factorization in integral
domains

LECTURE NOTES IN PURE AND APPLIED MATHEMATICS

Additional Volumes in Preparation

factorization in integral domains

edited by

Daniel D. Anderson
University of Iowa
Iowa City, Iowa

CRC Press
Taylor & Francis Group
Boca Raton London New York

CRC Press is an imprint of the
Taylor & Francis Group, an **informa** business

CRC Press
Taylor & Francis Group
6000 Broken Sound Parkway NW, Suite 300
Boca Raton, FL 33487-2742

First issued in hardback 2017

© 1997 by Taylor & Francis Group, LLC
CRC Press is an imprint of Taylor & Francis Group, an Informa business

No claim to original U.S. Government works

ISBN 13: 978-1-138-40178-5 (hbk)
ISBN 13: 978-0-8247-0032-4 (pbk)

Visit the Taylor & Francis Web site at
http://www.taylorandfrancis.com

and the CRC Press Web site at
http://www.

The Library of Congress Cataloging-in-Publication Data

Factorization in integral domains / edited by Daniel D. Anderson.
    p. cm. — (Lecture notes in pure and applied mathematics ; v. 189)
    Based on the proceedings of the conference entitled Factorization in Integral Domains, held Mar. 21, 1996, and of the special session in commutative ring theory, held Mar. 22-23, 1996, at the 909th Meeting of the American Mathematical Society in Iowa City.
    Includes bibliographical references and index.
    ISBN 0-8247-0032-5 (pbk. : alk. paper)
    1. Commutative rings—Congresses. 2. Integral domains (Ring theory)—Congresses. 3. Factorization (Mathematics)—Congresses. I. Anderson, Daniel D. II. Series.
QA251.3.F33   1997
512'.74—dc21

                                                        97-1911
                                                          CIP

Preface

This volume consists of the Proceedings of the conference entitled Factorization in Integral Domains held at The University of Iowa, Iowa City, and of the Special Session in Commutative Ring Theory organized by the editor which was held at the 909th Meeting of the American Mathematical Society in Iowa City.

In celebration of the State of Iowa's Sesquicentennial, the Department of Mathematics at The University of Iowa decided to have a number of conferences in conjunction with the regional AMS meeting being held in Iowa City. I decided to organize a conference around the topic of factorization in integral domains, a topic that has interested me for some time. I invited David F. Anderson, Scott Chapman, Alfred Geroldinger, Franz Halter-Koch, and Joe Mott to give one-hour survey talks on various aspects of factorization and divisibility. All five accepted my invitation, but Joe Mott was not able to attend and give his talk. However, the talk that Joe Mott was to have given is included in these Proceedings along with the other four talks that were given at the meeting. These five papers, the first five chapters, make up about half the total pages for these Proceedings. Over forty people attended the conference and special session, representing Austria, France, Israel, South Korea, and the United States. There were twenty-seven talks in the special session at the AMS meeting. Of the twenty other chapters in this volume, all were given by participants at the Iowa City meeting, but they are not necessarily from the talks given.

Let R be an atomic integral domain, that is, an integral domain in which every nonzero nonunit is a product of irreducible elements or atoms. Of course, R is a UFD precisely when the factorization into atoms is unique up to order and associates. The theory of nonunique factorizations in integral domains has its origins in the theory of algebraic number fields. Two recurrent themes are (1) how close can factorization into atoms be to being unique without the domain being a UFD and (2) how bad can factorization into atoms be in an atomic domain that is not a UFD. Many of the problems and results are combinatorial in nature and can be formulated for more general domains and for cancellative monoids. To date the strongest results are for Dedekind domains, or more generally, for Krull domains with nice divisor class groups and for one-dimensional Noetherian domains.

One of the first things to do when studying factorization into atoms is to look at the different lengths of factorizations. We say that a factorization $x = a_1 \cdots a_n$ of x into irreducibles has length n. Associated with each nonzero nonunit element x

iii

of R are the set of lengths $L(x) = \{n \mid x$ has a factorization into atoms of length $n\}$ and the elasticity $\rho(x) = \max\{L(x)\}/\min\{L(x)\}$ of x. The elasticity $\rho(R)$ of R is then defined to be $\sup\{\rho(x) \mid x$ is a nonzero nonunit of $R\}$. The article by David F. Anderson surveys results on elasticity. The sets of lengths of factorizations are surveyed in the article by Scott Chapman and Alfred Geroldinger and the article by Franz Halter-Koch. The article by Alfred Geroldinger discusses two relatively new invariants of factorizations: the catenary degree and the tame degree. The articles by Chapman, Geroldinger, and Halter-Koch give a good overview of the role of monoid-theoretic techniques involving divisor theory and block monoids. The article by Joe Mott surveys recent work on integral domains with a finite number of nonassociate atoms, integral domains with finitely generated groups of divisibility, and integral domains whose monoids of fractional ideals are finitely generated. These five chapters collect in one place for the first time results scattered throughout the literature. Also, the five survey articles each contain extensive bibliographies.

Besides factorization in integral domains, some of the topics covered in the twenty-five chapters are Prüfer domains, divided prime ideals, one-dimensional Noetherian rings, Henselian pullbacks, intersections of prime ideals, root closure, integral closure and complete integral closure, the ring of integer-valued polynomials, Hazlett's symbolic method for binary quantics, stable ideals and prime ideals in polynomial rings, and the genus class group.

I would like to thank The University of Iowa for its support of the conference, the invited hour speakers, the speakers at the special session, all the participants at the conference and the special session, and, of course, all those who contributed articles for these Proceedings. I would also like to thank the referees of papers in these Proceedings, others who helped with the conference and these Proceedings, the staff of Marcel Dekker including Maria Allegra, and finally, special thanks to our technical typist Brian Treadway.

Daniel D. Anderson

Contents

Contributors

Daniel D. Anderson, Department of Mathematics, University of Iowa, Iowa City, IA 52242

David F. Anderson, Department of Mathematics, University of Tennessee, Knoxville, TN 37996-1300

Joseph P. Brennan, Department of Mathematics, North Dakota State University, Fargo, ND 58105-5075

Jean-Luc Chabert, Département de Mathématiques, Université de Picardie, Saint-Quentin, France

Scott Chapman, Department of Mathematics, Trinity University, San Antonio, TX 78212-7200

Douglas L. Costa, Department of Mathematics, University of Virginia, Charlottesville, VA 22903-3199

Jim Coykendall, Department of Mathematics, Lehigh University, Bethlehem, PA 18015 (Current address: Department of Mathematics, North Dakota State University, Fargo, ND 58105-5075)

David E. Dobbs, Department of Mathematics, University of Tennessee, Knoxville, TN 37996-1300

Alfred Geroldinger, Institut für Mathematik, Karl-Franzens-Universität, A-8010 Graz, Austria

Robert Gilmer, Department of Mathematics, Florida State University, Tallahassee, FL 32306-3027

Robert Guralnick, Department of Mathematics, University of Southern California, Los Angeles, CA 90089-1113

Franz Halter-Koch, Institut für Mathematik, Karl-Franzens-Universität, A-8010 Graz, Austria

William Heinzer, Department of Mathematics, Purdue University, West Lafayette, IN 47907

David Lantz, Department of Mathematics, Colgate University, Hamilton, NY 13346

Rebecca L. Lewin, Department of Mathematics, University of Wisconsin-La Crosse, La Crosse, WI 54601

Aihua Li, Department of Mathematics, Loyola University, New Orleans, LA 70118

Thomas G. Lucas, Department of Mathematics, University of North Carolina at Charlotte, Charlotte, NC 28223-0001

Joe Mott, Department of Mathematics, Florida State University, Tallahassee, FL 32306-3027

Jaenam Park, Department of Mathematics, University of Tennessee, Knoxville, TN 37996-1300

Roy Quintero, Department of Mathematics, University of Iowa, Iowa City, IA 52242

Moshe Roitman, Department of Mathematics and Computer Science, University of Haifa, Mount Carmel, Haifa, Israel 31905

William W. Smith, Department of Mathematics, University of North Carolina, Chapel Hill, NC 27599-3250

Silvia Valdes-Leon, Department of Mathematics, University of Southern Maine, Portland, ME 04103

Roger Wiegand, Department of Mathematics, University of Nebraska, Lincoln, NE 68588-0323

Sylvia Wiegand, Department of Mathematics, University of Nebraska, Lincoln, NE 68588-0323

Janice Winner, Department of Mathematics, University of Kentucky, Lexington, KY 40506

Muhammad Zafrullah, MTA Teleport, Silver Spring, MD 20905

Elasticity of Factorizations in Integral Domains: A Survey

DAVID F. ANDERSON Department of Mathematics, The University of Tennessee, Knoxville, TN 37996, USA

1. INTRODUCTION

If R is a unique factorization domain (UFD), then any two factorizations of a nonzero nonunit of R into the product of irreducible elements have the same length. However, this need not be true for an arbitrary atomic domain (an integral domain is *atomic* if each nonzero nonunit is a product of irreducible elements (atoms)). Following Zaks [64], we say that an atomic domain R is a *half-factorial domain* (HFD) if whenever $x_1 \cdots x_m = y_1 \cdots y_n$ with each $x_i, y_j \in R$ irreducible, then $m = n$. Any UFD is obviously an HFD, but the converse fails since any Krull domain R with divisor class group $\text{Cl}(R) = \mathbb{Z}/2\mathbb{Z}$ is an HFD [65], but not a UFD. Recently there has been much activity on HFDs, other factorization properties weaker than unique factorization, and on invariants that measure different lengths of factorizations. In order to measure how far an atomic domain R is from being an HFD, we define $\rho(R) = \sup\{m/n \,|\, x_1 \cdots x_m = y_1 \cdots y_n, \text{ for } x_i, y_j \in R \text{ irreducible}\}$ if R is not a field, and $\rho(R) = 1$ if R is a field. Thus $1 \leq \rho(R) \leq \infty$, and $\rho(R) = 1$ if and only if R is an HFD. $\rho(R)$, called the *elasticity* of R, was introduced by Valenza [62] and has recently received considerable attention.

In this paper, we survey the major results on the elasticity of factorizations in integral domains. These results may be considered as generalizing results for HFDs. We include an extensive set of references, but we do not include complete proofs for some of the more technical results. Elasticity and factorization questions can also be investigated in the more general context of commutative cancellative monoids (see Section 7, [46], [55], and the two survey articles in these proceedings by S. Chapman and A. Geroldinger [32] and F. Halter-Koch [56]); however, we restrict our survey to integral domains.

In the second section, we include some elementary results on elasticity and give a brief history. In the third section, we introduce semi-length functions, one of the main tools for calculating $\rho(R)$. Much of the early investigation of factorization questions and elasticity centered on the class of Dedekind domains, or more generally, Krull domains. The elasticity of Krull domains is investigated in the fourth section. In the fifth section, we extend this study to more general classes of atomic domains such as Cohen-Kaplansky domains and weakly Krull domains. In the sixth section, we pass from the general theory to concrete, but often messy, calculations of $\rho(R)$, where R is a subring of either $K[X]$ or $K[[X]]$ for K a field. The final section includes two generalizations of elasticity for domains. The first is the elasticity $\rho(H)$ for H a commutative cancellative monoid, and the second is the elasticity $\rho_{\mathcal{F}}(R)$ of R relative to certain subsets $\mathcal{F}$ of irreducible elements of R. Several new observations are also sprinkled throughout this paper.

Throughout, R will denote an arbitrary atomic domain with R^* its set of nonzero elements, $U(R)$ its group of units, $\mathcal{I}(R)$ its set of irreducible elements, and $X(R)$ its set of height-one prime ideals. If R is a Krull domain, $\mathrm{Cl}(R)$ will be its divisor class group, and for $P \in X(R)$, its divisor class in $\mathrm{Cl}(R)$ will be denoted by $[P]$. We will repeatedly use the fundamental fact that for a Krull domain R, a nonunit $x \in R^*$ is irreducible if and only if in its v-factorization $xR = (P_1 \cdots P_n)_v$ with each $P_i \in X(R)$, no proper subproduct $(P_{i_1} \cdots P_{i_m})_v$ is principal. Let $\mathbf{Z}, \mathbf{Q}, \mathbf{R}$, and $\mathbf{C}$ be the ring of integers, and fields of rational numbers, real numbers, and complex numbers, respectively. Also, $\mathbf{Z}_+, \mathbf{Q}_+$, and $\mathbf{R}_+$ will denote the sets of nonnegative integers, nonnegative rational numbers, and nonnegative real numbers, respectively. As usual, $a/0 = \infty/0 = \infty/a = \infty + a = \infty$ for any positive $a \in \mathbf{R}$. We will use $\subset$ to denote proper inclusion. For other references and related results on factorization in integral domains, see [6], [7], [8], [32], [33], [34], or [56]. For general terminology see [48], and for basic facts on Krull domains see [39].

2 BASIC RESULTS AND HISTORY

In this section, we include some basic definitions and results. We
end with a short history of the study of elasticity. When studying
factorization properties, we will always assume that R is atomic, that
is, each nonzero nonunit of R is a product of irreducible elements.
Usually this will be automatic; in fact, usually stronger conditions will
hold, e.g., R will be either a Noetherian domain or a Krull domain.
Both Noetherian domains and Krull domains satisfy the ascending
chain condition on principal ideals (ACCP), and hence are atomic.
However, there are atomic domains which do not satisfy ACCP. The
first such example is due to Grams [51]; other examples have been
given by Zaks [66] and Roitman [59].

In an arbitrary atomic domain, a nonzero nonunit may have factorizations of different lengths. We thus introduce the following notation from [25]. For a nonzero nonunit x in an atomic domain R,
we define $l_R(x) = \inf\{n | x = x_1 \cdots x_n$, each $x_i \in R$ is irreducible$\}$,
$L_R(x) = \sup\{n | x = x_1 \cdots x_n$, each $x_i \in R$ is irreducible$\}$, and $l_R(x)$
$= L_R(x) = 0$ when $x \in U(R)$. Then $1 \le l_R(x) \le L_R(x) \le \infty$ for
each nonunit $x \in R^*$. Clearly $l_R(x)$ is always finite, but $L_R(x)$ may
be infinite. Next, define $\rho_R(x) = L_R(x)/l_R(x)$ for $x \in R^*$ a nonunit
and $\rho_R(x) = 1$ for $x \in U(R)$. Our original definition for $\rho(R)$ may
thus be recast as $\rho(R) = \sup\{\rho_R(x) | x \in R^*\}$.

Hence $1 \le \rho(R) \le \infty$, and $\rho(R) = 1$ if and only if R is an HFD.
Thus, $\rho(R)$ measures, in some sense, how far R is from being an
HFD. Later, we will give conditions on R so that $\rho(R)$ is finite or
rational. However, we first show that even for Dedekind domains,
$\rho(R)$ may be arbitrary. This result is from [1, Theorem 2.14] and uses
Claborn's construction of a Dedekind domain with specified relations
in its divisor class group ([38] or [39, Theorem 15.18]).

THEOREM 2.1. *Let $r \ge 1$ be a real number or $r = \infty$. Then there
is a Dedekind domain R with torsion divisor class group $Cl(R)$ such
that $\rho(R) = r$. Moreover, if r is rational, then we may choose $Cl(R)$
to be finite.* $\square$

We say that $\rho(R)$ is *realized by a factorization* if there are irreducible
$x_i, y_j \in R$ with $x_1 \cdots x_m = y_1 \cdots y_n$ and $\rho(R) = m/n$, i.e., $\rho(R) =
\rho_R(x) < \infty$ for some $x \in R^*$. If R is a Krull domain with $Cl(R)$ finite,
then $\rho(R)$ is necessarily realized by a factorization (see Theorem 4.2).

However, as noted in [1, page 233], for each rational number $q > 1$, the construction used for the proof of Theorem 2.1 yields a Dedekind domain R with infinite torsion divisor class group such that $\rho(R) = q$, but $\rho_R(x) < \rho(R)$ for each $x \in R^*$. Some other examples where $\rho(R)$ is rational, but $\rho(R)$ is not realized by a factorization, are given in Example 4.6 and Example 6.6.

There are several other realization results for $\rho(R)$. In [36, page 184], it is shown that each $q \in \mathbb{Q}$ with $q \geq 1$ can be realized as $\rho(R)$ for R a Dedekind domain with finite divisor class group. In [1, Example 3.3], a monoid domain construction was used to show that for any real number $r \geq 2$, there is an atomic one-dimensional quasilocal domain R with $\rho(R) = r$. For each $q \in \mathbb{Q} - \mathbb{Z}$ with $q \geq 2$, it is shown in [27, Theorem 2.5] that $\rho(R) = q$ for a suitable (power series) semigroup ring $R = K[[S]]$ over a field K, where $S = \langle n_1, \ldots, n_r \rangle$ is a numerical semigroup with $q = n_r/n_1$ (see Section 6). Also, if $n \geq 1$ is an integer, then there is a subring of $K[[X]]$ with $\rho(R) = n$ ([1, Example 3.5] or [24, Theorem 3.1]). Hence each $q \in \mathbb{Q}$ with $q \geq 2$ can be realized as $\rho(R)$ for R a subring of $K[[X]]$. In each case, $\rho(R)$ is realized by a factorization.

In [1], we defined R to be a *rationally bounded factorization domain* (RBFD) if $\rho(R) < \infty$. By the above comments, $\rho(R)$ may be infinite even for R a Dedekind domain. However, if $\rho(R) < \infty$, then R necessarily satisfies ACCP. In fact, R satisfies the stronger property that $L_R(x) < \infty$ for each $x \in R^*$; such a domain is called a *bounded factorization domain* (BFD) (see [6, Section 2]). Clearly BFDs satisfy ACCP, but the converse fails [6, Example 2.1]. Krull domains and Noetherian domains are BFDs [6, Proposition 2.2]; so by the above comments, the converse of Theorem 2.2 fails.

THEOREM 2.2 *Let R be an atomic integral domain. If $\rho(R) < \infty$, then R is a BFD; in particular, R satisfies ACCP.*
Proof: By definition, $L_R(x) \leq \rho(R) l_R(x) < \infty$ for each $x \in R^*$. Thus R is a BFD, and hence R satisfies ACCP. $\square$

Actually, we can do even better. As in [33], for an atomic domain R and positive integer n, we define $\mathcal{V}(n) = \{m | x_1 \cdots x_n = y_1 \cdots y_m$ for $x_i, y_j \in R$ irreducible $\}$, and let $\Phi(n) = |\mathcal{V}(n)|$. We say that R is *Φ-finite* if $\Phi(n) < \infty$ for each positive integer n. We then have the following result from [16, Theorem 1.2]. Examples given in [16] show that none of the implications is reversible (however, cf. Theorem 4.3

for the case when R is a Krull domain with $Cl(R) = \mathbb{Z}$).

THEOREM 2.3   *Let R be an atomic integral domain. Then $\rho(R) < \infty \Rightarrow R$ is Φ-finite $\Rightarrow R$ is a BFD.*
Proof: The first implication follows from the easily proved fact that $\Phi(n) \leq [(\rho(R)^2 - 1)/\rho(R)]n + 1$ [16, Lemma 1.1]. For the second implication, suppose that R is Φ-finite and $x = x_1 \cdots x_n$ with each $x_i \in R$ irreducible. Then $L_R(x) \leq \max \mathcal{V}(n) < \infty$; so R is a BFD. $\square$

Later, we will show that for a Krull domain R we have the following bound for $\rho(R)$ (in fact, we give a much better bound in Theorem 3.3). In particular, Theorem 2.4 yields that $\rho(R) < \infty$ when $Cl(R)$ is finite, and that R is an HFD when $Cl(R) = \mathbb{Z}/2\mathbb{Z}$.

THEOREM 2.4   *Let R be a Krull domain. Then $1 \leq \rho(R) \leq \max\{|Cl(R)|/2, 1\}$.*
Proof: See Theorem 3.3. $\square$

The above theorem also yields the following example from [1, Example 3.4].

EXAMPLE 2.5   Let K be a field and $R_n = K[X^n, XY, Y^n]$ for each integer $n \geq 1$. Then R_n is a two-dimensional Noetherian Krull domain with $Cl(R_n) = \mathbb{Z}/n\mathbb{Z}$ [13]. If $n = 1$, then of course R_n is a UFD; so suppose $n \geq 2$. By Theorem 2.4, $\rho(R_n) \leq n/2$. The factorization $(XY)^n = X^n Y^n$ with X^n, Y^n, and XY each irreducible in R_n shows that $\rho(R_n) \geq n/2$; so $\rho(R_n) = n/2$. In particular, R_n is an HFD if and only if either $n = 1$ or $n = 2$. $\square$

It is of interest to know how the elasticity behaves with respect to various ring-theoretic constructions. We next show that elasticity behaves well under the generalized "$D + M$" construction. This result may be used to construct domains with different ring-theoretic properties, but with the same elasticity. The quasilocal case is from [1, Example 3.7] (also, cf. [24, Remark 4.6(d)]). Note that for $R = k + M$ to be atomic, k must be a field [6, Proposition 1.2].

THEOREM 2.6   *Let T be an atomic integral domain of the form $T = K + M$, where K is a field contained in T and M is a nonzero maximal ideal of T. Let k be a subfield of K and $R = k+M$. Then R is* *atomic and $\rho(R) = \rho(T)$. Moreover, $\rho(R)$ is realized by a factorization if and only if $\rho(T)$ is realized by a factorization.*

Proof: Up to multiplication by an $\alpha \in K^*$ (resp., $\alpha \in k^*$), each element of T (resp., R) has the form m or $1 + m$ for some $m \in M$. Also, each of these elements is irreducible in R if and only if it is irreducible in T. All the statements in the theorem now follow easily from these observations. $\square$

The localization of an HFD need not be an HFD, [20, Theorem 5] and [21]. However, the localization of an HFD is again an HFD if we localize at a multiplicatively closed set generated by principal primes [7, Corollary 2.5]. If S is generated by all the principal primes of an atomic domain R, then R_S is atomic and has no principal primes [7, Corollary 1.4(d) and Corollary 1.7]. This shows that the prime elements play no role in determining $\rho(R)$ (cf. Theorem 3.1 and [23, Theorem 4.1]). The first part of our next theorem is from [21, Proposition 3.4].

THEOREM 2.7 *Let R be an atomic integral domain and S a multiplicatively closed subset of R generated by primes. Then R_S is atomic and $\rho(R_S) = \rho(R)$. Moreover, $\rho(R_S)$ is realized by a factorization if and only if $\rho(R)$ is realized by a factorization.*

Proof: By [7, Corollary 2.2], R_S is atomic. We may assume that R is not a UFD. Let $z = x_1 \cdots x_m = y_1 \cdots y_n$ with each $x_i, y_j \in R$ irreducible. Each factorization of z has the same number, say k, of prime factors in S. Let $x \in R$ be irreducible, but not prime. Then in R_S, zx^k is a product of m and n irreducible elements, so $\rho(R) \leq \rho(R_S)$. Conversely, factorizations of a $z \in R_S$ of lengths m and n yield factorizations in R of the form $sx_1 \cdots x_m = ty_1 \cdots y_n$ with $s, t \in S$ and each $x_i, y_j \in R - S$ irreducible. Thus $s = ut$ for some $u \in U(R)$; so $x_1 \cdots x_m$ is a product of m and n irreducible elements in R. Hence $\rho(R_S) \leq \rho(R)$, and thus $\rho(R_S) = \rho(R)$. The "moreover" statement follows easily by the above arguments. $\square$

How $\rho(R)$ behaves with respect to localization is investigated in more detail in [23]. In particular, $\rho(R_S) \leq \rho(R)$ and $\rho(R) = \max \{ \rho(R_S), \rho(R_T) \}$ when S is a splitting multiplicative set of R with complementary multiplicative set T ([23, Theorem 2.3], cf. [7]). Examples are also given to show that being realized by a factorization need not be preserved by localization.

Rather than consider the ratio of lengths of factorizations, we can also consider the difference of lengths of factorizations. More precisely, for each nonunit $x \in R^*$, define $\mu_R(x) = L_R(x) - l_R(x) = \sup\{m - n | x = x_1 \cdots x_m = y_1 \cdots y_n$, for $x_i, y_j \in R$ irreducible$\}$ and $\mu(x) = 0$ if $x \in U(R)$. Then $\mu(R) = \sup\{\mu_R(x) | x \in R^*\} = \sup\{m - n | x_1 \cdots x_m = y_1 \cdots y_n$, for $x_i, y_j \in R$ irreducible$\}$ gives a global measure of the difference in lengths of factorizations in R. Given a factorization $x = x_1 \cdots x_m = y_1 \cdots y_n$ with $x_i, y_j \in R$ irreducible, then the factorization $x^k = x_1^k \cdots x_m^k = y_1^k \cdots y_n^k$ has $k(m - n)$ as the difference of lengths of factorizations and $km/kn = m/n$ as the ratio of lengths of factorizations. Thus $\mu_R(x^n) \geq n\mu_R(x)$ for any integer $n \geq 1$. Hence, $\mu(R) = 0$ if R is an HFD and $\mu(R) = \infty$ if R is not an HFD. Thus $\mu(R)$ is not a very useful invariant.

It is, however, interesting to note that the limit $\overline{\mu}_R(x) = \lim_{n \to \infty} \mu_R(x^n)/n = \overline{L}_R(x) - \overline{l}_R(x)$ always exists (but may be infinite) since the limits $\overline{L}_R(x) = \lim_{n \to \infty} L_R(x^n)/n$ and $\overline{l}_R(x) = \lim_{n \to \infty} l_R(x^n)/n$ always exist [25, Theorem 3]. In fact, if R is a Krull domain, then $\overline{\mu}_R(x)$ is a nonnegative rational number by [44, Theorem 3] or [4, Theorem 13]. Moreover, in this case, $\overline{\mu}_R(x) = \overline{L}_R(x) - \overline{l}_R(x) = (\overline{L}_R(x)/\overline{l}_R(x) - 1)\overline{l}_R(x) = (\overline{\rho}_R(x) - 1)\overline{l}_R(x) \leq (\rho(R) - 1)\overline{l}_R(x)$, where $\overline{\rho}_R(x) = \lim_{n \to \infty} \rho_R(x^n)/n = \overline{L}_R(x)/\overline{l}_R(x)$. Other asymptotic results involving lengths of factorizations, $\rho(R)$, or the Φ-function are given in [4], [19], [22], [25], [36], [37], [44], and [53].

Finally, we discuss a little history about the development of $\rho(R)$. Much of it traces back to the observation of Carlitz [30] that (in our terminology) an algebraic number ring R is an HFD if and only R has class number $h_R \leq 2$ (recall that R is a UFD if and only if $h_R = 1$). Zaks then introduced HFDs in [64] and showed that any Krull domain R with $Cl(R) = \mathbb{Z}/2\mathbb{Z}$ is an HFD (cf. Corollary 3.4(c)). Carlitz's paper led Valenza [62] to define and study $\rho(R)$ for R an algebraic number ring ([63]). Independently, Steffan [61] showed that if R is a Dedekind domain (and not a PID) with finite divisor class group G, and $x_1 \cdots x_m = y_1 \cdots y_n$ for irreducible $x_i, y_j \in R$, then $1/q \leq m/n \leq q$, where $q = |G|/2$. In fact, he gave a better bound for q in terms of the Davenport constant $D(G)$ of G (see Section 3). One should note that the paper by Valenza was actually submitted November 19, 1980 and appeared in 1990, while the paper by Steffan was submitted March 20, 1985 and appeared in 1986. Then Anderson-Anderson investigated $\rho(R)$ systematically for arbitrary atomic domains in [1] and [2], Chapman-Smith [36] concentrated on Dedekind domains with torsion divisor class group, and the four combined forces in [4]. The paper of Narkiewicz [58] (submitted soon after [62] appeared) also observed the relationship between $\rho(R)$ and $D(\text{Cl}(R))$ when R is a Dedekind domain with finite divisor class group and each (nonzero) ideal class contains a prime ideal. Since then, there has been an avalanche of activity and generalizations on elasticity which will be detailed in this survey.

We close with another invariant for lengths of factorizations in atomic monoids defined by Geroldinger and Lettl in [46]. However, we state these only for an atomic integral domain R. For each integer $m \geq 1$, define $\mu_m(R) = \sup \{L_R(x) | x \in R^*$ has a factorization of length $m\}$ and $\mu_m^*(R) = \sup \{L_R(x) | x \in R^*$ and $l_R(x) = m\}$. Halter-Koch [55, Proposition 1] then showed that $\rho(R) = \sup\{\mu_m^*(R)/m | m \geq 1\} = \sup\{\mu_m(R)/m | m \geq 1\} = \lim_{m \to \infty} \mu_m(R)/m$.

3 SEMI-LENGTH FUNCTIONS

In this section, we study semi-length functions. They were introduced in [1] to determine upper and lower bounds for $\rho(R)$, and have proved quite useful in calculations (see Sections 4 and 6).

A function $\varphi : R^* \to \mathbb{R}_+$ is a *semi-length function* on R if (i) $\varphi(xy) = \varphi(x) + \varphi(y)$ for all $x, y \in R^*$, and (ii) $\varphi(x) = 0$ if and only if $x \in U(R)$. A semi-length function φ with $\text{im}\varphi \subseteq \mathbb{Z}$ is called a *length function*. To define a semi-length function φ, it suffices to define φ on $R^* - U(R)$ satisfying (i) and (ii) above, and then extend φ to R^* by defining $\varphi(x) = 0$ for all $x \in U(R)$ [1, Lemma 1.1]. As in [1], we define $M = M(R, \varphi) = \sup \{\varphi(x) | x \in R$ is irreducible $\}$, $m = m(R, \varphi) = \inf\{\varphi(x) | x \in R$ is irreducible $\}$, $M^* = M^*(R, \varphi) = \sup \{\varphi(x) | x \in R$ is irreducible, but not prime $\}$, $m^* = m^*(R, \varphi) = \inf \{\varphi(x) | x \in R$ is irreducible, but not prime $\}$, and $M^* = m^* = 1$ when R is a UFD. Clearly, if R is not a UFD, then $0 \leq m \leq m^* \leq M^* \leq M \leq \infty$, and $m < \infty$ and $0 < M$ for any integral domain R. We define φ to be a *bounded semi-length function* if $0 < m(R, \varphi)$ and $M(R, \varphi) < \infty$. By [1, Theorem 2.1], if φ is a semi-length function on R, then $1 \leq \rho(R) \leq M^*/m^*$, a key result on elasticity. This is an improvement on the more elementary inequality $1 \leq \rho(R) \leq M/m$, and again illustrates that principal primes play little role in studying lengths of factorizations.

THEOREM 3.1 *Let R be an atomic integral domain and φ a semi-length function on R. Then $1 \leq \rho(R) \leq M^*(R, \varphi)/m^*(R, \varphi)$. Moreover, if R has a bounded semi-length function, then $\rho(R) < \infty$.*

Proof. We may assume that R is not a UFD and that $0 < m^* = m^*(R, \varphi) \leq M^* = M^*(R, \varphi) < \infty$. Let $x = x_1 \cdots x_r = y_1 \cdots y_s$ be a factorization in R with each $x_i, y_j \in R$ irreducible and $r \geq s$. Since each factorization of x has the same prime factors (up to associates) and $r/s < (r - k)/(s - k)$ if $1 \leq k \leq s - 1$, we may assume that none of the x_i, y_j's is prime. Thus $rm^* \leq \varphi(x_1) + \cdots + \varphi(x_r) = \varphi(x) = \varphi(y_1) + \cdots + \varphi(y_s) \leq sM^*$, and hence $r/s \leq M^*/m^*$. Thus $\rho(R) \leq M^*/m^*$. The "moreover" statement is clear. $\square$

For Theorem 3.1 to be of any use, we need examples of semi-length functions. We next give several examples which we will use throughout this paper. Some more examples will be given in Section 6.

EXAMPLE 3.2 (a) Let R be an HFD. Then s_R defined by $s_R(x) = L_R(x)$ ($= l_R(x)$) for any $x \in R^*$ defines a length function on R. Moreover, if there is a length function φ with $\varphi(x) = 1$ if and only if x is irreducible, then R is an HFD [65, Lemma 1.3].

(b) Let R be a Krull domain with $\{v_P | P \in X(R)\}$ its set of essential discrete rank-one valuations. Define $V : R^* \to \mathbb{Z}_+$ by $V(x) = \sum v_P(x)$. (Thus $V(x) = n \geq 1$ if and only if $xR = (P_1 \cdots P_n)_v$ for some $P_i \in X(R)$). Then V defines a length function on R with $V(x) = 1$ if and only if x is prime. Note that $M^*(R, V) = M(R, V)$; while $m^*(R, V) = m(R, V)$ if and only if either R is a UFD or R has no principal primes.

(c) More generally, for a Krull domain R, let $\{r_P | P \in X(R)\}$ be any set of positive real numbers. Then $\varphi(x) = \sum r_P v_P(x)$ defines a semi-length function on R. In particular, if R is a UFD, then $m = m(R, \varphi) = \inf\{r_P\}$ and $M = M(R, \varphi) = \sup\{r_P\}$. Thus $m^* = M^* = 1$, but m and M may assume any real values such that $0 \leq m \leq M \leq \infty$, $m < \infty$, and $M > 0$.

(d) An important special case of part (c) above is when $Cl(R)$ is a torsion group. In this case, define $\mathcal{Z}_R(x) = \sum (n_P)^{-1} v_P(x)$, where n_P is the order of $[P]$ in $Cl(R)$. Note that $\mathcal{Z}_R(x) = 1$ if x is an irreducible element of the form $xR = (P^n)_v$. Thus $M^*(R, \mathcal{Z}_R) = M(R, \mathcal{Z}_R) \geq 1$ and $m^*(R, \mathcal{Z}_R) = m(R, \mathcal{Z}_R) \leq 1$. In [36], $\mathcal{Z}_R(x)$ was called the Zaks-Skula constant of x, but was defined for Dedekind domains using ideal classes rather than valuations. We will call $\mathcal{Z}_R$ the Zaks-Skula function of R. $\mathcal{Z}_R$ detects when R is an HFD (see Corollary 3.6(b)). The Zaks-Skula function in various forms has been used extensively in [34], [36], [60], and [65]. □

Semi-length functions on Krull domains were extensively investigated in [1] and [2]. It was shown [2, Theorem 1.2] that any semi-length function φ on a Krull domain R has the form $\varphi = \sum r_P v_P$ for some real numbers r_P. Moreover, if $\mathrm{Cl}(R)$ is a torsion group, then each $r_P > 0$ and is uniquely determined [1, Proposition 1.4]. However, if $\mathrm{Cl}(R)$ is not a torsion group, then the r_P's are not uniquely determined; for a specific example with $\mathrm{Cl}(R) = \mathbf{Z}$, see [2, Example 1.3].

We next define the *Davenport constant*, denoted by $D(G)$, of an abelian group G. As we will see, this constant comes up quite often in factorization problems. For G a finite abelian group, $D(G)$ is the least positive integer d such that for each sequence $S \subseteq G$ with $|S| = d$, some nonempty subseqence of S has sum 0. Equivalently, $D(G)$ is the length of the longest sequence of (not necessarily distinct) elements in G whose proper subsequences have nonzero sums. If G is infinite, then $D(G) = \infty$ in the sense that for each positive integer d, there is a sequence $S \subset G$ with $|S| = d$ and no nonempty subsequence of S has sum 0 [41, Proposition 2]. (In [61], $D(G)$ was denoted by $l(G)$, and in [62] it was denoted by $\sigma(G)$ and was called the sequential depth of G.) The importance of $D(G)$ is that if R is a Krull domain in which each (nonzero) divisor class contains a height-one prime ideal, then $D(\mathrm{Cl}(R))$ is the maximum number of height-one prime ideals (counted with multiplicity) which can occur in the prime ideal v-factorization of an irreducible element of R. For R the ring of integers in an algebraic number field, this was pointed out by H. Davenport (Midwestern Conference on Group Theory and Number Theory, Ohio State University, 1966). Also, observe that if each (nonzero) divisor class contains a height-one prime ideal, then $M^*(R, V) = D(\mathrm{Cl}(R))$.

We next state, without proof, some of the basic facts about $D(G)$ that we will use in this paper. First, $D(G) \leq |G|$, and we have equality when G is cyclic; $D(\mathbf{Z}/n\mathbf{Z}) = n$. However, $D(\mathbf{Z}/2\mathbf{Z} \oplus \mathbf{Z}/2\mathbf{Z}) = 3$. For $G = \mathbf{Z}/n_1\mathbf{Z} \oplus \cdots \oplus /n_r\mathbf{Z}$ with $1 < n_1 | \cdots | n_r$, define $M(G) = 1 + (n_1 - 1) + \cdots + (n_r - 1)$. It is always true that $D(G) \geq M(G)$, with equality if either $r \leq 2$ or G is a p-group, but not in general. In particular, for the p-group $G = (\mathbf{Z}/p\mathbf{Z})^{n_1} \oplus \cdots \oplus (\mathbf{Z}/p^r\mathbf{Z})^{n_r}$, we have $D(G) = n_1(p - 1) + n_2(p^2 - 1) + \cdots + n_r(p^r - 1) + 1$. Unfortunately, as yet there is no general formula for $D(G)$. The above and other facts about $D(G)$ are given in [31] and [47], two survey articles on the Davenport constant.

We next combine the above comments with Theorem 3.1 and Example 3.2(b) to obtain the "Fundamental theorem of elasticity for Krull domains" (as stated in [1, Theorem 2.2 and Corollary 2.3]). This result has been observed many different times. It is the Krull domain analogue of [61, Proposition 1] for Dedekind domains with finite divisor class group and [62, Proposition 4] for rings of integers in an algebraic number field. The Dedekind domain case of Corollary 3.4(b) is also due to Steffan [61, Proposition 2], and recovers the result of Carlitz [30] mentioned in Section 2. The Dedekind domain case has also been observed in [36, Lemma 1.2] and in [58] (when each ideal class contains a prime ideal). Corollary 3.4(c) is from [65, Theorem 1.4].

THEOREM 3.3 *Let R be a Krull domain which is not a UFD and $V = \sum v_P$. Then $1 \leq \rho(R) \leq M^*(R,V)/m^*(R,V) \leq M^*(R,V)/2 \leq D(Cl(R))/2 \leq |Cl(R)|/2$. In particular, a Krull domain R with finite divisor class group has $\rho(R) \leq \max\{|Cl(R)|/2, 1\} < \infty$.*
Proof. If R is not a UFD, then $m^*(R,V) \geq 2$. By Theorem 3.1, it suffices to show that $M^*(R,V) \leq D(Cl(R))$. We may assume that $Cl(R)$ is finite. Suppose that $s > D(Cl(R))$. Then for any $P_1, \ldots, P_s \in X(R)$, some proper subsum of $[P_1] + \cdots + [P_s]$ is 0 in $Cl(R)$. Thus if $s = V(x)$, then $xR = (P_1 \cdots P_s)_v$ is properly contained in a principal ideal and hence is not irreducible. Thus if x is irreducible, then $V(x) \leq D(Cl(R))$; so $M^*(R,V) = M(R,V) \leq D(Cl(R))$. $\square$

COROLLARY 3.4 *(a) Let R be a Krull domain with $Cl(R) = \mathbb{Z}/n_1\mathbb{Z} \oplus \cdots \oplus \mathbb{Z}/n_r\mathbb{Z}$ for $1 < n_1 | \cdots | n_r$. Then $M^*(R,V)/n_r \leq \rho(R) \leq M^*(R,V)/m^*(R,V)$. Thus $\rho(R) = M^*(R,V)/2$ when $Cl(R)$ is 2-elementary.*

(b) Let R be a Krull domain which is not a UFD. If each (nonzero) divisor class contains a height-one prime ideal, then $\rho(R) = M^*(R,V)/2 = D(Cl(R))/2$.

(c) A Krull domain R is an HFD if $Cl(R) = \mathbb{Z}/2\mathbb{Z}$. If each (nonzero) divisor class contains a height-one prime ideal, then R is an HFD if and only if $|Cl(R)| \leq 2$.

Proof: (a) Choose $x \in R$ irreducible with $t = V(x) = M^*(R, V)$. Then $xR = (P_1 \cdots P_t)_v$ for some $P_i \in X(R)$, and hence $x^{n_r} R = (P_1^{n_r} \cdots P_t^{n_r})_v$ with each $(P_i^{n_r})_v$ principal. Thus x^{n_r} has factorizations of length n_r and length at least t; so $M^*(R, V)/n_r = t/n_r \leq \rho(R)$. The other parts are clear.

(b) By Theorem 3.3, $\rho(R) \leq D(\mathrm{Cl}(R))/2$. Conversely, let $D(\mathrm{Cl}(R)) = r$. Then there is an irreducible $x \in R$ with $xR = (P_1 \cdots P_r)_v$ with each $P_i \in X(R)$. For each P_i, let $-[P_i] = [Q_i]$ for some $Q_i \in X(R)$. Then $(Q_1 \cdots Q_r)_v = yR$ with $y \in R$ irreducible. Also, each $(P_i Q_i)_v = z_i R$ with each $z_i \in R$ irreducible. Then the prime ideal v-factorization $((P_1 \cdots P_r)_v (Q_1 \cdots Q_r)_v)_v = ((P_1 Q_1)_v \cdots (P_r Q_r)_v)_v$ yields the factorization $xy = uz_1 \cdots z_r$ for some $u \in U(R)$. Hence $r/2 \leq \rho(R)$, and we have $\rho(R) = r/2 = D(\mathrm{Cl}(R))/2$.

(c) This follows easily from Theorem 3.3 and part (b) above. $\square$

Two special cases of Krull domains when each (nonzero) divisor class contains a height-one prime ideal are the ring of algebraic integers in a number field and a polynomial ring $R[X]$ over a Krull domain R [39, Theorem 14.3]. Thus if R is a Krull domain, then $\rho(R[X]) = \max\{D(\mathrm{Cl}(R))/2, 1\}$ by Corollary 3.4(b) since $\mathrm{Cl}(R[X]) = \mathrm{Cl}(R)$. Also, in this case, $\rho(R[X, X^{-1}]) = \rho(R[X, p/X]) = \rho(R[X])$ for any prime $p \in R$ by [3, Theorem 17]. For any finite abelian group G, there is a Dedekind HFD R with $\mathrm{Cl}(R) = G$ [65, Example 9]. This shows that the bounds in Theorem 3.3 are far from the best possible and that we may have $\rho(R) = 1$ and $\rho(R[X]) = n/2$ for any integer $n \geq 2$. In general, $R[X]$ need not be atomic when R is atomic [59, Example 5.1]. However, if both R and $R[X]$ are atomic, then clearly $\rho(R) \leq \rho(R[X])$.

For the Zaks-Skula semi-length function, we next obtain a lower bound for $\rho(R)$ [36, Corollary 1.7] (cf. [1, Corollary 2.5] and [62, Proposition 5]). Corollary 3.6(a) is from [2, Lemma 3.3] and Corollary 3.6(b) recovers [65, Theorem 3.3] and [60, Theorem 3.1].

THEOREM 3.5 *Let R be a Krull domain with torsion divisor class group, Zaks-Skula function $\mathcal{Z}_R$, $M^* = M^*(\mathcal{Z}_R, R)$, and $m^* = m^*(\mathcal{Z}_R, R)$. Then $\max\{M^*, 1/m^*\} \leq \rho(R)$.*

Proof: Suppose that $x \in R$ is irreducible and $xR = (P_1^{m_1} \cdots P_r^{m_r})_v$ for some $P_i \in X(R)$ and integers $m_i \geq 1$. Let $n_i = \|[P_i]\|$, $x_i R = (P_i^{n_i})_v$, $N = n_1 \cdots n_r$, and $N_i = N/n_i$. Since $x^N R = (P_1^{m_1 N} \cdots P_r^{m_r N})_v$, we have $x^N = u x_1^{m_1 N_1} \cdots x_r^{m_r N_r}$ for some $u \in U(R)$ and each $x_i \in R$

irreducible. Thus both $N/(m_1 N_1 + \cdots + m_r N_r)$ and $(m_1 N_1 + \cdots + m_r N_r)/N \leq \rho(R)$. However, $(m_1 N_1 + \cdots + m_r N_r)/N = m_1/n_1 + \cdots + m_r/n_r = \mathcal{Z}_R(x)$. The result now follows since $\mathcal{Z}_R(x)$ and $1/\mathcal{Z}_R(x) \leq \rho(R)$ for each irreducible $x \in R$. $\square$

COROLLARY 3.6 *Let R be a Krull domain with torsion divisor class group and Zaks-Skula function $\mathcal{Z}_R$, $M^* = M^*(\mathcal{Z}_R, R)$, and $m^* = m^*(\mathcal{Z}_R, R)$.*

(a) $(M^*/m^*)^{1/2} \leq \rho(R) \leq M^*/m^*$.
(b) *R is an HFD if and only if $M^* = m^* = 1$.*

Proof. (a) By Theorem 3.1 and Theorem 3.5, $\max\{M^*, 1/m^*\} \leq \rho(R) \leq M^*/m^*$. Hence $\rho(R) \leq M^*/m^* \leq \rho(R)^2$, and the result follows.

(b) This follows easily from Theorem 3.1 and Theorem 3.5. $\square$

There has also been some work on trying to define a Zaka-Skula function to detect when a Krull domain R with $\mathrm{Cl}(R) = \mathbb{Z}$ is an HFD in [34], [16], and [17].

The bounds given in Theorem 3.3 and Corollary 3.6(a) may not be very accurate. Claborn's construction was again used in [2, Theorem 3.1 and Theorem 3.4] to give the following result, which shows that essentially any (possible) thing can happen.

THEOREM 3.7 *(a) Let r and s be rational numbers with $1 \leq r \leq s$. Then there is a Dedekind domain R with finite divisor class group such that $\rho(R) = r$ and $M^*(R, V)/m^*(R, V) = s$.*

(b) Let r and s be real numbers or ∞ with $1 \leq r \leq s \leq r^2$. Then there is a Dedekind domain R with torsion divisor class group such that $\rho(R) = r$ and $M^*(R, \mathcal{Z}_R)/m^*(R, \mathcal{Z}_R) = s$. $\square$

4 KRULL DOMAINS

Much of the early work in factorization theory took place in the setting of algebraic number rings, more generally, in Dedekind domains or Krull domains. Usually results for Dedekind domains have direct analogues to Krull domains by just replacing products of maximal ideals in a Dedekind domain by v-products of height-one prime ideals in a Krull domain. In this case, we have already observed that bounds can be given for $\rho(R)$ in terms of the divisor class group $\mathrm{Cl}(R)$. However, $\rho(R)$ depends not on the group-theoretic properties of $\mathrm{Cl}(R)$, but rather on the distribution of the height-one prime ideals in the divisor classes (cf. Corollary 3.4(b)). Our next result is from [55] and uses the important concept of block monoids (see, for example, [32], [41], [42], [43], [55], and [56]).

THEOREM 4.1 *Let R be a Krull domain. Then $\rho(R)$ depends only on $G = Cl(R)$ and $S = \{g \in G | g = [P] \text{ for some } P \in X(R)\}$.* $\square$

Let G be an abelian group and $S \subseteq G - \{0\}$. We define $\{G, S\}$ to be a *realizable pair* if there is a Dedekind domain R with $\mathrm{Cl}(R) \cong G$ and $S = \{g \in G | g = [P] \text{ for some nonprincipal } P \in X(R)\}$. We will need two theorems of Grams [52, Corollaries 1.5 and 1.6] which characterize realizable pairs when either $G = \mathbb{Z}$ or G is a torsion abelian group. These characterizations are as follows: (i) $\{\mathbb{Z}, S\}$ is realizable if and only if S generates G and S contains both positive and negative elements of $\mathbb{Z}$ and, (ii) if G is torsion, then $\{G, S\}$ is realizable if and only if S generates G. These ideas extend in the obvious way to Krull domains. Let G be an abelian group and R_1 and R_2 be Krull domains with realizable pairs $\{G, S_1\}$ and $\{G, S_2\}$ respectively. Then $\rho(R_1) \leq \rho(R_2)$ if $S_1 \subseteq S_2$. Realizable pairs are also studied in [50], [57], and [60].

We have already observed in Theorem 2.1 that for Dedekind domains, $\rho(R)$ may be arbitrary. However, this is not the case if S is finite. Our next result [4, Theorem 10], conjectured in [1, page 233], is a special case of a more general result for atomic monoids proved in [4, Theorem 7]. This is another instance where suitable finiteness conditions imply that $\rho(R)$ is nice.

THEOREM 4.2 *Let R be a Krull domain such that only a finite number of divisor classes contain height-one prime ideals (thus $Cl(R)$ is finitely generated). Then $\rho(R)$ is rational and $\rho(R)$ is realized by a factorization. In particular, this is the case if $Cl(R)$ is finite.* $\square$

For R a Dedekind domain with finite divisor class group, how the distribution of prime ideals in the divisor classes affects various factorization properties has been studied extensively in [33], [34], [35], and [65].

There has been much less work on $\rho(R)$ when R is a Krull domain with non-torsion divisor class group. The case when $\text{Cl}(R) = \mathbb{Z}$ was studied in [35], [16], and [17]. We next summarize some of the main results; they are a very good illustration of how $\rho(R)$ is determined by the distribution of the height-one prime ideals in the divisor classes. If S is finite, then by Theorem 4.2, $\rho(R)$ is realized by a factorization. Our next result is from [16, Theorem 2.4], and improves Theorem 2.3 in the case R is a Krull domain with $\text{Cl}(R) = \mathbb{Z}$.

THEOREM 4.3 *Let R be a Krull domain with realizable pair $\{\mathbb{Z}, S\}$. Then the following statements are equivalent.*

(1) $\rho(R) < \infty$.
(2) R *is Φ-finite.*
(3) S *is either bounded above or below (or both).*

Proof: (1) $\Rightarrow$ (2) follows from Theorem 2.3. (2) $\Rightarrow$ (3) The proof of [16, Theorem 2.1] shows that $\Phi(2) = \infty$ if S is neither bounded below nor bounded above. (3) $\Rightarrow$ (1) We may assume that S is bounded below; the proof of the bounded above case is similar. Define $\varphi : R^* \to \mathbb{Z}_+$ by $\varphi(x) = \sum\{v_P(x)|[P] > 0\}$. By [16, claims 1 and 2, pages 104-105], φ is a bounded semi-length function on R. Hence $\rho(R) < \infty$ by Theorem 3.1. $\square$

With the notation of the above theorem, [16, Corollary 2.5] gives $\rho(R) \leq m(m+1)(m^2 + 3m - 2)/4$ if either $S \subseteq \{-m, -m+1, \dots\}$ or $S \subseteq \{\dots, m-1, m\}$ for m a positive integer.

By above, and using an automorphism argument, we may assume that $S = \{-m_1, -m_2, \ \dots, -m_t, n_1, n_2, \dots\}$, here and throughout, the m_i's and n_j's are positive integers. When $t = 1$, we have the following very satisfactory result for $\rho(R)$ from [16, Theorem 3.2 and Corollary 3.3].

THEOREM 4.4 *Let R be a Krull domain with realizable pair $\{\mathbb{Z}, S\}$, where $S = \{-m, n_1, n_2, \dots\}$. Then $\rho(R) < \infty$ and $\rho(R)$ is realized by a factorization.*

Proof: (Sketch) Let $\pi : \mathbb{Z} \to \mathbb{Z}/m\mathbb{Z}$ be the natural map with $\pi(n) = \overline{n}$ for each $n \in \mathbb{Z}$. Let $S' = \pi(S) - \{\overline{0}\}$. Then S' generates $\mathbb{Z}/m\mathbb{Z}$ since S generates $\mathbb{Z}$. Let R' be a Dedekind domain with realizable pair $\{\mathbb{Z}/m\mathbb{Z}, S'\}$. Then $\rho(R')$ is rational and $\rho(R')$ is realized by a factorization by Theorem 4.2. As in the proof of [16, Theorem 3.2], it may be shown that $\rho(R) = \rho(R')$ and $\rho(R)$ is realized by a factorization. □

COROLLARY 4.5. *Let R be a Krull domain with realizable pair $\{\mathbb{Z}, S\}$.*

    (a) *If $S = \{-2, n_1, n_2, \dots\}$, then R is an HFD.*

    (b) *If $m \geq 2$ and $S = \{-m, n_1, n_2, \dots\}$, then $\rho(R) \leq m/2$. Moreover, if $n_1, n_2, \dots$ forms a complete set of residues modulo m, then $\rho(R) = m/2$.*

    (c) *If $p \geq 2$ is prime and $S = \{-p, n_1, n_2, \dots\}$, then R is an HFD if and only if there is a fixed integer k with $1 \leq k \leq p-1$ such that each n_i is congruent to either 0 or k modulo p.*

Proof. Let R' be the Dedekind domain with realizable pair $\{R', S'\}$ and $\rho(R) = \rho(R')$ from the proof of Theorem 4.4. In part (a), $\mathrm{Cl}(R') = \mathbb{Z}/2\mathbb{Z}$, so R', and hence R, is an HFD by Corollary 3.4(c). In part (b), $\mathrm{Cl}(R') = \mathbb{Z}/m\mathbb{Z}$ and each nonzero divisor class contains a prime ideal. Hence $\rho(R) = \rho(R') = m/2$ by Corollary 3.4(b). Part (c) follows in a similar manner from [34, Corollary 3.3]. □

The $t = 2$ case is not so nice. In fact, our next example from [16, Example 3.4] shows that in this case $\rho(R)$ need not be realized by a factorization.

EXAMPLE 4.6 Let $S = \{-m, -1, n_1, n_2, \dots\}$ with infinitely many of the n_i's congruent to 1 modulo m. If R is a Dedekind domain with realizable pair $\{\mathbb{Z}, S\}$, then $\rho(R) = m$, but $\rho_R(R) < m$ for every nonunit $x \in R^*$ (see [16, Example 3.4] for details). □

However, we do have the following result from [17, Theorem 9]. For distinct primes $p_1, \dots, p_k$ and $m_1 = p_1^{a_1} p_2^{a_2} \cdots p_k^{a_k}$ and $m_2 = p_1^{b_1} p_2^{b_2} \cdots p_k^{b_k}$, we define $\langle\langle m_1, m_2 \rangle\rangle = p_1^{c_1} p_2^{c_2} \cdots p_k^{c_k}$, where $c_i = \max\{a_i, b_i\}$ if $a_i \neq b_i$ and $c_i = 0$ if $a_i = b_i$.

THEOREM 4.7 *Let R be a Krull domain with realizable pair $\{\mathbb{Z}, S\}$, where $S = \{-m_1, -m_2, n_1, n_2, \dots\}$ and $c = \langle\langle m_1, m_2 \rangle\rangle$. If $S_+ \subseteq \{c, 2c, 3c, \dots\}$, then $\rho(R)$ is rational and $\rho(R)$ is realized by a factorization.* □

Several other rather techincal conditions for a Krull domain R with $Cl(R) = \mathbb{Z}$ to be an HFD are given in [17].

For torsion-free abelian groups G other than $\mathbb{Z}$ not much is known about the elasticity of R when $Cl(R) = G$. We end with an interesting example from [16, Example 2.6].

EXAMPLE 4.8   Let $G = \mathbb{Z} \oplus \mathbb{Z}$ and $S = \{(-1,0), (1,0), (2,0), (3,0), \dots \} \cup \{(0,-1), (0,1), (0,2), (0,3), \dots \}$. Then $\{G, S\}$ is a realizable pair by [52]. Let R be a Dedekind domain with this pair. Any irreducible element $x \in R$ has the form $xR = P^n A_n$ or $xR = Q^n B_n$, where P, A_n, Q, and B_n are prime ideals from the ideal classes $(-1,0)$, $(n,0)$, $(0,-1)$, and $(0,n)$, respectively. A simple counting argument shows that R is an HFD, and hence $\rho(R) = 1$. However, $S' = S \cup \{(-1,1)\}$ is also a realizable pair; but if R' is a Dedekind domain with this pair, then $\rho(R') = \infty$. To see this, let A_n and B_n be as above (but prime ideals of R') and T a prime ideal of R' with $[T] = (-1,-1)$. For each $n \geq 1$, we have irreducibles x_n, y_n, z_n, and $w \in R'$ defined by $x_n R' = T^n B_1^n A_n$, $y_n R' = T^n A_1^n B_n$, $z_n R' = T^n A_n B_n$, and $wR' = TB_1 A_1$. Then for each $n \geq 1$, we have a factorization $x_n y_n = u_n w^n z_n$ for some $u_n \in U(R')$. Thus $\Phi(2) = \infty$, so $\rho(R') = \infty$ by Theorem 2.3. Hence by adding one additional element to the set S associated with an HFD, we obtain a new realizable pair and an associated Dedekind domain with infinite elasticity. Note that this cannot happen when $G = \mathbb{Z}$ by Theorem 4.2.   □

5   SPECIAL CLASSES OF DOMAINS

In this section, we investigate $\rho(R)$ for more general classes of integral domains. The simplest atomic domains are those with only a finite number of nonassociate atoms. Such atomic domains, called *Cohen-Kaplansky domains* (CK domains), were studied extensively in [10]. An integral domain R is a CK domain if and only if R is a one-dimensional semilocal Noetherian domain and for each non-principal maximal ideal of R, R/M is finite and R_M is analytically irreducible (i.e., its completion is a domain) [10, Theorem 2.4]. The next simplest atomic domains would be those domains in which almost all atoms are prime. Such atomic domains, called *generalized CK domains*, were introduced in [5]. A CK domain is certainly a generalized CK domain. However, for k a proper subfield of a finite field

K, $k + XK[[X]]$ and $k + XK[X]$ are each generalized CK domains, but $k + XK[X]$ is not a CK domain. Note that if R is a generalized CK domain and S is the multiplicatively closed set generated by all the principal primes of R, then R_S is an atomic CK domain. By Theorem 2.7, this usually reduces elasticity problems for generalized CK domains to CK domains. An integral domain R is a *weakly Krull domain* if $R = \cap\{R_P | P \in X(R)\}$ is of finite character. Clearly any Krull domain or one-dimensional Noetherian domain is a weakly Krull domain. Such domains, although not called weakly Krull domains there, were the subject of [11] (also, see [54] and [56]). As in [9], R is said to be a *weakly factorial domain* if each nonunit of R is a product of primary elements. A generalized CK domain is also weakly factorial [5, Corollary 5]. Moreover, a weakly Krull domain R is weakly factorial if and only if $Cl_t(R) = 0$ [12, Theorem]. Recall that $Cl_t(R)$, the *t-class group* of R, is the group of t-invertible t-ideals of R modulo its subgroup of principal fractional ideals. This definition is due to Bouvier and Zafrullah and was introduced in [28]. When R is a Krull domain $Cl_t(R) = Cl(R)$, while $Cl_t(R) = Pic(R)$ for any one-dimensional integral domain R. For these and other results, the reader is referred to [28] or [14]. First we show that $\rho(R)$ is always realized by a factorization when R is a generalized CK domain. Some specific calculations will be given in the next section; also see [56, Section 8].

THEOREM 5.1 *Let R be a generalized CK domain. Then $\rho(R)$ is rational and $\rho(R)$ is realized by a factorization.*
Proof: This is stated for R a CK domain in [4, Corollary 8]. As observed in [19, Theorem 3.5], an elementary generalization of [4, Theorem 7] establishes the result for generalized CK domains. Theorem 5.1 also follows from [4, Theorem 7] via Theorem 2.7. □

THEOREM 5.2 *Let R be an atomic weakly Krull domain. Then $\rho(R) \leq D(Cl_t(R)) \sup\{\rho(R_P) | P \in X(R)\}$. In particular, if R is a one-dimensional Noetherian domain, then $\rho(R) \leq D(Pic(R)) \cdot \sup\{\rho(R_P) | P \in X(R)\}$.*
Proof: This is [1, Theorem 2.14] and the paragraph after the proof. The "in particular" statement holds since $Pic(R) = Cl_t(R)$ for any one-dimensional integral domain R. □

Our next corollary is from [1, Corollary 2.15] (also, see [23, Corollary

2.5]).

COROLLARY 5.3 *Let R be an atomic weakly factorial domain. Then $\rho(R) = \sup\{\rho(R_P)|P \in X(R)\}$.*

Proof: We have already observed that an atomic weakly factorial domain is weakly Krull and $\mathrm{Cl}_t(R) = 0$. Theorem 5.2 thus gives "$\leq$". For the reverse inequality, let $P \in X(R)$ and $0 \neq z \in P_P$. Then $zR_P \cap R = yR$ for some $y \in R^*$ [12, Theorem (6)], and $L_R(y) = L_{R_P}(z)$ and $l_R(y) = l_{R_P}(z)$ by [12, Theorem (5)]. Thus $\rho_R(y) = \rho_{R_P}(z)$; so $\sup\{\rho(R_P)|P \in X(R)\} \leq \rho(R)$ and we have equality. $\square$

Somewhat stronger estimates are given in [55, Theorem 5] for the case when R is a one-dimensional Noetherian domain with integral closure $\overline{R}$ a finitely generated R-module. In particular, we have the following result [55, Corollary 4].

THEOREM 5.4 *Let R be an order in an algebraic number field and $\overline{R}$ its integral closure.*

(a) *If for some prime ideal P of R there is more than one prime ideal of $\overline{R}$ lying above P, then $\rho(R) = \infty$.*

(b) *If for every prime ideal P of R there is exactly one prime ideal of $\overline{R}$ lying above P, then $\rho(R)$ is rational and $\rho(R)$ is realized by a factorization.* $\square$

It was also conjectured [1, page 231] that $1 \leq \rho(R) \leq \max\{|\mathrm{Cl}_t(R)|/2, 1\}\sup\{\rho(R_P)|P \in X(R)\}$ when R is an atomic weakly Krull domain. This conjecture is shown to be false in [55, Example, p. 384]. Let $R = \mathbb{Z}[3i] \subset \mathbb{C}$. It is shown that $\rho(R) = 3/2$, while the stated conjecture would imply that $\rho(R) = 1$, a contradiction. Another counterexample to the conjecture may be obtained using $R = \mathbb{Z}/p\mathbb{Z}[X^2, X^3]$. In this case, $\mathrm{Cl}_t(R) = \mathrm{Pic}(R) = \mathbb{Z}/p\mathbb{Z}$, and thus the right-hand side in the conjecture becomes $(p/2)(3/2) = 3p/4$. However, by Theorem 6.1 of the next section, $\rho(R) = (p+2)/2$. Thus the conjecture holds precisely when $p \geq 4$. Hence $\mathbb{Z}/2\mathbb{Z}[X^2, X^3]$ and $\mathbb{Z}/3\mathbb{Z}[X^2, X^3]$ are also counterexamples to the conjecture.

We end this section with a characterization of one-dimensional local domains (local means Noetherian quasilocal) with finite elasticity from [1, Theorem 2.12] (also cf. [1, Theorem 2.7 and Theorem, page

227]). Some specific examples will be given in the next section. [1, Example 3.3] shows that the Noetherian hypothesis is needed.

THEOREM 5.5 *The following statements are equivalent for a one-dimensional local domain R with maximal ideal M.*

(1) $\rho(R) < \infty$.

(2) R is analytically irreducible.

(3) For every DVR $(D, \pi D)$ centered on M with associated valuation v, $M^*(R, v|_R) < \infty$.

(4) There is a DVR $(D, \pi D)$ centered on M with associated valuation v such that $M^*(R, v|_R) < \infty$.

(5) Every semi-length function on R is bounded.

(6) There is a bounded semi-length function on R. Moreover, if $\rho(R) < \infty$, then $\rho(R)$ is rational. $\square$

6   SUBRINGS OF $K[X]$ or $K[[X]]$

In this section, we give some very specific calculations for $\rho(R)$ when R is a subring of either $K[X]$ or $K[[X]]$ for K a field. Clearly neither $K[X^2, X^3]$ nor $K[[X^2, X^3]]$ is an HFD since $X^6 = (X^2)^3 = (X^3)^2$; in fact, $\rho(R) \geq 3/2$ for each. We first compute $\rho(K[X^2, X^3])$ as in [19, Theorem 2.4], and the Davenport constant somewhat surprisingly appears. This is probably the first nontrivial calculation of $\rho(R)$ for a class of domains other than Krull domains.

THEOREM 6.1 *Let K be a field and $R = K[X^2, X^3]$. Then $\rho(R) = (D(K) + 2)/2$. Thus $\rho(R) < \infty$ if and only if K is a finite field. Moreover, if K is finite, then $\rho(R)$ is realized by a factorization. In particular, if $|K| = p^n$, then $\rho(R) = (n(p - 1) + 3)/2$.*
Proof: (Sketch). Define a semi-length function φ on $K[X]$ by $\varphi(aX^n f_1 \cdots f_m) = n + m$, where $0 \neq a \in K$ and each $f_i \in K[X]$ is irreducible with $f_i(0) = 1$. First assume that K is a finite field, and hence $D(K) < \infty$. Let $f = aX^n f_1 \cdots f_m \in R$ as above be irreducible. If $n = 0$, then $m \leq D(K)$; and if $n = 2$ or $n = 3$, then $m \leq D(K) - 1$. Thus $M^* = M^*(R, \varphi) \leq 3 + D(K) - 1 = D(K) + 2$. Clearly $m^* = m^*(R, \varphi) = 2$ since X^2 is irreducible, but not prime, in R. Thus $\rho(R) \leq M^*/m^* \leq (D(K) + 2)/2$ by Theorem 3.1. Suppose

that $D(K) = n$. Let $a_1, \ldots, a_{n-1} \in K$ with no subsum 0. Then $f = X^3(1+a_1X)\cdots(1+a_{n-1}X)$ and $g = X^3(1-a_1X)\cdots(1-a_{n-1}X)$ are each irreducible in R since any product of the $(1 + a_iX)$'s (or $(1 - a_iX)$'s) has nonzero X coefficient. However, $fg = (X^2)^3(1 - a_1^2X^2)\cdots(1 - a_{n-1}^2X^2)$ is a product of 2 and $D(K) + 2$ irreducible elements in R. Hence $(D(K) + 2)/2 \le \rho(R)$ and we have equality. This factorization also shows that $\rho(R)$ is realized by a factorization. If $|K| = p^n$, then $K \cong (\mathbb{Z}/p\mathbb{Z})^n$ as an abelian group; so $D(K) = n(p-1)+1$. Hence $\rho(R) = (D(K)+2)/2 = (n(p-1)+3)/2$. The above proof also shows that if K is infinite (and hence also $D(K) = \infty$), then $\rho(R) = \infty$. □

However, $\rho(K[[X^2, X^3]]) = 3/2$ [1, Example 3.6]. This is essentially because in the above proof, each $1 - a_i^2X^2$ is a unit in $K[[X^2, X^3]]$.

The correct setting for these examples is the semigroup ring $K[S]$ or (power series) semigroup ring $K[[S]]$, for S a numerical semigroup. $\rho(K[S])$ and $\rho(K[[S]])$ have been investigated in a series of papers [19], [22], [26], and [27]. We next recall a few facts about numerical semigroups and semigroup rings.

A *numerical semigroup* S is an additive submonoid of $\mathbb{Z}_+$ with $\mathbb{Z}_+ - S$ finite. It is well known that $S = \langle n_1, \ldots, n_r \rangle$ for a unique minimal set of generators with $n_1 < \cdots < n_r$ and $\gcd(n_1, \ldots, n_r) = 1$. Conversely, given positive integers $n_1 < \cdots < n_r$ with $\gcd(n_1, \ldots, n_r) = 1$, $S = \langle n_1, \ldots, n_r \rangle$ is a numerical semigroup. When we write $S = \langle n_1, \ldots, n_r \rangle$, we we will always mean that $\{n_1, \ldots, n_r\}$ is such a minimal generating set for S. The *Frobenius number* of a numerical semigroup S is $g(S) = \max\{n \in \mathbb{Z}_+ | n \in S\}$ (let $g(\mathbb{Z}_+) = 0$). Thus $X^n f \in K[S]$ (resp., $K[[S]]$) for any $n > g(S)$ and $f \in K[X]$ (resp., $K[[X]]$). Note that any nonzero submonoid T of $\mathbb{Z}_+$ is isomorphic to a numerical semigroup (just divide out by the gcd of the elements of T); thus any subring of $K[X]$ (resp., $K[[X]]$) generated by monomials over K is isomorphic to $K[S]$ (resp., $K[[S]]$) for a suitable numerical semigroup S. For $S = \langle n_1, \ldots, n_r \rangle$, $K[S] = K[X^{n_1}, \ldots, X^{n_r}] \subseteq K[X]$ and $K[[S]] = K[[X^{n_1}, \ldots, X^{n_r}]] \subseteq K[[X]]$, and $X^{n_1}, \ldots, X^{n_r}$ are the only irreducible monomials in either ring. Each ring is also one-dimensional with finitely generated integral closure, $K[X]$ or $K[[X]]$, respectively. We are interested in the case when S is a proper submonoid of $\mathbb{Z}_+$, i.e., when $n_1 \ge 2$ and $r \ge 2$ (equivalently, $K[S] \ne K[X]$ and $K[[S]] \ne K[[X]]$). Note that in this case, $K[[S]]$ has no prime elements. Also, $K[[S]]$ is local with $U(K[[S]]) = \{f \in K[[S]] | f(0) \ne 0\}$

and analytically irreducible (cf. [1, Theorem 2.12]). Moreover, $K[S]$ is a generalized CK domain and $K[[S]]$ is a CK domain when K is finite. General references for numerical semigroups are [40] and [49].

The two parts of our next theorem are from [26, Theorem 2.4] and [27, Theorem 2.3], respectively. In Theorem 6.2(a), $G(K, S)$ is an abelian group associated with K and S. It turns out that $G(K, S) = \text{Pic}(K[S])$ (this may be seen using the Mayer-Vietoris exact sequence for (U, Pic)). (This observation is due to F. Halter-Koch, see [56, Theorem 8.1].) As in Theorem 6.1, these calculations are based on estimates using semi-length functions. For the $K[S]$ case, we use a semi-length function φ similar to that used in the proof of Theorem 6.1, namely $\varphi(aX^n f_1 \cdots f_m) = 2n/n_1 + m$. For the $K[[S]]$ case, we use the usual order valuation on $K[[X]]$. Finally, observe that $\rho(K[S]) < \infty$ (in fact, $K[S]$ is Φ-finite) if and only if K is a finite field; moreover, in this case $\rho(K[S])$ is realized by a factorization [19, Theorem 3.5] (cf. Theorem 5.1).

THEOREM 6.2 *Let K be a field and $S = \langle n_1, \dots, n_r \rangle$ a numerical semigroup. Then*

(a) $n_r/n_1 \le \rho(K[S]) \le (n_1 D(G(K, S)) + 2g(S) + n_1)/(2n_1)$.
(b) $n_r/n_1 \le \rho(K[[S]]) \le (g(S) + n_1)/n_r$.

Moreover, when $g(S) + n_1 = n_r$, each right-hand side inequality is an equality. $\square$

Note that $g(S) + n_1 \ge n_r$ always holds. This case was investigated in [26, Section 3], where we said that a numerical semigroup $S = \langle n_1, \dots, n_r \rangle$ satisfies *property* $(*)$ if $g(S) + n_1 = n_r$. We now review some facts about such semigroups from [26]. First, suppose that $r = 2$; so $S = \langle n_1, n_2 \rangle$ with $n_1 \ge 2$. Then $g(S) = n_1 n_2 - n_1 - n_2$ [40]. Thus $g(S) + n_1 = n_2$ if and only if $n_1 = r = 2$. It is clear that $n_1 \ge r$ for any $S = \langle n_1, \dots, n_r \rangle$ which satisfies property $(*)$ since $n_1, \dots, n_r$ have distinct residues modulo n_1 because $\{n_1, \dots, n_r\}$ is a minimal generating set for S. However, when $r = 3$, it is no longer true that $S = \langle n_1, n_2, n_3 \rangle$ satisfies property $(*)$ if and only if $n_1 = r = 3$. The easiest such example is $S = \langle 4, 5, 11 \rangle$. Here, $r = 3$, $n_1 = 4$, $n_3 = 11$, and $g(S) = 7$. Thus $g(S) + n_1 = 11 = n_3$; so $S = \langle 4, 5, 11 \rangle$ satisfies property $(*)$. (Other examples are given in [26, Example 3.5].) However, S does satisfy property $(*)$ when $n_1 = r$ [26, Proposition 3.6]. This generalizes the $S = \langle n, n+1, \dots, 2n-1 \rangle$

case from [1, Example 3.6] and [22, Theorem 2.4]. Our next corollary follows directly from the above observations and Theorem 6.2.

COROLLARY 6.3   *(a) If K is a finite field, then $\rho(K[X^n, X^{n+1}, \ldots, X^{2n-1}]) = (nD(G(K,S)) + 3n - 2)/2n$.*
  (b) For any field K, $\rho(K[[X^n, X^{n+1}, \ldots, X^{2n-1}]]) = (2n-1)/n$.   □

We next give some specific examples (also, see [56, Section 8]). They show that $\rho(K[[S]])$ may be strictly between the bounds in Theorem 6.2.(b) when S does not satisfy $(*)$ and $\rho(K[[S]])$ may depend on the field K.

EXAMPLE 6.4   (a) ([26, Example 3.4]) Let $S = \langle 2, 5 \rangle = \{0, 2, 4, 5, 6, \ldots\}$. Then $r = 2$, $n_1 = 2$, $n_2 = 5$, and $g(S) = 3$ and $K[S] = K[X^2, X^5]$. Let char $K = p$ and $|K| = p^n$. Then $G = G(K, S) \cong (\mathbb{Z}/p\mathbb{Z})^{2n}$ if $p \neq 3$ and $G \cong (\mathbb{Z}/9\mathbb{Z})^n$ if $p = 3$. Thus $D(G) = 2n(p - 1) + 1$ if $p \neq 3$ and $D(G) = 8n + 1$ if $p = 3$. Hence $\rho(K[X^2, X^5]) = (5 + 2n(p-1))/2$ if $K = GF(p^n)$ with $p \neq 3$; $(5+8n)/2$ if $K = GF(3^n)$; and ∞ if K is infinite.
  (b) ([27, Theorem 3.5]) Let K be a field and $n > 3$ an integer with $n \not\equiv 0 \pmod 3$. Then $\rho(K[[X^3, X^n]]) = 3$ if $n = 5$ and K is quadratically closed (a field K is *quadratically closed* if each $f(X) \in K[X]$ of degree 2 has a root in K). Otherwise, $\rho(K[[X^3, X^n]]) = 2n/3$.
  (c) ([27, Theorem 3.7(c)]) Let K be a field. Then $\rho(K[[X^4, X^6, X^7]]) = 3$.   □

Next, we look at two other classes of subrings of either $K[X]$ or $K[[X]]$ which were studied in [24]. The rings in the next several results are atomic by [24, Theorem 2.1]. Example 6.6 from [24, Example 3.3] gives a very simple example where $\rho(R)$ is not realized by a factorization.

THEOREM 6.5.   *Let $R_0 \subseteq R_1 \subseteq R_2 \subseteq \cdots$ be an ascending sequence of subrings of a field K with R_0 a field, $K = \cup R_n$, and $R = \prod R_n X^n = \{\sum a_n X^n \in K[[X]] | a_n \in R_n\}$.*
  (a) If each $R_n \subset K$, then $\rho(R) = \infty$.
  (b) Suppose that $R_{n-1} \subset R_n = K$ for some $n \geq 1$. Then $\rho(R) = n$. Moreover, $\rho(R)$ is realized by a factorization if and only if R_{n-1} is not a valuation domain with quotient field K.

Proof: (a) and the first part of (b) are from [24, Theorem 3.1]. The "moreover" statement in (b) is from [24, Theorem 3.2]. □

EXAMPLE 6.6   Let t be an indeterminate over $\mathbb{Q}$ and let $R = \mathbb{Q} + \mathbb{Q}[t]_{(t)}X + X^2\mathbb{Q}(t)[[X]]$ and $S = \mathbb{Q} + \mathbb{Q}[t]X + X^2\mathbb{Q}(t)[[X]]$ be subrings of $\mathbb{Q}(t)[[X]]$. By Theorem 6.5, $\rho(R) = \rho(S) = 2$. Moreover, $\rho(R)$ is not realized by a factorization since $\mathbb{Q}[t]_{(t)}$ is a valuation domain with quotient field $\mathbb{Q}(t)$. However, $\rho(S)$ is realized by the factorization $(aX^2)(a^{-1}X^2) = X^4$, where $a = (t+1)/(t-1) \in \mathbb{Q}(t)$. □

THEOREM 6.7   *Let $R_0 \subseteq R_1 \subseteq R_2 \subseteq \cdots$ be an ascending sequence of subrings of a field K with R_0 a field, $K = \cup R_n$, and $R = \oplus R_n X^n = \{\sum a_n X^n \in K[X] | a_n \in R_n\}$.*

  (a) *If each $R_n \subset K$, then $\rho(R) = \infty$.*
  (b) *If $R_0 \subseteq R_1 = K$, then $\rho(R) = 1$ (R is an HFD).*
  (c) *Suppose that $R_{n-1} \subset R_n = K$ for some $n \geq 2$. Then $\rho(R) < \infty$ if and only K is a finite field. Moreover, in this case $\rho(R) = (2n + D(Pic(R)) - 1)/2$ and $\rho(R)$ is realized by a factorization.*

Proof: (a), (b), and the first part of (c) are from [24, Theorem 4.1]. The "moreover" statement in (c) is from [24, Theorem 4.4 and Remark 4.6(d)]. □

EXAMPLE 6.8   ([24, Remark 4.6(c)]) Let $GF(p^n) = F \subset K = GF(p^{mn})$ with $m > 1$. Let $A = F + X^2 K[X]$ and $B = F + FX + X^2 K[X]$. Then $\rho(A) = (mn(p-1)+3)/2$ by Theorem 6.1 and Theorem 2.6, and $\rho(B) = (n(m-1)(p-1)+4)/2$ by Theorem 6.7(c). □

We close this section by showing that $\rho(\text{Int}(\mathbb{Z})) = \infty$, where $\text{Int}(\mathbb{Z}) = \{f \in \mathbb{Q}[X] | f(\mathbb{Z}) \subseteq \mathbb{Z}\}$ is the *ring of integer-valued polynomials*. $\text{Int}(\mathbb{Z})$ is a two-dimensional Prüfer domain which satisfies ACCP and has no prime elements [15, Theorem 3.2]. It is well known that $\text{Int}(\mathbb{Z})$ is a free abelian group on the set of binomial polynomials $\{B_n(X)\}$, where $B_0(X) = 1$, and $B_n(X) = X(X-1)\cdots(X-n+1)/n!$ for each integer $n \geq 1$. That $\rho(\text{Int}(\mathbb{Z})) = \infty$ was first proved in [29, Theorem 1.6]; we give a stronger result from [15, Theorem 2.4].

THEOREM 6.9   *In $Int(\mathbb{Z})$, $\Phi(k) = \infty$ for each integer $k \geq 2$. Thus $Int(\mathbb{Z})$ is not Φ-finite, and hence $\rho(Int(\mathbb{Z})) = \infty$.*

Proof: For each integer $n \geq 2$, $nB_n(X) = B_{n-1}(X)(X - (n-1))$ with both of the right-hand side factors irreducible in $\text{Int}(\mathbf{Z})$. Since n can be chosen with an arbitrary number of prime factors and any prime element in $\mathbf{Z}$ is irreducible in $\text{Int}(\mathbf{Z})$, we have $\Phi(2) = \infty$. Thus $\Phi(k) = \infty$ for each integer $k \geq 2$, and hence $\rho(\text{Int}(\mathbf{Z})) = \infty$ by Theorem 2.3. $\square$

More generally, let R be an integral domain with quotient field K. We define $\text{Int}(R) = \{f \in K[X] | f(R) \subseteq R\}$ to be the *ring of integer-valued polynomials* over R, and for $\emptyset \neq S \subseteq R$, we define $\text{Int}(S, R) = \{f \in K[X] | f(S) \subseteq R\}$, the *ring of polynomials integer-valued over S*. Several calculations for $\rho(\text{Int}(R))$ and $\rho(\text{Int}(S, R))$ and other results concerning factorization in $\text{Int}(R)$ or $\text{Int}(S, R)$ are given in [15] and [29]. In particular, $\text{Int}(S, R)$ is not atomic when S is finite [15, Proposition 1.1]. We know of no example where $\text{Int}(R) \neq R[X]$ and $\rho(\text{Int}(R)) < \infty$. We next state [15, Proposition 1.7], which includes Theorem 6.9 as a special case.

THEOREM 6.10 *Let R be an integral domain with quotient field K and S a (infinite) subset of R such that*

   i) *$Int(S, R)$ is atomic,*
  ii) *there is a discrete valuation v on K, and*
 iii) *there is a principal prime ideal P such that $|R/P| < \infty$.*

Then $Int(S, R)$ is not Φ-finite, and hence $\rho(Int(S, R)) = \infty$. $\square$

7   GENERALIZATIONS

In this short concluding section, we indicate two ways in which the elasticity $\rho(R)$ of an integral domain R can be generalized. The first generalization concerns commutative cancellative monoids, and the second F-sets.

First, the monoid case. Let H be a commutative cancellative monoid written multiplicatively with $U(H)$ its group of invertible elements. As for domains, we can define the usual terms such as atomic monoid and lengths of factorizations. We say that H is *atomic* if each nonunit $x \in H$ is a product of irreducible elements. Then the *elasticity* of H is $\rho(H) = \sup\{m/n | x_1 \cdots x_m = y_1 \cdots y_n, \text{ each } x_i, y_j \in H \text{ irreducible}\}$. Thus $\rho(R) = \rho(H)$, where $H = R^*$ is the monoid of nonzero elements of R under multiplication. $\rho(H)$ was first defined in [4], where many of the main results were stated in terms of atomic monoids. This idea has been extensively investigated in [55]; also see [32] and [56] for survey articles. This monoid approach is more general and often reveals the combinatorial nature of factorization problems. There is also an extensive literature on divisibility in monoids (cf. [46]).

Secondly, as in [15], we can consider just certain, rather than all, ratios of lengths of factorizations in R. Let R be an atomic integral domain and $\mathcal{I}(R)$ its set of irreducible elements. A nonempty subset $\mathcal{F} \subseteq \mathcal{I}(R)$ is called a *factorization set* (F-set) if whenever some factorization of an $x \in R$ has an irreducible factor from $\mathcal{F}$, then any factorization of x has an irreducible factor from $\mathcal{F}$. F-sets were introduced and studied extensively in [18]. Let $\mathcal{F}$ be an F-set of R and $x \in R$ a nonzero nonunit. In analogy for domains, we then define $l_{\mathcal{F}}(x) = \inf\{n | x = x_1 \cdots x_k \text{ with } x_i \in \mathcal{I}(R) \text{ and } n \text{ of the factors from } \mathcal{F}\}$, $L_{\mathcal{F}}(x) = \sup\{n | x = x_1 \cdots x_k \text{ with } x_i \in \mathcal{I}(R) \text{ and } n \text{ of the factors from } \mathcal{F}\}$, $R_{\mathcal{F}} = \{x \in R | x \text{ is a nonzero nonunit with } l_{\mathcal{F}}(x) \neq 0\}$, and $\rho_{\mathcal{F}}(R) = \sup\{L_{\mathcal{F}}(x)/l_{\mathcal{F}}(x) | x \in R_{\mathcal{F}}\}$. Then $\rho(R) = \rho_{\mathcal{I}(R)}(R)$. These concepts were introduced in [15, Section 3] and were used to investigate factorizations in $\text{Int}(\mathbf{Z})$. In particular, it was shown that $\rho_{\mathcal{F}}(\text{Int}(\mathbf{Z})) = \infty$ for any F-set $\mathcal{F}$ of $\text{Int}(\mathbf{Z})$ [15, Theorem 3.4]. $\rho_{\mathcal{F}}(R)$ has also been studied in [23].

REFERENCES

1. D.D. Anderson and D.F. Anderson, *Elasticity of factorizations in integral domains*, J. Pure Appl. Algebra 80 (1992), 217-235.

2. D.D. Anderson and D.F. Anderson, *Elasticity of factorizations in integral domains, II*, Houston J. Math. 20 (1994), 1-15.

3. D.D. Anderson and D.F. Anderson, *The ring $R[X, r/X]$*, Lecture Notes in Pure and Applied Mathematics, vol. 171, Marcel Dekker, New York, 1995, chapter 8, 95-113.

4. D.D. Anderson, D.F. Anderson, S. Chapman, and W.W. Smith, *Rational elasticity of factorizations in Krull domains*, Proc. Amer. Math. Soc. 117 (1993), 37-43.

5. D.D. Anderson, D.F. Anderson, and M. Zafrullah, *Atomic domains in which almost all atoms are prime*, Comm. Algebra 20 (1992), 1447-1462.

6. D.D. Anderson, D.F. Anderson, and M. Zafrullah, *Factorization in integral domains*, J. Pure Appl. Algebra 69 (1990), 1-19.

7. D.D. Anderson, D.F. Anderson, and M. Zafrullah, *Factorization in integral domains, II*, J. Algebra 152 (1992), 78-93.

8. D.D. Anderson, D.F. Anderson, and M. Zafrullah, *Rings between $D[X]$ and*

$K[X]$, Houston J. Math. **17** (1991), 109-129.

9. D.D. Anderson and L.A. Mahaney, *On primary factorizations*, J. Pure Appl. Algebra **54** (1988), 141-154.

10. D.D. Anderson and J.L. Mott, *Cohen-Kaplansky domains: integral domains with a finite number of irreducible elements*, J. Algebra **148** (1992), 17-41.

11. D.D. Anderson, J.L. Mott, and M. Zafrullah, *Finite character representations for integral domains*, Boll. U. M. I. **16-B** (1992), 613-630.

12. D.D. Anderson and M. Zafrullah, *Weakly factorial domains and groups of divisibility*, Proc. Amer. Math. Soc. **109** (1990), 907-913.

13. D.F. Anderson, *Subrings of $k[X,Y]$ generated by monomials*, Canad. J. Math. **30** (1978), 215-224.

14. D.F. Anderson, *A general theory of class groups*, Comm. Algebra **16** (1988), 805-847.

15. D.F. Anderson, P.-J. Cahen, S. Chapman, and W.W. Smith, *Some factorization properties of the ring of integer-valued polynomials*, Lecture Notes in Pure and Applied Mathematics, vol. 171, Marcel Dekker, New York, 1995, chapter 10, 125-142.

16. D.F. Anderson, S. Chapman, and W.W. Smith, *Some factorization properties of Krull domains with infinite cyclic divisor class group*, J. Pure and Apppl. Algebra **96** (1994), 97-112.

17. D.F. Anderson, S. Chapman, and W.W. Smith, *On Krull half-factorial domains with infinite cyclic divisor class group*, Houston J. Math. **20** (1994), 561-570.

18. D.F. Anderson, S. Chapman, and W.W. Smith, *Factorization sets and half-factorization sets in integral domains*, J. Algebra **178** (1995), 92-121.

19. D.F. Anderson, S. Chapman, F. Inman, and W.W. Smith, *Factorization in $K[X^2, X^3]$*, Arch. Math. **61** (1993), 521-528.

20. D.F. Anderson, S. Chapman, and W.W. Smith, *Overrings of half-factorial domains*, Canad. Math. Bull. **37** (1994), 437-442.

21. D.F. Anderson, S. Chapman, and W.W. Smith, *Overrings of half-factorial domains, II*, Comm. Algebra **23** (1995), 3961-3976.

22. D.F. Anderson and S. Jenkens, *Factorization in $K[X^n, X^{n+1}, \ldots, X^{2n-1}]$*, Comm. Algebra **23** (1995), 2561-2576.

23. D.F. Anderson, G.-I. Kim, H.-J. Oh, and J. Park, *Splitting multiplicative sets and elasticity*, preprint.

24. D.F. Anderson and J. Park, *Factorization in subrings of $K[X]$ or $K[[X]]$*, these Proceedings.

25. D.F. Anderson and P. Pruis, *Length functions on integral domains*, Proc. Amer. Math. Soc. **113** (1991), 933-937.

26. D.F. Anderson and C. Scherpenisse, *Factorization in $K[S]$*, Lecture Notes in Pure and Applied Mathematics, vol. 185, Marcel Dekker, New York, 1997, 45-56.

27. D.F. Anderson and J. Winner, *Factorization in $K[[S]]$*, these Proceedings.

28. A. Bouvier, *Le groupe des classes d'un anneau intégré*, 107ème Congres National des Societes Savantes, Brest, Fasc. **IV** (1982), 85-92.

29. P.-J. Cahen and J.-L. Chabert, *Elasticity for integer-valued polynomials*, J. Pure Appl. Algebra **103** (1995), 303-311.

30. L. Carlitz, *A characterization of algebraic number fields with class number two*, Proc. Amer. Math. Soc. 11 (1960), 391-392.

31. S. Chapman, *On the Davenport constant, the cross number, and their applications in factorization theory*, Lecture Notes in Pure and Applied Mathematics, vol. 171, Marcel Dekker, New York, 1995, chapter 14, 167-190.

32. S. Chapman and A. Geroldinger, *Krull domains and monoids, their sets of lengths and associated combinatorial problems*, these Proceedings.

33. S. Chapman and W.W. Smith, *Factorization in Dedekind domains with finite class group*, Israel J. Math. 71 (1990), 65-95.

34. S. Chapman and W. W. Smith, *On the HFD, CHFD, and k-HFD properties in Dedekind domains*, Comm. Algebra 20 (1992), 1955-1987.

35. S. Chapman and W. W. Smith, *On the k-HFD property in Dedekind domains with small class group*, Mathematika 39 (1992), 330-340.

36. S. Chapman and W.W. Smith, *An analysis using the Zaks-Skula constant of element factorizations in Dedekind domains*, J. Algebra 159 (1993), 176-190.

37. S. Chapman and W.W. Smith, *On the lengths of factorizations of elements in an algebraic number ring*, J. Number Theory 43 (1993), 24-30.

38. L. Claborn, *Specified relations in the ideal group*, Michigan Math. J. 15 (1968), 249-255.

39. R. M. Fossum, *The Divisor Class Group of a Krull Domain*, Springer, New York, 1973.

40. R. Fröberg, G. Gottlieb, and R. Häagkvist, *On numerical semigroups*, Semigroup Forum 35 (1987), 63-83.

41. A. Geroldinger, *Über nicht-eindeutige Zerlegungen in irreduzible Elements*, Math. Z. 197 (1988), 505-529.

42. A. Geroldinger, *T-block monoids and their arithmetical applications to noetherian domains*, Comm. Algebra 22 (1994), 1603-1615.

43. A. Geroldinger and F. Halter-Koch, *Non-unique factorizations in block semigroups and arithmetical applications*, Math. Slovaca (1992), 641-661.

44. A. Geroldinger and F. Halter-Koch, *On the asymptotic behaviour of lengths of factorizations*, J. Pure Appl. Algebra 77 (1992), 239-252.

45. A. Geroldinger and F. Halter-Koch, *Realization theorems for semigroups with divisor theory*, Semigroup Forum 44 (1992), 229-237.

46. A. Geroldinger and G. Lettl, *Factorization problems in semigroups*, Semigroup Forum 40 (1990), 23-38.

47. A. Geroldinger and R. Schneider, *On Davenport's constant*, J. Combin. Theory Ser. A 61 (1992), 147-152.

48. R. Gilmer, *Multilpicative Ideal Theory*, Marcel Dekker, New York, 1972.

49. R. Gilmer, *Commutative Semigroup Rings*, Chicago Lectures in Mathematics, University of Chicago Press, Chicago, 1984.

50. R. Gilmer, W. Heinzer, and W.W. Smith, *On the distribution of prime ideals within the ideal class group*, Houston J. Math. 22 (1996), 51-59.

51. A. Grams, *Atomic domains and the ascending chain condition for principal ideals*, Math. Proc. Cambridge Philos. Soc. 75 (1974), 321-329.

52. A. Grams, *The distribution of prime ideals of a Dedekind domain*, Bull. Austral. Math. Soc. 11 (1974), 429-441.

53. F. Halter-Koch, *On the asymptotic behaviour of the number of distinct factorizations into irreducibles*, Ark. Mat. 31 (1993), 297-305.

54. F. Halter-Koch, *Divisor theories with primary elements and weakly Krull domains*, Boll. U. M. I. 9-B (1995), 417-441.

55. F. Halter-Koch, *Elasticity of factorizations in atomic monoids and integral domains*, J. Théorie des Nombres de Bordeaux 7 (1995), 367-385.

56. F. Halter-Koch, *Finitely generated monoids, finitely primary monoids and factorization properties of integral domains*, these Proceedings.

57. D. Michel and J.-L. Steffan, *Répartition des idéaux premiers parmi les classes d'idéaux dans un anneau de Dedekind et équidécomposition*, J. Algebra 98 (1986), 82-94.

58. W. Narkiewicz, *A note on elasticity of factorizations*, J. Number Theory 51 (1995), 46-47.

59. M. Roitman, Polynomial extensions of atomic domains, J. Pure Appl. Algebra 87 (1993), 187-199.

60. L. Skula, *On c-semigroups*, Act. Arith. 31 (1976), 247-257.

61. J. L. Steffan, *Longuers des diecompositions en produits d'ieliements irrieductibles dans un anneau de Dedekind*, J. Algebra 102 (1986), 229-236.

62. R. J. Valenza, *Elasticity of factorizations in number fields*, J. Number Theory 36 (1990), 212-218.

63. R. J. Valenza, *Personal communication* (March 1, 1996).

64. A. Zaks, *Half-factorial domains*, Bull. Amer. Math. Soc. 82 (1976), 721-724.

65. A. Zaks, *Half-factorial domains*, Israel J. Math. 37 (1980)), 281-302.

66. A. Zaks, *Atomic rings without a.c.c. on principal ideals*, J. Algebra 74 (1982), 223-231.

Finitely Generated Monoids, Finitely Primary Monoids, and Factorization Properties of Integral Domains

Franz Halter-Koch
Institut für Mathematik, Karl-Franzens-Universität, A-8010 Graz, Austria.

1. MONOIDS AND INTEGRAL DOMAINS

The theory of non-unique factorizations in integral domains has its origins in the theory of algebraic number fields. It became an independent area of research by the work of W. Narkiewicz about thirty years ago; see [29], Ch. 9. It soon turned out that most problems and results in the theory are combinatorial in nature and can be formulated for more general integral domains and for cancellative monoids.

In this survey article, we give an overview of results which can be derived in finitely generated and finitely primary monoids. In our opinion, the language of monoids is the adequate framework for the formulation and investigation of most factorization problems, even if the main interest lies in the applications for integral domains. This is not primarily a question of utmost generality but of usefulness. In many cases, the factorization properties of an integral domain have their counterparts in suitably constructed monoids (see section 5) and it is much easier to deal with them in those monoids instead of considering the integral domains themselves.

Throughout this paper, a monoid H is assumed to be commutative and cancellative. Unless otherwise stated, we write H multiplicatively and denote by $1 \in H$ its neutral element. Only the monoids $\mathbf{N_0}$, $\mathbf{N_0^s}$ (for $s \in \mathbf{N}$) and $\mathbf{N_0^{(I)}}$ (for any set I) will be written additively. If H_1, H_2 are monoids, we denote by $H_1 \times H_2$ their direct product, and we view H_1 and H_2 as submonoids of $H_1 \times H_2$. If $(H_\lambda)_{\lambda \in \Lambda}$ is a family of monoids, we define their *coproduct* by

$$\coprod_{\lambda \in \Lambda} H_\lambda = \left\{ (x_\lambda)_{\lambda \in \Lambda} \in \prod_{\lambda \in \Lambda} H_\lambda \;\middle|\; x_\lambda = 1 \text{ for almost all } \lambda \in \Lambda \right\}.$$

For a monoid H, we denote by $H^\times$ the group of invertible elements of H, and we call H *reduced* if $H^\times = \{1\}$. We use the notions of divisibility theory in a monoid H as in [25], section 2.14. Since associated elements have the same divisibility properties, we may pass from H to $H/H^\times$ and assume that H is reduced whenever this is convenient.

For any set P, we denote by $\mathcal{F}(P)$ the free abelian monoid generated by P, and we write the elements of $\mathcal{F}(P)$ in the form

$$a = \prod_{p \in P} p^{v_p(a)},$$

where $v_p(a) \in \mathbf{N_0}$, $v_p(a) = 0$ for almost all $p \in P$. A monoid H is factorial if and only if $H = \mathcal{F}(P) \times H^\times$, where P is a full system of repesentatives of non-associated primes of H.

For a subset M of a monoid H, we denote by $[M]$ the submonoid of H generated by M. The irreducible elements of H are called *atoms*, $\mathcal{A}(H)$ denotes the set of all atoms of H, and H is called *atomic* if $H = [\mathcal{A}(H) \cup H^\times]$.

For an integral domain R, we denote by $R^\bullet = R \setminus \{0\}$ the multiplicative monoid of R. We call R atomic if $R^\bullet$ is atomic. More general, if $\mathbf{P}$ is any factorization property, we say that R has property $\mathbf{P}$ if and only if the monoid $R^\bullet$ has property $\mathbf{P}$. The reduced monoid $\mathcal{H}(R) = R^\bullet/R^\times$ can be identified with the monoid of non-zero principal ideals of R.

2. SOME FINITENESS PROPERTIES OF FACTORIZATIONS

Throughout this section, let H be a reduced atomic monoid, $H \neq H^\times$ and $\mathcal{A} = \mathcal{A}(H)$ the set of atoms of H. The canonical epimorphism

$$\pi: \begin{cases} \mathbb{N}_0^{(\mathcal{A})} & \to & H \\ (n(u))_{u \in \mathcal{A}} & \mapsto & \prod_{u \in \mathcal{A}} u^{n(u)} \end{cases}$$

is called the *factorization homomorphism* of H. The monoid $\mathcal{Z}(H) = \mathcal{N}_0^{(\mathcal{A})}$ is called the *factorization monoid of* H. For any $a \in H$, the elements $\mathbf{n} \in \pi^{-1}(a)$ are called the *factorizations of* a, and for $\mathbf{n} \in \pi^{-1}(a)$ we call

$$|\mathbf{n}| = \sum_{u \in \mathcal{A}} n(u) \in \mathbb{N}_0$$

the *length* of $\mathbf{n}$. For $a \in H$, we denote by

$$L(a) = L^H(a) = \{|\mathbf{n}| \mid \mathbf{n} \in \pi^{-1}(a)\} \subset \mathbb{N}_0$$

the set of lengths of (factorizations of) a and by

$$\mathcal{L}(H) = \{L(a) \mid a \in H\}$$

the system of all sets of lengths of factorizations of elements of H.

For $a \in H \setminus \{1\}$, we call

$$\varrho(a) = \varrho^H(a) = \frac{\sup L(a)}{\inf L(a)} \in \mathbb{Q}_{\geq 1} \cup \{\infty\}$$

the *elasticity* of a, and we denote by

$$\varrho(H) = \sup \{\varrho(a) \mid 1 \neq a \in H\} \in \mathbb{R}_{\geq 1} \cup \{\infty\}$$

the *elasticity of* H. We say that H *has accepted elasticity* if $\varrho(H) = \varrho(a)$ holds for some $a \in H$. The elasticity is a first rather coarse measure for the deviation of H from being factorial; see the survey article of D. F. Anderson [8] and the references given there.

For $a \in H$, we denote by

$$\mathbf{f}(a) = \mathbf{f}^H(a) = \#\pi^{-1}(a) \in \mathbb{N} \cup \{\infty\}$$

the number of factorizations of a.

Lemma 2.1. *Let H be a reduced atomic monoid.*
  i) *H is factorial if and only if $\mathbf{f}(a) = 1$ for all $a \in H$.*
  ii) *If H is not factorial, then $\sup\{\mathbf{f}(a) \mid a \in H\} = \infty$.*

Proof. i) is obvious.

  ii) If H is not factorial, then there is some $a \in H$ such that $\mathbf{f}(a) \geq 2$. Suppose that $\mathbf{m}, \mathbf{n} \in \pi^{-1}(a)$, $\mathbf{m} \neq \mathbf{n}$. For $0 \leq j \leq k$, we have $j\mathbf{m} + (k-j)\mathbf{n} \in \pi^{-1}(a^k)$, and consequently $\mathbf{f}(a^k) \geq k+1$. $\square$

Lemma 2.1 has a counterpart for lengths. Recall that H is called *half-factorial* if $\#L(a) = 1$ for all $a \in H$ (equivalently: any two factorizations of an element $a \in H$ have the same length).

Lemma 2.2. *Let H be an atomic monoid.*
  i) *H is half-factorial if and only if $\varrho(H) = 1$.*
  ii) *If H is not half-factorial, then $\sup\{\#L(a) \mid a \in H\} = \infty$.*

Proof. i) is obvious by definition, and ii) is proved as in Lemma 2.1. $\square$

Next we discuss two finiteness theorems concerning the number of factorizations.

Theorem 2.3. *Let H be a reduced atomic monoid and $a \in H$. Then $\mathbf{f}(a) < \infty$ if and only if the number of atoms dividing a is finite.*

Theorem 2.4. *Let H be a reduced atomic monoid, $a \in H$, and let $\mathcal{A}(a)$ be the set of all $u \in \mathcal{A}$ dividing some power a^n of a.*
  i) *If $\mathcal{A}(a)$ is finite, then*

$$\mathbf{f}(a^n) = An^d + O(n^{d-1})$$

  holds for some $A \in \mathbb{Q}_{>0}$ and $d \in \mathbb{N}_0$.
  ii) *If $\mathcal{A}(a)$ is infinite, then*

$$\mathbf{f}(a^n) \gg n^d$$

  holds for all $d \in \mathbb{N}$.

Theorem 2.3 is proved in [20], Theorem 2, and Theorem 2.4 is in [21], Theorems 1 and 2. Below we shall present proofs of Theorems

2.3 and 2.4 in the case where H is finitely generated (since H is reduced, this means that $\mathcal{A}$ is finite) in order to illustrate the methods. Our proof of Theorem 2.4 will be much simpler than that in [21]. However, it only shows that A is a positive real number and not that it is rational. We formulate the necessary tools as Lemmas 2.5 and 2.6.

Lemma 2.5 (Dickson's Finiteness Theorem). *Let* $M \subset \mathbb{N}_0^s$ *be a set whose points are pairwise incomparable under* $\leq$. *Then* M *is finite.*

Proof. [30], Satz 12. □

Lemma 2.6 (Lattice points in expanding polyhedra). *Let* E *be a real-euclidean vector space of dimension* $d \in \mathbb{N}$, $P \subset E$ *a compact polyhedron and* $\Gamma \subset E$ *a complete lattice. Then*

$$\#(tP \cap \Gamma) = At^d + O(t^{d-1}),$$

where

$$A = \frac{\text{vol}(P)}{\text{vol}(\Gamma)} \in \mathbb{R}_{\geq 0}$$

is the ratio between the volume of P *and the volume of a fundamental domain of* Γ.

Proof. [27], Ch. VI, §2, Theorem 2. □

Proof of Theorems 2.3 and 2.4 for finitely generated H. Suppose that $\mathcal{A} = \{u_1, \ldots, u_s\}$. Then the factorization homomorphism of H has the form

$$\pi : \begin{cases} \mathbb{N}_0^s & \to \quad H \\ \mathbf{k} = (k_1, \ldots k_s) & \mapsto u_1^{k_1} \cdot \ldots \cdot u_s^{k_s} . \end{cases}$$

For $a \in H$, the points $\mathbf{n} \in \pi^{-1}(a)$ are pairwise incomparable, and therefore $\mathbf{f}(a) = \#\pi^{-1}(a) < \infty$ by Lemma 2.5.

For the proof of Theorem 2.4, let Q be a quotient group of H and $\overline{\pi} : \mathbb{Z}^s \to Q$ the unique group homomorphism extending π. Then $\Gamma = \text{Ker}(\overline{\pi}) \subset \mathbb{Z}^s$ is a complete lattice in the real-euclidean vector space $E = \mathbb{R}\Gamma$ satisfying $\Gamma \cap \mathbb{N}_0^s = \{0\}$ since H is reduced.

If $a \in H$ and $\mathbf{k} \in \pi^{-1}(a)$, then $P = \{\mathbf{v} \in E \mid \mathbf{v} \geq -\mathbf{k}\}$ is a compact polyhedron, and for $n \in \mathbb{N}$, we obtain

$$\#\pi^{-1}(a^n) = \#\left[(n\mathbf{k} + \Gamma) \cap \mathbb{N}_0^s\right] = \#\{\mathbf{v} \in \Gamma \mid \mathbf{v} \geq -n\mathbf{k}\} = \#(\Gamma \cap nP).$$

Now the assertion follows from Lemma 2.6.   □

Next we show how Theorems 2.3 and 2.4 apply to ring theory. An integral domain R is called an *FF-domain (finite factorization domain)* if R is atomic and for every $a \in R^\bullet$ there exist (up to associates) only finitely many atoms dividing a; R is called an *SFF-domain (strong finite factorization domain)* if R is atomic and for every $a \in R^\bullet$ there exist (up to associates) only finitely many atoms dividing some power a^n of a.

Corollary 2.7.
   i) *If R is an FF-domain, then we have $\mathbf{f}(a) < \infty$ for all $a \in R^\bullet$.*
   ii) *If R is an SFF-domain and $a \in R^\bullet$, then $\mathbf{f}(a^n) = An^d + O(n^{d-1})$ holds for some $A \in \mathbb{Q}_{>0}$ and $d \in \mathbb{N}_0$.*

Proof. For the proof of i) resp. ii), we consider the finitely generated reduced monoid H consisting of all $\alpha \in \mathcal{H}(R) = R^\bullet/R^\times$ dividing a resp. some power a^n of a. Applying Theorems 2.3 and 2.4 for H, yields the assertion.   □

The notion of FF-domains was introduced in [3], that of SFF-domains in [21] (note that in [7] another notion of SFF-domains is used). We note the following criteria for an integral domain R to be an FF-domain or an SFF-domain.

Proposition 2.8.
   i) *Every Krull domain is an SFF-domain.*
   ii) *Let R be a noetherian domain whose integral closure $\overline{R}$ is a finitely generated R-module. Then R is an FF-domain if and only if $(\overline{R}^\times : R^\times) < \infty$.*

Proof. i) [17], Theorem 6.
   ii) [20], Theorem 7.   □

Example. Let $\overline{K}/K$ be an algebraic field extension and $R = K + X\overline{K}[X]$. Then we have $\overline{R} = \overline{K}[X]$ and $\overline{R}^{\times}/R^{\times} = \overline{K}^{\times}/K^{\times}$. Hence R is an FF-domain if and only if $\overline{K}$ is finite.

The above example is from [3]. Proposition (2.8) ii) was generalized in [7]. Concerning the SFF-property for noetherian domains, we have only a criterion in the one-dimensional case. We start with two preparations.

A monoid H is said to have *nearly unique factorization* if H is atomic, and up to associates, almost all atoms are prime. For an integral domain R, let $\mathcal{I}(R)$ be the monoid of invertible (integral) ideals of R. The prime elements of $\mathcal{I}(R)$ are the invertible prime ideals of R, and if R is noetherian, then every atom of $\mathcal{I}(R)$ is primary.

Proposition 2.9. *Let R be a one-dimensional noetherian domain such that its integral closure $\overline{R}$ is a finitely generated R-module.*

i) *If $\mathcal{I}(R)$ has nearly unique factorization, then R is an SFF-domain.*

ii) *$\mathcal{I}(R)$ has nearly unique factorization if and only if the canonical homomorphism $\mathrm{spec}\,(\overline{R}) \to \mathrm{spec}\,(R)$ is injective and $R/\mathfrak{p}$ is finite for all maximal ideals $\mathfrak{p} \supset [R : \overline{R}]$.*

iii) *If $\mathcal{I}(R)$ does not have nearly unique factorization and $\mathrm{Pic}\,(R)$ is finite, then R is not an SFF-domain.*

Proof. i) [17], Theorem 7.

ii) [19], Satz 2, together with [23], Proposition 7.

iii) If $\mathcal{I}(R)$ does not have nearly unique factorization, then there exists some maximal ideal $\mathfrak{p}$ of R such that $\mathfrak{p} \supset [R : \overline{R}]$, and there exists an infinite sequence of distinct irreducible invertible $\mathfrak{p}$-primary ideals, say $(\mathfrak{q}_n)_{n \geq 1}$. Since $\mathrm{Pic}\,(R)$ is finite, we may assume that all $\mathfrak{q}_n$ lie in the same class. Let $\mathfrak{a} \in \mathcal{I}(R)$ be such that $\mathfrak{a}\mathfrak{q}_1 \in \mathcal{H}(R)$ but $\mathfrak{a}_0\mathfrak{q}_1 \notin \mathcal{H}(R)$ for all proper divisors $\mathfrak{a}_0$ of $\mathfrak{a}$. Then we have $\mathfrak{a}\mathfrak{q}_n = u_n R$ for all $n \in \mathbb{N}$, where the u_n are pairwise non-associated atoms of R. Since all $\mathfrak{q}_n$ are $\mathfrak{p}$-primary, there exist integers $k_n \in \mathbb{N}$ such that $\mathfrak{q}_1^{k_n} \subset \mathfrak{q}_n$ and consequently $\mathfrak{q}_1^{k_n} = \mathfrak{q}_n\mathfrak{c}_n$ for some $\mathfrak{c}_n \in \mathcal{I}(R)$. From

$$(\mathfrak{a}\mathfrak{q}_1)^{k_n} = (\mathfrak{a}\mathfrak{q}_n)(\mathfrak{a}^{k_n-1}\mathfrak{c}_n)$$

we see that $u_n \mid u_1^{k_n}$ for all $n \in \mathbb{N}$, and therefore R is not an SFF-domain. $\square$

Example. Let K be a field and

$$R = \{f \in K[X] \mid f(0) = f(1)\}.$$

We have $\overline{R} = K[X]$, $[R : \overline{R}] = X(X-1)\overline{R}$, the elements $u_n = X^n(X-1) \in R$ are atoms, and since $u_n \mid u_1^{n+1}$ for all $n \in \mathbb{N}$, we see that R is not an SFF-domain. Since $\operatorname{Pic}(R) \simeq K$, this example is covered by Proposition 2.9 only if K is finite.

3. THE STRUCTURE OF SETS OF LENGTHS

If an atomic monoid is not half-factorial, then Lemma 2.2 implies that there are arbitrary large sets of lengths. For monoids of arithmetical interest, the structure of these large sets of lengths was investigated by A. Geroldinger in [11], [12] and [14]. In this section we discuss the case of finitely generated monoids.

Definition. A subset $L \subset \mathbb{Z}$ is called an *almost arithmetical progression bounded by* $M \in \mathbb{N}$, if

$$
\begin{aligned}
L = \{x_1, \ldots, x_\alpha, \, & y + \delta_1, \ldots, y + \delta_\eta, y + d, \\
& y + \delta_1 + d, \ldots, y + \delta_\eta + d, y + 2d, \\
& \cdots \\
& y + \delta_1 + (k-1)d, \ldots, y + \delta_\eta + (k-1)d, y + kd, \\
& z_1, \ldots, z_\beta\},
\end{aligned}
$$

where $\alpha, \beta, \eta, k, d \in \mathbb{N}_0$, $x_1 < \cdots < x_\alpha < y \leq y + kd < z_1 < \cdots < z_\beta$, $0 < \delta_1 < \cdots < \delta_\eta < d$ and $\max\{\alpha, \beta, d\} \leq M$.

Theorem 3.1. *Let H be a finitely generated reduced monoid. Then there exists some $M \in \mathbb{N}$ such that every $L \in \mathcal{L}(H)$ is an almost arithmetical progression bounded by M.*

Clearly, Theorem 3.1 holds for all monoids H for which the associated reduced monoid $H/H^\times$ is finitely generated. Thus it applies for atomic integral domains possessing (up to associates) only finitely many atoms. Such integral domains are called *Cohen-Kaplansky domains* and were investigated in [5]. In section 7 we shall see how Theorem 3.1 applies for Krull domains with finite class group and thus in particular for rings of integers in algebraic number fields. In fact, this is the case in which Theorem 3.1 was proved first in [11]. In this case, the quantities α, β, d and $\delta_1, \ldots, \delta_\eta$ have arithmetical significance and allow combinatorial descriptions; see [9]. In section 5, we shall see that Theorem 3.1 continues to hold for a large class of atomic monoids which are not finitely generated. A precise description of the class of monoids for which Theorem 3.1 holds is not known.

It was observed in [22] that Theorem 3.1 is in fact a result on lattice points in polyhedra and has a formulation within the theory of linear inequalities. We present this formulation in the following Proposition 3.2 and we show subsequently how Theorem 3.1 follows from it.

Proposition 3.2. *Let* $A \in \mathbb{Z}^{m \times s}$ *be an integral* $(m \times s)$-*matrix of rank* s, $\Lambda < \mathbb{Z}^s$ *a subgroup such that*

$$\{\mathbf{x} \in \Lambda \mid A\mathbf{x} \geq 0\} = \{0\}$$

and $\varphi \in \mathrm{Hom}(\mathbb{Z}^s, \mathbb{Z})$. *Then there exists some* $M \in \mathbb{N}$ *such that, for any* $\mathbf{k} \in \mathbb{Z}^s$, *the set*

$$\{\varphi(\mathbf{x}) \mid \mathbf{x} \in \mathbf{k} + \Lambda, \ A\mathbf{x} \geq 0\}$$

is an almost arithmetical progression bounded by M.

Proof of Theorem 3.1 by means of Proposition 3.2. In order to deduce Theorem 3.1 from Proposition 3.2, suppose that $H = [u_1, \ldots, u_s]$ is reduced, where $u_1, \ldots, u_s$ are distinct atoms of H, and let $\pi : \mathbb{N}_0^s \to H$ be the factorization homomorphism. Let $\overline{\pi} : \mathbb{Z}^s \to Q$ be the extension of π to an epimorphism of the quotient groups, set $\Lambda = \mathrm{Ker}(\overline{\pi}) \subset \mathbb{Z}^s$, and apply Proposition 3.2 with the unit matrix $A = I_s$ and $\varphi(\mathbf{x}) = |\mathbf{x}| = x_1 + \cdots + x_s$. If $a \in H$ and $a = \pi(\mathbf{k})$, then

$$L(a) = \{\varphi(\mathbf{x}) \mid \mathbf{x} \in \mathbf{k} + \Lambda, \ \mathbf{x} \geq 0\}$$

is an almost arithmetical progression bounded by some M which only depends on H. $\square$

Theorem 3.3. *Every finitely generated monoid has accepted elasticity.*

Theorem 3.3 was first proved in [2]. Below we shall present a fresh proof of this result using geometrical ideas. A. Geroldinger [12] showed that every finitely generated monoid has tame factorizations and that every monoid with tame factorizations has finite elasitcity. It should be noted that Theorem 3.3 is independent from Theorem 3.1. Indeed, Theorem 3.1 is also true for finitely primary monoids which usually have infinite elasticity (see sections 4 and 5 below).

The following Lemma contains the geometrical background for our proof of Theorem 3.3.

Lemma 3.4. *Let $W \subset \mathbb{Q}^s$ be a subspace such that $W \cap \mathbb{Q}^s_{\geq 0} = \{0\}$.*
  i) *For every $m \in \mathbb{Q}^s_{\geq 0}$, the set*

$$W(\mathbf{m}) = \{\mathbf{x} \in W \mid \mathbf{x} \geq -\mathbf{m}\}$$

  is a polytope, and there exists a constant $B > 0$ depending only on W such that $|\mathbf{x}| \leq B|\mathbf{m}|$ holds for all $\mathbf{m} \in \mathbb{Q}^s_{\geq 0}$ and $\mathbf{x} \in W(\mathbf{m})$.
  ii) *The set*

$$P = \{(\mathbf{k}, \mathbf{m}) \in \mathbb{Q}^{2s}_{\geq 0} \mid \mathbf{k} - \mathbf{m} \in W, |\mathbf{m}| = 1\}$$

  is a polytope.

Proof. We make freely use of the theory of polyhedra as explained in [31], ch.8.

  i) By definition, $W(\mathbf{m})$ is a polyhedron with characteristic cone $\{\mathbf{x} \in W \mid \mathbf{x} \geq 0\} = \{0\}$, and therefore $W(\mathbf{m})$ is a polytope. Every vertex of $W(\mathbf{m})$ is the unique solution of a system of linear equations of the form

$$A\mathbf{x} = 0, \quad x_j = -m_j \ (j \in J)$$

where A is a rational matrix such that $W = \{\mathbf{x} \in \mathbb{Q}^s \mid A\mathbf{x} = 0\}$ and $J \subset \{1, \ldots, s\}$. Consequently there exists some constant $B > 0$ (depending only on A and hence on W) such that $|\mathbf{p}| \leq B|\mathbf{m}|$ holds for every vertex $\mathbf{p}$ of $W(\mathbf{m})$. Since $W(\mathbf{m})$ is the convex hull of its vertices, the inequality holds for all $\mathbf{p} \in W(\mathbf{m})$.

ii) By definition, P is a polyhedron. If $(\mathbf{k}, \mathbf{m}) \in P$, then $\mathbf{k} = \mathbf{m} + \mathbf{w}$ for some $\mathbf{w} \in W(\mathbf{m})$, and **i)** implies $|\mathbf{w}| \leq B|\mathbf{m}| = B$. Consequently,

$$|(\mathbf{k}, \mathbf{m})| \leq |\mathbf{k}| + |\mathbf{m}| \leq B + 2\,,$$

and therefore P is bounded. $\square$

Proof of Theorem 3.3. Suppose that $H = [u_1, \ldots, u_s]$ is reduced, where $u_1, \ldots, u_s$ are distinct atoms of H, and let $\pi \colon \mathbb{N}_0^s \to H$ be the factorization homomorphism. Let $\overline{\pi} \colon \mathbb{Z}^s \to Q$ be the extension of π to an epimorphism of the quotient groups, set $\Lambda = \mathrm{Ker}\,(\overline{\pi}) \subset \mathbb{Z}^s$ and $W = \mathbb{Q}\Lambda \subset \mathbb{Q}^s$. Then we have $W \cap \mathbb{Q}_{\geq 0}^s = \{0\}$ (since H is reduced), and by Lemma 3.4 the set

$$P = \{(\mathbf{k}, \mathbf{m}) \in \mathbb{Q}_{\geq 0}^{2s} \mid \mathbf{k} - \mathbf{m} \in W\,, \ |\mathbf{m}| = 1\}$$

is a polytope. The linear functional ϕ, defined by $\phi(\mathbf{k}, \mathbf{m}) = |\mathbf{k}|$, takes its maximum in one of the vertices of P, and thus there exists some $(\mathbf{k}^*, \mathbf{m}^*) \in P$ such that

$$|\mathbf{k}^*| = \max\{|\mathbf{k}| \mid (\mathbf{k}, \mathbf{m}) \in P\}\,.$$

We assert that there exists some $a^* \in H$ such that $\varrho(H) = \varrho(a^*) = |\mathbf{k}^*|$. Let $q \in \mathbb{N}$ be such that $q\mathbf{k}^*, q\mathbf{m}^* \in \mathbb{N}_0^s$ and $q(\mathbf{k}^* - \mathbf{m}^*) \in \Lambda$. If $a^* = \pi(q\mathbf{k}^*)$, then

$$|q\mathbf{k}^*| \in L(a^*)\,, \ |q\mathbf{m}^*| \in L(a^*) \quad \text{and} \quad \varrho(a^*) \geq \frac{|q\mathbf{k}^*|}{|q\mathbf{m}^*|} = |\mathbf{k}^*|\,.$$

It remains to prove that $\varrho(a) \leq |\mathbf{k}^*|$ holds for all $a \in H$. Indeed, if $a \in H$ and $\mathbf{k}, \mathbf{m} \in \pi^{-1}(a)$ are such that $|\mathbf{k}| = \max L(a)$ and $|\mathbf{m}| = \min L(a) = m \in \mathbb{N}$, then we have $(m^{-1}\mathbf{k}, m^{-1}\mathbf{m}) \in W$ and consequently

$$|\mathbf{k}^*| \geq |m^{-1}\mathbf{k}| = \frac{|\mathbf{k}|}{|\mathbf{m}|} = \varrho(a)\,. \quad \square$$

4. FINITELY PRIMARY MONOIDS

Definition. A monoid H is called *finitely primary (of rank $s \in \mathbb{N}$ and exponent $\alpha \in \mathbb{N}$)* if H is a submonoid of a factorial monoid F with s pairwise non-associated prime elements $p_1, \ldots, p_s$,

$$H \subset F = [p_1, \ldots, p_s] \times F^\times$$

such that the following two conditions are fulfilled:

1) $(p_1 \cdot \ldots \cdot p_s)^\alpha F \subset H$.
2) If $\varepsilon p_1^{\alpha_1} \cdot \ldots \cdot p_s^{\alpha_s} \in H$ (where $\varepsilon \in F^\times$ and $\alpha_i \geq 0$), then either $\alpha_1 = \cdots = \alpha_s = 0$, $\varepsilon \in H^\times$ or $\alpha_1 \geq 1, \ldots, \alpha_s \geq 1$.

For $a = \varepsilon p_1^{\alpha_1} \cdot \ldots \cdot p_s^{\alpha_s} \in H$ (where $\varepsilon \in F^\times$ and $\alpha_i \geq 0$), we set

$$\omega(a) = \min\{\alpha_1, \ldots, \alpha_s\}.$$

If H is a finitely primary monoid of rank s and exponent α and F is as above, then $F = \hat{H}$ is the complete integral closure of H, see [15], Theorem 1. In particular, F is (up to isomorphism) uniquely determined by H, and s and α are invariants of H.

Finitely primary monoids were introduced in [23] in order to investigate the arithmetic of one-dimensional noetherian domains. Their algebraic and arithmetical properties were studied in [15]. The following Proposition collects some simple facts showing that the arithmetic of finitely primary monoids is quite different from that of finitely generated monoids.

Proposition 4.1. *Let H be a finitely primary monoid of rank s and exponent α.*

   i) *H is atomic, and for every $a \in H$, $\max L(a) \leq \omega(a)$.*
   ii) *If $s \geq 2$, then we have $\min L(a) \leq 2\alpha$ for every $a \in H$.*
   iii) *$\varrho(H) < \infty$ if and only if $s = 1$.*
   iv) *$H/H^\times$ is finitely generated if and only if $s = 1$ and $(\hat{H}^\times : H^\times) < \infty$.*

Proof. Suppose that $H \subset F = [p_1, \ldots, p_s] \times F^\times$ as in the definition above.

   i) If $a \in H$, then we have $\omega(c) < \omega(a)$ for every proper divisor c of a, and therefore $a = a_1 \cdot \ldots \cdot a_n$ (where $a_i \in H \setminus H^\times$) implies $n \leq \omega(a)$.

ii) We set $a = \varepsilon p_1^{\alpha_1} \cdot \ldots \cdot p_s^{\alpha_s}$ (where $\varepsilon \in F^\times$, $\alpha_i \geq 0$) and we may assume that $\omega(a) = \min\{\alpha_1, \ldots, \alpha_s\} \geq 2\alpha$. Then we have $a = a'a''$, where

$$a' = \varepsilon p_1^\alpha p_2^{\alpha_2 - \alpha} \cdot \ldots \cdot p_s^{\alpha_s - \alpha}, \quad a'' = p_1^{\alpha_1 - \alpha}(p_2 \cdot \ldots \cdot p_s)^\alpha,$$

and $\max L(a') \leq \alpha$, $\max L(a'') \leq \alpha$ implies $\min L(a) \leq 2\alpha$.

iii) If $s \geq 2$, then ii) implies $\varrho(H) = \infty$. If $s = 1$, then every atom $u \in H$ satisfies $\omega(u) \leq 2\alpha - 1$ and since ω is a semilength function, we infer $\varrho(H) \leq 2\alpha - 1$, see [1].

iv) $H/H^\times$ is finitely generated if and only if H possesses up to associates only finitely many atoms.

If $s = 1$ and $\hat{H}^\times/H^\times = \{\varepsilon_1 H^\times, \ldots, \varepsilon_n H^\times\}$, then every atom of H is of the form $p_1^\beta \varepsilon_i \varepsilon$, where $1 \leq \beta < 2\alpha$, $1 \leq i \leq n$ and $\varepsilon \in H^\times$, and among these elements there are only finitely many non-associates.

If $\hat{H}^\times/H^\times = \{\varepsilon_j H^\times \mid j \in J\}$ is infinite, then $\{p_1^\alpha \varepsilon_j \mid j \in J\}$ is an infinite set of pairwise non-associated elements with $\max L(p_1^\alpha \varepsilon_j) = \alpha$, and therefore there are infinitely many non-associated atoms dividing them.

If $s \geq 2$, then $\{p_1^n(p_2 \cdot \ldots \cdot p_s)^\alpha \mid n \geq \alpha\}$ is an infinite set of pairwise non-associated elements satisfying $\max L\big(p_1^n(p_2 \cdot \ldots \cdot p_s)^\alpha\big) = \alpha$ and we argue as before. $\square$

Usually, finitely primary monoids do not contain prime elements. We deduce a slightly more general result. A monoid H is called *primary* if, for all $a \in H$ and $b \in H \setminus H^\times$, there exists some $n \in \mathbb{N}$ such that $a \mid b^n$.

Proposition 4.2. *Let H be a monoid.*
  i) *If H is finitely primary, then H is primary.*
  ii) *If H is primary and contains a prime element p, then $H = [p] \times H^\times$.*

Proof. i) follows from the very definition of finitely primary monoids.

ii) See [15], Proposition 5 (the proof given there works without the assumption that H is atomic). $\square$

Next we recall from [15] which integral domains R have finitely primary multiplicative monoids $R^\bullet$.

Proposition 4.3. *Let R be an integral domain. Then $R^\bullet$ is finitely primary if and only if R is quasilocal, $\dim(R) = 1$, the complete integral closure $\hat{R}$ of R is a semilocal principal ideal domain, and $f\hat{R} \subset R$ holds for some $f \in R^\bullet$.*

If these conditions are fulfilled, then $s = \#\max(\hat{R})$ is the rank of $R^\bullet$.

Proof. [15], Theorem 2. Observe that $\hat{R}^\bullet = \widehat{R^\bullet}$ and that $\#\max(\hat{R})$ is the number of non-associated prime elements of $\hat{R}^\bullet$. $\square$

Corollary 4.4. *Let R be a local noetherian domain. Then $R^\bullet$ is finitely primary if and only if $\dim(R) = 1$ and the integral closure $\overline{R}$ of R is a finitely generated R-module.*

Proof. Obvious from Proposition 4.3. $\square$

Proposition 4.5. *Let R be an integral domain. Then $R^\bullet$ is primary if and only if R possesses exactly one non-zero prime ideal $\mathfrak{p}$. If $\mathfrak{p}$ is principal, then R is a discrete valuation ring.*

Proof. By [24], Theorem 4.1 and Proposition 4.2. $\square$

5. DIVISOR HOMOMORPHISMS AND BLOCKS

Definition. A monoid homomorphism $\varphi\colon H \to D$ is called a *divisor homomorphism* if $u, v \in H$ and $\varphi(u)\,|\,\varphi(v)$ implies $u\,|\,v$. A submonoid $H \subset D$ is called *saturated* if the injection $H \hookrightarrow D$ is a divisor homomorphism.

For a systematic theory of divisor homomorphisms see [18]. Note that $\varphi\colon H \to D$ is a divisor homomorphism if and only if the associated homomorphism $\overline{\varphi}\colon H/H^\times \to D/D^\times$ is a divisor homomorphism. If D is reduced, then a monoid homomorphism $\varphi\colon H \to D$ is a divisor homomorphism if and only if it splits in the form

$$\varphi\colon H \to H/H^\times \xrightarrow{\sim} \varphi(H) \hookrightarrow D,$$

and $\varphi(H)$ is a saturated submonoid of D. The importance of divisor homomorphisms comes from the fact that they allow us to deduce arithmetical properties of H from that of D. We shall see this in detail in the next section, where we deal with Krull and weakly Krull domains.

Next we introduce class groups. Let $\varphi \colon H \to D$ be a monoid homomorphism. We define an equivalence relation $\sim$ on D by means of $a \sim b$ if $a\varphi(u) = b\varphi(v)$ holds for some $u, v \in H$, and we denote by $[a] = [a]_\varphi$ the equivalence class of a. In fact, $\sim$ is a congruence relation, and the set of equivalence classes becomes an additive (cancellative) monoid by means of $[a] + [b] = [ab]$ whose quotient group $\mathcal{C}(\varphi)$ is called the *class group of* φ.

Lemma 5.1. *A monoid homomorphism* $\varphi \colon H \to D$ *is a divisor homomorphism if and only if the following two conditions are fulfilled.*
1) *For all* $z \in H$, $\varphi^{-1}(\varphi(z)) \subset zH^\times$.
2) $\varphi(H) = \{a \in D \mid [a]_\varphi = 0 \in \mathcal{C}(\varphi)\}$.

Proof. Let first φ be a divisor homomorphism. If $z \in H$ and $u \in \varphi^{-1}(\varphi(z))$, then $\varphi(u) = \varphi(z)$ implies $u \mid z$ and $z \mid u$, whence $u \in zH^\times$. Morever, $[\varphi(z)] = [1] = 0$. Suppose now that $a \in D$ satisfies $[a] = 0 = [1]$. Then we have $a\varphi(u) = \varphi(v)$ for some $u, v \in H$, and $\varphi(u) \mid \varphi(v)$ implies $u \mid v$. If $z \in H$ satisfies $v = uz$, then $\varphi(v) = \varphi(u)\varphi(z)$ implies $\varphi(z) = a \in \varphi(H)$.

Let now **1)**, **2)** be satisfied, and let $u, v \in H$ be such that $\varphi(u) \mid \varphi(v)$. If $a \in D$ satisfies $\varphi(v) = a\varphi(u)$, then we have $[a] = 0$ and hence $a = \varphi(z)$ for some $z \in H$. Now $\varphi(v) = \varphi(uz)$ implies $v \in \varphi^{-1}(\varphi(uz)) \subset uzH^\times$, and hence $u \mid v$. Thus φ is a divisor homomorphism. $\square$

For our next result we have to recall the definition of Davenport's constant $\mathcal{D}(G_0)$ for a subset G_0 of an additive abelian group G. $\mathcal{D}(G_0) \in \mathbb{N}_0 \cup \{\infty\}$ is the supremum of all $d \in \mathbb{N}_0$ such that there exists a sequence $(g_1, \ldots, g_d)$ in G_0 satisfying

$$\sum_{j=1}^{d} g_j = 0 \quad \text{and} \quad \sum_{j \in J} g_j \neq 0 \quad \text{for all} \quad \emptyset \neq J \subsetneq \{1, \ldots, d\}.$$

It was proved in [11], Proposition 2, that $\#G_0 < \infty$ implies $\mathcal{D}(G_0) < \infty$. For more information concerning $\mathcal{D}(G_0)$, we refer to [9].

The following result shows the significance of Davenport's constant for the transfer of factorization properties by means of divisor homomorphisms; see also [14], Lemma 4.4.

Proposition 5.2. *Let* D *be an atomic monoid and suppose that* $L^D(a)$ *is finite for all* $a \in D$. *Let* $\varphi \colon H \to D$ *be a divisor homomorphism. Then* H *is atomic, and for all* $z \in H$,

$$\max L^H(z) \leq \max L^D\big(\varphi(z)\big) \leq \big[\max L^H(z)\big]\mathcal{D}(G_0),$$

where $G_0 \subset \mathcal{C}(\varphi)$ *is the set of all classes containing atoms of* D.

Proof. We may assume that D is reduced, H is a saturated submonoid of D and $\varphi = (H \hookrightarrow D)$. If $z \in H$ and $z = z_1 \cdot \ldots \cdot z_k$, where $z_1, \ldots, z_k \in H$, then we may factor each z_i into atoms of D and obtain $k \leq \max L^H(z)$. Therefore H is atomic, and $\max L^H(z) \leq \max L^D(z)$. Let now $z = u_1 \cdot \ldots \cdot u_d$ be a factorization of z into atoms of D such that $d = \max L^D(z)$. Then we have $[u_j] \in G_0$ and $[u_1] + \cdots + [u_d] = 0$. By definition of $\mathcal{D}(G_0)$, there exists a partition $\{1, \ldots, d\} = J_1 \uplus \cdots \uplus J_m$ such that, for all $\mu \in \{1, \ldots, m\}$,

$$\#J_\mu \leq \mathcal{D}(G_0) \quad \text{and} \quad \sum_{j \in J_\mu} [u_j] = 0.$$

If

$$z_\mu = \prod_{j \in J_\mu} u_j,$$

then $[z_\mu] = 0$ implies $z_\mu \in H$ by Lemma 5.1, and $z = z_1 \cdot \ldots \cdot z_m$ implies $m \leq \max L^H(z)$. Therefore we obtain

$$d = \sum_{\mu=1}^m \#J_\mu \leq \big[\max L^H(z)\big]\mathcal{D}(G_0). \quad \square$$

Atomic monoids D satisfying $\#L^D(a) < \infty$ for all $a \in D$ are called *BF-monoids (bounded factorization monoids)* and where investigated in [20]. Proposition 5.2 implies in particular that a saturated submonoid of a BF-monoid is again a BF-monoid; see [20], Theorem 3 for a more general result. Surprisingly, a saturated submonoid of an atomic monoid need not be atomic, even if the class group of the injection is finite. This is shown by the following example.

Example 5.3. Let Q be the free abelian group with basis

$$\{v_n, v'_n \mid n \in \mathbb{N}_0\},$$

and let D be the submonoid generated by the elements

$$v_n, \; v'_n \quad \text{and} \quad u_n = (v_{n+1}v'_{n+1})^{-1}(v_n v'_n) \; (n \geq 0).$$

It is easily checked that all elements v_n, v'_n and u_n are atoms of D (whence D is atomic), and that the element $v_0 v'_0$ possesses the factorizations

$$v_0 v'_0 = v_n v'_n \prod_{\nu=0}^{n-1} u_\nu \quad (n \geq 0)$$

and no others. Let $\varphi \colon Q \to \mathbb{Z}/2\mathbb{Z}$ be the group homomorphism defined by $\varphi(v_n) = \varphi(v'_n) = 1 + 2\mathbb{Z}$, and set $H = \mathrm{Ker}(\varphi) \cap D \subset D$. Then H is a saturated submonoid of D, the class group of $H \hookrightarrow D$ is $\mathbb{Z}/2\mathbb{Z}$, and H is not atomic. Indeed, $v_0 v'_0 \in H$ possesses no factorizations into atoms of H (note that we know all factorizations of $v_0 v'_0$ into atoms of D).

The following simple but important Lemma characterizes a class of monoid homomorphisms which preserve lengths of factorizations.

Lemma 5.4 (Transfer Lemma). *Let* $\beta \colon H \to B$ *be a surjective homomorphism of reduced monoids with the following two properties.*

(T1) $\beta^{-1}(1) = \{1\}$.

(T2) *If* $z \in H$ *and* $\beta(z) = bc$, *where* $b, c \in B$, *then there exist* $x, y \in H$ *such that* $z = xy$, $\beta(z) = b$ *and* $\beta(y) = c$.

Then H *is atomic if and only if* B *is atomic, and for all* $z \in H$, $L^H(z) = L^B(\beta(z))$. *In particular,* $u \in H$ *is an atom of* H *if and only if* $\beta(u)$ *is an atom of* B.

Proof. We prove first that $u \in H$ is an atom of H if and only if $\beta(u)$ is an atom of B. If $u \in H$ is an atom and $\beta(u) = bc$, where $b, c \in B$, then **(T2)** implies $u = xy$, where $x, y \in H$, $\beta(x) = b$ and $\beta(y) = c$. Since u is an atom, we infer $x = 1$ or $y = 1$ and consequently $b = 1$ or $c = 1$. Hence $\beta(u)$ is also an atom. If $\beta(u)$ is an atom of B and $u = xy$, where $x, y \in H$, then $\beta(u) = \beta(x)\beta(y)$, which implies $\beta(x) = 1$ or $\beta(y) = 1$. By **(T1)** we infer $x = 1$ or $y = 1$, and hence u is an atom.

Now suppose that H is atomic. If $a \in B$ and $z \in \beta^{-1}(a)$, then $z = u_1 \cdot \ldots \cdot u_r$ where $u_1, \ldots, u_r \in H$ are atoms, and therefore $\beta(z) = \beta(u_1) \cdot \ldots \cdot \beta(u_r)$. By the above, the elements $\beta(u_i)$ are atoms of B, which implies that B is atomic and also that $L^H(z) \subset L^B(\beta(z))$.

If B is atomic and $z \in H$, we choose a factorization $\beta(z) = b_1 \cdot \ldots \cdot b_m$ into atoms $b_i \in B$. By (T2), there exist $u_1, \ldots, u_m \in H$ such that $\beta(u_j) = b_j$ for all $j \in \{1, \ldots m\}$ and $z = u_1 \cdot \ldots \cdot u_m$. By the above, the elements u_j are atoms of H. Consequently, H is atomic, and $L^B(\beta(z)) \subset L^H(z)$. $\square$

Now we show how to attach to a divisor homomorphism $\varphi \colon H \to D$ a monoid homomorphism $\beta \colon H \to B$ satisfying the assumptions of the Transfer Lemma.

If D is a reduced atomic monoid and P is the set of prime elements of D, then every $a \in D$ has a unique decomposition in the form

$$a = p_1 \cdot \ldots \cdot p_n b \,,$$

where $n \in \mathbb{N}_0$, $p_1, \ldots, p_n \in P$, $b \in D$, and there is no $p \in P$ dividing b. Consequently,

$$(*) \qquad\qquad D = \mathcal{F}(P) \times T \,,$$

where $T \subset D$ is the submonoid of all $b \in D$ satisfying $p \nmid b$ for all $p \in P$. We shall refer to $(*)$ as the *canonical decomposition of D*; see [13], Lemma 2.

Let now $D = \mathcal{F}(P) \times T$ be the canonical decomposition of a reduced atomic monoid D, let $\varphi \colon H \to D$ be a divisor homomorphism and $G_0 = \{g \in \mathcal{C}(\varphi) \mid g \cap P \neq \emptyset\}$ the set of all classes of $\mathcal{C}(\varphi)$ containing primes. We define a homomorphism $\iota \colon \mathcal{F}(G_0) \times T \to \mathcal{C}(\varphi)$ by means of

$$\iota(g_1 \cdot \ldots \cdot g_n t) = g_1 + \cdots + g_n + [t]$$

(where $g_1, \ldots, g_n \in G_0$, $t \in T$), and we set

$$B = \iota^{-1}(0) \subset \mathcal{F}(G_0) \times T \,.$$

By construction, B is a saturated submonoid of $\mathcal{F}(G_0) \times T$. We call B the *block monoid attached to φ* and write $B = \mathcal{B}(\varphi)$. Next we define a homomorphism $\overline{\beta} \colon \mathcal{F}(P) \times T \to \mathcal{F}(G_0) \times T$ by means of

$$\overline{\beta}(p_1 \cdot \ldots \cdot p_n t) = [p_1] \cdot \ldots \cdot [p_n] t$$

(where $p_1, \ldots, p_n \in P$, $t \in T$). For every $c \in D$, $\iota \circ \overline{\beta}(c) = [c] \in$ $\mathcal{C}(\varphi)$ is the class containing c. In particular, by Lemma 5.1 we have $\overline{\beta}(c) \in B$ if and only if $c \in \varphi(H)$. The homomorphism

$$\beta = \overline{\beta} \circ \varphi \colon H \to B$$

is called the *block homomorphism attached to* φ, and we obtain the commutative diagram

$$
\begin{array}{ccccc}
H & \overset{\varphi}{\to} & D = \mathcal{F}(P) \times T & \overset{\pi}{\to} & \mathcal{C}(\varphi) \\
\beta \downarrow & & \downarrow \overline{\beta} & & \| \\
B & \hookrightarrow & \mathcal{F}(G_0) \times T & \overset{\iota}{\to} & \mathcal{C}(\varphi)
\end{array}
$$

where $\pi(c) = [c]$ for all $c \in D$. In particular, we see that $\mathcal{C}(\varphi)$ is also the class group of $B \hookrightarrow \mathcal{F}(G_0) \times T$.

Theorem 5.5. *Let* $\varphi \colon H \to D$ *be a divisor homomorphism of reduced monoids, suppose that* D *is atomic, and let* $\beta \colon H \to B$ *be the block homomorphism attached with* φ. *Then* β *satisfies the assumptions of the Transfer Lemma.*

Proof. Let all notations be as above. If $z \in H$ and $1 = \beta(z) = \overline{\beta}(\varphi(z))$, then $\varphi(z) = 1$ and hence $z = 1$, since φ is a divisor homomorphism.

For the proof of **(T2)**, let $z \in H$ and $b_1, b_2 \in B$ be such that $\beta(z) = b_1 b_2$. If $b_i = \overline{\beta}(c_i)$, where $c_i \in D$, then we infer $c_i \in \varphi(H)$, say $c_i = \varphi(x_i)$, where $x_i \in H$. Since $\beta(x_i) = \overline{\beta} \circ \varphi(x_i) = b_i$, the assertion follows. $\square$

The construction of block monoids and block homomorphisms was first done in [13]. The special case $T = \{1\}$, $D = \mathcal{F}(P)$ is of particular importance. In this case,

$$\mathcal{B}(\varphi) = \mathcal{B}(G_0) = \{g_1 \cdot \ldots \cdot g_n \in \mathcal{B}(G_0) \mid g_1 + \cdots + g_n = 0\}$$

is the ordinary block monoid introduced by W. Narkiewicz [28]. In general, $\mathcal{B}(G_0) \subset \mathcal{B}(\varphi)$ is a divisor-closed submonoid (i. e., if $B \in \mathcal{B}(G_0)$, then all divisors of B in $\mathcal{B}(\varphi)$ also belong to $\mathcal{B}(G_0)$). In particular, $\mathcal{B}(G_0) \subset \mathcal{B}(\varphi)$ is saturated.

6. ARITHMETIC OF WEAKLY KRULL DOMAINS

For an integral domain R, let $\mathfrak{X}(R)$ be the set of all prime ideals of height one. Following [4], we call R a *weakly Krull domain* if

$$R = \bigcap_{\mathfrak{p} \in \mathfrak{X}(R)} R_{\mathfrak{p}},$$

and this intersection is of finite character (which means that for every $a \in R^{\bullet}$ we have $a \in R_{\mathfrak{p}}^{\times}$ for all but finitely many $\mathfrak{p} \in \mathfrak{X}(R)$). If R is weakly Krull then the canonical homomorphism

$$\varphi \colon R^{\bullet} \to D = \coprod_{\mathfrak{p} \in \mathfrak{X}(R)} R_{\mathfrak{p}}^{\bullet}/R_{\mathfrak{p}}^{\times}$$

is a divisor homomorphism whose class group $G = \mathcal{C}(\varphi)$ is called the *class group of* R. Its block monoid $B = \mathcal{B}(\varphi)$ is called the *block monoid of* R and its block homomorphism $\beta \colon R^{\bullet} \to B$ is called the *block homomorphism of* R.

We shall identify G with the t-class group of R. We refer to [6], section 2 for the elementary properties of t-invertible t-ideals needed in the sequel. Let K be a quotient field of R, $\mathfrak{F}(R)$ the group of t-invertible fractional t-ideals of R (inside K) and $\mathfrak{H}(R)$ the subgroup of non-zero fractional principal ideals. Then the t-class group of R is defined by

$$\mathcal{C}_t(R) = \mathfrak{F}(R)/\mathfrak{H}(R).$$

For $\mathfrak{a} \in \mathfrak{F}(R)$, we denote by $[\mathfrak{a}] = \mathfrak{a}\mathfrak{H}(R)$ the class containing $\mathfrak{a}$. A non-zero fractional t-ideal $\mathfrak{a}$ of R is t-invertible if and only if it is locally free, which means $\mathfrak{a}R_{\mathfrak{p}} = u_{\mathfrak{p}}R_{\mathfrak{p}}$ for all $\mathfrak{p} \in \mathfrak{X}(R)$ (for some $u_{\mathfrak{p}} \in R_{\mathfrak{p}}$). Since $u_{\mathfrak{p}}$ is uniquely determined modulo $R_{\mathfrak{p}}^{\times}$ and

$$\mathfrak{a} = \bigcap_{\mathfrak{p} \in \mathfrak{X}(R)} \mathfrak{a}R_{\mathfrak{p}}$$

holds for all $\mathfrak{a} \in \mathfrak{F}(R)$, we obtain an isomorphism

$$\Phi \colon \mathfrak{F}(R) \xrightarrow{\sim} \coprod_{\mathfrak{p} \in \mathfrak{X}(R)} K^{\times}/R_{\mathfrak{p}}^{\times},$$

given by $\Phi(\mathfrak{a}) = (u_\mathfrak{p} R_\mathfrak{p}^\times)_{\mathfrak{p} \in \mathfrak{X}(R)}$ if $\mathfrak{a} R_\mathfrak{p} = u_\mathfrak{p} R_\mathfrak{p}$ for all $\mathfrak{p} \in \mathfrak{X}(R)$. Under this isomorphism,

$$\Phi\big(\mathfrak{H}(R)\big) = \Delta(K^\times/R^\times),$$

where

$$\Delta: \begin{cases} K^\times/R^\times & \to & \coprod\limits_{\mathfrak{p} \in \mathfrak{X}(R)} K^\times/R_\mathfrak{p}^\times \\ aR^\times & \to & (aR_\mathfrak{p}^\times)_{\mathfrak{p} \in \mathfrak{X}(R)} \end{cases}$$

denotes the diagonal embedding. The canonical homomorphism

$$\begin{cases} D & \to & G \\ \alpha & \mapsto & [\alpha] \end{cases}$$

extends to an epimorphism

$$\theta: \coprod\limits_{\mathfrak{p} \in \mathfrak{X}(R)} K^\times/R_\mathfrak{p}^\times \to G$$

of the quotient group with kernel $\Delta(K^\times/(R^\times))$. Thus we obtain the following commutative diagram with exact rows and vertical isomorphisms:

$$\begin{array}{ccccccccc} 1 & \to & \mathfrak{H}(R) & \hookrightarrow & \mathfrak{F}(R) & & \to & \mathcal{C}_t(R) & \to & 1 \\ & & \downarrow & & \downarrow \Phi & & & \downarrow \overline{\Phi} & & \\ 1 & \to & K^\times/R^\times & \overset{\Delta}{\to} & \coprod\limits_{\mathfrak{p} \in \mathfrak{X}(R)} K^\times/R_\mathfrak{p}^\times & \overset{\theta}{\to} & G & \to & 0. \end{array}$$

The isomorphism $\overline{\Phi}$ is given by $\overline{\Phi}([\mathfrak{a}]) = [\Phi(\mathfrak{a})]$ for all t-invertible t-ideals $\mathfrak{a} \lhd R$. Since

$$\coprod\limits_{\mathfrak{p} \in \mathfrak{X}(R)} K^\times/R_\mathfrak{p}^\times = \operatorname{Im}(\Delta)\, D,$$

we infer $\theta(D) = G$.

For every subset $\Omega \subset \mathfrak{X}(R)$, we consider the embedding

$$\coprod\limits_{\mathfrak{p} \in \Omega} K^\times/R_\mathfrak{p}^\times \hookrightarrow \coprod\limits_{\mathfrak{p} \in \mathfrak{X}(R)} K^\times/R_\mathfrak{p}^\times,$$

given by inserting 1 at all components $\mathfrak{p} \in \mathfrak{X}(R) \setminus \Omega$. In particular, for $\Omega = \{\mathfrak{p}_0\} \subset \mathfrak{X}(R)$, we obtain the embedding

$$K^\times / R_{\mathfrak{p}_0}^\times \hookrightarrow \coprod_{\mathfrak{p} \in \mathfrak{X}(R)} K^\times / R_{\mathfrak{p}}^\times$$

into the $\mathfrak{p}_0$-component.

Let P be the set of all $\mathfrak{p} \in \mathfrak{X}(R)$ for which $R_{\mathfrak{p}}$ is a discrete valuation ring. For $\mathfrak{p} \in P$, let $\pi_{\mathfrak{p}} \in R_{\mathfrak{p}}$ be a prime element for $\mathfrak{p}$. Every $\mathfrak{p} \in \mathfrak{X}(R)$ is a t-ideal, and if $\mathfrak{p} \in P$, then $\mathfrak{p}$ is t-invertible. Indeed, we have

$$\mathfrak{p} R_{\mathfrak{q}} = \begin{cases} \pi_{\mathfrak{p}} R_{\mathfrak{p}}, & \text{if } \mathfrak{q} = \mathfrak{p}, \\ R_{\mathfrak{q}}, & \text{if } \mathfrak{q} \neq \mathfrak{p}. \end{cases}$$

We assert that $\{\pi_{\mathfrak{p}} R_{\mathfrak{p}}^\times \mid \mathfrak{p} \in P\} \subset D$ is the set of all prime elements of D. Indeed, prime elements are non-trivial in exactly one component, and thus the assertion follows from Proposition 4.5.

The map

$$\psi_0 : \begin{cases} \mathcal{F}(P) & \to & \coprod_{\mathfrak{p} \in P} R_{\mathfrak{p}}^\bullet / R_{\mathfrak{p}}^\times \\ \prod_{\mathfrak{p} \in P} \mathfrak{p}^{n_{\mathfrak{p}}} & \mapsto & (\pi_{\mathfrak{p}}^{n_{\mathfrak{p}}} R_{\mathfrak{p}}^\times)_{\mathfrak{p} \in P} \end{cases}$$

is an isomorphism. If

$$T = \coprod_{\mathfrak{p} \in \mathfrak{X}(R) \setminus P} R_{\mathfrak{p}}^\bullet / R_{\mathfrak{p}}^\times,$$

then

$$D = \coprod_{\mathfrak{p} \in P} R_{\mathfrak{p}}^\bullet / R_{\mathfrak{p}}^\times \times T$$

is the canonical decomposition of D, ψ_0 extends to an isomorphism

$$\psi = \psi_0 \times \mathrm{id}_T : \mathcal{F}(P) \times T \to D,$$

and we set

$$\iota = \theta \circ \psi : \mathcal{F}(P) \times T \to G.$$

Clearly, ι is a surjective homomorphism, $\iota \mid T = \theta \mid T$, and for $\mathfrak{p} \in P$, we have

$$\iota(\mathfrak{p}) = \theta(\pi_{\mathfrak{p}} R_{\mathfrak{p}}^\times) = \theta \circ \Phi(\mathfrak{p}) = \overline{\Phi}([\mathfrak{p}]).$$

Thus the set G_0 of all classes $g \in G$ containing primes of D corresponds to the set of classes in $\mathcal{C}_t(R)$ containing prime t-ideals under $\overline{\Phi}$.

We shall henceforth identify

$$D = \mathcal{F}(P) \times T \quad \text{and} \quad G = \mathcal{C}_t(R)$$

by means of Φ and $\overline{\Phi}$. Then the divisor homomorphism $\varphi \colon R^\bullet \to D$ is given by

$$\varphi(a) = \left(\prod_{\mathfrak{p} \in P} \mathfrak{p}^{v_\mathfrak{p}(a)} \right) \left(a R_\mathfrak{p}^\times \right)_{\mathfrak{p} \in \mathfrak{X}(R) \setminus P},$$

where $v_\mathfrak{p}$ is the discrete valuation associated with $R_\mathfrak{p}$. The canonical homomorphism

$$\iota \colon \begin{cases} \mathcal{F}(P) \times T & \to & D \\ \alpha & \mapsto & [\alpha] \end{cases}$$

is given by

$$\iota \left(\prod_{\mathfrak{p} \in P} \mathfrak{p}^{n_\mathfrak{p}} t \right) = \sum_{\mathfrak{p} \in P} n_\mathfrak{p} [\mathfrak{p}] + \theta(t).$$

The block monoid of R has the form

$$\mathcal{B}(R) = \{ g_1 \cdot \ldots \cdot g_n t \in \mathcal{F}(G_0) \times T \mid g_1 + \cdots + g_n + \theta(t) = 0 \},$$

$\mathcal{B}(G_0) \subset \mathcal{B}(R) \subset \mathcal{F}(G_0) \times T$ are saturated submonoids, and the block homomorphism $\beta \colon R^\bullet \to \mathcal{B}(R)$ is given by

$$\beta(a) = \left(\prod_{\mathfrak{p} \in P} [\mathfrak{p}]^{v_\mathfrak{p}(a)} \right) t, \quad \text{where} \quad t = (a R_\mathfrak{p}^\times)_{\mathfrak{p} \in \mathfrak{X}(R) \setminus P} \in T.$$

We shall refer to this explicit descriptions later on. Let us now discuss the arithmetical consequences. First of all, we shall apply Theorem 3.1 for Krull domains.

Theorem 6.1. *Let R be a Krull domain, and suppose that the set G_0 of all classes containing primes is a finite subset of the class group of R. Then there exists some $M \in \mathbb{N}$ depending only on G_0 such that, for all $a \in R^\bullet$, $L(a)$ is an almost arithmetical progression bounded by M.*

Proof. If R is a Krull domain, then $P = \mathfrak{X}(R)$, $T = \{1\}$ and $\mathcal{B}(R) = \mathcal{B}(G_0)$. If G_0 is finite, then $\mathcal{B}(G_0)$ is a finitely generated monoid by [11], Proposition 2. Therefore Theorem 3.1 applies for $\mathcal{B}(G_0)$, and the proof is completed by Theorem 5.5 and Lemma 5.4. □

A. Geroldinger succeeded in proving an analog to Theorem 6.1 for a class of weakly Krull domains. In fact, he proved the arithmetical result for a suitable class of monoids and then he applied it to the block monoid. We present this monoid-theoretical result and its application to weakly Krull domains. Up to now, it is the most impressive application of block homomorphisms and the Transfer Lemma.

Theorem 6.2. *Let* $D = D_1 \times \cdots \times D_n$ *be a finite product of finitely primary monoids* $D_1, \ldots, D_n$, *and let* $\varphi \colon H \to D$ *be a divisor homomorphism with finite class group. Then* H *is atomic, and there exists some* $M \in \mathbb{N}$ *such that, for all* $a \in H$, $L(a)$ *is an almost arithmetical progression bounded by* M.

Proof. [14], Theorem 6.2. □

Definition. A weakly Krull domain R is said to be *of finite type* if its integral closure $\overline{R}$ is a Krull domain and a finitely generated R-module.

Krull domains and noetherian weakly Krull domains R whose integral closure $\overline{R}$ is a finitely generated R-module are the most important examples of weakly Krull domains of finite type. Note that a noetherian domain is weakly Krull if and only if every prime ideal of depth one has height one; see [16], Lemma 7.6. In particular, a one-dimensional noetherian domain R is weakly Krull of finite type if and only if its integral closure $\overline{R}$ is a finitely generated R-module.

Lemma 6.3. *Let* R *be a weakly Krull domain of finite type and* $E = \mathfrak{X}(R) \setminus P$ *the set of all* $\mathfrak{p} \in \mathfrak{X}(R)$ *for which* $R_\mathfrak{p}$ *is not a discrete valuation ring. Then* E *is finite, and for all* $\mathfrak{p} \in E$, $R_\mathfrak{p}^\bullet$ *is finitely primary.*

Proof. Since $\overline{R}$ is a finitely generated R-module, there exists some $f \in R^\bullet$ such that $f\overline{R} \subset R$. If $\mathfrak{p} \in \mathfrak{X}(R)$ and $f \notin \mathfrak{p}$, then $R_\mathfrak{p} = \overline{R}_\mathfrak{p}$ is a discrete valuation ring, since $\overline{R}$ is a Krull domain, and consequently $\mathfrak{p} \notin E$. Thus we have

$$E \subset \{\mathfrak{p} \in \mathfrak{X}(R) \mid f \in \mathfrak{p}\},$$

and this set is finite. For $\mathfrak{p} \in E$, $\overline{R}_\mathfrak{p} = \overline{R_\mathfrak{p}}$ is a finitely generated R-module and hence a semilocal Krull domain. Therefore it is a principal ideal domain, and the assertion follows from Proposition 4.3. $\square$

Theorem 6.4. *Let R be a weakly Krull domain of finite type whose class group G is finite. Then there exists some $M \in \mathbb{N}$ such that, for all $a \in R^\bullet$, $L(a)$ is an almost arithmetical progression bounded by M.*

Proof. By Lemma 5.4 and Theorem 5.5, it suffices to prove the assertion for $\mathcal{B}(R)$. Let G_0 be the set of all classes containing primes $\mathfrak{p} \in P$. Then $\mathcal{B}(R)$ is a saturated submonoid of $\mathcal{F}(G_0) \times T$, where

$$T = \coprod_{\mathfrak{p} \in \mathfrak{X}(R) \setminus P} R_\mathfrak{p}^\bullet / R_\mathfrak{p}^\times$$

is a finite product of finitely primary monoids by Lemma 6.3. Since G is the class group of $\mathcal{B}(R) \hookrightarrow \mathcal{F}(G_0) \times T$, the assertion follows from Theorem 6.2. $\square$

7. DESCRIPTION OF THE CLASS GROUP

The main goal of this section is to make the description of the class group of weakly Krull domains of finite type more explicit. The results will enable us to calculate sets of lengths and the elasticity of one-dimensional noetherian domains using their block monoids in section 8.

Let R be a weakly Krull domain of finite type with quotient field K, and keep all notions introduced in the preceding section. Let $\overline{R}$ be the integral closure of R in K. For a prime ideal $\mathfrak{P} \in \mathfrak{X}(\overline{R})$, let $v_\mathfrak{P} : K \to \mathbb{Z} \cup \{\infty\}$ be the normalized valuation associated with $\mathfrak{P}$. Let $E = \mathfrak{X}(R) \setminus P$ be the finite set of all $\mathfrak{p} \in \mathfrak{X}(R)$ for which $R_\mathfrak{p}$ is not a discrete valuation ring. For $\mathfrak{p} \in \mathfrak{X}(R)$, let $\overline{R}_\mathfrak{p} = (R \setminus \mathfrak{p})^{-1}\overline{R}$ be the localization of $\overline{R}$ at $\mathfrak{p}$. Then $\overline{R}_\mathfrak{p}$ is the integral closure of $R_\mathfrak{p}$ in K, and $R_\mathfrak{p} = \overline{R}_\mathfrak{p}$ holds if and only if $\mathfrak{p} \in P$. In this case $\mathfrak{p}\overline{R}$ is the only prime ideal of $\overline{R}$ above $\mathfrak{p}$, and $\overline{R}_{\mathfrak{p}\overline{R}} = \overline{R}_\mathfrak{p} = R_\mathfrak{p}$.

Let now $\mathfrak{p} \in \mathfrak{X}(R)$ be arbitrary, and let $\mathfrak{P}_1, \ldots, \mathfrak{P}_s$ be the prime ideals of $\overline{R}$ lying above $\mathfrak{p}$. Then $\overline{R}_\mathfrak{p}$ is a semilocal principal ideal domain, and $\mathfrak{P}_1\overline{R}_\mathfrak{p}, \ldots, \mathfrak{P}_s\overline{R}_\mathfrak{p}$ are its maximal ideals. Since

$$(\overline{R}_\mathfrak{p})_{\mathfrak{P}_i\overline{R}_\mathfrak{p}} = \overline{R}_{\mathfrak{P}_i}$$

holds for all $i \in \{1, \ldots, s\}$, we obtain

$$(*) \qquad\qquad \overline{R}_\mathfrak{p} = \bigcap_{i=1}^{s} \overline{R}_{\mathfrak{P}_i}$$

Lemma 7.1. *If* $\mathfrak{p} \in \mathfrak{X}(R)$ *and* $\mathfrak{P}_1, \ldots, \mathfrak{P}_s$ *are the prime ideals of* $\overline{R}$ *above* $\mathfrak{p}$, *then the map*

$$\omega_\mathfrak{p} : \begin{cases} K^\times/\overline{R}_\mathfrak{p}^\times & \to & \coprod\limits_{i=1}^{s} K^\times/\overline{R}_{\mathfrak{P}_i}^\times \\ a\overline{R}_\mathfrak{p}^\times & \mapsto & (a\overline{R}_{\mathfrak{P}_1}^\times, \ldots, a\overline{R}_{\mathfrak{P}_s}^\times) \end{cases}$$

is an isomorphism.

Proof. It follows from $(*)$ that

$$\overline{R}_\mathfrak{p}^\times = \bigcap_{i=1}^{s} \overline{R}_{\mathfrak{P}_i}^\times,$$

and therefore $\omega_\mathfrak{p}$ is a monomorphism. To prove surjectivity, let $a_1, \ldots, a_s \in K^\times$ be given. By the weak approximation theorem, there exists some $a \in K^\times$ such that $v_{\mathfrak{P}_i}(a) = v_{\mathfrak{P}_i}(a_i)$ and hence $a^{-1}a_i \in \overline{R}_{\mathfrak{P}_i}^\times$ for all $i \in \{1, \ldots, s\}$. This implies

$$\omega_\mathfrak{p}(a\overline{R}_\mathfrak{p}^\times) = (a_1\overline{R}_{\mathfrak{P}_1}^\times, \ldots, a_s\overline{R}_{\mathfrak{P}_s}^\times). \quad \square$$

Lemma 7.2. *For every finite subset* $F \subset \mathfrak{X}(R)$,

$$\coprod_{\mathfrak{p}\in\mathfrak{X}(R)} K^\times/R_\mathfrak{p}^\times = \mathrm{Im}\,(\Delta) \coprod_{\mathfrak{p}\in\mathfrak{X}(R)\backslash F} R_\mathfrak{p}^\bullet/R_\mathfrak{p}^\times,$$

and consequently

$$G = \theta\Big(\coprod_{\mathfrak{p}\in\mathfrak{X}(R)\backslash F} R_\mathfrak{p}^\bullet/R_\mathfrak{p}^\times \Big).$$

Proof. We may assume that $E \subset F$. Let

$$\alpha = (a_{\mathfrak{p}} R_{\mathfrak{p}}^{\times})_{\mathfrak{p} \in \mathfrak{X}(R)} \in \coprod_{\mathfrak{p} \in \mathfrak{X}(R)} K^{\times}/R_{\mathfrak{p}}^{\times}$$

be given. Let $f \in R$ be such that $f\overline{R} \subset R$. By the approximation theorem (see [10], Theorem 3.13), there exists some $a \in K^{\times}$ such that $v_{\mathfrak{P}}(aa_{\mathfrak{p}} - 1) > v_{\mathfrak{P}}(f)$ holds for all $\mathfrak{p} \in F$ and all $\mathfrak{P} \in \mathfrak{X}(\overline{R})$ above $\mathfrak{p}$. Then we have $aa_{\mathfrak{p}} \in \overline{R}_{\mathfrak{p}}^{\times}$ and $aa_{\mathfrak{p}} \in 1 + f\overline{R}_{\mathfrak{p}} \subset R_{\mathfrak{p}}$, which implies $aa_{\mathfrak{p}} \in R_{\mathfrak{p}}^{\times}$ for all $\mathfrak{p} \in F$. There are only finitely many primes $\mathfrak{p} \in P$ such that $v_{\mathfrak{p}\overline{R}}(aa_{\mathfrak{p}}) < 0$, say $\mathfrak{p}_1, \ldots, \mathfrak{p}_n$ and $v_{\mathfrak{p}_i \overline{R}}(aa_{\mathfrak{p}_i}) = -k_i < 0$. Since $\mathfrak{p}_1, \ldots, \mathfrak{p}_n \notin F$, we may apply [26], Theorem 81 to obtain some $x \in R$ satisfying

$$x \in \left(\bigcap_{i=1}^{n} \mathfrak{p}_i^{k_i} \right) \setminus \left(\bigcup_{\mathfrak{p} \in F} \mathfrak{p} \right),$$

which implies $xaa_{\mathfrak{p}} \in R_{\mathfrak{p}}^{\bullet}$ for all $\mathfrak{p} \in \mathfrak{X}(R)$ and $xaa_{\mathfrak{p}} \in R_{\mathfrak{p}}^{\times}$ for all $\mathfrak{p} \in F$. Consequently, we obtain

$$\alpha = \Delta\big((xa)^{-1}\big)\,(xaa_{\mathfrak{p}})_{\mathfrak{p} \in \mathfrak{X}(R)} \in \mathrm{Im}\,(\Delta) \coprod_{\mathfrak{p} \in \mathfrak{X}(R) \setminus F} R_{\mathfrak{p}}^{\bullet}/R_{\mathfrak{p}}^{\times}. \quad \square$$

Remark. In the language of ideals, Lemma 7.2 asserts that, given any finite subset $F \subset \mathfrak{X}(R)$, every class $g \in G$ contains t-invertible t-ideals $\mathfrak{a}$ such that $\mathfrak{a} \not\subset \mathfrak{p}$ for all $\mathfrak{p} \in F$.

Corollary 7.3. $G = [G_0]$.

Proof. If $g \in G$, then $g = \theta(\alpha)$ for some

$$\alpha = (a_{\mathfrak{p}} R_{\mathfrak{p}}^{\times})_{\mathfrak{p} \in P} \in \coprod_{\mathfrak{p} \in P} R_{\mathfrak{p}}^{\bullet}/R_{\mathfrak{p}}^{\times}$$

If $\mathfrak{p} \in P$, then $\mathfrak{p}R_{\mathfrak{p}} = \pi_{\mathfrak{p}} R_{\mathfrak{p}}$ and $a_{\mathfrak{p}} R_{\mathfrak{p}}^{\times} = \pi_{\mathfrak{p}}^{n_{\mathfrak{p}}} R_{\mathfrak{p}}^{\times}$ for some $\pi_{\mathfrak{p}} \in R_{\mathfrak{p}}$ and $n_{\mathfrak{p}} \in \mathbb{N}_0$. This implies

$$g = \sum_{\mathfrak{p} \in P} n_{\mathfrak{p}} [\mathfrak{p}] \in [G_0]. \quad \square$$

Next we compare the class groups $G = \mathcal{C}_t(R)$ and $\overline{G} = \mathcal{C}_t(\overline{R})$. Lemma 7.1 yields an isomorphism

$$\coprod_{\mathfrak{P} \in \mathfrak{X}(\overline{R})} K^\times / \overline{R}_{\mathfrak{P}}^\times = \coprod_{\mathfrak{p} \in \mathfrak{X}(R)} \coprod_{\substack{\mathfrak{P} \in \mathfrak{X}(\overline{R}) \\ \mathfrak{P} \cap R = \mathfrak{p}}} K^\times / \overline{R}_{\mathfrak{P}}^\times \xrightarrow{\sim} \coprod_{\mathfrak{p} \in \mathfrak{X}(R)} K^\times / \overline{R}_{\mathfrak{p}}^\times ,$$

by means of which the exact sequence for $\overline{G}$ takes the form

$$1 \to K^\times / \overline{R}^\times \xrightarrow{\overline{\Delta}} \coprod_{\mathfrak{p} \in \mathfrak{X}(R)} K^\times / \overline{R}_{\mathfrak{p}}^\times \xrightarrow{\overline{\theta}} \overline{G} \to 0;$$

here $\overline{\Delta}$ and $\overline{\theta}$ have the same meaning for $\overline{R}$ as Δ and θ have for R. Combining the exact sequences for G and $\overline{G}$, we obtain the following commutative diagram with exact rows and columns

$$
\begin{array}{ccccccccc}
 & & 1 & & 1 & & 0 & & \\
 & & \downarrow & & \downarrow & & \downarrow & & \\
1 & \longrightarrow & \overline{R}^\times / R^\times & \longrightarrow & \coprod_{\mathfrak{p} \in \mathfrak{X}(R)} \overline{R}_{\mathfrak{p}}^\times / R_{\mathfrak{p}}^\times & \longrightarrow & \mathcal{M} & \longrightarrow & 0 \\
 & & \downarrow & & \downarrow & & \downarrow & & \\
1 & \longrightarrow & K^\times / R^\times & \xrightarrow{\Delta} & \coprod_{\mathfrak{p} \in \mathfrak{X}(R)} K^\times / R_{\mathfrak{p}}^\times & \xrightarrow{\theta} & G & \longrightarrow & 0 \\
 & & \downarrow & & \downarrow & & \downarrow & & \\
1 & \longrightarrow & K^\times / \overline{R}^\times & \xrightarrow{\overline{\Delta}} & \coprod_{\mathfrak{p} \in \mathfrak{X}(R)} K^\times / \overline{R}_{\mathfrak{p}}^\times & \xrightarrow{\overline{\theta}} & \overline{G} & \longrightarrow & 0 \\
 & & \downarrow & & \downarrow & & \downarrow & & \\
 & & 1 & & 1 & & 0 & &
\end{array}
$$

Here the upper vertical arrows are injections, and

$$\mathcal{M} = \theta \Big(\coprod_{\mathfrak{p} \in \mathfrak{X}(R)} \overline{R}_{\mathfrak{p}}^\times / R_{\mathfrak{p}}^\times \Big) \subset G.$$

The epimorphism $G \to \overline{G}$ is induced from the extension of ideals from R to $\overline{R}$.

For $\mathfrak{p} \in P$, let $\pi_{\mathfrak{p}} \in R_{\mathfrak{p}}$ be a prime element for $\mathfrak{p}$. For $\mathfrak{p} \in E$, let $\pi_{\mathfrak{p},1} \dots, \pi_{\mathfrak{p},s(\mathfrak{p})}$ be non-associated prime elements of $\overline{R}_{\mathfrak{p}}$ such that $\pi_{\mathfrak{p},1}\overline{R}_\mathfrak{p},\ldots,\pi_{\mathfrak{p},s(\mathfrak{p})}\overline{R}_\mathfrak{p}$ are the maximal ideals of $\overline{R}_\mathfrak{p}$. Then every $\xi \in K^\times/R_\mathfrak{p}^\times$ has a unique representation

$$\xi = \pi_{\mathfrak{p},1}^{k_1} \cdot \ldots \cdot \pi_{\mathfrak{p},s(\mathfrak{p})}^{k_{s(\mathfrak{p})}} u R_\mathfrak{p}^\times,$$

where $k_1,\ldots,k_{s(\mathfrak{p})} \in \mathbb{Z}$ and $u R_\mathfrak{p}^\times \in \overline{R}_\mathfrak{p}^\times/R_\mathfrak{p}^\times$. We have $\xi \in \overline{R}_\mathfrak{p}^\bullet/R_\mathfrak{p}^\times$ if and only if $k_1,\ldots,k_{s(\mathfrak{p})} \in \mathbb{N}_0$, and if $\xi \in R_\mathfrak{p}^\bullet/R_\mathfrak{p}^\times$ then either $k_1 = \cdots = k_{s(\mathfrak{p})} = 0$, $u \in R_\mathfrak{p}^\times$ or $k_1,\ldots,k_{s(\mathfrak{p})} \in \mathbb{N}$. If morever

$$[R_\mathfrak{p}:\overline{R}_\mathfrak{p}] = \pi_{\mathfrak{p},1}^{e_{\mathfrak{p},1}} \cdot \ldots \cdot \pi_{\mathfrak{p},s(\mathfrak{p})}^{e_{\mathfrak{p},s(\mathfrak{p})}} \overline{R}_\mathfrak{p},$$

then $k_1 \geq e_{\mathfrak{p},1},\ldots,k_{s(\mathfrak{p})} \geq e_{\mathfrak{p},s(\mathfrak{p})}$ implies $\xi \in R_\mathfrak{p}^\bullet/R_\mathfrak{p}^\times$. In particular, $R_\mathfrak{p}^\bullet/R_\mathfrak{p}^\times$ is finitely primary of rank $s(\mathfrak{p})$, and $\overline{R}_\mathfrak{p}^\bullet/R_\mathfrak{p}^\times \simeq \mathbb{N}_0^{s(\mathfrak{p})} \times \overline{R}_\mathfrak{p}^\times/R_\mathfrak{p}^\times$ is its complete integral closure. Now the group homomorphism

$$\iota\colon \mathcal{F}(P) \times T \to G$$

is given as follows: For $\mathfrak{p} \in P$, we have $\iota(\mathfrak{p}) = [\mathfrak{p}]$. If

$$t = \left(\pi_{\mathfrak{p},1}^{k_{\mathfrak{p},1}} \cdot \ldots \cdot \pi_{\mathfrak{p},s(\mathfrak{p})}^{k_{\mathfrak{p},s(\mathfrak{p})}} u_\mathfrak{p} R_\mathfrak{p}^\times\right)_{\mathfrak{p} \in E} \in T,$$

then

$$\iota(t) = \sum_{\mathfrak{p}\in E}\left(\theta(u_\mathfrak{p} R_\mathfrak{p}^\times) + \sum_{i=1}^{s(\mathfrak{p})} k_{\mathfrak{p},i}\theta(\pi_{\mathfrak{p},i})\right) \in G.$$

If R is a one-dimensional noetherian domain, then the kernel $\mathcal{M}$ of the canonical epimorphism $G \to \overline{G}$ has a global description by means of the conductor $\mathfrak{f} = [R:\overline{R}]$.

Lemma 7.4. *Let R be a one-dimensional noetherian domain whose integral closure $\overline{R}$ is a finitely generated R-module, and let $\mathfrak{f} = [R:\overline{R}]$ be its conductor.*

i) *We have $(R/\mathfrak{f})^\times = (\overline{R}/\mathfrak{f})^\times \cap (R/\mathfrak{f})$, and the map*

$$\iota\colon \begin{cases} (\overline{R}/\mathfrak{f})^\times/(R/\mathfrak{f})^\times & \to & \coprod_{\mathfrak{p}\in\mathfrak{X}(R)} \overline{R}_\mathfrak{p}^\times/R_\mathfrak{p}^\times \\ (u+\mathfrak{f})(R/\mathfrak{f})^\times & \mapsto & (u R_\mathfrak{p}^\times)_{\mathfrak{p}\in\mathfrak{X}(R)} \end{cases}$$

is an isomorphism.

ii) *The epimorphism*

$$\theta \Big| \coprod_{\mathfrak{p} \in \mathfrak{X}(R)} \overline{R}_\mathfrak{p}^\times / R_\mathfrak{p}^\times : \coprod_{\mathfrak{p} \in \mathfrak{X}(R)} \overline{R}_\mathfrak{p}^\times / R_\mathfrak{p}^\times \longrightarrow \mathcal{M}$$

induces an isomorphism

$$(\overline{R}/\mathfrak{f})^\times / (R/\mathfrak{f})^\times \eta(\overline{R}^\times) \xrightarrow{\sim} \mathcal{M} \,,$$

where $\eta : \overline{R} \to \overline{R}/\mathfrak{f}$ *is the natural epimorphism.*

Proof. Since $R \subset \overline{R}$ is an integral extension, the same is true for $R/\mathfrak{f} \subset \overline{R}/\mathfrak{f}$, and hence $(R/\mathfrak{f})^\times = (\overline{R}/\mathfrak{f})^\times \cap (R/\mathfrak{f})$ by [26], Theorem 15.

If $\mathfrak{p} \in P$, then $\overline{R}_\mathfrak{p}^\times = R_\mathfrak{p}^\times$, and therefore we have in fact

$$\coprod_{\mathfrak{p} \in \mathfrak{X}(R)} \overline{R}_\mathfrak{p}^\times / R_\mathfrak{p}^\times = \coprod_{\mathfrak{p} \in E} \overline{R}_\mathfrak{p}^\times / R_\mathfrak{p}^\times \,.$$

If $\mathfrak{p} \in E$, then $\mathfrak{f}\overline{R}_\mathfrak{p}$ is contained in all maximal ideals of $\overline{R}_\mathfrak{p}$. Therefore, the natural map

$$\eta_\mathfrak{p} : \overline{R}_\mathfrak{p}^\times \to (\overline{R}_\mathfrak{p}/\mathfrak{f}\overline{R}_\mathfrak{p})^\times$$

is surjective and satisfies $\eta_\mathfrak{p}(R_\mathfrak{p}^\times) = (R_\mathfrak{p}/\mathfrak{f}R_\mathfrak{p})^\times$. Hence it induces an isomorphism

$$\overline{R}_\mathfrak{p}^\times / R_\mathfrak{p}^\times \xrightarrow{\sim} (\overline{R}_\mathfrak{p}/\mathfrak{f}\overline{R}_\mathfrak{p})^\times / (R_\mathfrak{p}/\mathfrak{f}R_\mathfrak{p})^\times \,.$$

The Chinese remainder theorem induces isomorphisms

$$\overline{R}/\mathfrak{f} \xrightarrow{\sim} \coprod_{\mathfrak{p} \in \mathfrak{X}(R)} \overline{R}_\mathfrak{p}/\mathfrak{f}\overline{R}_\mathfrak{p} \,, \quad R/\mathfrak{f} \xrightarrow{\sim} \coprod_{\mathfrak{p} \in \mathfrak{X}(R)} R_\mathfrak{p}/\mathfrak{f}R_\mathfrak{p} \,,$$

and therefore **i)** follows. The isomorphism

$$\iota^{-1} : \coprod_{\mathfrak{p} \in \mathfrak{X}(R)} \overline{R}_\mathfrak{p}^\times / R_\mathfrak{p}^\times \longrightarrow (\overline{R}/\mathfrak{f})^\times / (R/\mathfrak{f})^\times$$

maps $\Delta(\overline{R}^\times/R^\times)$ onto $\eta(\overline{R}^\times)(R/\mathfrak{f})^\times/(R/\mathfrak{f})^\times$. Hence we obtain the exact sequence

$$1 \to \overline{R}^\times/R^\times \to \coprod_{\mathfrak{p}\in\mathfrak{X}(R)} \overline{R}_\mathfrak{p}^\times/R_\mathfrak{p}^\times \to (\overline{R}/\mathfrak{f})^\times/(R/\mathfrak{f})^\times\eta(\overline{R}^\times) \to 1\,,$$

which proves the Lemma. $\square$

8. SETS OF LENGTHS AND ELASTICITY OF ONE-DIMENSIONAL NOETHERIAN DOMAINS: EXAMPLES

Throughout this section, let R be a one-dimensional noetherian domain whose integral closure $\overline{R}$ is a finitely generated R-module, and we suppose that the conductor $\mathfrak{f} = [R:\overline{R}]$ is $\mathfrak{p}$-primary for some $\mathfrak{p} \in \mathfrak{X}(R)$. We continue to use all notions introduced in sections 6 and 7.

We do one example with $s(\mathfrak{p}) > 1$, a second one with $s(\mathfrak{p}) = 1$, where the prime divisor of $\mathfrak{p}$ in $\overline{R}$ is not principal, and finally we consider in more generality the case where $s(\mathfrak{p}) = 1$ and $\overline{R}$ is a principal ideal domain.

Example 1: $R = \mathbb{Z}[\sqrt{-7}]$.

The integral closure of R is

$$\overline{R} = \mathbb{Z}\left[\frac{1 + \sqrt{-7}}{2}\right],$$

and $\mathfrak{f} = \mathfrak{p} = 2\overline{R} = (\pi\overline{R})(\pi'\overline{R})$, where

$$\pi = \frac{1 + \sqrt{-7}}{2}, \quad \pi' = \frac{1 - \sqrt{-7}}{2}.$$

$\overline{R}$ is a principal ideal domain, and since

$$(\overline{R}/2\overline{R})^\times \simeq (\overline{R}/\pi\overline{R})^\times \times (\overline{R}/\pi'\overline{R})^\times = \{1\}\,,$$

Lemma 7.4 implies $\mathcal{M} = G = \{0\}$. Hence the block monoid $B = \mathcal{B}(R)$ is given by

$$B = \mathcal{F}(\{0\}) \times R_\mathfrak{p}^\bullet/R_\mathfrak{p}^\times\,,$$

and its elements have the form

$$0^m \pi^k \pi'^{k'} R_{\mathfrak{p}}^{\times},$$

where $m \in \mathbf{N}_0$ and $(k, k') \in B_0 = (\mathbf{N} \times \mathbf{N}) \cup \{(0,0)\}$. Therefore B is isomorphic to $\mathbf{N}_0 \times B_0$, and its system of sets of lengths is given by

$$\mathcal{L}(B) = \{m + L \mid m \in \mathbf{N}_0, \ L \in \mathcal{L}(B_0)\}.$$

Thus it remains to calculate $\mathcal{L}(B_0)$. The atoms of B_0 are the elements $(1, k)$ and $(k, 1)$ for all $k \in \mathbf{N}$. If $(k, l) \in B_0$ and $l \geq k \geq 2$, then there is exactly one factorization of length 2 of (k, l), namely

$$(k, l) = (1, l - 1) + (k - 1, 1).$$

Any other factorization of (k, l) contains at least 3 atoms which can be replaced by 2 atoms by means of the relations

$$(1, a) + (1, b) + (1, c) = (1, a + b + c - 1) + (2, 1)$$
$$(1, a) + (1, b) + (c, 1) = (1, a + b) + (c + 1, 1)$$
$$(1, a) + (b, 1) + (c, 1) = (1, a + 1) + (b + c, 1)$$
$$(a, 1) + (b, 1) + (c, 1) = (a + b + c - 1, 1) + (1, 2).$$

Therefore we obtain

$$L\big((k, l)\big) = \{2, 3, \ldots, k - 1, k\},$$

and the factorization of length k is given by

$$(k, l) = (k - 1)(1, 1) + (1, l - k + 1).$$

The ring $\mathbf{Z}[\sqrt{-7}]$ has infinite elasticity and is a frequently used example in factorization theory; see [17], Example 11 and [21], Example 4.

Example 2: $R = \mathbf{Z}[\sqrt{-20}]$.

The integral closure of R is $\overline{R} = \mathbf{Z}[\sqrt{-5}]$, $\mathfrak{f} = \mathfrak{z} = 2R \in \mathfrak{X}(R)$, and $2\overline{R}_{\mathfrak{z}} = \pi^2 \overline{R}_{\mathfrak{z}}$, where $\pi = 1 + \sqrt{-5}$. It is well known that the class group $\overline{G}$ of $\overline{R}$ is a group with 2 elements. From

$$\overline{R}_{\mathfrak{z}}^{\times} / R_{\mathfrak{z}}^{\times} \simeq (\overline{R}/2\overline{R})^{\times} / (R/2R)^{\times}$$

we see that $\overline{R}_{\mathfrak{z}}^{\times}/R_{\mathfrak{z}}^{\times}$ is also a group with 2 elements, generated by

$$\varepsilon = (2 + \sqrt{-5})R_{\mathfrak{z}}^{\times} \in \overline{R}_{\mathfrak{z}}^{\times}/R_{\mathfrak{z}}^{\times} .$$

Since $\overline{R}^{\times} = R^{\times} = \{\pm 1\}$, it follows that

$$\theta \colon \overline{R}_{\mathfrak{z}}^{\times}/R_{\mathfrak{z}}^{\times} \to \mathcal{M}$$

is an isomorphism, which implies $\#\mathcal{M} = 2$ and $\#G = 4$. To prove that G is cyclic, we consider the decomposition $3R = \mathfrak{q}\mathfrak{q}'$, where $\mathfrak{q} = (3, 1 - 4\sqrt{-5})$ and $\mathfrak{q}' = (3, 1 + 4\sqrt{-5})$. Now $\mathfrak{q}^2 = (9, 1 - 4\sqrt{-5})$, $(R : \mathfrak{q}^2) = 9$, and 9 has no representation in the form $x^2 + 20y^2$ with $x, y \in \mathbb{Z}$. Therefore $\mathfrak{q}^2$ is not principal and G is cyclic of order 4, generated by $g = [\mathfrak{q}]$. In particular, we have $2g = [\mathfrak{q}^2] = \theta(\varepsilon) \in \mathcal{M}$ (indeed, $\mathfrak{q}^2\overline{R} = (2 + \sqrt{-5})\overline{R}$). We have

$$\overline{R}_{\mathfrak{z}}^{\bullet}/R_{\mathfrak{z}}^{\times} = \{\pi^k \varepsilon^a \mid k \in \mathbb{N}_0, \ a = 0, 1\},$$

and obviously $\pi^k \varepsilon^a \in R_{\mathfrak{z}}^{\bullet}/R_{\mathfrak{z}}^{\times}$ for $k \geq 2$, $\pi\varepsilon^0 = (1 + \sqrt{-5})R_{\mathfrak{z}}^{\times} \notin R_{\mathfrak{z}}^{\bullet}/R_{\mathfrak{z}}^{\times}$ and $\pi\varepsilon^1 = (-3 + 3\sqrt{-5})R_{\mathfrak{z}}^{\times} \notin R_{\mathfrak{z}}^{\bullet}/R_{\mathfrak{z}}^{\times}$. Thus we obtain

$$R_{\mathfrak{z}}^{\bullet}/R_{\mathfrak{z}}^{\times} = \{\pi^k \varepsilon^a \mid k \geq 2, \ a = 0, 1 \quad \text{or} \quad k = a = 0\}.$$

In order to calculate $\theta(\pi R_{\mathfrak{z}}^{\times})$, we consider the element

$$\alpha = (a_{\mathfrak{p}} R_{\mathfrak{p}}^{\times})_{\mathfrak{p} \in \mathfrak{X}(R)} \in \coprod_{\mathfrak{p} \in \mathfrak{X}(R)} K^{\times}/R_{\mathfrak{p}}^{\times} ,$$

defined by $a_{\mathfrak{z}} = \pi = 1 + \sqrt{-5}$ and $a_{\mathfrak{p}} = 1$ if $\mathfrak{p} \neq \mathfrak{z}$. Since $\mathfrak{q}\overline{R}_{\mathfrak{q}} = (1 - \sqrt{-5})\overline{R}_{\mathfrak{q}}$ and $\mathfrak{q}'\overline{R}_{\mathfrak{q}'} = (1 + \sqrt{-5})\overline{R}_{\mathfrak{q}'}$, we obtain

$$\frac{a_{\mathfrak{p}}}{1 + \sqrt{-5}} \in R_{\mathfrak{p}}^{\times} \quad \text{for all} \quad \mathfrak{p} \neq \mathfrak{q}'$$

and

$$\frac{a_{\mathfrak{q}'}}{a + \sqrt{-5}} = \pi^{-1}, \quad v_{\mathfrak{q}'\overline{R}}(\pi) = 1,$$

which implies

$$\theta(\pi R_{\mathfrak{z}}^{\times}) = \theta(\alpha) = \theta\left(\Delta\left(\frac{1}{1 + \sqrt{-5}}\right)\alpha\right) = \theta(\pi^{-1}R_{\mathfrak{q}'}^{\times}) = -[\mathfrak{q}'] = [\mathfrak{q}] = g.$$

Therefore the block monoid $B = \mathcal{B}(R)$ consists of all elements of the form

$$0^m g^b (2g)^c (3g)^d (\pi^k \varepsilon^a) \in \mathcal{F}(G) \times T,$$

where $m, b, c, d, k \in \mathbb{N}_0$, $a \in \{0, 1\}$, either $k \geq 2$ or $k = a = 0$ and

$$b + 2c + 3d + k + 2a \equiv 0 \mod 4.$$

B contains 12 atoms which are not prime, namely

$$g^4, \, g(3g), \, (2g)^2, \, g^2(2g)$$
$$g^2(\pi^2), \, (2g)(\pi^2), \, g(\pi^3), \, (2g)(3g)(\pi^3),$$
$$(\pi^2 \varepsilon), \, g(2g)(\pi^3 \varepsilon), \, g^3(\pi^3 \varepsilon), \, (3g)(\pi^3 \varepsilon).$$

An easy but lengthy calculation shows that all sets of lengths are of the form

$$\{k, k+1, \ldots, k+l\}$$

for some $k \in \mathbb{N}$, $l \in \mathbb{N}_0$.

Finally we calculate the elasticity of R, $\varrho(R) = \varrho(B)$ using semilength functions, see [1]. We summarize the needed results on semilength functions as follows.

A semilength function $f \colon B \to \mathbb{N}_0$ is a monoid homomorphism satisfying $f^{-1}(0) = \{1\}$. If

$$M^*(f) = \sup \{f(x) \mid x \in B \text{ is an atom, but not prime}\},$$
$$m^*(f) = \inf \{f(x) \mid x \in B \text{ is an atom, but not prime}\},$$

then

$$\varrho(B) \leq \frac{M^*(f)}{m^*(f)}.$$

In our case we use the semilength function

$$f\big(0^m g^b (2g)^c (3g)^d (\pi^k a)\big) = b + c + d + k.$$

The above list of atoms yields $M^*(f) = 5$, $m^*(f) = 2$, and hence

$$\varrho(B) \leq \frac{5}{2}.$$

The factorization

$$\left[(2g)(3g)(\pi^3)\right]^2 \left[g(2g)(\pi^3\varepsilon)\right]^2 = \left[g(3g)\right]^2 \left[(2g)^2\right]^2 (\pi^2)^6$$

shows that in fact

$$\varrho(R) = \varrho(B) = \frac{5}{2}.$$

Special case: $s(\mathfrak{p}) = 1$ and $\overline{R}$ is a principal ideal domain

First we recall some facts on numerical semigroups. By a *numerical semigroup* S we mean an (additive) submonoid $S \subset \mathbb{N}_0$ for which $\mathbb{N}_0 \setminus S$ is a finite set. Such a numerical semigroup has the form

$$S = [n_1, \ldots, n_r],$$

where $1 \leq n_1 < \cdots < n_r$, $\gcd(n_1, \ldots, n_r) = 1$ and either $n_1 = r = 1$ or $n_1 \geq 2$. We always suppose that r is chosen minimal, which implies that $n_1, \ldots, n_r$ are the atoms of S.

Now let R be a one-dimensional noetherian domain such that its integral closure $\overline{R}$ is a principal ideal domain and $[R:\overline{R}] = \pi^e \overline{R}$ for some prime element π of $\overline{R}$. We set $\mathfrak{p} = \pi\overline{R} \cap R \in \mathfrak{X}(R)$. Then $[R:\overline{R}]$ is a $\mathfrak{p}$-primary R-ideal, and $s(\mathfrak{p}) = 1$. Since $\overline{R}$ is a principal ideal domain, we have $G = \mathcal{M}$ and $\theta(\pi R_\mathfrak{p}^\times) = 0$. For $k \in \mathbb{N}_0$, we set

$$\mathcal{U}_k = \{\varepsilon R_\mathfrak{p}^\times \in \overline{R}_\mathfrak{p}^\times / R_\mathfrak{p}^\times \mid \pi^k \varepsilon \in R_\mathfrak{p}\}.$$

Since $[R_\mathfrak{p}:\overline{R}_\mathfrak{p}] = \pi^e \overline{R}_\mathfrak{p}$ and $\overline{R}_\mathfrak{p}^\times \cap R_\mathfrak{p} = R_\mathfrak{p}^\times$, we obtain $\mathcal{U}_k = \overline{R}_\mathfrak{p}^\times / R_\mathfrak{p}^\times$ if and only if $k \geq e$, and $\mathcal{U}_0 = \{1\}$. Morever, $\mathcal{U}_k \mathcal{U}_l \subset \mathcal{U}_{k+l}$ for all $k, l \in \mathbb{N}_0$, and

$$S = \{k \in \mathbb{N}_0 \mid \mathcal{U}_k \neq \emptyset\} \subset \mathbb{N}_0$$

is a numerical semigroup containing all natural numbers $k \geq e$. As above, we write S in the form $S = [n_1, \ldots, n_r]$, where $1 \leq n_1 < \cdots < n_r$ are the atoms of S.

The elements of the block monoid $B = \mathcal{B}(R)$ have a unique representation in the form

$$g_1 \cdots \cdots g_n (\pi^k \varepsilon) \in \mathcal{F}(G_0) \times R_\mathfrak{p}^\bullet / R_\mathfrak{p}^\times,$$

where $g_1, \ldots, g_n \in G_0$, $k \in \mathbb{N}_0$, $\varepsilon \in \mathcal{U}_k$ and $g_1 + \cdots + g_n + \theta(\varepsilon) = 0$.
In order to estimate the elasticity of R, we consider the semilength
function $f \colon B \to \mathbb{N}_0$, defined by

$$f\big(g_1 \cdot \ldots \cdot g_n(\pi^k \varepsilon)\big) = n + \frac{2k}{n_1},$$

and we obtain

$$\varrho(R) = \varrho(B) \le \frac{M^*(f)}{m^*(f)},$$

where $M^*(f)$ resp. $m^*(f)$ is the supremum resp. infimum of all
numbers $f(x)$ for atoms $x \in B$ which are not prime; see Example
2 above.

Suppose now that $x = g_1 \cdot \ldots \cdot g_n(\pi^k \varepsilon) \in B$ is an atom which is
not prime. We have to distinguish two cases.

CASE 1: $\pi^k \varepsilon = 1$. In this case, $g_1 \cdot \ldots \cdot g_n$ is an irreducible block
in $\mathcal{B}(G_0)$, and therefore $f(x) = n \le \mathcal{D}(G_0)$.

CASE 2: $\pi^k \varepsilon \ne 1$. If $k \ge e + n_1$ and $\varepsilon_1 \in \mathcal{U}_{n_1}$, then the
equation

$$x = g_1 \cdot \ldots \cdot g_n(\pi^{n_1} \varepsilon_1)(\pi^{k-n_1} \varepsilon_1^{-1} \varepsilon)$$

shows that x is not an atom of B. Thus we have $k \le e + n_1 - 1$,
and clearly $g_1 \cdot \ldots \cdot g_n$ contains no irreducible block, which implies
$n \le \mathcal{D}(G_0) - 1$,

$$f(x) = n + \frac{2k}{n_1} \le \mathcal{D}(G_0) + 1 + \frac{2(e-1)}{n_1},$$

and consequently

$$M^*(f) \le \mathcal{D}(G_0) + 1 + \frac{2(e-1)}{n_1}.$$

Using the trivial estimate $m^*(f) \ge 2$, we obtain

$$\varrho(R) = \varrho(B) \le \frac{\mathcal{D}(G_0) + 1}{2} + \frac{e-1}{n_1}.$$

Under additional assumptions equality can be obtained. Suppose that
$G_0 = -G_0$, $n_r = e - 1 + n_1$, and that there exists some $\eta \in \mathcal{U}_{n_1}$ such
that $\theta(\eta) \in G$ has finite order. Let $g_1 \cdot \ldots \cdot g_D \in \mathcal{B}(G_0) \subset B$ be an

irreducible block of length $D \le \mathcal{D}(G_0)$. Then $(-g_1) \cdot \ldots \cdot (-g_D)$ is also an irreducible block. Since $\theta \colon \overline{R}_{\mathfrak{p}}^\times / R_{\mathfrak{p}}^\times \to G$ is surjective, there exists some $\varepsilon \in \overline{R}_{\mathfrak{p}}^\times / R_{\mathfrak{p}}^\times$ satisfying $\theta(\varepsilon) = g_D$, and since $n_r \ge e$, the elements

$$u = g_1 \cdot \ldots \cdot g_{D-1}(\pi^{n_r}\varepsilon) \quad \text{and} \quad u' = (-g_1) \cdot \ldots \cdot (-g_{D-1})(\pi^{n_r}\varepsilon^{-1})$$

are atoms of B. If $N \in \mathbb{N}$ satisfies $N\theta(g) = 0$, then the factorization

$$(uu')^{Nn_1} = [g_1(-g_1)]^{Nn_1} \cdot \ldots \cdot [g_{D-1}(-g_{D-1})]^{Nn_1} (\pi_{n_1}\eta)^{2Nn_r}$$

shows that

$$\varrho(B) \ge \frac{Nn_1(D-1) + 2Nn_r}{2Nn_1}.$$

If $\mathcal{D}(G_0) = \infty$, this implies $\varrho(B) = \infty$. If $\mathcal{D}(G_0) < \infty$, we set $D = \mathcal{D}(G_0)$ and conclude

$$\varrho(B) = \frac{\mathcal{D}(G_0) + 1}{2} + \frac{e - 1}{n_1}.$$

We summarize the result of our considerations in the following Theorem.

Theorem 8.1. Let R be a one-dimensional noetherian domain whose integral closure $\overline{R}$ is a principal ideal domain satisfying $[R : \overline{R}] = \pi^e \overline{R}$, where π is a prime element of $\overline{R}$ and $e \in \mathbb{N}$. We set $\mathfrak{p} = \pi \overline{R} \cap R \in \mathfrak{X}(R)$. Let S be the numerical semigroup consisting of all $k \in \mathbb{N}_0$ such that $\pi^k \eta \in R_{\mathfrak{p}}$ holds for some $\eta \in \overline{R}_{\mathfrak{p}}^\times$, and set $S = [n_1, \ldots, n_r]$, where $1 \le n_1 < \cdots < n_r$ are the atoms of S. Let G be the class group of R and G_0 the set of all classes containing prime ideals. Then we have

$$\varrho(R) \le \frac{\mathcal{D}(G_0) + 1}{2} + \frac{e - 1}{n_1}.$$

If $G_0 = -G_0$, $n_r = e - 1 + n_1$, and if there exists some $\eta \in \mathcal{U}_{n_1}$ such that $\theta(\eta) \in G$ has finite order, then equality holds.

D. F. Anderson and his co-authors proved several special cases of Theorem 7.1, in particular for monoid algebras $K[S]$ and $K[[S]]$; see [8] for an overview.

Example 3: $R = \mathbb{Z}[\sqrt{8}]$.

The integral closure of R is $\overline{R} = \mathbb{Z}[\sqrt{2}]$, and $\mathfrak{f} = \mathfrak{p} = 2R = 2\overline{R} = (\sqrt{2}\,\overline{R})^2$. Since $\overline{R}_\mathfrak{p}^\times/R_\mathfrak{p}^\times \simeq (\overline{R}/2\overline{R})^\times/(R/2R)^\times$, we see that $\overline{R}_\mathfrak{p}^\times/R_\mathfrak{p}^\times$ is a group with 2 elements, generated by $\varepsilon = (1+\sqrt{2})R_\mathfrak{p}^\times$. Since $1+\sqrt{2} \in \overline{R}^\times$, the map $\Delta\colon \overline{R}^\times/R^\times \to \overline{R}_\mathfrak{p}^\times/R_\mathfrak{p}^\times$ is surjective, and $\mathcal{M} = G = 0$. Since $\sqrt{2} \notin R_\mathfrak{p}$, $\sqrt{2}(1+\sqrt{2}) \notin R_\mathfrak{p}$ and $(1+\sqrt{2})^2 \in R_\mathfrak{p}$, $B = \mathcal{B}(R) = \mathcal{F}(\{0\}) \times R_\mathfrak{p}^\bullet/R_\mathfrak{p}^\times$ is the block monoid of R, and its elements are of the form

$$0^m \sqrt{2}^n \varepsilon^a,$$

where $m, n \in \mathbb{N}_0$, $a \in \{0,1\}$ and either $n \geq 2$ or $n = a = 0$. Consequently, we obtain $S = [2,3]$, $r = n_1 = 2$, $n_2 = 3$, $e = 2$, $G_0 = G = \{0\}$, $\mathcal{D}(G_0) = 1$ and hence

$$\varrho(R) = \frac{3}{2}$$

(which, of course, also could be immediately verified). B has 4 atoms which are not prime, namely $2, 2\varepsilon, 2\sqrt{2}$ and $2\sqrt{2}\varepsilon$ (we omit the factor $\varepsilon^0 = R_\mathfrak{p}^\times$). The defining relations are

$$2^2 = (2\varepsilon)^2, \quad 2^3 = (2\sqrt{2})^2 = (2\sqrt{2}\varepsilon)^2,$$
$$2^2(2\varepsilon) = (2\sqrt{2})(2\sqrt{2}\varepsilon), \quad 2(2\sqrt{2}) = (2\varepsilon)(2\sqrt{2}\varepsilon).$$

The only prime element of B is 0. The map

$$f\colon \begin{cases} B & \to \mathbb{N}_0 \times [2,3] \\ 0^m\sqrt{2}^n\varepsilon & \mapsto \quad (m,n) \end{cases}$$

is length-preserving. Therefore

$$\mathcal{L}(B) = \{m + L_0 \mid L_0 \in \mathcal{L}([2,3])\},$$

which implies that $\mathcal{L}(B)$ consists of all sets

$$\{a, a+1, \ldots, b\},$$

where $a, b \in \mathbb{N}_0$, $a \leq b \leq \frac{3}{2}a$.

It is worth mentioning that $\mathcal{L}(B)$ coincides with $\mathcal{L}(\mathcal{B}(C_3))$. Since $\mathcal{B}(C_3)$ is a Krull monoid, we see, how little the structure of sets of lengths tells us about the structure of the underlying monoid.

Example 4: $R = \mathbf{Z}[3i]$, $i = \sqrt{-1}$.

The integral closure of R is $\overline{R} = \mathbf{Z}[i]$, and $\mathfrak{f} = \mathfrak{p} = 3R = 3\overline{R} \in \mathfrak{X}(R)$. The group $(\overline{R}/3\overline{R})^\times$ is cyclic of order 8, generated by $(1-i)+3\overline{R}$. Since $(R/3R)^\times = <-1+3R>$, the group

$$\overline{R}_\mathfrak{p}^\times / R_\mathfrak{p}^\times \simeq (\overline{R}/3\overline{R})^\times / (R/3R)^\times$$

is cyclic of order 4, generated by $\varepsilon = (1-i)R_\mathfrak{p}^\times \in \overline{R}_\mathfrak{p}^\times / R_\mathfrak{p}^\times$. Since $\Delta(\overline{R}^\times / R^\times) = < iR_\mathfrak{p}^\times > = < \varepsilon^2 >$, the class group $G = \mathcal{M}$ is of order 2, generated by $g = \theta(\varepsilon)$. By Corollary 7.3, $G = [G_0]$, and consequently $G = G_0$. Hence the block monoid $B = \mathcal{B}(R)$ of R consists of all elements

$$0^m g^b (3^k \varepsilon^a) \in \mathcal{F}(G) \times R_\mathfrak{p}^\bullet / R_\mathfrak{p}^\times,$$

where $m, b \in \mathbf{N}_0$, either $k \in \mathbf{N}$, $a \in \{0,1\}$ or $k = a = 0$, and $b + a \equiv 0 \mod 2$. B possesses exactly 3 atoms which are not prime, namely

$$3, \quad g^2 \quad \text{and} \quad g(3\varepsilon),$$

and 0 is the only prime element of B. There is one generating relation, namely

$$[g(3\varepsilon)]^2 = 3^2 g^2,$$

and therefore again $\mathcal{L}(B)$ consists of all sets

$$\{a, a+1, \ldots, b\},$$

where $a, b \in \mathbf{N}_0$, $a \le b \le \frac{3}{2}a$.

Example 5: $R = \mathbb{F}_2[X^2, X^5]$.

The integral closure of R is $\overline{R} = \mathbb{F}_2[X]$, $\mathfrak{f} = X^4\overline{R}$, $\mathfrak{p} = X\overline{R} \cap R$, $\pi = X$, $e = 4$, $S = [2,5] \subset \mathbf{N}_0$, $r = n_1 = 2$ and $n_2 = 5$. The group $(\overline{R}/X^4\overline{R})^\times$ is of order 8, generated by $\alpha = (1+X)+X^4\overline{R}$ and $\beta = (1+X^3)+X^4\overline{R}$, where $\alpha^4 = \beta^2 = 1+X^4\overline{R}$ and $\alpha^2 \in$

$(R/X^4\overline{R})^\times$. The isomorphism $\overline{R}_\mathfrak{p}^\times/R_\mathfrak{p}^\times \simeq (\overline{R}/X^4\overline{R})^\times/(R/X^4\overline{R})^\times$ of Lemma 7.4 shows that

$$\overline{R}_\mathfrak{p}^\times/R_\mathfrak{p}^\times = <u,v>,$$

where $u = (1+X)R_\mathfrak{p}^\times$, $v = (1+X^3)R_\mathfrak{p}^\times$ and $u^2 = v^2 = 1$. Since $\overline{R}^\times = R^\times = \mathbb{F}_2^\times$, the map $\theta : \overline{R}_\mathfrak{p}^\times/R_\mathfrak{p}^\times \to \mathcal{M} = G$ is an isomorphism. Therefore

$$G = <g,h> \simeq C_2 \oplus C_2,$$

where $g = \theta(u)$ and $h = \theta(v)$. Since g contains the prime ideal $(1+X)\overline{R} \cap R$, h contains the prime ideal $(1+X^2+X^3)\overline{R} \cap R$, and $g+h$ contains the prime ideal $(1+X+X^3)\overline{R} \cap R$, we see that $G = G_0$ (more general, it follows from the prime ideal theorem in algebraic function fields, that $G = G_0$ holds whenever R is an order in an algebraic function field in one variable over a finite field). Theorem 8.1 implies

$$\varrho(R) = \frac{7}{2}.$$

Next we calculate the sets $\mathcal{U}_k$. Clearly, $\mathcal{U}_0 = \{1\}$, $\mathcal{U}_1 = \mathcal{U}_3 = \emptyset$ and $\mathcal{U}_k = \overline{R}_\mathfrak{p}^\times/R_\mathfrak{p}^\times$ for $k \geq 4$. Since

$$\mathcal{U}_2 = \{u^i v^j \mid i,j = 0,1;\ X^2 u^i v^j \subset R_\mathfrak{p}\},$$

we see that $\mathcal{U}_2 = \{1,v\}$. Hence the block monoid $B = \mathcal{B}(R)$ consists of all elements of the form

$$0^m g^b h^c (g+h)^d (X^k u^\varepsilon v^{\varepsilon'}),$$

where $m,b,c,d \in \mathbb{N}_0$, $\varepsilon,\varepsilon' \in \{0,1\}$, $b+d+\varepsilon \equiv c+d+\varepsilon' \equiv 0 \mod 2$ and either $k = \varepsilon = \varepsilon' = 0$ or $k = 2$, $\varepsilon = 0$ or $k \geq 4$. There are 11 atoms which are not prime, namely

$$g^2,\ h^2,\ gh(g+h),\ X^2,\ X^5,\ h(X^2 v),\ g(g+h)(X^2 v),$$
$$g(X^5 u),\ h(X^5 v),\ (g+h)(X^5 uv),\ gh(X^5 uv).$$

This time, the sets of lengths are not always of the simple form as in the examples above. For example, if $a \geq 2$, then

$$L\big(g^3(X^{10a}u)\big) =$$
$$\{2a,\ 2a+1,\ 2a+3,\ 2a+4,\ 2a+6,\ 2a+7,\ 2a+9,$$
$$\dots,\ 5a-9,\ 5a-8,\ 5a-6,\ 5a-5,\ 5a-2\}$$

is already a rather general almost arithmetical progression.

REFERENCES

[1] D. D. Anderson and D. F. Anderson, *Elasticity of factorizations in integral domains*, J. Pure Appl. Algebra **80**: 217–235 (1992).

[2] D. D. Anderson, D. F. Anderson, S. Chapman and W. W. Smith, *Rational elasticity of factorizations in Krull domains*, Proc. AMS **117**: 37–43 (1993).

[3] D. D. Anderson, D. F. Anderson and M. Zafrullah, *Factorization in integral domains*, J. Pure Appl. Algebra **69**: 1–19 (1990).

[4] D. D. Anderson, D. F. Anderson and M. Zafrullah, *Atomic domains in which almost all atoms are prime*, Comm. Alg. **20**: 1447–1462 (1992).

[5] D. D. Anderson and J. L. Mott, *Cohen-Kaplansky domains: integral domains with a finite number of irreducible elements*, J. Algebra **148**: 17–41 (1992).

[6] D. D. Anderson, J. Mott and M. Zafrullah, *Finite character representations for integral domains*, Bollettino U.M.I. (**7**) **6-B**: 613–630 (1992).

[7] D. D. Anderson and B. Mullins, *Finite factorization domains*, Proc. AMS **124**: 389–396 (1996).

[8] D. F. Anderson, *Elasticity of factorizations in integral domains, a survey*, these Proceedings.

[9] S. T. Chapman and A. Geroldinger, *Krull domains and monoids, their sets of lengths and associated combinatorial problems*, these Proceedings.

[10] O. Endler, *Valuation Theory*, Springer, 1972.

[11] A. Geroldinger, *Über nicht-eindeutige Zerlegungen in irreduzible Elemente*, Math. Z. **197**: 505–529 (1988).

[12] A. Geroldinger, *On the arithmetic of certain not integrally closed noetherian integral domains*, Comm. Algebra **19**: 685–698 (1991).

[13] A. Geroldinger, *T-block monoids and their arithmetical applications to certain integral domains*, Comm. Algebra **22**: 1603–1615 (1994).

[14] A. Geroldinger, *Chains of factorizations and sets of lengths*, J. Algebra.

[15] A. Geroldinger, *On the structure and arithmetic of finitely primary monoids*, Czech. Math. J..

[16] A. Geroldinger, *Chains of factorizations in weakly Krull domains*, Colloq. Math..

[17] A. Geroldinger and F. Halter-Koch, *On the asymptotic behaviour of lengths of factorizations*, J. Pure Appl. Algebra **77**: 239–252 (1992).

[18] A. Geroldinger and F. Halter-Koch, *Arithmetical theory of monoid homomorphisms*, Semigroup Forum **48**: 333–362 (1994).

[19] F. Halter-Koch, *Zur Zahlen- und Idealtheorie eindimensionaler noetherscher Integritätsbereiche*, J. Algebra **136**: 103–108 (1991).

[20] F. Halter-Koch, *Finiteness theorems for factorizations*, Semigroup Forum
44: 112–117 (1992).

[21] F. Halter-Koch, *On the asymptotic behaviour of the number of distinct factorizations into irreducibles*, Arkiv f. Math. **31**: 297–305 (1993).

[22] F. Halter-Koch, *Über Längen nicht-eindeutiger Faktorisierungen und Systeme linearer diophantischer Ungleichungen*, Abh. Math. Sem. Univ. Hamburg **63**: 265–276 (1993).

[23] F. Halter-Koch, *Elasticity of factorizations in atomic monoids and integral domains*, J. Th. Nomb. Bordeaux **7**: 367–385 (1995).

[24] F. Halter-Koch, *Divisor theories with primary elements and weakly Krull domains*, Boll. UMI **9-B**: 417–441 (1995).

[25] N. Jacobson, *Basic Algebra I*, W. H. Freeman & Co., 1985.

[26] I. Kaplansky, *Commutative Rings*, Allyn and Bacon, 1970.

[27] S. Lang, *Algebraic Number Theory*, Addison-Wesley, 1970.

[28] W. Narkiewicz, *Finite abelian groups and factorization problems*, Coll. Math. **42**: 319–330 (1979).

[29] W. Narkiewicz, *Elementary and Analytic Theory of Algebraic Numbers*, Springer, 1990.

[30] L. Rédei, *Theorie der endlich erzeugbaren kommutativen Halbgruppen*, Physica-Verlag, 1963.

[31] A. Schrijver, *Theory of linear and integer programming*, John Wiley & Sons, 1986.

Krull Domains and Monoids, Their Sets of Lengths, and Associated Combinatorial Problems

Scott Chapman, Trinity University, Department of Mathematics, San Antonio, Texas 78212-7200

Alfred Geroldinger, Institut für Mathematik, Karl-Franzens-Universität, 8010 Graz, Austria

1. INTRODUCTION

This is a survey article on the theory of non-unique factorization. This field has its origins in algebraic number theory. Today, problems involving the factorization of elements into irreducibles are studied in general integral domains using a huge variety of techniques (see [A-A-Z1], [A-A-Z2] and [A-A-Z3]). In this paper, we consider factorization properties of Krull domains, including integrally closed noetherian domains and rings of integers in algebraic number fields. We restrict our interest to sets of lengths and the invariants derived from them. Our results are valid mainly for Krull domains with finite divisor class group.

Although our emphasis and the main applications of the theory of non-unique factorization lies in ring theory, this paper is written in the language of monoids. The reason for this is not for higher generality, but usefulness and simplicity. We start in chapter 2 with

Krull monoids by discussing their sets of lengths and various constants which control these sets. We gather in this chapter many results which have appeared in the literature using different notation and terminology. In chapter 3, we study block monoids. These are suitably constructed monoids which provide the opportunity to understand and describe more clearly the invariants which have been previously introduced. Having these monoids at our disposal, we can reduce ring theoretical problems to problems in finitely generated monoids and are able to apply geometrical methods. There is another striking example of the usefulness of this approach. Results for Krull monoids can be applied to investigate one-dimensional noetherian domains R which are not integrally closed, but have non-zero conductor $\mathrm{Ann}_R(\overline{R}/R)$. In chapter 4, we give a short account of the analytic aspects of non-unique factorization. Chapter 5 is devoted to the investigation of the combinatorial problems which arise during the work in chapters 2, 3 and 4.

2. KRULL MONOIDS

We divide this chapter into two sections. In Section 2.1, we develop the algebraic properties of Krull monoids and provide a wide array of examples. In Section 2.2 we begin discussion of sets of lengths and define several combinatorial constants which play a key role in describing the arithmetic of a Krull monoid.

2.1 Definition and Examples of Krull monoids

Throughout this paper, a monoid is a commutative and cancellative semigroup with unit element. Our main interest lies in monoids which are multiplicative monoids of integral domains. Let R be an integral domain. Then $R^{\bullet} = R \setminus \{0\}$ denotes its multiplicative monoid, $R^{\times} = R^{\bullet \times}$ the unit group of R, $\mathcal{P}(R)$ the set of maximal ideals of R, $\mathcal{I}(R)$ the multiplicative monoid of integral invertible ideals of R (with usual ideal multiplication as composition) and $\mathcal{H}(R) \subseteq \mathcal{I}(R)$ the submonoid of principal ideals. Furthermore, $\mathcal{H}(R) \simeq R^{\bullet}/R^{\times}$, the embedding $\mathcal{H}(R) \hookrightarrow \mathcal{I}(R)$ is a homomorphism and $\mathcal{I}(R)/\mathcal{H}(R) = \mathrm{Pic}(R)$ is just the Picard group of the domain R.

For a non-zero ideal $\mathfrak{f} \subseteq R$ we set

$$\mathcal{I}_{\mathfrak{f}}(R) = \{I \in \mathcal{I}(R) \,|\, I + \mathfrak{f} = R\}$$

and
$$\mathcal{H}_f(R) = \mathcal{I}_f(R) \cap \mathcal{H}(R) \,.$$

Clearly, $\mathcal{I}_f(R)$ (resp. $\mathcal{H}_f(R)$) is a submonoid of $\mathcal{I}(R)$ (resp. $\mathcal{H}(R)$).

We use the standard notions of divisibility theory as developed in [Ja; section 2.14] or in [Gi; chapter 1]. Furthermore, our notation is consistent with F. Halter-Koch's survey article in this volume [HK9]. For the convenience of the reader, we briefly recall some concepts. If not stated otherwise, monoids will be written multiplicatively.

For a family of monoids $(H_p)_{p \in P}$

$$\coprod_{p \in P} H_p = \{(a_p)_{p \in P} \in \prod_{p \in P} H_p \,|\, a_p = 1 \quad \text{for almost all} \quad p \in P\}$$

denotes the coproduct of the H_p. For every $Q \subseteq P$ we view $\coprod_{p \in Q} H_p$ as a submonoid of $\coprod_{p \in P} H_p$. If all H_p are infinite cyclic (i.e., $H_p \simeq (\mathbb{N}_0, +)$), then $\coprod_{p \in P} H_p$ is the free abelian monoid with basis P and will be denoted by $\mathcal{F}(P)$. In this case, every $a \in \mathcal{F}(P)$ has a unique representation

$$a = \prod_{p \in P} p^{v_p(a)}$$

with $v_p(a) \in \mathbb{N}_0$ and $v_p(a) = 0$ for almost all $p \in P$. Furthermore,

$$\sigma(a) = \sum_{p \in P} v_p(a) \in \mathbb{N}_0$$

is called the *size* of a.

Let D be a monoid. Then $D^\times$ denotes the group of invertible elements of D. D is called reduced, if $D^\times = \{1\}$. Clearly, $D_{\mathrm{red}} = D/D^\times$ is reduced. $\mathcal{Q}(D)$ denotes a quotient group of D with $D \subseteq \mathcal{Q}(D)$. The *root closure* $\widetilde{D}$ of D and the *complete integral closure* $\widehat{D}$ of D are defined by

$$\widetilde{D} = \{x \in \mathcal{Q}(D) \,|\, x^n \in D \quad \text{for some} \quad n \in \mathbb{N}\}$$

and

$$\widehat{D} = \{x \in \mathcal{Q}(D) \,|\, \text{ there exists some } c \in D \text{ such that } cx^n \in D \text{ for all } n \in \mathbb{N}\} \,.$$

D is called *root closed*, if $D = \tilde{D}$ and *completely integrally closed* if $D = \hat{D}$. Clearly,

$$D \subseteq \tilde{D} \subseteq \hat{D} \subseteq \mathcal{Q}(D).$$

A submonoid $H \subseteq D$ is called *saturated*, if $a, b \in H$, $c \in D$ and $a = bc$ implies that $c \in H$ (equivalently, $H = D \cap \mathcal{Q}(H)$).

A monoid homomorphism $\varphi : H \to D$ induces a monoid homomorphism $\varphi_{\text{red}} : H_{\text{red}} \to D_{\text{red}}$ and a group homomorphism $\mathcal{Q}(\varphi) : \mathcal{Q}(H) \to \mathcal{Q}(D)$.

Let $\varphi : H \to D$ be a monoid homomorphism. Then φ is called a *divisor homomorphism*, if $a, b \in H$ and $\varphi(a) \,|\, \varphi(b)$ implies that $a \,|\, b$. The following conditions are equivalent (cf. [G-HK2; Lemma 2.6]).

   i) φ is a divisor homomorphism,
   ii) φ_{red} is a divisor homomorphism,
   iii) φ_{red} is injective and $\varphi(H) \subseteq D$ is saturated.

Definition 2.1. A divisor homomorphism $\varphi : H \to D$ into a free abelian monoid D is called a *divisor theory* (for H), if for all $\alpha \in D$ there are $a_1, \ldots, a_n \in H$ such that $\alpha = \gcd\{\varphi(a_1), \ldots \varphi(a_n)\}$. The quotient group $C(H) = \mathcal{Q}(D)/\mathcal{Q}(\varphi)(\mathcal{Q}(H))$ is called the *divisor class group* of H.

Dedekind domains serve as a classic example of a divisor theory. For, if R is a Dedekind domain, then $R^{\bullet}$ has a divisor theory $\varphi : R^{\bullet} \to \mathcal{I}(R)$ given by $\alpha \to \alpha R$. The prime divisors of $\mathcal{I}(R)$ are just the prime ideals of R and the divisor class group is the usual ideal class group of R. The definition of a divisor theory given above goes back to Skula (see [Sk1]). For a survey on monoids with divisor theory the reader is referred to [HK1].

Let H be a monoid. It follows directly from the definition that H admits a divisor theory if and only if the reduced monoid H_{red} admits a divisor theory. Recall the following two facts.

   i) if H admits a divisor homomorphism into a free abelian group, then H admits a divisor theory.
   ii) if $\varphi : H \to D$ and $\varphi' : H \to D'$ are divisor theories for H, then there exists a monoid isomorphism $\phi : D \to D'$ such that $\varphi' = \phi \circ \varphi$. In particular, the divisor class group of H just depends on H (and not on φ).

Both i) and ii) can be proved using the theory of divisorial ideals in H. Proofs may be found in the book of Gundlach ([Gu; chapter 9]). Using the same methods one can also show if H admits a divisor theory, then the canonical homomorphism $\partial : H \to \mathcal{I}_v(H)$ into the monoid of integral divisorial ideals of H is a divisor theory.

An alternate proof of i) and ii) involves defining families of monoid homomorphisms. A family $(\varphi_p : H \to \mathbb{N}_0)_{p \in P}$ of monoid homomorphisms is a *defining family for* H, if

$$H = \bigcap_{p \in P} \mathcal{Q}(\varphi_p)^{-1}(\mathbb{Z})$$

and the intersection is of finite character (cf. [G-HK2; HK2]). If H has a defining family of the above type, it has a defining family of essential surjective homomorphisms $(\psi_p : H \to \mathbb{N}_0)_{p \in P}$ (cf. [HK2]).

We summarize our discussion in the following theorem. The remaining equivalences can be found in [G-HK1; Theorem 1] and [Cho; Proposition 2].

Theorem 2.2. *For a monoid H following conditions are equivalent:*
1. *H admits a divisor theory.*
2. *The canonical homomorphism $\partial : H \to \mathcal{I}_v(H)$ into the monoid of integral divisorial ideals is a divisor theory.*
3. *H is completely integrally closed and satisfies the ascending chain condition on divisorial ideals.*
4. *$H = H^\times \times T$ and T is a saturated submonoid of a free abelian monoid.*
5. *H admits a divisor homomorphism into a free abelian monoid.*
6. *H has a defining family $(\varphi_p : H \to \mathbb{N}_0)_{p \in P}$.*
7. *H has a defining family $(\psi_p : H \to \mathbb{N}_0)_{p \in P}$ where all ψ_p are essential and surjective.*

Definition 2.3. A monoid H satisfying the equivalent conditions of the previous theorem is called a *Krull monoid.*

The notion of a Krull monoid was introduced by L. Chouinard in [Cho]. It is important for our purposes due to the following result which was stated by Skula in [Sk2] and first proved by Krause in [Kr] (a simple proof appears in [HK2; Satz 5]).

Theorem 2.4. *An integral domain R is a Krull domain if and only if its multiplicative monoid $R^{\bullet}$ is a Krull monoid. If R is Krull, then the divisor class group of the Krull monoid $R^{\bullet}$ is the usual divisor class group of the Krull domain R.*

We present a series of examples of Krull monoids which are not multiplicative monoids of Krull domains, but none the less are of interest.

a) Krull rings Let R be a Marot ring. Then R is a Krull ring (in the sense of [Hu]) if and only if the multiplicative monoid of regular elements of R is a Krull monoid (cf. [HK7]).

b) Hilbert monoids The monoid structures of several classical objects in commutative algebra and number theory are represented in the following construction. We will start in a very abstract way and specialize to well known examples which at first may seem unrelated. Let R be a Dedekind domain. Then $\mathcal{I}(R)$ is a free abelian monoid with basis $\mathcal{P}(R)$ (i. e., $\mathcal{I}(R) = \mathcal{F}(\mathcal{P}(R))$). Let Γ_0 be a monoid, $\Gamma \subseteq \Gamma_0$ a submonoid and $\pi : R^{\bullet} \to \Gamma_0$ a monoid homomorphism. Then

$$H = R_{\Gamma,\pi} = \pi^{-1}(\Gamma) \subseteq R^{\bullet}$$

is a submonoid. Suppose $\Gamma \subseteq \Gamma_0^{\times}$ is a subgroup. Then $H^{\times} = H \cap R^{\times}$,

$$H/H^{\times} \simeq \{aR \,|\, a \in H\} \subseteq \mathcal{H}(R).$$

and $H \hookrightarrow R^{\bullet}$ is a divisor homomorphism, since $H = \mathcal{Q}(H) \cap R^{\bullet}$. Therefore H admits a divisor theory (i. e., H is a Krull monoid).

Let $\{1\} \subseteq \Gamma' \subseteq \Gamma \subseteq \Gamma_0^{\times}$ be subgroups. Then

$$R_{\{1\},\pi} \subseteq R_{\Gamma',\pi} \subseteq R_{\Gamma,\pi} \subseteq R_{\Gamma_0^{\times},\pi},$$

$R_{\Gamma,\pi} \subseteq R_{\Gamma_0^{\times},\pi}$ is saturated and there is a natural epimorphism

$$\rho : \mathcal{C}(R_{\Gamma',\pi}) \to \mathcal{C}(R_{\Gamma,\pi}).$$

Conversely, let $H \subseteq R_{\Gamma_0^{\times},\pi}$ be a saturated submonoid with $R_{\{1\},\pi} \subseteq H$. We show that H is of the form $R_{\Gamma,\pi}$ for some subgroup $\Gamma \subseteq \Gamma_0^{\times}$. Since $H \subseteq R_{\Gamma_0^{\times},\pi}$ is saturated, we have

$$H = \mathcal{Q}(H) \cap R_{\Gamma_0^{\times},\pi}.$$

Let $\mathcal{Q}(\pi) : \mathcal{Q}(R^{\bullet}) \to \mathcal{Q}(\Gamma_0)$ be the extension of π to the quotient groups. Then $R_{\{1\},\pi} \subseteq H$ implies that $\mathrm{Ker}\big(\mathcal{Q}(\pi)\big) \subseteq \mathcal{Q}(H)$ and thus $\mathcal{Q}(H) = \mathcal{Q}(\pi)^{-1}\big(\mathcal{Q}(\pi)\mathcal{Q}(H)\big)$. Clearly,

$$\Gamma = \mathcal{Q}(\pi)\big(\mathcal{Q}(H)\big) \subseteq \Gamma_0^{\times}$$

and

$$H = \mathcal{Q}(\pi)^{-1}(\Gamma) \cap \pi^{-1}(\Gamma_0^{\times}) = \pi^{-1}(\Gamma).$$

After these preliminaries we consider monoids Γ_0 of arithmetical interest. Let

$$\mathfrak{f}^{*} = \mathfrak{f}\,\omega_1\ldots\omega_m$$

be a cycle of R. Hence, $\mathfrak{f}^{*}$ is a formal product of an ideal $\mathfrak{f} \in \mathcal{I}(R)$ and m distinct ring monomorphisms $\omega_1,\ldots,\omega_m : R \to \mathbb{R}$. For $1 \le i \le m$ we set $\sigma_i = \mathrm{sgn} \circ \omega_i$ where $\mathrm{sgn} : \mathbb{R} \to \{-1,0,1\}$ denotes the signum function.

We say that $a,b \in R$ are congruent modulo $\mathfrak{f}^{*}$, if $a \equiv b \bmod \mathfrak{f}$ and $\sigma_i(a) = \sigma_i(b)$ for $1 \le i \le m$. This defines a congruence relation on R. For every $a \in R$ we denote by $[a]$ the congruence class containing a and by $R/\mathfrak{f}^{*}$ the set of all congruence classes. Clearly, $R/\mathfrak{f}^{*}$ is a (not necessarily cancellative) semigroup with $[a][b] = [ab]$ for all $a,b \in R$. Let $\pi : R^{\bullet} \to R/\mathfrak{f}^{*} = \Gamma_0$ denote the canonical epimorphism and let $\Gamma \subseteq \Gamma_0^{\times}$ be a subgroup. Then

$$R_{\mathfrak{f}^{*},\Gamma} = \pi^{-1}(\Gamma) = \{a \in R^{\bullet} \,|\, [a] \in \Gamma\}$$

is a Krull monoid, called the *Hilbert monoid* associated to $\mathfrak{f}^{*}$ and Γ. We may identify the reduced Hilbert monoid $(R_{\mathfrak{f}^{*},\Gamma})_{\mathrm{red}}$ with

$$\mathcal{H}_{\mathfrak{f}^{*},\Gamma}(R) = \{aR \,|\, a \in R^{\bullet},\ [a] \in \Gamma\} \subseteq \mathcal{H}(R).$$

Furthermore, the embedding

$$\mathcal{H}_{\mathfrak{f}^{*},\Gamma}(R) \to \mathcal{I}_{\mathfrak{f}}(R)$$

is a divisor theory (cf. [HK2; Proof of Satz 7]). If $\Gamma = (R/\mathfrak{f}^{*})^{\times}$, we simply write $\mathcal{H}_{\mathfrak{f}^{*}}(R)$ instead of $\mathcal{H}_{\mathfrak{f}^{*},(R/\mathfrak{f}^{*})^{\times}}(R)$.

This construction was first illustrated by F. Halter-Koch in [HK2] and generalizes the original examples of D. Hilbert. Clearly,

$$(R/\mathfrak{f})^{\times} = \{a + \mathfrak{f} \in R/\mathfrak{f} \,|\, a \in R,\ aR + \mathfrak{f} = R\}.$$

In [HK2; Hilfssatz 2 and Satz 7] it was verified that

$$(R/\mathfrak{f}^*)^\times = \{[a] \in R/\mathfrak{f}^* \mid a \in R, \ aR + \mathfrak{f} = R\}$$

and that there is an exact sequence of groups

$$0 \to (\mathbb{Z}/2\mathbb{Z})^m \to (R/\mathfrak{f}^*)^\times \to (R/\mathfrak{f})^\times \to 1 .$$

Hence we have that

$$
\begin{aligned}
\mathcal{H}_{\mathfrak{f}^*}(R) &= \{aR \mid a \in R, \ [a] \in (R/\mathfrak{f}^*)^\times\} \\
&= \{aR \mid a \in R, \ aR + \mathfrak{f} = R\} \\
&= \{aR \mid a \in R, \ [a] \in (R/\mathfrak{f})^\times\} = \mathcal{H}_{\mathfrak{f}}(R) .
\end{aligned}
$$

We consider the following simple cases.

i) If $m = 0$, $\mathfrak{f} = (1) = R$, $\mathfrak{f}^* = \mathfrak{f} = R$, then $\mathcal{H}_{\mathfrak{f}}(R) = \mathcal{H}(R)$.

ii) Let $R = \mathbb{Z}$ and $\omega_1 : \mathbb{Z} \hookrightarrow \mathbb{R}$ be the embedding $\mathfrak{f} = f\mathbb{Z}$ for some $f \in \mathbb{N}$. Then

$$R_{\mathfrak{f}^*,\{1\}} = \{a \in \mathbb{Z} \mid a > 0, \ a \equiv 1 \bmod f\} = 1 + f\mathbb{N},$$

the classical Hilbert monoid.

iii) Let R be the ring of integers in an algebraic number field K, $\omega_1, \ldots, \omega_{r_1} : K \to \mathbb{R}$ the real embeddings of K, $\mathfrak{f} = (R)$ and $\mathfrak{f}^* = \omega_1 \ldots \omega_{r_1}$. Then

$$H = R_{\mathfrak{f}^*,\{1\}} = \{a \in R^\bullet \mid \omega_i(a) > 0 \quad 1 \le i \le r\}$$

is the monoid of totally positive algebraic integers in K. Its divisor class group $\mathcal{C}(H)$ is called the ideal class group in the narrow sense (cf. [Na2; p.94]).

Let K be a global field (i. e., either an algebraic number field or an algebraic function field in one variable over a finite field). Let $S(K)$ denote the set of all non-archimedean places and for some $\nu \in S(K)$ let R_ν be the corresponding valuation domain. For a finite subset $S \subset S(K)$, $S \ne \emptyset$ in the function field case,

$$R = R_S = \bigcap_{\nu \in S(K) \backslash S} R_\nu \subseteq K$$

is called the holomorphy ring of K associated with S. R is a Dedekind domain with quotient field K. A cycle $\mathfrak{f}^*$ of R (in the sense of class field theory) is a cycle

$$\mathfrak{f}^* = \mathfrak{f}\omega_1 \ldots \omega_m$$

with $\mathfrak{f} \in \mathcal{I}(R)$, $m \geq 0$ and $\omega_1, \ldots, \omega_m : K \to \mathbb{R}$ real embeddings ($m = 0$ in the function field case). Then

$$\mathcal{H}_{\mathfrak{f}^*,\{1\}} = \{aR \mid a \in R, \ a \equiv 1 \bmod \mathfrak{f}^*\}$$

is the *principal ray modulo* $\mathfrak{f}^*$. Its divisor class group

$$\mathcal{C}(\mathcal{H}_{\mathfrak{f}^*,\{1\}}) = \mathcal{I}_\mathfrak{f}(R)/\mathcal{H}_{\mathfrak{f}^*,\{1\}}(R)$$

is a finite abelian group, called the *ray class group modulo* $\mathfrak{f}$. It gives rise to the following sequence of finite abelian groups:

$$0 \to \mathcal{H}_\mathfrak{f}(R)/\mathcal{H}_{\mathfrak{f}^*,\{1\}}(R) \to \mathcal{I}_\mathfrak{f}(R)/\mathcal{H}_{\mathfrak{f}^*,\{1\}}(R) \to \mathcal{I}_\mathfrak{f}(R)/\mathcal{H}_\mathfrak{f}(R) = \mathcal{I}(R)/\mathcal{H}(R) \to 0.$$

c) **Submonoids of orders in Dedekind domains** Let R be a Dedekind domain and $\mathfrak{o} \subseteq R$ an order in R. Thus, $\mathfrak{o}$ is one-dimensional noetherian, R is the integral closure of $\mathfrak{o}$ (in some quotient field of $\mathfrak{o}$) and R is a finitely generated $\mathfrak{o}$-module. Let

$$\mathfrak{f} = \mathrm{Ann}_\mathfrak{o}(R/\mathfrak{o}) = \{a \in \mathfrak{o} \mid aR \subseteq \mathfrak{o}\}$$

denote the conductor of $\mathfrak{o}$. If $\mathfrak{o} \neq R$, then $\mathfrak{o}$ is not integrally closed. Thus $\mathfrak{o}$ is not a Krull domain and hence $\mathcal{H}(\mathfrak{o})$ fails to be a Krull monoid. However, $\mathcal{H}(\mathfrak{o})$ contains a divisor closed submonoid that is Krull which yields information on the factorization properties of $\mathcal{H}(\mathfrak{o})$. This submonoid can be described as follows. The extension of ideals

$$\psi : \mathcal{I}_\mathfrak{f}(\mathfrak{o}) \quad \to \quad \mathcal{I}_\mathfrak{f}(R)$$
$$I \quad \mapsto \quad IR$$

is a monoid isomorphism with

$$\psi^{-1}(J) = J \cap \mathfrak{o} \quad \text{for all} \quad J \in \mathcal{I}_\mathfrak{f}(R)$$

(cf. [G-HK-K; §3]). Therefore $\mathcal{I}_{\mathfrak{f}}(\mathfrak{o})$ is free abelian. The embedding

$$\mathcal{H}_{\mathfrak{f}}(\mathfrak{o}) \hookrightarrow \mathcal{I}_{\mathfrak{f}}(\mathfrak{o})$$

is a divisor theory and hence $\mathcal{H}_{\mathfrak{f}}(\mathfrak{o})$ is a Krull monoid.

Note that in general, $\psi(\mathcal{H}_{\mathfrak{f}}(\mathfrak{o})) \neq \mathcal{H}_{\mathfrak{f}}(R)$.

d) **Block monoids** Let G be an additively written abelian group and $G_0 \subseteq G$ a nonempty subset. Then

$$\mathcal{B}(G_0) = \left\{ \prod_{g \in G_0} g^{n_g} \in \mathcal{F}(G_0) \mid \sum_{g \in G} n_g g = 0 \right\} \subseteq \mathcal{F}(G_0)$$

is called the *block monoid over* G_0. Clearly, the embedding $i :$ $\mathcal{B}(G_0) \hookrightarrow \mathcal{F}(G_0)$ is a divisor homomorphism and hence $\mathcal{B}(G_0)$ is a Krull monoid. Block monoids are the appropriate tool for studying factorization questions in Krull domains and will be discussed further in chapter 3.

e) **Root closed finitely generated monoids** Krull monoids are completely integrally closed and hence root closed. For finitely generated monoids the converse holds.

Proposition 2.5. *Let H be a monoid.*
1. *H is finitely generated if and only if $\widetilde{H}$ is finitely generated.*
2. *Suppose that H is finitely generated. Then H is a Krull monoid if and only if $H = \widetilde{H}$.*

Proof. See [Le], [HK6; Theorem 5] and [G-HK2; Proposition 6.1]. □

Let H be a finitely generated monoid such that $\widetilde{H}$ is reduced. By the above theorem, $\widetilde{H}$ is a saturated submonoid of a finitely generated free abelian monoid. Changing to additive notation, we may suppose that

$$H \subseteq (\mathbb{N}^s, +) \subseteq (\mathbb{Z}^s, +) \subseteq (\mathbb{Q}^s, +)$$

which allows us to study H by geometrical methods. For example, it turns out that

$$\widetilde{H} = \mathrm{cone}(H) \cap \mathbb{Z}^s$$

where $\text{cone}(H)$ denotes the convex cone generated by H. It was this geometrical point of view which was used in the proof of Theorem 2.5.

Hence, finitely generated monoids $H \subseteq \mathbb{Z}^s$ with $H = \text{cone}(H) \cap \mathbb{Z}^s$ are Krull monoids. In particular, this is the case for the set of solutions of linear diophantine inequalities. Let $m, s \in \mathbb{N}$, $A \in M_{m,s}(\mathbb{Z})$ and

$$H = \{x \in \mathbb{Z}^s \,|\, Ax \geq 0\} \subseteq \mathbb{Z}^s.$$

Then H is a root closed monoid which is finitely generated by [HK6; Theorem 1] and thus is Krull.

2.2 Sets of lengths in Krull monoids

Let H be a monoid. An element $u \in H \setminus H^\times$ is called *irreducible* (or an *atom*), if for all $a, b \in H$ $u = ab$ implies that $a \in H^\times$ or $b \in H^\times$. Let $\mathcal{A}(H)$ denote the set of atoms of H. If $\rho : H \to H_{\text{red}}$ is the canonical epimorphism, then $\rho(\mathcal{A}(H)) = \mathcal{A}(H_{\text{red}})$ and $\rho^{-1}(\mathcal{A}(H_{\text{red}})) = \mathcal{A}(H)$. The free abelian monoid

$$\mathcal{Z}(H) = \mathcal{F}(\mathcal{A}(H_{\text{red}}))$$

with basis $\mathcal{A}(H_{\text{red}})$ is called the *factorization monoid* of H. Furthermore, the canonical homomorphism

$$\pi = \pi_H : \mathcal{Z}(H) \to H_{\text{red}}$$

is called the *factorization homomorphism* of H and for $a \in H$ the elements of

$$\mathcal{Z}_H(a) = \mathcal{Z}(a) = \pi^{-1}(aH^\times) \subseteq \mathcal{Z}(H)$$

are called *factorizations* of a. We say that H is *atomic* if π is surjective (equivalently, H is generated by $\mathcal{A}(H) \cup H^\times$). So, when studying factorizations we may restrict ourselves to reduced monoids.

Let H be a reduced atomic monoid. For an element $z = \prod_{u \in \mathcal{A}(H)} u^{n_u} \in \mathcal{Z}(H)$,

$$\sigma(z) = \sum_{u \in \mathcal{A}(H)} n_u \in \mathbb{N}_0$$

is called the *length of the factorization* z. For $a \in H$,

$$L_H(a) = L(a) = \{\sigma(z) \mid z \in \mathcal{Z}(a)\} \subseteq \mathbb{N}_0$$

denotes the *set of lengths* of a. Furthermore, we call

$$\mathcal{L}(H) = \{L(a) \mid 1 \neq a \in H\}$$

the *system of sets of lengths* of H. $\mathcal{L}(H)$ is a subset of the power set of $\mathbb{N}$. By definition we have

   i) $L(a) = \{0\}$ if and only if $a = 1$.
   ii) $L(a) = \{1\}$ if and only if $a \in \mathcal{A}(H)$.

We say that H is *half-factorial* if for all $a \in H$ any two factorizations of a have the same length. Such monoids gained interest after Carlitz [Ca] showed (using different terminology) that a ring of integers in an algebraic number field is half-factorial if and only if its class number is less than or equal to two. Further results concerning domains that are half-factorial can be found in [Sk2] [Z1] and [Z2]. We will return to this topic later in chapter 5.

Lemma 2.6. *Let H be a reduced atomic monoid.*
   1. *Then the following conditions are equivalent:*
      i)   *H is half-factorial.*
      ii) *$\#L(a) = 1$ for all $a \in H$.*
      iii) *$\mathcal{L}(H) = \{\{n\} \mid n \in \mathbb{N}\}$.*
   2. *If H is not half-factorial, then for every $N \in \mathbb{N}$ there exists some $a \in H$ with $\#L(a) \geq N$.*

Proof. The proof of **1.** is obvious. For the proof of **2.**, see [HK9; Lemma 2.2]. $\square$

Let H be a Krull monoid with divisor theory $\varphi : H \to D$. We define

$$\mathcal{D}(H, D) = \sup\{\sigma(\varphi(u)) \mid u \in \mathcal{A}(H)\} \in \mathbb{N} \cup \{\infty\}.$$

Hence, $\mathcal{D}(H, D)$ is the maximum number of prime divisors of D dividing the image of some irreducible element of H. The following lemma lists some basic facts relating these concepts.

Lemma 2.7. *Let H be a Krull monoid with divisor theory φ : $H \to \mathcal{F}(P)$ and divisor class group G.*

1. *H is atomic.*
2. *An element $a \in H$ is prime in H if and only if $\varphi(a) \in P$.*
3. *H is a finite-factorization monoid (i. e., all sets $\mathcal{Z}(a)$ are finite).*
4. *All sets $L \in \mathcal{L}(H)$ are finite.*
5. *The following statements are equivalent:*
   - i) *H is factorial.*
   - ii) *$\mathcal{D}(H, D) = 1$.*
   - iii) *$\#G = 1$.*

Proof. See [HK1; Korollar 2 and Satz 10] and [HK5; Corollary 3]. $\square$

We restrict ourselves to arithmetical questions dealing with lengths of factorizations. For information on other arithmetical invariants, the reader is referred to [Ge10]. We introduce some arithmetical invariants which help describe the structure of sets of lengths. We list some of their most elementary properties but offer a more thorough treatment after the introduction of block monoids.

We divide our discussion into four problem areas. Throughout, let H be a reduced atomic monoid.

a) The μ_k - functions We compare the minimum and the supremum of sets of lengths. Let $k \in \mathbb{N}$. We define the following three invariants:

$$\mu_k(H) = \sup\{\sup L \mid \min L \le k, \ L \in \mathcal{L}(H)\},$$

$$\mu'_k(H) = \sup\{\sup L \mid k \in L, \ L \in \mathcal{L}(H)\} \text{ and}$$

$$\mu_k^*(H) = \sup\{\sup L \mid \min L = k, \ L \in \mathcal{L}(H)\},$$

using the convention $\sup \emptyset = 0$ if there is no $L \in \mathcal{L}(H)$ with $\min L = k$. By definition we have

$$\mu_k^*(H) \le \mu'_k(H) \le \mu_k(H).$$

Let $a \in H$ with $\min L(a) = l \le k$. Choose some $u \in \mathcal{A}(H)$. Then $k \in L(au^{k-l})$ and

$$\sup L(au^{k-l}) \ge \sup L(a) + (k - l),$$

which implies $\mu'_k(H) \ge \mu_k(H)$. Thus $\mu'_k(H) = \mu_k(H)$.

Lemma 2.8. *Let H be a reduced atomic monoid and $k \in \mathbb{N}$. If $\mu_k(H) < \infty$, then $\mu_k^*(H) = \mu_k(H)$.*

Proof. Suppose $\mu_k(H) < \infty$ and let $a \in H$ be given with $\min L(a) \leq k$ and $\max L(a) = \mu_k(H)$. For some $u \in \mathcal{A}(H)$ we set $b = au^{k-\min L(a)}$. Then

$$\min L(b) \leq \min L(a) + k - \min L(a) = k$$

and

$$\mu_k(H) \geq \max L(b) \geq \max L(a) + k - \min L(a) \geq \mu_k(H),$$

which implies that $k = \min L(a)$ and hence $\mu_k^*(H) = \mu_k(H)$. $\square$

The invariant $\mu_k(H)$ was introduced in [G-L] and special aspects of the μ_k-function have been investigated in various terminology. We point out the relationship between these concepts, but first gather some elementary properties of $\mu_k(H)$.

Lemma 2.9. *Let H be a reduced atomic monoid.*
  1. *$\mu_1(H) = 1$ and $k + l \leq \mu_k(H) + \mu_l(H) \leq \mu_{k+l}(H)$ for all $k, l \in \mathbb{N}$.*
  2. *If $\mu_k(H) = k$ for some $k \in \mathbb{N}$, then $\mu_l(H) = l$ for all $1 \leq l \leq k$.*
  3. *H is half-factorial if and only if $\mu_k(H) = k$ for all $k \in \mathbb{N}$.*

Proof. The proofs of **1.** and **3.** are clear by the definition. To verify **2.**, suppose $\mu_k(H) = k$ for some $k \in \mathbb{N}$. Then for every $1 \leq l < k$ we have that

$$l \leq \mu_l(H) \leq \mu_k(H) - \mu_{k-l}(H) \leq k - (k - l) = l. \quad \square$$

An atomic monoid H is said to be *k-half-factorial* for some $k \in \mathbb{N}$, if $\mu_k(H) = k$. This property has been investigated in a series of papers (see [Ch-S3],[Ch-S4] and [Ch-S6]) and the interested reader is referred there for specific constructions. The following lemma shows that in finitely generated monoids there exists a constant $k \in \mathbb{N}$ such that k-half-factoriality implies half-factoriality. Some efforts have been made to determine the minimal $k \in \mathbb{N}$ with this property. We return to this problem in section 3.1 and chapter 5.

Proposition 2.10. *Let H be a finitely generated monoid. Then there exists some $k \in \mathbb{N}$ such that H is half-factorial if and only if H is k-half-factorial.*

Proof. We may suppose that H is reduced. Clearly, if H is half-factorial, then it is m-half-factorial for all $m \in \mathbb{N}$. So it remains to show the converse.

Let H be generated by $u_1, \ldots, u_s \in H$ and consider the set

$$S = \{(\mathbf{m}, \mathbf{n}) \in \mathbb{N}_0^{2s} \mid \prod_{i=1}^{s} u_i^{m_i} = \prod_{i=1}^{s} u_i^{n_i} \neq 1\}.$$

By Dickson's Theorem (see [Re; Satz 2]) the set of minimal points $S_0 \subseteq S$ is finite. For $\mathbf{m} \in \mathbb{N}_0^s$ we set

$$|\mathbf{m}| = \sum_{i=1}^{s} m_i.$$

Obviously, it is sufficient to verify that if

$$|\mathbf{m}| = |\mathbf{n}| \text{ for all } (\mathbf{m}, \mathbf{n}) \in S_0 \qquad (*)$$

then H is half-factorial.

Suppose that $(*)$ holds and assume to the contrary that H is not half-factorial. Then there exists some $a \in H$ with

$$a = \prod_{i=1}^{s} u_i^{m_i} = \prod_{i=1}^{s} u_i^{n_i},$$

$$\min L(a) = |\mathbf{m}| < |\mathbf{n}|$$

and $|\mathbf{m}|$ minimal. Then $(\mathbf{m}, \mathbf{n}) \notin S_0$ and hence there exists some $(\mathbf{m}', \mathbf{n}') \in S_0$ with $\mathbf{m}' \leq \mathbf{m}$ and $\mathbf{n}' \leq \mathbf{n}$. But then we have

$$\prod_{i=1}^{s} u_i^{m_i - m_i'} = \prod_{i=1}^{s} u_i^{n_i - n_i'} \in H$$

and $|\mathbf{m} - \mathbf{m}'| < |\mathbf{n} - \mathbf{n}'|$, a contradiction to the minimality of $|\mathbf{m}|$. $\square$

A further arithmetical concept closely related with the μ_k-function is the concept of elasticity. Let H be an atomic monoid. The *elasticity* $\varrho(H)$ of H is defined as

$$\varrho(H) = \sup\left\{ \frac{\sup L}{\min L} \,\Big|\, L \in \mathcal{L}(H) \right\} \in \mathbb{R}_{\geq 1} \cup \{\infty\} \,.$$

A detailed study of elasticity is contained in an article by David Anderson in this volume ([An1]). It is easy to see that

$$\varrho(H) = \lim_{k\to\infty} \frac{\mu_k(H)}{k}$$

(cf. [HK8; Proposition 1]).

In section 5.3 we shall review what is known concerning $\mu_k(H)$ for Krull monoids H.

b) Distances of successive lengths Our next topic deals with possible distances of successive lengths of factorizations for elements $a \in H$. For a finite set $L = \{x_1, \cdots, x_r\} \subseteq \mathbb{Z}$ with $x_1 < \cdots < x_r$ we set

$$\Delta(L) = \{x_i - x_{i-1} \mid 2 \leq i \leq r\} \subseteq \mathbb{N} \,,$$

and

$$\Delta(H) = \bigcup_{L \in \mathcal{L}(H)} \Delta(L) \subseteq \mathbb{N} \,.$$

Hence, $\Delta(H)$ is the set of distances of successive lengths. The proof of the following property of $\Delta(H)$ is not difficult and can be found in [Ge6; Lemma 3].

Lemma 2.11. *Let H be a reduced atomic monoid. Then $\min\Delta(H) = \gcd\Delta(H)$.*

By definition we have that H is half-factorial if and only if $\Delta(H) = \emptyset$. A monoid H is called *d-congruence half-factorial*, if for all $a \in H$ and any two factorizations $z, z' \in \mathcal{Z}(a)$ we have

$$\sigma(z) \equiv \sigma(z') \mod d$$

(cf. [Ch-S2]). Hence H is d-congruence half-factorial with $d \in \mathbb{N}$ minimal if and only if $\min\Delta(H) = d$. We will consider the d-congruence half-factorial property, as well as the set $\Delta(H)$, in more detail in section 5.2.

c) **Structure of sets of lengths** An atomic monoid is either half-factorial or sets of lengths can become arbitrarily large. Hence, an obvious question is to describe such sets for Krull monoids. While in general these sets may not be perfect arithmetic progressions, they are almost arithmetic in the following sense.

Definition 2.12. A subset $L \subset \mathbb{Z}$ is called an *almost arithmetical progression bounded by* $M \in \mathbb{N}$, if

$$L = \{x_1, \ldots, x_\alpha, \quad y + \delta_1, \quad \ldots, \quad y + \delta_\eta, \quad y + d,$$
$$y + \delta_1 + d, \quad \ldots, \quad y + \delta_\eta + d, \quad y + 2d,$$
$$\vdots$$
$$y + \delta_1 + (k-1)d, \quad \ldots, \quad y + \delta_\eta + (k-1)d, \quad y + kd,$$
$$z_1, \ldots, z_\beta\},$$

where $\alpha, \beta, \eta, k, d \in \mathbb{N}_0$, $x_1 < \cdots < x_\alpha < y \leq y + kd < z_1 < \cdots < z_\beta$, $0 < \delta_1 < \cdots < \delta_\eta < d$ and $\max\{\alpha, \beta, d\} \leq M$.

The following Theorem solves the characterization question and can be found in [Ge2, Satz 1].

Theorem 2.13. *Let* H *be a Krull monoid which has only finitely many divisor classes containing prime divisors. Then there exists some* $M \in \mathbb{N}$ *such that every* $L \in \mathcal{L}(H)$ *is an almost arithmetical progression bounded by* M.

This result is sharp in the sense that all the parameters in Definition 2.12 are necessary (see [Ge9] for an example which illustrates this). Hence, the result raises further questions concerning the parameters of almost arithmetical progressions. The only invariant which has been investigated thus far deals with possible distances in arithmetical progressions. To be more precise, for H as in Theorem 2.13 let

$$\Delta_1(H)$$

denote the set of all $d \in \mathbb{N}$ such that for all $N \in \mathbb{N}$ there exists some $L \in \mathcal{L}(H)$ with $\#L \geq N$ and with

$$L = \{x_1, \ldots, x_\alpha, y, y + d, \ldots, y + kd, z_1, \ldots z_\beta\}$$

where $x_1 < \cdots < z_\beta, \alpha \leq M$ and $\beta \leq M$ with M as in Theorem 2.13. Notice that $\Delta_1(H) \subseteq \Delta(H)$, and that there are examples of differences which appear in $\Delta(H)$ which do not appear as a distance in arbitrarily long almost arithmetic progressions. We return to $\Delta_1(H)$ in section 5.2.

d) Systems of sets of lengths In algebraic number theory, the ideal class group G is considered as a measure for the deviation of the ring of integers R from being a unique factorization domain. G is thought to determine the arithmetic of R. A definition of a type of arithmetic equivalence is given by F. Halter-Koch in [HK1; Korollar 4].

In [Na2; problem 32], W. Narkiewicz asked for an arithmetical characterization of the ideal class group of a ring of integers in an algebraic number field. Various answers to this question have been given by J. Kaczorowski, F. Halter-Koch, D.E. Rush, A. Czogala, U. Krause and others (see [Ge6 and Ge7 for a survey]). However, it is still unknown if it is possible to characterize the ideal class group by using only lengths of factorizations. To be more precise, we formulate the following problem:

Problem: Let H and H' be Krull monoids with finite divisor class groups G and G' such that each divisor class contains a prime divisor. If $\#G \geq 4$, does $\mathcal{L}(H) = \mathcal{L}(H')$ imply that $G = G'$?

We shall not consider this problem further but add the following remarks.

Remark. 1. Suppose that H and H' are the multiplicative monoids of the rings of algebraic integers R and R'. A positive answer to the Problem implies that sets of lengths completely determine the arithmetic of a ring of integers.

2. The analogous question for arbitrary orders in algebraic number fields ("do sets of lengths determine the arithmetic of an arbitrary order") has a negative answer (see [HK9; Example 3]).

3. Clearly, the assumption that each class contains a prime divisor is necessary for obtaining a positive answer.

4. The Problem is answered positively for cyclic groups in [Ge4].

3. BLOCK MONOIDS

Let G be an additively written abelian group and $\emptyset \neq G_0 \subseteq G$ a subset. Let $< G_0 >$ denote the subgroup and $[G_0]$ the submonoid generated by G_0. If we define the *content homomorphism*

$$\iota : \mathcal{F}(G_0) \quad \to \quad G$$
$$\prod_{g \in G} g^{n_g} \quad \mapsto \quad \sum_{g \in G_0} n_g g,$$

then

$$\mathcal{B}(G_0) = \{S \in \mathcal{F}(G_0) \,|\, \iota(S) = 0\}$$

is called the *block monoid* over G_0. It is a Krull monoid, since the embedding $\mathcal{B}(G_0) \hookrightarrow \mathcal{F}(G_0)$ is a divisor homomorphism. We also have the following (see [HK1; Satz 4 and Korollar] for a proof).

Proposition 3.1. *Let G be an abelian group and G_0 a nonempty subset. Then the following holds:*

1. *The embedding $\mathcal{B}(G_0) \hookrightarrow \mathcal{F}(G_0)$ is a divisor theory if and only if $< G_0 > = [G_0 \setminus \{g\}]$ for every $g \in G_0$.*

2. *If $\#G \leq 2$, then $\mathcal{B}(G)$ is factorial.*

3. *If $\#G \geq 3$, then $\mathcal{B}(G) \hookrightarrow \mathcal{F}(G)$ is a divisor theory with class group (isomorphic to) G and each class contains exactly one prime divisor.*

3.1 The block monoid associated to a Krull monoid

Let H be a reduced Krull monoid, $\varphi : H \to D = \mathcal{F}(P)$ its divisor theory and $\pi : D \to \mathcal{C}(H) = G$ the canonical epimorphism onto its divisor class group. Let $G_0 \subseteq G$ denote the set of classes containing prime divisors (i. e., $G_0 = \{g \in G \,|\, g \cap P \neq \emptyset\}$). We define a monoid epimorphism

$$\overline{\beta} : \mathcal{F}(P) \quad \to \quad \mathcal{F}(G_0)$$
$$p \quad \mapsto \quad [p]$$

which maps each prime divisor onto its divisor class. This induces a monoid epimorphism

$$\beta = \overline{\beta} \circ \varphi : H \to \mathcal{B}(G_0)$$

and we obtain the following commutative diagram

$$
\begin{array}{ccccc}
H & \overset{\varphi}{\to} & D = \mathcal{F}(P) & \overset{\pi}{\to} & G \\
\downarrow \beta & & \downarrow \overline{\beta} & & \| \\
\mathcal{B}(G_0) & \hookrightarrow & \mathcal{F}(G_0) & \overset{\iota}{\to} & G
\end{array} .
$$

$\beta : H \to \mathcal{B}(G_0)$ is called the *block homomorphism* and $\mathcal{B}(G_0)$ the *block monoid* associated with the Krull monoid H (resp. with the divisor theory $\varphi : H \to D$). The significance of this construction can be seen from the following lemma.

Lemma 3.2. *Let* H *be a reduced Krull monoid,* $a \in H$ *and* $\beta : H \to \mathcal{B}(G_0)$ *the corresponding block homomorphism. Then we have*

1. *a is irreducible in* H *if and only if* $\beta(a)$ *is irreducible in* $\mathcal{B}(G_0)$, $\beta(\mathcal{A}(H)) = \mathcal{A}(\mathcal{B}(G_0))$ *and* $\beta^{-1}(\mathcal{A}(\mathcal{B}(G_0))) = \mathcal{A}(H)$.
2. $L_H(a) = L_{\mathcal{B}(G_0)}(\beta(a))$ *and* $\mathcal{L}(H) = \mathcal{L}(\mathcal{B}(G_0))$.
3. $\mathcal{D}(H, D) = \mathcal{D}(\mathcal{B}(G_0), \mathcal{F}(G_0))$.
4. $\varrho(H) = \varrho(\mathcal{B}(G_0))$, $\Delta(H) = \Delta(\mathcal{B}(G_0))$, $\Delta_1(H) = \Delta_1(\mathcal{B}(G_0))$ *and* $\mu_k(H) = \mu_k(\mathcal{B}(G_0))$ *for every* $k \in \mathbb{N}$.

Proof. The proofs of **3.** and **4.** are immediate consequences of **1.** and **2.** Proofs of **1.** and **2.** may be found in [Ge2; Proposition 1]. Alternatively, notice that β satisfies the assumption of the Transfer Lemma in [HK9]. $\square$

Lemma 3.2 states that sets of lengths (and hence all invariants dealing with lengths of factorizations) in a Krull monoid may be studied in the associated block monoid. Let us mention an important application: it is sufficient to prove the structure theorem for sets of lengths in Krull monoids (Theorem 2.13) for block monoids. We give a further example to illustrate how the Lemma 3.2 works.

Proposition 3.3. *Let* H *be a Krull monoid with divisor class group* G *and let* $G_0 \subseteq G$ *denote the set of classes containing prime divisors. Suppose that* G_0 *is finite. Then there exists a constant* $k \in \mathbb{N}$ *such that* H *is half-factorial if and only if* H *is* k-*half-factorial.*

Proof. By Lemma 3.2 it sufficient to prove the assertion for $\mathcal{B}(G_0)$ instead of H. Since G_0 is finite, $\mathcal{B}(G_0)$ is finitely generated (cf. [Ge2; Proposition 2]). Now Proposition 2.10 implies the result. $\square$

Block monoids were introduced by W. Narkiewicz in [Na1] and first used systematically in [Ge2]. Consequently, block monoids have been attached to arbitrary divisor homomorphisms $\varphi : H \to D$ (cf. [Ge8]). For an overview the reader is referred to [HK9].

We use the following notation throughout the remainder of this article.

Notation: $\mathcal{L}(G_0) = \mathcal{L}(\mathcal{B}(G_0))$, $\Delta(G_0) = \Delta(\mathcal{B}(G_0)), \Delta_1(G_0) = \Delta_1(\mathcal{B}(G_0))$,
$$\mathcal{D}(G_0) = \mathcal{D}(\mathcal{B}(G_0), \mathcal{F}(G_0)), \mu_k(G_0) = \mu_k(\mathcal{B}(G_0)) \quad \text{and}$$
$\varrho(G_0) = \varrho(\mathcal{B}(G_0))$.

For an abelian group G and a subset $\emptyset \neq G_0 \subseteq G$, $\mathcal{D}(G_0)$ is called the *Davenport constant* of G. By definition we have

$$\mathcal{D}(G_0) = \sup\{\sigma(B) \mid B \in \mathcal{A}(\mathcal{B}(G_0))\}$$

Davenport's constant plays a central role in factorization theory (cf. [Ch]). Its properties will be discussed in section 5.1.

3.2 Realization theorems

As noted above, it is easy to see that block monoids admit a divisor theory. However, it is surprising that conversely every reduced Krull monoid is isomorphic to a block monoid. A proof of the following can be found in [G-HK1; Theorem 2].

Theorem 3.4. *For a monoid H the following conditions are equivalent:*

1. H *is a Krull monoid.*
2. *There exists an abelian group G and a subset $\emptyset \neq G_0 \subseteq G$ such that $H \simeq H^\times \times \mathcal{B}(G_0)$.*

Further realization theorems for Krull monoids as arithmetically closed submonoids of special type are derived in [G-HK1; section 4].

By Lemma 3.2, sets of lengths in a Krull monoid just depend on the pair (G, G_0). So one might ask for which pairs (G, G_0) there exists a Krull monoid H with divisor class group (isomorphic) G such that G_0 is the set of classes containing prime divisors. If there is such an H, we say that the pair (G, G_0) is *realizable* by H.

Theorem 3.5. *Let G be an abelian group, $(m_g)_{g \in G}$ a family of cardinal numbers and $G_0 = \{g \in G | m_g \neq 0\}$. Then the following conditions are equivalent:*

1. *There exists a Krull monoid H with divisor theory $\varphi \colon H \to \mathcal{F}(\mathcal{P})$, divisor class group C and a group isomorphism $\psi \colon G \to C$ such that $m_g = \operatorname{card}(\mathcal{P} \cap \psi(g))$ for all $g \in G$.*

2. *$G = [G_0]$, and for all $g \in G_0$ with $m_g = 1$ we have $G = [G_0 \setminus \{g\}]$.*

Proof. [HK1; Satz 5] $\square$

Building on the work of L. Claborn, A. Grams and L. Skula characterized pairs (G, G_0) which are realizable by Dedekind domains.

Theorem 3.6. *Let G be an abelian group and $G_0 \subseteq G$ a nonempty subset. Then the following conditions are equivalent:*

1. *There exists a Dedekind domain R with ideal class group G such that G_0 is the set of classes containing prime ideals.*

2. *$G = [G_0]$.*

Proof. See [Gr; Theorem 1.4]. The result is also proved independently in [Sk2; Theorem 2.4]. $\square$

4. ARITHMETICAL KRULL MONOIDS

Let R be the ring of integers in an algebraic number field and P some factorization property. Then one might ask for an asymptotic formula for the number of elements $\alpha \in R$ (counted up to associates) satisfying property P and with norm $N(\alpha)$ bounded by x. The prototype of these questions is to count the primes $p \in \mathbb{N}$ with $p \leq x$. Such quantitative aspects of non-unique factorizations in algebraic number fields were first considered by E. Fogels in the forties. Systematic investigations were started by W. Narkiewicz in the sixties (see [Na2; Chapter 9]).

Quantitative investigations of phenomena of non-unique factorizations are interesting mainly for holomorphy rings in global fields (including rings of integers in algebraic number fields and in algebraic function fields in one variable over a finite field). However, it has turned out that most of the results can be derived for very general structures in the setting of abstract analytic number theory. It was this axiomatic procedure which recently allowed the extension of results from principal orders to arbitrary orders in global fields (see [G-HK-K]). A further advantage of the axiomatic method is that it allows us to describe and investigate the combinatorial structures which are responsible for the various phenomena of non-unique factorization. Abstract analytic number theory is carefully presented in the monograph by J. Knopfmacher [Kn], who introduced the notion of an arithmetical formation. Our definition will be slightly different.

Definition 4.1. 1. A *norm (function)* $|\cdot|$ on a reduced monoid H is a monoid homomorphism $|\cdot| : H \to (\mathbb{N}, \cdot)$ satisfying $|a| = 1$ if and only if $a = 1$.

2. An *arithmetical Krull monoid* (*an arithmetical formation*) consists of a triple $[D, H, |\cdot|]$ where $D = \mathcal{F}(P)$ is a free abelian monoid, $H \subseteq D$ a saturated submonoid with finite divisor class group G and a norm $|\cdot| : D \to \mathbb{N}$ satisfying the following axiom: for every $g \in G$

$$\sum_{p \in P \cap g} |p|^{-s} = \frac{1}{\#G} \log \frac{1}{s-1} + h_g(s)$$

where $h_g(s)$ is regular in the half-plane $\mathrm{Re}(s) \geq 1$ and in some neighborhood of $s = 1$.

Examples and Remarks. 1. Let K be an algebraic number field and R a holomorphy ring in K (e.g., the ring of algebraic integers in K). For every ideal $I \in \mathcal{I}(R)$ we set $|I| = \#(R/I)$. Then $|\cdot| : \mathcal{I}(R) \to \mathbb{N}$ is a norm function.

Let $\mathfrak{f}^* = \mathfrak{f}\omega_1 \ldots \omega_m$ be a cycle of R with $\mathfrak{f} \in \mathcal{I}(R)$, $m \geq 0$ and real embeddings $\omega_1, \ldots, \omega_m : K \to \mathbb{R}$. Let $\pi : R^\bullet \to (R/\mathfrak{f}^*)^\times$ denote the canonical epimorphism, $\Gamma \subseteq (R/\mathfrak{f}^*)^\times$ a subgroup and $H = \mathcal{H}_{\mathfrak{f}^*, \Gamma}(R)$ the reduced Hilbert monoid associated with $\mathfrak{f}^*$ and Γ (see chapter 2). Then Chebotarev's density theorem implies that

$$[\mathcal{I}_\mathfrak{f}(R), H, |\cdot|]$$

is an arithmetical Krull monoid (see [HK4; Proposition 3]).

2. Let K be an algebraic number field, R its ring of integers and $\mathfrak{o} \subseteq R$ an order with conductor $\mathfrak{f}$. For every $I \in \mathcal{I}_\mathfrak{f}(\mathfrak{o})$ we have

$(\mathfrak{o} : I) = (R : IR)$ and hence $|\cdot| : \mathcal{I}_{\mathfrak{f}}(\mathfrak{o}) \to \mathbb{N}$ is a norm function. The embedding

$$\mathcal{H}_{\mathfrak{f}}(\mathfrak{o}) \hookrightarrow \mathcal{I}_{\mathfrak{f}}(\mathfrak{o})$$

is a divisor theory with divisor class group isomorphic to $\mathrm{Pic}(\mathfrak{o})$. Thus

$$[\mathcal{H}_{\mathfrak{f}}(\mathfrak{o}),\ \mathcal{I}_{\mathfrak{f}}(\mathfrak{o}),\ |\cdot|]$$

is an arithmetical Krull monoid (cf. [G-HK-K; Prop. 3 and Remark after Definition 4]).

3. Suppose that in the above definition we just require that the functions $h_g(s)$ are regular in the open half-plane $\mathrm{Re}(s) > 1$. Then examples **1.** and **2.** can be carried out not only in algebraic number fields but in global fields. However, the asymptotic results are weaker. Apart from this, further variants of the analytical axiom have been studied. However, all axioms are modelled on the concrete situation in algebraic number fields or algebraic function fields (in one variable over a finite constant field). The reader is referred to [G-HK-K] or [G-K].

Let $[D, H, |\cdot|]$ be an arithmetical formation. For a subset $Z \subseteq H$ let

$$Z(x) = \#\{a \in H \mid |a| \leq x,\ a \in Z\}$$

denote the associated counting function. Quantitative theory of non-unique factorizations studies $Z(x)$ for subsets Z of arithmetical interest. We restrict ourselves to subsets Z which give information on sets of lengths. Among others, the following sets have been studied for every $k \in \mathbb{N}$

$$M_k(H) = M_k = \{a \in H \mid \max L(a) \leq k\},$$
$$G_k = \{a \in H \mid \#L(a) \leq k\}$$

and

$$P = \{a \in H \mid L(a) \text{ is an arithmetical progression with distance } 1\}.$$

We say a subset $L \subset \mathbb{Z}$ is an arithmetical progression with distance 1, if $L = \{x, x+1, \ldots, x+m\}$ for some $x \in \mathbb{Z}$ and $m \in \mathbb{N}_0$. Moreover, note that

$$M_1 = \{a \in H \mid a \text{ is irreducible}\}.$$

For all these sets Z , the asymptotic behaviour of $Z(x)$ is of the form

$$Z(x) \sim C x (\log x)^{-A} (\log \log x)^{B} \qquad (*)$$

with $C \in \mathbb{R}_{>0}$, $0 \le A \le 1$ and $B \in \mathbb{N}_0$. As usual, $f \sim g$ for two real valued functions f and g means that

$$\lim_{x \to \infty} \frac{f(x)}{g(x)} = 1 .$$

We give a brief outline of a proof of formula $(*)$. We do it in a manner which allows us to obtain combinatorial descriptions of the involved exponents A and B. We proceed in three steps.

a) **Block monoids** Due to the following lemma, it is sufficient to study the sets Z in the associated block monoid.

Lemma 4.2. *Let H be a Krull monoid with divisor class group G such that each class contains a prime divisor and let $\beta : H \to \mathcal{B}(G)$ denote the block homomorphism. Then for each set Z we have*

$$Z(H) = \beta^{-1}\big(Z(\mathcal{B}(G))\big)$$

(i. e., $Z(H) = \{a \in H \mid \beta(a) \in Z(\mathcal{B}(G))\}$).

Proof. This follows immediately from Lemma 3.2. $\square$

b) **Combinatorial part** Let G be a finite abelian group and $Z = Z(\mathcal{B}(G)) \subseteq \mathcal{B}(G)$. Our aim is to reveal the structure of Z. For this we introduce the following combinatorial tool.

For a subset $Q \subseteq G$ and a function $\sigma : G \setminus Q \to \mathbb{N}_0$ we set

$$\Omega(Q, \sigma) = \{S \in \mathcal{F}(G) \mid v_g(S) = \sigma(g) \text{ for all } g \in G \setminus Q\} .$$

and

$$|\sigma| = \sum_{g \in G \setminus Q} \sigma(g) \in \mathbb{N}_0 .$$

Lemma 4.4. *Let* $Z = M_k$ *or* $Z = G_k$ *for some* $k \in \mathbb{N}$. *Then there exist finitely many pairs* $(G_1, \sigma_1), \ldots, (G_r, \sigma_r)$ *such that*

$$Z = \bigcup_{i=1}^{r} \Omega(G_i, \sigma_i). \qquad (**)$$

Further we have:
  i) *the representation* $(**)$ *is unique.*
  ii) *if* $Z = M_k$, *then* $G_1 = \cdots = G_r = \emptyset$.
  iii) *if* $Z = G_k$, *then* $\Delta(G_1) = \cdots = \Delta(G_r) = \emptyset$.

Proof. The existence and uniqueness of the representation $(**)$ is proved in [HK4; Proposition 9].

ii) Suppose that $\Omega(G_0, \sigma_0) \subseteq M_k$. If $g \in G_0$, then $B = (g^{\mathrm{ord}(g)})^{k+1} \in \mathcal{B}(G) \cap \Omega(G_0, \sigma_0)$ but $B \notin M_k$. This shows that $G_0 = \emptyset$.

iii) Suppose that $\Omega(G_0, \sigma_0) \subseteq G_k$. If $\Delta(G_0) \neq \emptyset$, there exists some $B \in \mathcal{B}(G_0)$ with $\#L(B) \geq 2$. Thus $B^{k+1} \in \Omega(G_0, \sigma_0) \cap \mathcal{B}(G)$ but $B^{k+1} \notin G_k$. This shows that $\Delta(G_0) = \emptyset$. $\square$

The previous result gives rise to the following definition.

Definition 4.5. Let G be a finite abelian group.
  1. For every $k \in \mathbb{N}$ let $\mathcal{D}_k(G) = \sup\{\sigma(B) \mid B \in \mathcal{B}(G), \mathrm{max}L(B) \leq k\}$.
  2. $\eta(G) = \max\{\#G_0 \mid \Delta(G_0) = \emptyset\}$.

Clearly, $\mathcal{D}_1(G) = \mathcal{D}(G)$ is just Davenport's constant of G. The next lemma gives information on the structure of the set P.

Lemma 4.6. *Let* $G' = G \setminus \{0\}$ *and* $A^* = \prod_{g \in G'} g$. *Then for all* $A \in \mathcal{B}(G)$ *with* $A^* | A$ *we have* $A \in P$ *i. e.,*

$$\mathcal{B}(G) \setminus P \subseteq \bigcup_{g \in G'} \Omega(G \setminus \{g\}, \sigma_g)$$

with $\sigma_g : \{g\} \to \mathbb{N}_0$ *and* $\sigma_g(g) = 0$ *for all* $g \in G'$.

Proof. See [Ge9]. $\square$

c) **Analytical part** The above combinatorial results show that it is sufficient to study the asymptotic behaviour of functions $\Omega(Q,\sigma)(x)$ in order to obtain results for $Z(x)$. A function of the form $\Omega(Q,\sigma)(x)$ was first investigated by P. Remond for algebraic number fields (for citations and historical remarks the reader is referred to Narkiewicz's book [Na2; chapter 9]). The main analytical tool for these investigations is a Tauberian theorem of Ikehara-Delange. For the proof of the following lemma, see [HK4; Proposition 10].

Lemma 4.7. Let $Q \subseteq G$ be a subset and $\sigma : G \setminus Q \to \mathbb{N}_0$ a function with $|\sigma| > 0$ if $Q = \emptyset$. Then, for x tending to infinity,

$$\#\{a \in D \mid \beta(a) \in \Omega(Q,\sigma), \ |a| \le x\} \sim Cx(\log x)^{-\eta}(\log \log x)^d$$

where

$$\eta = \frac{\#(G \setminus Q)}{\#G} \quad and \quad d = \begin{cases} |\sigma| & Q \neq \emptyset \\ |\sigma| - 1 & Q = \emptyset. \end{cases}$$

Theorem 4.8. Let $[D, H, | \cdot |]$ be an arithmetical Krull monoid. Then we have for every $k \in \mathbb{N}$

1. $M_k(x) \sim Cx(\log x)^{-1}(\log \log x)^{D_k(G)}$ for some $C \in \mathbb{R}_{>0}$.
2. $G_k(x) \sim C(\log x)^{\frac{\eta(G) - \#G}{\#G}}(\log \log x)^B$ for some $C \in \mathbb{R}_{>0}$ and some $B \in \mathbb{N}_0$.
3. $P(x) = H(x) + O\left(\frac{x}{(\log x)^{1/\#G}}\right)$, in particular

$$\lim_{x \to \infty} \frac{P(x)}{H(x)} = 1.$$

Proof. See [HK3], [Ge3], [Ge2; Satz 2] and [G-K; Theorem 7]. $\square$

5. COMBINATORIAL PROBLEMS IN ABELIAN GROUPS

In chapters 2 and 4, we discussed arithmetical questions in Krull domains and defined invariants which describe their arithmetic. As explained in chapter 3, all these notions depend solely on the divisor class group of the domain and the distribution of prime divisors in the divisor classes. This final chapter is devoted to the investigation of these group theoretical constants. The resulting problems belong to the zero sum area, a part of additive group theory or combinatorial number theory.

Let G be an abelian group. We use additive notation throughout this section. Let $S = (g_1, \ldots, g_l)$ be a finite sequence of elements of G. Usually, one says that S is a *zero sequence*, if $\sum_{i=1}^l g_i = 0$ and that S is a *minimal zero sequence*, if $\sum_{i \in I} g_i \neq 0$ for each proper subset $\emptyset \neq I \subset \{1, \ldots, l\}$. To be consistent with the previous chapters, we view S as an element of the free abelian monoid $\mathcal{F}(G)$ and use multiplicative notation. Hence,

$$S = \prod_{i=1}^l g_i = \prod_{g \in G} g^{v_g(S)} \in \mathcal{F}(G)$$

and $\sigma(S) = l$ denotes the number of elements of S. By definition, S is a zero sequence if and only if S is a block and S is a minimal zero sequence if and only if S is an irreducible block. We define

$$-S = \prod_{i=1}^l (-g_i) \, ,$$

and, for every subset $G_0 \subseteq G$ set $-G_0 = \{-g \,|\, g \in G_0\}$. The cyclic group of order $n \in \mathbb{N}$ will be denoted by C_n and

$$C_n = \{\bar{a} = a + n\mathbb{Z} \,|\, 0 < a < n\} = \{\bar{1}, \ldots, \overline{n-1}\} \, .$$

For a real number $x \in \mathbb{R}$ let $[x] \in \mathbb{Z}$ be the smallest integer $d \in \mathbb{Z}$ with $d \leq x$.

5.1 On $D_k(G)$

Throughout section 5.1, let G is a finite abelian group and suppose that $G = \bigoplus_{i=1}^r C_{n_i}$ where $n_1 \mid n_2 \mid \cdots \mid n_r$. Then n_r is the exponent of the group and r is the maximal p-rank of G. We call r the rank of G. If k is a positive integer, set

$$M_k(G) = \sum_{i=1}^{r-1} (n_i - 1) + kn_r$$

and $M_1(G) = M(G)$.

Proposition 5.1. *Let* k *be a positive integer and* G *be as above.*

1. $M_k(G) \leq \mathcal{D}_k(G) \leq k\mathcal{D}(G) \leq kn_r(1 + \log \frac{\#G}{n_r})$.
2. *If* $G_1 \subsetneq G$ *is a proper subgroup, then* $\mathcal{D}_k(G_1) < \mathcal{D}_k(G)$.
3. *If* $G = G' \oplus G''$, *then* $\mathcal{D}_k(G') + \mathcal{D}(G'') - 1 \leq \mathcal{D}_k(G)$.

Proof. The upper bound for $\mathcal{D}(G)$ was first shown in [E-K]. For a simplified proof see [A-G-P; Theorem 1.1]. Proofs of all other assertions may be found in [HK3]. □

Proposition 5.2. *Let* k *be a positive integer and* G *be as above.*

1. *If rank* $r \leq 2$, *then* $M_k(G) = \mathcal{D}_k(G)$.
2. *If* G *is a p-group, then* $M(G) = \mathcal{D}(G)$.

Proof. The assertions for $\mathcal{D}(G)$ were derived independently by Olson and Kruyswijk (see [Ol] and [E-K]). Halter-Koch proved **1.** for arbitrary k in [HK3]. □

There are infinitely many groups of rank four with $M(G) < \mathcal{D}(G)$ (see [G-S]). It is still an open problem whether there are groups of rank three with $M(G) < \mathcal{D}(G)$. For recent results for groups of rank three the reader is referred to [Ga].

5.2 On $\Delta(G_0)$

Proposition 5.3. *Let* G *be a finite abelian group.*

1. *Let* $\emptyset \neq G_0 \subseteq G$ *be a subset. If* $\mathcal{D}(G_0) \geq 3$, *then* $\Delta(G_0) \subseteq \{1, \ldots, \mathcal{D}(G_0) - 2\}$. *If* $G_0 = -G_0$, *then* $\{\mathrm{ord}(g) - 2 | g \in G, \mathrm{ord}(g) > 2\} \subseteq \Delta(G_0)$.
2. $\Delta(G) = \emptyset$ *if and only if* $\#G \leq 2$. $\Delta(G) = \{1\}$ *if and only if* $G \in \{C_3, C_3^2, C_2^2\}$. *In all other cases we have* $\#\Delta(G) \geq 2$ *and* $1 = \min \Delta(G)$.
3. $\Delta(C_n) = \{1, \ldots, n - 2\}$ *for every* $n \geq 3$.

Proof. The proofs can be found in [Ge2; Proposition 3 and the Example prior to Proposition 4] and [G-K; Proposition 4]. □

Suppose that G is a torsion abelian group and $G_0 \subseteq G$ a subset. We say that G_0 is *half-factorial*, if $\Delta(G_0) = \emptyset$ (equivalently, the block monoid $\mathcal{B}(G_0)$ is half-factorial). Our first aim is to gather some results on half-factorial subsets. For a sequence $S = g_1 g_2 \cdots g_t$ of elements in G we set

$$k(S) = \sum_{i=1}^{t} \frac{1}{\mathrm{ord}(g_i)}.$$

$k(S)$ is known as the *cross number of* S. For properties of this invariant and its relevance in factorization theory the reader is referred to [Ch] and [Ch-G].

The following result was obtained independently in both [Sk2; Theorem1] and [Z1; Proposition 1].

Proposition 5.4. *Let G be an abelian torsion group and $G_0 \subseteq G$. The following statements are equivalent:*
1. *G_0 is half-factorial.*
2. *$k(B) = 1$ for all irreducible blocks $B \in \mathcal{B}(G_0)$.*

Proof. Assume that $k(B) = 1$ for all irreducible blocks. If $U_1, \ldots, U_r$, $V_1, \ldots, V_s$ are irreducible blocks in $\mathcal{B}(G_0)$ and $U_1 \cdots U_r = V_1 \cdots V_s$, then $k(U_1 \cdots U_r) = k(V_1 \cdots V_s)$ implies that $r = s$.

Assume that $k(B) \neq 1$ for some irreducible block $B = g_1 \cdots g_l \in \mathcal{B}(G_0)$. Let $m = \mathrm{lcm}\{\mathrm{ord}(g_1), \ldots, \mathrm{ord}(g_l)\}$, $n_i \mathrm{ord}(g_i) = m$ and $U_i = g_i^{\mathrm{ord}(g_i)}$ be an irreducible block in $\mathcal{B}(G_0)$ for each $1 \leq i \leq l$. Since

$$k(B) = \frac{n_1 + \cdots + n_l}{m} \neq 1$$

we have

$$B^m = U_1^{n_1} \cdots U_l^{n_l}.$$

Thus B^m has factorizations of two different lengths. $\square$

Proposition 5.5. *Let G be a direct sum of cyclic groups. Then there exists a half-factorial subset G_0 such that $\langle G_0 \rangle = G$.*

Proof. Since G is a free $\mathbb{Z}$-module, it has a basis G_0, which has the required property. $\square$

The question of whether Proposition 5.5 holds for all abelian groups is still open. The arithmetical relevance of this problem lies in its combination with the realization theorems in section 3.2. Given an abelian group G which is a direct sum of cyclic groups, Proposition

5.5 thus provides a half-factorial Dedekind domain with class group G. A discussion of some other classes of abelian groups for which the assertion of 5.5 holds can be found in [M-S].

Proposition 5.6. *Let n be a positive integer, p a prime and $G_0 \subseteq C_{p^n}$ a nonempty subset of C_{p^n}. Then G_0 is half-factorial if and only if there exists an automorphism φ of C_{p^n} such that $\varphi(G_0) \subseteq \{\overline{1}, \overline{p}, \dots, \overline{p^{n-1}}\}$.*

Proof. See [Sk2; Prop. 3.4] and [Z1;Corollary 5]. □

We characterize small half-factorial subsets of cyclic groups.

Proposition 5.7. *Let n be a positive integer and $G_0 \subseteq C_n$ with $\overline{0} \notin G_0$.*

1. *Suppose $G_0 = \{a + n\mathbb{Z}, b + n\mathbb{Z}\}$ and set*

$$n_1 = \frac{n \gcd(a, b, n)}{\gcd(a, n) \gcd(b, n)}.$$

Then G_0 is half-factorial if and only if $n_1 \leq 2$ or

$$\frac{a}{\gcd(a, n)} \equiv \frac{b}{\gcd(b, n)} \pmod{n_1}.$$

2. *Suppose $G_0 = \{\overline{1}, \overline{a}, \overline{b}\}$ with $1 \leq a, b < n$. Then G_0 is half-factorial if and only if $a \mid n$ and $b \mid n$.*

Proof. The proof of 1. is [Ge1; Proposition 5]. For the proof of 2., see [Ch-S1; Theorem 3.8]. □

Example. There is no analogue for the case $\#G_0 = 4$. Consider $G = C_{30}$ and $G_0 = \{\overline{1}, \overline{6}, \overline{10}, \overline{15}\}$. If

$$B = \overline{15} \cdot \overline{10} \cdot \overline{10} \cdot \overline{6} \cdot \overline{6} \cdot \overline{6} \cdot \overline{6} \cdot \overline{1}$$

then $k(B) = 2$ and hence $\Delta(G_0) \neq \emptyset$ by Proposition 5.4 [Ch-S1, Example 11].

In chapter 4 we defined

$$\eta(G) = \max\{\#G_0 \mid G_0 \subseteq G, G_0 \text{ half-factorial}\}.$$

This invariant was first studied by J. Sliwa [Sl], and then by J. Kaczorowski and the first author. However, very little is known about $\eta(G)$. We restate one result from [G-K; section 13].

Proposition 5.8. *For a prime* p *and an integer* $r \in \mathbb{N}$ *we have*

$$1 + (r - 2[\tfrac{r}{2}]) + p[\tfrac{r}{2}] \leq \eta(C_p^r) \leq 1 + [p\tfrac{r}{2}] \ .$$

Next we study subsets G_0 of finite abelian groups G which are not half-factorial. As pointed out in section 2.2, a central point is to determine $\min \Delta(G_0)$. There is an algorithm for doing so [Gel; Proposition 3].

Proposition 5.9. *Let* $G_0 = \{g_1, \ldots, g_m\} \subseteq G$ *where* G *is finite abelian and* $\Delta(G_0) \neq 0$. *Let* $B_1, B_2, \ldots, B_\psi$ *be the irreducible blocks in* $\mathcal{B}(G_0)$ *and suppose that* $d = \min \Delta(G_0)$. *Then* d *is the solution of the following linear, integral optimization problem: minimize*

$$\sum_{i=1}^{\psi} x_i$$

under the restrictions

$$\sum_{i=1}^{\psi} v_{g_j}(B_i) \cdot x_i = 0$$

for every $j \in \{1, \ldots, m\}$ *and*

$$\sum_{i=1}^{\psi} x_i > 0$$

where $x_i \in \mathbb{Z}$ *for every* $i \in \{1, \ldots, \psi\}$.

For a cyclic group G and $G_0 \subseteq G$ containing a generator, $\min \Delta(G_0)$ can be determined easily from the irreducible blocks in $\mathcal{B}(G_0)$ [Gel; Proposition 7].

Proposition 5.10. *Let* $n \geq 3$ *and* $G_0 \subseteq C_n$ *a subset with* $\bar{1} \in G_0$ *and* $\Delta(G_0) \neq \emptyset$. *Then*

$$\min \Delta(G_0) = \gcd\{\tfrac{1}{n} \sum_{i=1}^{n-1} i \cdot v_{\bar{i}}(B) - 1 \,|\, \bar{0} \neq B \in \mathcal{B}(G_0) \ \text{irreducible}\} \ .$$

Example. We consider some special cases of the last result.

   i. $\min \Delta(\{\bar{1}, \overline{n-1}\}) = n - 2$.

  ii. If n is odd then,

     a. $\min \Delta(\{\bar{1}, \overline{\frac{n+1}{2}}\}) = 1$.

     b. $\min \Delta(\{\bar{1}, \overline{\frac{n-1}{2}}\}) = \frac{n-3}{2}$.

Alternate calculations for these values can be found in [Ch-S2; Theorem 4] and [Ch-S1; Theorems 4.5 and 4.6].

If in addition G_0 consists of only two elements, then there is an explicit formula for $\min \Delta(G_0)$ not involving irreducible blocks of $\mathcal{B}(G_0)$. This result is obtained by methods of diophantine approximation in [Ge1; Theorem 1].

Proposition 5.11. *Let* $n \geq 3$, $a \in \{2, \ldots, n-1\}$ *with* $\gcd(a, n) = 1$ *and* $l \in \{1, \ldots, n-1\}$ *such that* $al + 1 \equiv 0 \pmod{n}$. *Suppose the continued fraction expansion of* $\frac{n}{l} = [a_0; a_1, \ldots, a_m]$. *Then*

$$\min \Delta(\{\bar{1}, \bar{a}\}) = \gcd\{a - 1, \frac{al + 1}{n} - 1, a_0 - 1, a_2, \ldots, a_{2s}\}$$

where s *is determined explicitly.*

Proposition 5.3 indicates that for any finite abelian group G with $\#G \geq 3$ we have $\min \Delta(G) = 1$. The next result shows that, apart from very few exceptions, there are always subsets $G_0 \subseteq G$ with $\min \Delta(G_0) > 1$. The proof can be found in [Ch-S1; Corollary 4.14].

Proposition 5.12. *Let* G *be a non-trivial finite abelian group. Then there exists a nonempty subset* $G_0 \subseteq G$ *with* $\min \Delta(G_0) > 1$ *if and only if* $G \notin \{C_2, C_3, C_2 \oplus C_2, C_3 \oplus C_3\}$.

We close with a result on $\Delta_1(G)$. The proof can be found in [Ge1; Propositions 1 and 2].

Proposition 5.13. *Let* G *be a finite abelian group. Setting*

$$S = \{\min \Delta(G_0) \mid \emptyset \neq G_0 \subseteq G, \Delta(G_0) \neq \emptyset\}$$

we obtain

$$S \subseteq \Delta_1(G) \subseteq \{d \mid d|s \text{ for some } s \in S\}.$$

5.3 On $\mu_k(G)$

We freely use the elementary properties of $\mu_k(G)$ mentioned in section 2.2.

Proposition 5.14. *Let $k \in \mathbb{N}$ and G be a non-trivial finite abelian group.*

1. $\mu_{2k}(G) = kD(G)$.
2. $kD(G) + 1 \leq \mu_{2k+1}(G) \leq kD(G) + \left\lceil \frac{D(G)}{2} \right\rceil$.
3. $\mu_{2k-1}(G) + D(G) \leq \mu_{2k+1}(G)$.
4. *Let $m \in \mathbb{N}$ such that*

$$\mu_{2m+1}(G) - mD(G) = \max\left\{\mu_{2r+1}(G) - rD(G) \mid r \in \mathbb{N}\right\}.$$

Then

$$\mu_{2m+2i+1}(G) = \mu_{2m+1}(G) + iD(G)$$

for all $i \geq 1$.

Proof. Set $G' = G\backslash\{0\} \neq \emptyset$ and let $U \in \mathcal{B}(G)$ be irreducible with $\sigma(U) = D(G)$.

For 1. and 2., we obviously have $\max L((-U)^k U^k) = kD(G)$ and hence $\mu_{2k}(G) \geq kD(G)$. Similarly, $\max L((-U)^k U^{k+1}) \geq kD(G) + 1$ and thus $\mu_{2k+1}(G) \geq kD(G) + 1$. To prove the remaining inequality, let $B \in \mathcal{B}(G')$ be given with $\min L(B) = \ell \in \{2k, 2k+1\}$. Then

$$2 \max L(B) \leq \sigma(B) \leq D(G)\min L(B) = \ell D(G)$$

implies

$$\max L(B) \leq \left\lceil \frac{\ell D(G)}{2} \right\rceil.$$

For 3., let $B \in \mathcal{B}(G')$ with $\min L(B) \leq 2k-1$ and $\max L(B) = \mu_{2k-1}(G)$. Then $\min L(B(-U)U) \leq 2k+1$ and $\max L(B) \geq D(G) + \mu_{2k-1}(G)$. This implies $\mu_{2k+1}(G) \geq D(G) + \mu_{2k-1}(G)$.

For 4., induction on 3. implies that, for all $i \geq 1$,

$$\mu_{2m+2i+1}(G) \geq \mu_{2m+1}(G) + iD(G).$$

By definition of m we infer that

$$\mu_{2m+2i+1}(G) - (m+i)D(G) \leq \mu_{2m+1}(G) - mD(G),$$

which implies the assertion. $\square$

Part 1. above improves a result found in [Ch-S5; Lemma 4], where it is shown using different notation that $\mu_{2kD(G)}(G) = kD(G)^2$ for every integer $k \geq 1$. For elementary 2-groups we are able to explicitly compute $\mu_\ell(G)$ for all positive integers ℓ.

Proposition 5.15. *Let $G = C_2^r$ be an elementary 2-group of rank $r \geq 1$ and let $\ell \geq 2$. Then*

$$\mu_\ell(G) = \left[\frac{\ell D(G)}{2}\right].$$

Proof. By Proposition 5.14, the assertion holds if ℓ is even. So let ℓ be odd. Proposition 5.14 part 2. implies that $\mu_\ell(G) \leq \left[\frac{\ell D(G)}{2}\right]$. By Proposition 5.14 part 4, it is sufficient to verify that

$$\mu_3(G) \geq D(G) + \left[\frac{D(G)}{2}\right].$$

Let $e_1, \ldots, e_r$ be a generating system of G and $e_0 = \sum_{i=1}^r e_i$. Note that by Proposition 5.2 we have $D(G) = r + 1$. We give 3 irreducible blocks U_1, U_2 and U_3 such that

$$\max L(U_1 U_2 U_3) = D(G) + \left[\frac{D(G)}{2}\right].$$

Set $U_1 = \prod_{i=0}^r e_i$.

Case 1 Suppose $r = 2k + 1$. Then set

$$U_2 = \left(\prod_{i=1}^{k+1} e_i\right) \left(\prod_{i=1}^k e_i + e_{i+k+1}\right) \left(\sum_{i=k+1}^{2k+1} e_i\right)$$

and

$$U_3 = \left(\prod_{i=k+2}^{2k+1} e_i\right) \left(\prod_{i=1}^k e_i + e_{i+k+1}\right) \left(\sum_{i=k+1}^{2k+1} e_i\right) e_0.$$

Case 2 Suppose $r = 2k$. Then set

$$U_2 = \left(\prod_{i=1}^k e_i\right) \left(\prod_{i=1}^k e_i + e_{i+k}\right) \left(\sum_{i=k+1}^{2k} e_i\right)$$

and

$$U_3 = \left(\prod_{i=1}^k e_i + e_{i+k}\right) \left(\prod_{i=k+1}^{2k} e_i\right) \left(\sum_{i=1}^k e_i\right). \quad \square$$

Proposition 5.16. *Let G be a bounded abelian torsion group with exponent m and $G_0 \subseteq G$ a nonempty subset. Then $\mu_m(G_0) = m$ if and only if $\mu_k(G_0) = k$ for all positive integers k.*

Proof. Suppose that $\mu_k(G_0) \neq k$ for some integer k. Then by Proposition 5.4 there exists some irreducible block

$$B = \prod_{i=1}^{l} g_i \in \mathcal{B}(G_0)$$

with cross number $k(B) \neq 1$. Thus

$$B^m = \left(\prod_{i=1}^{l} g_i \right)^m = \prod_{i=1}^{l} \left(g_i^{\text{ord}(g_i)} \right)^{\frac{m}{\text{ord}(g_i)}},$$

where $m \neq mk(B)$, and hence $\mu_m(G_0) \neq m$. □

In general, the exponent of the group is the minimal possible m such that the above result holds. This can be seen from the following example.

Example. Let $k \geq 4$ be given. Then there exists a finite abelian group G (depending on k) and a subset G_0 for which $\mu_k(G_0) = k$ but $\mu_{k+1}(G_0) \neq k+1$. The argument runs as follows.

Let $m > k+1$ and set $G = \oplus_{i=1}^{k} C_m$. If $e_1, \ldots, e_k$ is a $\mathbb{Z}$-module basis for G and $e_0 = -\sum_{i=1}^{k} e_i$, then $G_0 = \{e_1, \ldots, e_k, e_0\}$ has the required property (see [Ch-S3; Examples 2.6 - 2.9]).

In special cases, the value m obtained in Proposition 5.16 can be improved, as the following result demonstrates.

Proposition 5.17. *Suppose that G is a non-trivial finite abelian group and $G_0 \subseteq G$ a nonempty subset which satisfies any of the following conditions:*
  1. *$G = C_{p^n}$ for p a prime integer and n a positive integer.*
  2. *$G = C_{pq}$ for distinct prime integers p and q.*
  3. *$\#G \leq 15$.*
  4. *G is cyclic and G_0 contains a generator.*

Then $\mu_2(G_0) = 2$ if and only if $\mu_k(G_0) = k$ for all integers $k \geq 1$.

Proof. The proofs of 1., 2. and 3. are slight modifications of [Ch-S3; Theorem 3.2], [Ch-S3; Corollary 3.5] and [Ch-S4; Theorem 1]. For 4. see [Ch-S6; Theorem 1]. □

REFERENCES

[A-G-P] W.R. Alford, A. Granville and C. Pomerance, *There are infinitely many Carmichael numbers*, Ann. Math. **140**: 703–722 (1994).

[A-A-Z1] D.D. Anderson, D.F. Anderson and M. Zafrullah, *Factorization in integral domains*, J. Pure Applied Algebra **69**: 1-19 (1990).

[A-A-Z2] D.D. Anderson, D.F. Anderson and M. Zafrullah, *Rings between D[X] and K[X]* , Houston J. Math. **17**: 109-129 (1991).

[A-A-Z3] D.D. Anderson, D.F. Anderson and M. Zafrullah, *Factorization in integral domains, II*, J. Algebra **152**: 78-93 (1992).

[An1] D. F. Anderson, *Elasticity of factorization in integral domains, a survey*, these Proceedings.

[Ca] L. Carlitz, *A characterization of algebraic number fields with class number two*, Proc. Amer. Math. Soc. **11**: 391–392 (1960).

[Ch] S. Chapman, *On the Davenport's constant, the Cross number, and their application in factorization theory*, in Zero-dimensional commutative rings, Lecture Notes in Pure Appl. Math., Marcel Dekker **171**: 167–190 (1995).

[Ch-G] S. Chapman and A. Geroldinger, *On cross numbers of minimal zero sequences*, Australasian J. Comb..

[Ch-S1] S. Chapman and W.W. Smith, *Factorization in Dedekind domains with finite class group*, Israel J. Math. **71**: 65–95 (1990).

[Ch-S2] S. Chapman and W.W. Smith, *On a characterization of algebraic number fields with class number less than three*, J. Algebra **135**: 381–387 (1990).

[Ch-S3] S. Chapman and W.W. Smith, *On the HFD, CHFD, and the k-HFD properties in Dedekind domains*, Comm. Algebra **20**: 1955–1987 (1992).

[Ch-S4] S. Chapman and W.W. Smith, *On the k-HFD property in Dedekind domains with small class group*, Mathematika **39**: 330–340 (1992).

[Ch-S5] S. Chapman and W.W. Smith, *On the lengths of factorizations of elements in an algebraic number ring*, J. Number Theory **43**: 24–30 (1993).

[Ch-S6] S. Chapman and W.W. Smith, *Finite cyclic groups and the k-HFD property*, Colloq. Math. **70**: 219–226 (1996).

[Cho] L. Chouinard, *Krull semigroups and divisor class groups*, Canad. J. Math. **19**: 1459–1468 (1981).

[E-K] P. van Emde Boas and D. Kruyswijk, *A combinatorial problem on finite abelian groups III*, Report ZW-1969-008, Math. Centre Amsterdam.

[Ga] W. Gao, *On Davenport's constant of finite abelian groups with rank three*, preprint.

[Ge1]   A. Geroldinger, *On non-unique factorizations into irreducible elements II*, Coll. Math. Soc. J. Bolyai, Number Theory, Budapest **51**:723–757 (1987).

[Ge2]   A. Geroldinger, *Über nicht-eindeutige Zerlegungen in irreduzible Elemente*, Math. Z. **197**: 505–529 (1988).

[Ge3]   A. Geroldinger, *Ein quantitatives Resultat über Faktorisierungen verschiedener Länge in algebraischen Zahlkörpern*, Math. Z. **205**: 159–162 (1990).

[Ge4]   A. Geroldinger, *Systeme von Längenmengen*, Abh. Math. Sem. Univ. Hamburg **60**: 115–130 (1990).

[Ge5]   A. Geroldinger, *Arithmetical characterizations of divisor class group*, Arch. Math. **54**: 455–464 (1990).

[Ge6]   A. Geroldinger, *On the arithmetic of certain not integrally closed noetherian integral domains*, Comm. Algebra **19**: 685–698 (1991).

[Ge7]   A. Geroldinger, *Arithmetical characterizations of divisor class group II*, Acta Math. Univ. Comenianæ **61**: 193–208 (1992).

[Ge8]   A. Geroldinger, *T-block monoids and their arithmetical applications to certain integral domains*, Comm. Algebra **22**: 1603–1615 (1994).

[Ge9]   A. Geroldinger, *Factorization of algebraic integers*, Springer Lecture Notes in Mathematics **1380**, 63–74.

[Ge10]   A. Geroldinger, *The catenary degree and tameness of factorizations in weakly Krull domains*, these Proceedings.

[G-HK1]   A. Geroldinger and F. Halter-Koch, *Realization theorems for semigroups with divisor theory*, Semigroup Forum **44**: 229–237 (1992).

[G-HK2]   A. Geroldinger and F. Halter-Koch, *Arithmetical theory of monoid homomorphisms*, Semigroup Forum **48**: 333–362 (1994).

[G-HK-K] A. Geroldinger, F. Halter-Koch and J. Kaczorowski, *Non-unique factorizations in orders of global fields*, J. Reine Angew. Math. **459**: 89–118 (1995).

[G-K]   A. Geroldinger and J. Kaczorowski, *Analytic and arithmetic theory of semigroups with divisor theory*, Sém. Théorie d. Nombres Bordeaux **4**: 199–238 (1992).

[G-L]   A. Geroldinger and G. Lettl, *Factorization problems in semigroups*, Semigroup Forum **40**: 23–38 (1990).

[G-S]   A. Geroldinger and R. Schneider, *On Davenport's constant*, J. Comb. Theory Series A **61**: 147-152 (1992).

[Gi]   R. Gilmer, *Commutative semigroup rings*, The University of Chicago Press (1984).

[Gr]   A. Grams, *The distribution of prime ideals of a Dedekind domain*, Bull. Austral. Math. Soc. **11**: 429–441 (1974).

[Gu]   K. B. Gundlach, *Einführung in die Zahlentheorie*, B. I. Hochschultaschenbücher **Bd. 772** (1972).

[HK1] F. Halter-Koch, *Halbgruppen mit Divisorentheorie*, Expo. Math. 8: 27–66 (1990).

[HK2] F. Halter-Koch, *Ein Approximationssatz für Halbgruppen mit Divisorentheorie*, Result. Math. 19: 74–82 (1991).

[HK3] F. Halter-Koch, *A generalization of Davenport's constant and its arithmetical applications*, Colloq. Math. 63: 203–210 (1992).

[HK4] F. Halter-Koch, *Chebotarev formations and quantitative aspects of non-unique factorizations*, Acta Arith. 62: 173–206 (1992).

[HK5] F. Halter-Koch, *Finiteness theorems for factorizations*, Semigroup Forum 44: 112–117 (1992).

[HK6] F. Halter-Koch, *The integral closure of a finitely generated monoid and the Frobenius problem in higher dimensions, Semigroups, edited by C. Bonzini, A. Cherubini and C. Tibiletti*, World Scientific (1993), 86 – 93.

[HK7] F. Halter-Koch, *A characterization of Krull rings with zero divisors*, Archivum Math. (Brno) 29: 119–122 (1993).

[HK8] F. Halter-Koch, *Elasticity of factorizations in atomic monoids and integral domains*, J. Théorie des Nombres Bordeaux 7: 367–385 (1995).

[HK9] F. Halter-Koch, *Finitely generated monoids, finitely primary monoids and factorization properties of integral domains*, these Proceedings.

[Hu] J. Huckaba, *Commutative rings with zero divisors*, Marcel Dekker (1988).

[Ja] N. Jacobson, *Basic Algebra I*, W. H. Freeman & Co. (1985).

[Kn] J. Knopfmacher, *Abstract Analytic Number Theory*, North-Holland, 1975.

[Kr] U. Krause, *On monoids of finite real character*, Proc. Amer. Math. Soc. 105: 546-554 (1989).

[Le] G. Lettl, *Subsemigroups of fintely generated groups with divisor theory*, Mh. Math. 106: 205–210 (1988).

[M-S] D. Michel and J. Steffan, *Répartition des idéaux premiers parmi les classes d'idéaux dans un anneau de Dedekind et équidécomposition*, J. Algebra 98: 82–94 (1986).

[Na1] W. Narkiewicz, *Finite abelian groups and factorization problems*, Coll. Math 42: 319–330 (1979).

[Na2] W. Narkiewicz, *Elementary and analytic theory of algebraic numbers*, Springer (1990).

[Ol] J. Olson, *A combinatorial problem on finite abelian groups I and II*, J. Number Theory 1: 8–11, 195–199 (1969).

[Re] L. Rédei, *Theorie der endlich erzeugbaren kommutativen Halbgruppen*, Physica Verlag (1963).

[Sk1] L. Skula, *Divisorentheorie einer Halbgruppe*, Math. Z. 114: 113–120 (1970).

[Sk2]   L. Skula, *On c-semigroups*, Acta Arith. **31**: 247–257 (1976).

[Sl]   J. Sliwa, *Remarks on factorizations in number fields*, Acta Arith. **46** (1982).

[Z1]   A. Zaks, *Half-factorial domains*, Bull. Amer. Math. Soc. **82**: 721–723 (1976).

[Z2]   A. Zaks, *Half factorial domains*, Isr. J. Math. **37**: 281–302 (1980).

The Catenary Degree and Tameness of Factorizations in Weakly Krull Domains

Alfred Geroldinger, Institut für Mathematik, Karl-Franzens-Universität, 8010 Graz, Austria

1. THE ARITHMETIC OF WEAKLY KRULL DOMAINS

Weakly Krull domains were introduced in [A-M-Z]. We briefly recall those properties which will be crucial for the study of their arithmetic. For proofs the reader is referred to [Ge5; section 7], [HK3] and [HK2; section 6].

Let R be an integral domain. Then $R^{\bullet} = R \setminus \{0\}$ denotes the multiplicative monoid of R, $R^{\times}$ the unit group and $R^{\#} = R^{\bullet}/R^{\times}$ the reduced multiplicative monoid. Clearly, $R^{\#}$ is isomorphic to $\mathcal{H}(R)$, the monoid of principal integral ideals. R is said to be *weakly Krull*, if

$$R = \bigcap_{\mathfrak{p} \in \mathfrak{X}(R)} R_{\mathfrak{p}},$$

where $\mathfrak{X}(R)$ denotes the set of height-one prime ideals and the intersection is of finite character. Let R be a weakly Krull domain and $\mathfrak{p} \in \mathfrak{X}(R)$. Then $R_{\mathfrak{p}}^{\bullet}$ and $R_{\mathfrak{p}}^{\#}$ are finitely primary monoids, whose rank equals the number of prime ideals of the complete integral closure $\widehat{R_{\mathfrak{p}}}$ of $R_{\mathfrak{p}}$. Clearly, $R_{\mathfrak{p}}$ is a discrete valuation ring if and only if $R_{\mathfrak{p}}^{\#}$ is isomorphic to $(\mathbf{N}_0, +)$. The monoid $\mathcal{I}_t(R)$ of integral

t-invertible t-ideals (equipped with t-multiplication) is isomorphic to $\coprod_{\mathfrak{p} \in \mathfrak{X}(R)} R_\mathfrak{p}^\#$. Let $P \subseteq \mathfrak{X}(R)$ denote the set of prime ideals $\mathfrak{p}$ for which $R_\mathfrak{p}$ is a discrete valuation ring and set $\Omega = \mathfrak{X}(R) \setminus P$. Then

$$\coprod_{\mathfrak{p} \in \mathfrak{X}(R)} R_\mathfrak{p}^\# \simeq \mathcal{F}(P) \times T$$

where $\mathcal{F}(P)$ is the free abelian monoid with basis P and $T = \coprod_{\mathfrak{p} \in \Omega} R_\mathfrak{p}^\#$.

Since $\mathcal{H}(R) \subseteq \mathcal{I}_t(R)$ is saturated, $H = R^\bullet / R^\times \simeq \mathcal{H}(R)$ is a saturated submonoid of a coproduct D of finitely primary monoids such that the class group of $H \hookrightarrow D$ (in the sense of [HK2; section 5]) is isomorphic to the t-class group of R.

Krull domains and noetherian domains, for which every prime ideal of depth one has height one, are weakly Krull. Furthermore, the above set Ω is finite for Krull domains and for noetherian weakly Krull domains R whose integral closure $\overline{R}$ is a finite R-module (e. g., for orders in Dedekind domains).

In the theory of non-unique factorizations mainly sets of lengths, and associated invariants as the elasticity, are used to describe the arithmetic of a non-factorial domain (cf. the three survey articles [An], [HK2] and [Ch-G] in this volume). Indeed, in Krull domains with finite divisor class group where each class contains a prime divisor sets of lengths carry a great part of arithmetical information (cf. [Ch-G; section 2.2d]). However, in more general domains sets of lengths alone are a too weak measure for describing the arithmetic (cf. the examples in [HK2]).

In this article we discuss invariants as the catenary degree and the tame degree, which control the non-uniqueness of factorizations not by their lengths but in a direct way. The following examples of weakly Krull domains show some surprising behaviour of these invariants:

i) for every $\varrho \in \mathbf{N}$ there exists a Dedekind domain with elasticity ϱ and infinite catenary degree (see [Ge7; section 3]).

ii) orders in algebraic number fields have finite catenary degree. The elasticity and the tame degree might be finite or infinite (see Theorems 2.7 and 3.6).

iii) there is a one-dimensional noetherian domain which is not an FF-domain but has tame degree 2 (see the example after Proposition 3.3).

This paper follows the survey articles [HK2] and [Ch-G] in this volume. The reader is referred to them for questions of notation and for standard notions in the theory of non-unique factorizations (e. g., atomic, BF- and FF-monoids and domains, sets of lengths, finitely generated and finitely primary monoids, saturated submonoids, block monoids and class groups). Furthermore, we use the language of monoids right from the beginning. The reasons for such a semigroup theoretical procedure are discussed in [HK2] and [Ch-G]. Our main results are formulated for saturated submonoids of coproducts of finitely primary monoids. Hence, they apply to weakly Krull domains, as explained above.

In sections 2 and 3 we introduce the main arithmetical concepts: chains of factorizations, catenary degree, tame degree and local tameness of factorizations. The presentation in these sections is expository. The main results (Theorem 2.7 and Theorem 3.6) are from [Ge5] and [Ge7]. In the case of Krull domains with finite class group where each class contains a prime divisor we derive explicit lower and upper bounds for the catenary and the tame degree (Theorems 2.10, 3.8 and 3.9).

In section 4 we prove that in certain weakly Krull domains sets of lengths of large elements are arithmetical progressions with distance 1 (Theorem 4.4). This result and the analytical machinery from [G-HK-K] imply that (in certain orders of global fields) the density of these elements equals 1 (Theorem 5.4).

2. THE CATENARY DEGREE

Let H be a reduced atomic monoid, $H \neq H^\times$ and $\mathcal{A}(H)$ the set of atoms of H. The free abelian monoid

$$\mathcal{Z}(H) = \mathcal{F}(\mathcal{A}(H))$$

with basis $\mathcal{A}(H)$ is called the *factorization monoid* of H. The canonical epimorphism

$$\pi \colon \mathcal{Z}(H) \to H$$

is the *factorization homomorphism* of H. For every $a \in H$, the elements of $\mathcal{Z}(a) = \pi^{-1}(a)$ are the *factorizations* of a.

In order to study these sets $\mathcal{Z}(a)$ we define a *distance function*

$$d\colon \mathcal{Z}(H) \times \mathcal{Z}(H) \to \mathbf{N}_0$$

by

$$d(z, z') = \max\left\{ \sigma\left(\frac{z}{\gcd(z, z')}\right), \ \sigma\left(\frac{z'}{\gcd(z, z')}\right) \right\} \in \mathbf{N}_0$$

for two factorizations z and $z' \in \mathcal{Z}(H)$. This means, if $z = u_1 \ldots u_l v_1 \ldots v_m$ and $z' = u_1 \ldots u_l w_1 \ldots w_n$ with $u_i, v_j, w_k \in \mathcal{A}(H)$ such that $\{v_j | 1 \leq j \leq m\} \cap \{w_k | 1 \leq k \leq n\} = \emptyset$, then $d(z, z') = \max\{m, n\}$. If $z, z' \in \mathcal{Z}(a)$ for some $a \in H$ and $z \neq z'$, then $d(z, z') \geq 2$.

The following lemma shows that d has indeed the expected properties of a distance function (cf. [Ge7; Lemma 3.1]).

Lemma 2.1. *Let H be a reduced atomic monoid. Then the distance function $d\colon \mathcal{Z}(H) \times \mathcal{Z}(H) \to \mathbf{N}_0$ satisfies the following properties for all $k \in \mathbf{N}$ and all $x, x', y, y', z \in \mathcal{Z}(H)$:*
  1. $d(x, y) = 0$ *if and only if* $x = y$,
  2. $d(x, y) = d(y, x)$,
  3. $d(x, y) \leq d(x, z) + d(z, y)$,
  4. $d(x^k, y^k) = k d(x, y)$,
  5. $d(xy, x'y') \leq d(x, x') + d(y, y')$,
  6. $d(x, y) = d(xz, yz) \leq d(x, yz)$.

Lemma 2.2. *Let H be a reduced atomic monoid.*
  1. H *is factorial if and only if* $\#\mathcal{Z}(a) = 1$ *for all* $a \in H$.
  2. *Suppose that H is not factorial. Then for every $k \in \mathbf{N}$ there exists some $a \in H$ and factorizations $z, z' \in \mathcal{Z}(a)$ such that* $\#\mathcal{Z}(a) \geq k$ *and* $d(z, z') \geq k$.

Proof. **1.** is clear. To verify **2.** let $k \in \mathbf{N}$ be given. Since H is not factorial, there exists some $a \in H$ and factorizations $x, y \in \mathcal{Z}(a)$ with $d(x, y) > 1$. For $0 \leq i \leq k$ set $z_i = x^{k-i} y^i$. Then

$$x^k = z_0, z_1, \ldots, z_k = y^k$$

are pairwise distinct factorizations of a^k and $d(z_0, z_k) = k d(x, y) > k$. $\square$

The above proof shows that a^k has factorizations $z = x^k$ and $z' = y^k$ with $d(z, z') > k$. However, there are factorizations between such that

$$d(z_{i-1}, z_i) = d(x^{k-i+1}y^{i-1}, x^{k-i}y^i) = d(x, y) \leq N$$

holds for $1 \leq i \leq k$ and some N independent on k. This behaviour motivates the following definition.

Definition 2.3. Let H be a reduced atomic monoid.
1. Let $a \in H$, $z, z' \in Z(a)$ and $N \in \mathbb{N}_0$. We say that there is an *N-chain (of factorizations) from z to z'*, if there exist factorizations $z = z_0, z_1, \ldots, z_k = z' \in Z(a)$ such that $d(z_{i-1}, z_i) \leq N$ for $1 \leq i \leq k$.
2. The *catenary degree*

$$c_H(H') = c(H') \in \mathbb{N}_0 \cup \{\infty\}$$

of a subset $H' \subseteq H$ is the minimal $N \in \mathbb{N}_0 \cup \{\infty\}$ such that for every $a \in H'$ and each two factorizations $z, z' \in Z(a)$ there exists an N-chain from z to z'. For simplicity, we write $c(a)$ instead of $c(\{a\})$.

The following lemma gathers some elementary properties of the catenary degree, which follow immediately from the definition.

Lemma 2.4. *Let H be a reduced atomic monoid.*
1. *H is factorial if and only if $c(H) = 0$.*
2. *For every $a \in H$ and each two factorizations $z, z' \in Z(a)$ we have $2 + |\sigma(z) - \sigma(z')| \leq d(z, z')$.*
3. *For every subset $H' \subseteq H$ we have $2 + \sup \Delta(H') \leq c(H')$. In particular, $c(H) = 2$ implies that H is half-factorial; furthermore, if $c(H) = 3$, then all sets of lengths $L \in \mathcal{L}(H)$ are arithmetical progressions with distance 1.*

In his survey article F. Halter-Koch pointed out the significance of finitely generated and finitely primary monoids for factorization theory in integral domains. Monoids of both classes have finite catenary degree. The next result was first proved in [Ge2]; it is also a consequence of the forthcoming Propositions 3.3 and 3.4.

Proposition 2.5. *Every finitely generated monoid has finite cate-nary degree.*

A proof of the following proposition can be found in [Ge4; Theorem 3]

Proposition 2.6. *Let H be a finitely primary monoid with exponent $\alpha \in \mathbb{N}$. Then H has finite catenary degree $c(H) \leq max\{3, 4\alpha - 2\}$.*

Theorem 2.7. *Let $H \subseteq D = \coprod_{i \in I} D_i$ be a saturated submonoid with bounded class group G. Suppose that all D_i are finitely primary of some exponent $\alpha \in \mathbb{N}$ and that $\mathcal{D}(G_1) < \infty$ with $G_1 = \{g \in G \mid g \cap \mathcal{U}(D) \neq \emptyset\}$. Then H has finite catenary degree. More precisely, we have*

$$c(H) \leq (\alpha + \beta)\mathcal{D}(G_1)\big[4\beta + \alpha + (\alpha + \beta)\mathcal{D}(G_1)(2\alpha - 1)\big]$$

where $\beta = \alpha \exp(G)$.

Proof. see [Ge5; Theorem 5.4] □

For Krull domains and monoids a better upper bound for the cate-nary degree can be obtained. For this we need block monoids (cf. [HK2; section 5]) which turn out to be a powerful tool also for study-ing chains of factorizations. We cite a short version of [Ge5; Proposition 4.2].

Proposition 2.8. *Let $H \subseteq D = \mathcal{F}(P) \times T$ be a saturated atomic submonoid of a reduced atomic monoid D and let $\mathcal{B}$ denote the associated block monoid. Then $c(\mathcal{B}) \leq c(H) \leq \max\{2, c(\mathcal{B})\}$.*

For an abelian group G and a subset $G_0 \subseteq G$ we write $c(G_0)$ instead of $c(\mathcal{B}(G_0))$.

Theorem 2.9. *Let H be a Krull monoid with divisor class group G and $G_0 \subseteq G$ the set of classes containing prime divisors. Then $c(H) \leq \mathcal{D}(G_0)$.*

Proof. Obviously, the assertion holds if H is factorial. Suppose H is not factorial. Then G is not the trivial group and since G_0 generates G, there exists an element $g \in G_0$ with $\text{ord}(g) \geq 2$. Therefore $\mathcal{D}(G_0) \geq 2$.

Since by [G-L; Proposition 3] $c(G_0) \leq \mathcal{D}(G_0)$ holds, the assertion follows from the previous proposition. For an alternate proof cf. [Ge5; Proposition 4.3]. □

The determination of $c(G)$ is a purely combinatorial, group-theoretical problem. We use the same notations as in [Ch-G; section 5]. In particular, we have the following two size functions. Let $A = \prod_{i=1}^{l} g_i \in \mathcal{B}(G) \subseteq \mathcal{F}(G)$ and $Z = U_1 \ldots U_k \in \mathcal{Z}(\mathcal{B}(G))$ a factorization of A. Then, as defined above, $\sigma(Z) = \sigma_{\mathcal{Z}(\mathcal{B}(G))}(Z) = k$ denotes the length of the factorization Z and $\sigma(A) = \sigma_{\mathcal{F}(G)}(A) = l$ denotes the size of A. Hence, if A is irreducible, then $A = U_1 = Z$, $\sigma_{\mathcal{Z}(\mathcal{B}(G))}(U_1) = 1$ and $\sigma_{\mathcal{F}(G)}(U_1) = l$. Nevertheless, we shall write σ without indices. Its meaning will become clear from the context.

Theorem 2.10. *Let* $G = C_{n_1} \oplus \cdots \oplus C_{n_r}$ *be a finite abelian group with* $1 < n_1 \mid \ldots \mid n_r$. *Then we have*
1. $\max\{n_r, 1 + \sum_{i=1}^{r}\left[\frac{n_i}{2}\right]\} \leq c(G) \leq \mathcal{D}(G)$.
2. $c(G) = \mathcal{D}(G)$ *holds if and only if* G *is either cyclic or an elementary 2-group.*

Corollary 2.11. *Let* $G = C_2 \oplus C_{2n}$ *for some* $n \in \mathbf{N}$. *Then* $c(G) = 2n$.

Proof. By Theorem 2.10 we have

$$2n = \exp(G) \leq c(G) \leq \mathcal{D}(G) - 1 = 2n . \quad \square$$

We start with the preparations for the proof of Theorem 2.10. For an element $S = \prod_{i=1}^{l} g_i \in \mathcal{F}(G)$ let

$$\Sigma(S) = \left\{\sum_{j \in J} g_j \in G \mid \phi \neq J \subseteq \{1, \ldots, l\}\right\}$$

denote the set of sums of all divisors $1 \neq S'$ of S .

Lemma 2.12. *Let* G *be a finite abelian group,* $H \subset G$ *a proper subgroup and* $U = \prod_{i=0}^{l} g_i \in \mathcal{B}(G)$ *an irreducible block with* $1 + l = \mathcal{D}(G)$. *Then we have*
1. $\mathcal{D}(H) < \mathcal{D}(G)$ *and* $< g_1, \ldots, g_l > = G$.

2. $\Sigma(\prod_{i=1}^{l} g_i) = G \setminus \{0\}$.

Proof. **1.** Let $V = \prod_{j=1}^{k} h_j$ be an irreducible block in H with $k = \mathcal{D}(H)$ and let $g \in G \setminus H$. Then

$$W = g(h_1 - g) \prod_{j=2}^{k} h_j$$

is an irreducible block in $\mathcal{B}(G)$ with $\sigma(W) > \mathcal{D}(H)$, whence $\mathcal{D}(G) > \mathcal{D}(H)$. Since

$$G' = \langle g_0, \ldots, g_l \rangle = \langle g_1, \ldots, g_l \rangle \subseteq G$$

is a subgroup with $\mathcal{D}(G') = \mathcal{D}(G)$, the first assertion implies that $G' = G$.

2. Since for every $g \in G \setminus \{0\}$

$$S = g \prod_{i=1}^{l} g_i$$

contains a subsequence with sum zero, the assertion follows. $\quad\square$

Lemma 2.13. *Let G be a finite abelian group which is neither cyclic nor an elementary 2-group. Then every product of two irreducible block $U, V \in \mathcal{B}(G)$ with $\sigma(U) = \sigma(V) = \mathcal{D}(G)$ allows a factorization of the form*

$$UV = W_1 \ldots W_k$$

with irreducible blocks $W_1, \ldots, W_k, \sigma(W_1) < \mathcal{D}(G)$ and $k < \mathcal{D}(G)$.

Proof. Let $U, V \in \mathcal{B}(G)$ be two irreducible blocks with $\sigma(U) = \sigma(V) = \mathcal{D}(G)$. We divide the proof into two cases.

Case 1. $V \neq -U$. First we assume that all irreducible blocks $W \in \mathcal{B}(G)$ with $W \mid UV$ have length $\sigma(W) = \mathcal{D}(G)$. Let $g \in G$ with $g \mid U$ and set $V = \prod_{i=1}^{l} h_i$ with $l = \mathcal{D}(G)$. For every $i \in \{1, \ldots, l\}$ and every $J_i = \{1, \ldots, l\} \setminus \{i\}$ consider

$$S_i = g \prod_{j \in J_i} h_j \in \mathcal{F}(G).$$

Since $\sigma(S_i) = \mathcal{D}(G)$, S_i contains an irreducible block S_i' and clearly $S_i' \mid S_i \mid UV$. This implies that $S_i' = S_i$ is a block. Hence

$$h_i = -\sum_{j \in J_i} h_j = g$$

and thus $\mathcal{D}(G) = l = \text{ord}\,(g)$. Therefore $H = <g> = G$ by the previous Lemma, a contradiction.

Thus there exists an irreducible block $W = W_1$ with $W \mid UV$ and $\sigma(W) < \mathcal{D}(G)$. This gives a factorization

$$UV = W_1 \ldots W_k$$

with irreducible blocks $W_2, \ldots, W_k$. Now, $\sigma(W_1) = \cdots = \sigma(W_k) = 2$ would imply that $W_i = -a_i \cdot a_i$ for some $a_i \in G$ and that $U = \prod_{i=1}^k a_i = -V$. Thus $\sigma(W_i) > 2$ for some $i \in \{1, \ldots, k\}$ which implies that $k < \mathcal{D}(G)$.

Case 2. $V = -U$. Since G is not an elementary 2-group, there exists an element $g_0 \in G$ with $\text{ord}(g_0) > 2$ and $g_0 \mid U$ (see the previous lemma). Set

$$U = g_0^m \prod_{i=1}^l g_i$$

with $g_0 \notin \{g_1, \ldots, g_l\}$. Again by Lemma 2.12 it follows that $l \geq 2$, since G is not cyclic. Consider

$$W' = (-g_0)^m \prod_{i=1}^l g_i \in \mathcal{F}(G).$$

Then $W' \mid -UU$ and since $\sigma(W') = \mathcal{D}(G)$, there is an irreducible block $W \in \mathcal{B}(G)$ dividing W'. Because U is irreducible, it follows that $W \nmid \prod_{i=1}^l g_i$ and hence $-g_0 \mid W$. Since $g_0 \notin \{g_1, \ldots, g_l\}$ and $\text{ord}(g_0) > 2$, we infer that $W \neq -g_0 \cdot g_0$ and thus $\sigma(W) > 2$. Obviously, it suffices to show that $\sigma(W) < \mathcal{D}(G)$.

Assume to the contrary that $\sigma(W) = \mathcal{D}(G)$. Then $W = W' = (-g_0)^m \prod_{i=1}^l g_i$. Since U and W are blocks, it follows that $2mg_0 = 0$ and hence $m > 1$. Set

$$S = g_0^m \prod_{i=1}^{l-1} g_i.$$

Then by the previous lemma we have $\Sigma(S) = G \setminus \{0\}$ and thus $(m+1)g_0 \in \Sigma(S)$. This means that

$$(m+1)g_0 = rg_0 + \sum_{i \in I} g_i$$

with $0 \le r \le m$ and $I \subseteq \{1, \ldots, l-1\}$. Assume $r = 0$; then

$$0 = 2mg_0 = (m-1)g_0 + \sum_{i \in I} g_i \in \Sigma(S),$$

a contradiction. Thus $r \ge 1$ and

$$T = (-g_0)^{m+1-r} \prod_{i \in I} g_i$$

is a block with $T \mid W$ and $T \ne W$, a contradiction to the irreducibility of W. $\square$

Proof of Theorem 2.10. **1.** The right inequality in **1.** follows from Theorem 2.9.

Let $e_1, \ldots, e_r$ be a generating system of G with $\mathrm{ord}(e_i) = n_i$ for $1 \le i \le r$. Then

$$A = (e_r^{n_r})(-e_r^{n_r}) = \prod_{i=1}^{n_r} (-e_r \cdot e_r) \in \mathcal{B}(G)$$

has exactly the two given factorizations. This shows that $c(G) \ge n_r = \exp(G)$.

Next, set $k_i = \left[\frac{n_i}{2}\right]$ for $1 \le i \le r$ and let $e_0 = -\sum_{i=1}^{r} k_i e_i$. Then

$$U = e_0 \prod_{i=1}^{r} e_i^{k_i}$$

is an irreducible block. Consider

$$A = -U \cdot U = (-e_0 \cdot e_0) \prod_{i=1}^{r} (-e_i \cdot e_i)^{k_i}.$$

Since $L(A) = \{2, 1 + \sum_{i=1}^{r} k_i\}$, it follows that $c(G) \ge 1 + \sum_{i=1}^{r} k_i$.

2. If G is cyclic or an elementary 2-group, then **1.** implies that $c(G) = \mathcal{D}(G)$ holds.

Now suppose that G is neither cyclic nor an elementary 2-group. We show that for every $A \in \mathcal{B}(G)$ and any two factorizations Z, Z' of A there exists a $(\mathcal{D}(G) - 1)$-chain of factorizations from Z to Z'. We argue by induction on $\sigma(A)$. Clearly, the assertion holds for all blocks A with $\sigma(A) \leq \mathcal{D}(G) - 1$.

Let $A \in \mathcal{B}(G)$ be given with $\sigma(A) \geq \mathcal{D}(G)$ and let $Z = U_1 \ldots U_r$ and $Z' = V_1 \ldots V_s$ be two factorizations of A. If $r \leq \mathcal{D}(G) - 1$ and $s \leq \mathcal{D}(G) - 1$, then $d(Z, Z') \leq \mathcal{D}(G) - 1$. So suppose that $r \geq \mathcal{D}(G)$.

If $\sigma(V_1) = \sigma(V_2) = \mathcal{D}(G)$, the previous lemma gives us a factorization

$$Z'' = W_1 \ldots W_k V_3 \ldots V_s$$

of A with $\sigma(W_1) < \mathcal{D}(G)$ and $d(Z', Z'') = \max\{2, k\} < \mathcal{D}(G)$. So it means no restriction to suppose that $\sigma(V_1) < \mathcal{D}(G)$. Then we have

$$r - 1 \geq \mathcal{D}(G) - 1 \geq \sigma(V_1).$$

Hence after a suitable renumeration we obtain $V_1 \mid U_1 \ldots U_{r-1}$ and hence $U_1 \ldots U_{r-1} = V_1 X_1 \ldots X_t$ with irreducible blocks $X_1, \ldots, X_t$. By induction hypothesis there is a $(\mathcal{D}(G) - 1)$-chain from $U_1 \ldots U_{r-1}$ to $V_1 X_1 \ldots X_t$ and thus a $(\mathcal{D}(G) - 1)$-chain from Z to $V_1 X_1 \ldots X_t U_r$.

Again by induction hypothesis there is a $(\mathcal{D}(G) - 1)$-chain from $X_1 \ldots X_t U_r$ to $V_2 \ldots V_s$ which induces a $(\mathcal{D}(G) - 1)$-chain from $V_1 X_1 \ldots X_t U_r$ to Z' and the assertion is proved. $\square$

3. TAMENESS OF FACTORIZATIONS

The concept of tameness of factorizations was introduced in [Ge2]. In [Ge7] it was refined to a notion of local tameness, which turned out to be a main conceptional tool in the proofs of structure theorems for sets of lengths. Its usefulness will become obvious in the proofs of Theorems 4.3 and 4.4 in the next section.

Definition 3.1. Let H be a reduced atomic monoid.

1. The *tame degree* $t_H(H', X) = t(H', X)$ of a subset $H' \subseteq H$ with respect to a subset $X \subseteq \mathcal{Z}(H)$ is the minimum of all $N \in \mathbf{N}_0 \cup \{\infty\}$ with the following property: for all $a \in H'$, $z \in \mathcal{Z}(a)$ and

$x \in X$ with $\pi(x) \mid a$ (in H) there exists a factorization $z' \in Z(a)$ with $x \mid z'$ (in $Z(H)$) and $d(z, z') \leq N$.

For $x \in Z(H)$ we write $t(H', x)$ instead of $t(H', \{x\})$.

2. H is called *locally tame*, if $t_H(H, u) < \infty$ for all $u \in \mathcal{A}(H)$. H is said to be *tame*, if the *tame degree* $t(H) = t(H, \mathcal{A}(H)) < \infty$.

Remarks. Let H be a reduced atomic monoid.

1. Let $u \in \mathcal{A}(H)$. Then $t(H, u) = 0$ if and only if u is prime. Therefore, H is factorial if and only if $t(H) = 0$.

2. For every $H' \subseteq H$ and every $X \subseteq Z(H)$ we have $t(H', X) \leq \sup\{\sup L(a) \mid a \in H'\}$.

3. Let $x = u_1 \ldots u_r \in Z(H)$. An inductive argument shows that $t(H, x) \leq \sum_{i=1}^{r} t(H, u_i) \leq \sigma(x) t(H)$.

4. If $H = \coprod_{i \in I} H_i$, then $\mathcal{A}(H) = \cup_{i \in I} \mathcal{A}(H_i)$ and hence $t(H) = \sup\{t(H_i) \mid i \in I\}$.

5. Let $\varphi \colon H \to D$ be a monoid epimorphism onto a reduced atomic monoid D with $\varphi(\mathcal{A}(H)) \subseteq \mathcal{A}(D) \cup \{1\}$. Then φ has a canonical extension to $\varphi \colon Z(H) \to Z(D)$ and for $z, z' \in Z(H)$ we have $d(\varphi z, \varphi z') \leq d(z, z')$. Therefore, $t_D(\varphi H', \varphi X) \leq t_H(H', X)$ for all $H' \subseteq H$ and $X \subseteq Z(H)$.

Lemma 3.2. *Let H be a reduced atomic monoid which is not factorial. Then for every $a \in H$ and each two factorizations $y, z \in Z(a)$ we have $\sigma(z) \leq \sigma(y) t(H)$.*

Proof. Let $a \in H$ and $y, z \in Z(a)$. Since H is not factorial, we have $t(H) \geq 1$. We suppose that $\gcd(y, z) = 1$, from which the general case follows immediately. By definition of tameness there exists a factorization $z' \in Z(a)$ with $y \mid z'$ and $d(z, z') \leq t(H, y)$. Since $y \in Z(a)$ and $z' \in Z(a)$, $y \mid z'$ implies that $y = z'$ and thus

$$\sigma(z) \leq \max\{\sigma(y), \sigma(z)\} = d(z, y) \leq t(H, y).$$

By the previous remark it follows that

$$t(H, y) \leq \sigma(y) t(H). \quad \square$$

Proposition 3.3. *Let H be a tame monoid. Then H is a BF-monoid with finite elasticity and finite catenary degree. Moreover, if H is not factorial, then*

$$\max\{\varrho(H),\ c(H)\} \le t(H).$$

Proof. Suppose that H is not factorial. Let $a \in H$, $y \in \mathcal{Z}(a)$ with $\sigma(y) = \min L(a)$ and $z \in \mathcal{Z}(a)$ arbitrary. Then, by Lemma 3.2

$$\sigma(z) \le \min L(a) \cdot t(H)$$

which implies that

$$\varrho(a) = \frac{\sup L(a)}{\min L(a)} \le t(H).$$

Thus

$$\varrho(H) = \sup\{\varrho(a) \mid a \in H\} \le t(H).$$

Since H has finite elasticity, it is a BF-monoid.

We show that $c(a) \le t(H)$ holds for all $a \in H$, which implies that $c(H) \le t(H)$. We proceed by induction on $\max L(a)$. Clearly, the assertion is true for all $a \in H$ with $\max L(a) \le t(H)$. Let $a \in H$ and $z, z' \in \mathcal{Z}(a)$ two distinct factorizations. Suppose $z' = uy$ where $u \in \mathcal{A}(H)$ such that $u \nmid z$ and $y \in \mathcal{Z}(b)$ for some $b \in H$. There exists a factorization $z'' = ux \in \mathcal{Z}(a)$ with $d(z, z'') \le t(H)$. Since $\max L(b) < \max L(a)$, induction hypothesis gives a chain of factorizations $x = x_0,\ x_1, \ldots, x_k = y$ with $d(x_{i-1},\ x_i) \le t(H)$ for $1 \le i \le k$. Hence there is a $t(H)$-chain of factorizations from z to z'. $\square$

The following example shows that tame monoids need not be FF-monoids.

Example. Let us consider the domain $R = \mathbb{R} + X\mathbb{C}[X]$, which is a well studied example in the theory of non-unique factorizations. The arithmetic of domains of this type was first investigated in [A-A-Z1]. R is a one-dimensional noetherian domain with integral closure $\overline{R} = \mathbb{C}[X]$. Furthermore,

$$A = \{e^{i\varphi} X \mid 0 \le \varphi < \pi\} \cup \{p = 1 + Xf \mid p \in \mathbb{C}[X] \text{ irreducible}\}$$

is a system of pairwise non-associated representatives of irreducible elements of R. Hence $\mathcal{Z}(R) = \mathcal{Z}(R^\bullet/R^\times)$ is the free abelian monoid with basis A. This shows immediately that $t(R) = 2$ which implies that $c(R) = 2$ and that R is half-factorial. However, R is not an FF-domain.

Proposition 3.4. *Every finitely generated monoid is tame.*

Proof. Let H be a finitely generated monoid. Without restriction we may suppose that H is reduced. Obviously it is sufficient to show that H is locally tame. Set $\mathcal{A}(H) = \{u_1, \ldots, u_s\}$ and choose some $i \in \{1, \ldots, s\}$. By Dickson's theorem the set

$$S_i = \{\mathbf{k} \in \mathbf{N}_0^s \mid u_i \mid \prod_{j=1}^s u_j^{k_j}\} \subseteq \mathbf{N}_0^s$$

has only finitely many minimal points. Hence

$$H_i = \{\prod_{i=1}^s u_i^{m_i} \in H \mid \mathbf{m} \text{ is minimal in } S_i\}$$

is finite.

Let $a \in H$ with $u_i \mid a$ and let $z = u_1^{l_1} \ldots u_s^{l_s}$ be a factorization of a. Then there is some minimal point $\mathbf{m} \in S_i$ with $\mathbf{m} \leq \mathbf{l}$ and

$$z = u_1^{m_1} \ldots u_s^{m_s} \prod_{i=1}^s u_i^{l_i - m_i}.$$

Therefore, $t(H, u_i) = t(H_i, u_i)$ and $t(H_i, u_i) < \infty$ since H_i is finite. $\square$

Proposition 3.5. *Let H be a finitely primary monoid of rank s and exponent α. Then H is locally tame. H is tame if and only if $s = 1$. In this case, $t(H) \leq (3\alpha - 1)(2\alpha - 1)$.*

Proof. See [Ge7; Lemma 4.6 and Proposition 4.7]. $\square$

Theorem 3.6. *Let* $H \subseteq D = \coprod_{i \in I} D_i$ *be a saturated submonoid with bounded class group* G. *Suppose that all* D_i *are finitely primary of some exponent* $\alpha \in \mathbf{N}$ *and that* $\mathcal{D}(G_1) < \infty$ *with* $G_1 = \{g \in G \mid g \cap \mathcal{U}(D) \neq \emptyset\}$. *Then* H *is locally tame and the following conditions are equivalent:*

1. *All* D_i *have rank 1.*
2. H *is tame.*
3. H *has finite elasticity.*

Proof. H is locally tame by [Ge7; Proposition 4.7]. Suppose that all D_i are of rank 1. Then by Proposition 3.5 we infer that

$$t(D) = \sup\{t(D_i) \mid i \in I\} \leq (3\alpha - 1)(2\alpha - 1).$$

Then D is tame and hence H is tame by [Ge7; Proposition 4.5]. Therefore, the elasticity of H is finite by Proposition 3.3.

Conversely, suppose that for some $i \in I$

$$D_i \subseteq \widehat{D_i} = [p_1, \ldots, p_s] \times \widehat{D_i}^{\times}$$

has rank $s \geq 2$. Let $m \geq 2$. Then the elements

$$b_1 = (p_1 \ldots p_s)^{\alpha \exp(G)},$$

$$b_2 = \left(p_1 p_2^{m-1} \ldots p_s^{m-1}\right)^{\alpha \exp(G)}$$

and

$$b_3 = \left(p_1^{m-1} p_2 \ldots p_s\right)^{\alpha \exp(G)}$$

are in H and $\max L_H(b_i) \leq \alpha \exp(G)$ for $1 \leq i \leq 3$. Then for

$$a = b_1^m = b_2 b_3$$

we have $\sup L_H(a) \geq m$ and $\min L_H(a) \leq 2\alpha \exp(G)$ which implies that

$$t(H) \geq \varrho(H) \geq \varrho(a) \geq \frac{1}{2\alpha \, \exp(G)} \cdot m. \quad \square$$

Theorem 3.6 implies in particular, that Krull monoids with finite divisor class group have finite tame degree. Again block monoids make it possible to obtain sound lower and upper bounds for the tame degree.

Proposition 3.7. *Let* $H \subseteq D = \mathcal{F}(P) \times T$ *a saturated atomic submonoid of a reduced atomic monoid* D *with class group* G. *Let* $G_0 = \{[p] \in G \mid p \in P\}$, $\beta : H \to \mathcal{B}$ *the block homomorphism and* $\beta : \mathcal{Z}(H) \to \mathcal{Z}(\mathcal{B})$ *its canonical extension.*
Then for every subset $\emptyset \neq H' \subseteq H$ *and every* $u \in \mathcal{A}(H)$

$$t_{\mathcal{B}}\big(\beta(H'), \beta(u)\big) \leq t_H(H', u) \leq \max\{t_{\mathcal{B}}\big(\beta(H'), \beta(u)\big), \mathcal{D}(G_0) + \vartheta\}$$

where $\vartheta \in \{0,1\}$ *and* $\vartheta = 0$ *if* $T = \{1\}$. *In particular, if* H *is (locally) tame, then* $\mathcal{B}$ *is (locally) tame.*

Proof. The left inequality follows from Remark 5 after Definition 3.1. To obtain the right inequality, let $\emptyset \neq H' \subseteq H$, $a \in H'$, $z \in \mathcal{Z}(a)$ and $u \in \mathcal{A}(H)$ with $u \mid a$. Without restriction we may suppose that $u \nmid z$ in $\mathcal{Z}(H)$. We set $Z = \beta(z)$ and $U = \beta(u)$.
Case 1. $U \nmid Z$. There exists some $Z' \in \mathcal{Z}\big(\beta(a)\big)$ with $d(Z, Z') \leq t_{\mathcal{B}}\big(\beta(H'), \beta(u)\big)$ and $U \mid Z'$ in $\mathcal{Z}(\mathcal{B})$. Say

$$Z = YV_1 \ldots V_r \quad \text{and} \quad Z' = YU_1 \ldots U_s$$

with V_i, $U_j \in \mathcal{A}(\mathcal{B})$, $\{V_1, \ldots, V_r\} \cap \{U_1, \ldots, U_s\} = \emptyset$, $d(Z, Z') = \max\{r, s\}$ and $U_1 = U$.

We may choose $y \in \mathcal{Z}(H)$ and $u_2, \ldots, u_s, v_1, \ldots, v_r \in \mathcal{A}(H)$ such that $\beta(y) = Y$, $\beta(u_i) = U_i$, $\beta(v_j) = V_j$ and $z = yv_1 \ldots v_r$, $z' = yuu_2 \ldots u_s \in \mathcal{Z}(a)$. Clearly $d(z, z') = d(Z, Z')$.
Case 2. $U \mid Z$. Say

$$z = v_1 \ldots v_r, \quad Z = V_1 \ldots V_r$$

with $v_i \in \mathcal{A}(H), \beta(v_i) = V_i$ for $1 \leq i \leq k$ and $V_1 = U$. Suppose $u = p_1 \ldots p_k \alpha$ with $p_i \in P$ and $\alpha \in T$. Then $\alpha \mid v_1$ in $\mathcal{F}(P) \times T$. Without restriction we may suppose that

$$u \mid v_1 v_2 \ldots v_{k+\vartheta} \quad \text{in} \quad \mathcal{F}(P) \times T$$

By exchanging primes belonging to the same classes we obtain

$$v_1 v_2 \ldots v_{k+\vartheta} = uu_2 \ldots u_{k+\vartheta}$$

with $u_i \in \mathcal{A}(H)$ such that $\beta(u_i) = \beta(v_i)$ for $1 \leq i \leq k + \vartheta$.
Setting $z' = u_1 \ldots u_{k+\vartheta} v_{k+\vartheta+1} \ldots v_r$ we infer that

$$d(z, z') \leq k + \vartheta \leq \mathcal{D}(G_0) + \vartheta. \quad \square$$

Let G be an abelian group and $G_0 \subseteq G$ a subset. In the sequel we write $t(G_0)$ instead of $t(\mathcal{B}(G_0))$. The next result sharpens Proposition 4.5 in [Ge7].

Theorem 3.8. *Let H be a Krull monoid with divisor class group G and let $G_0 \subseteq G$ denote the set of classes containing prime divisors. Then we have*

1. *$t(G_0) \leq t(H) \leq \max\{t(G_0), \mathcal{D}(G_0)\}$.*
2. *$t(G_0) \leq 1 + \frac{\mathcal{D}(G_0)(\mathcal{D}(G_0)-1)}{2}$.*
3. *If $G_0 = -G_0$ and $\mathcal{D}(G_0) > 2$, then $\mathcal{D}(G_0) \leq t(G_0) = t(H)$.*

Proof. **1.** follows immediately from the previous proposition.

2. Let $A \in \mathcal{B}(G_0)$, $U = \prod_{i=1}^{k} g_i \in \mathcal{B}(G_0)$ irreducible and $Z = V_1 \ldots V_r$ a factorization of A. If $k = 1$, then U is prime and hence $t(\mathcal{B}(G_0), U) = 0$. Therefore suppose that $k \geq 2$. After a suitable renumeration we have $U | V_1 \ldots V_l$ where $\sigma(V_i) \geq 2$ for $1 \leq i \leq l \leq \min\{k, r\}$. Then

$$V_1 \ldots V_l = UW$$

for some $W \in \mathcal{B}(G_0)$ with

$$\sigma(W) = \sum_{i=1}^{l} \sigma(V_i) - \sigma(U) \leq \sigma(U)(\mathcal{D}(G_0) - 1)$$

$$\leq \mathcal{D}(G_0)(\mathcal{D}(G_0) - 1).$$

Let Y be a factorization of W. Then

$$\sigma(Y) \leq \frac{1}{2}\sigma(W) \leq \frac{1}{2}\mathcal{D}(G_0)(\mathcal{D}(G_0) - 1).$$

Obviously,

$$Z' = UYV_{l+1} \ldots V_r$$

is a factorization of A and

$$t(\mathcal{B}(G_0), U) \leq d(Z, Z') \leq \max\{l, 1 + \sigma(Y)\} \leq 1 + \frac{1}{2}\mathcal{D}(G_0)(\mathcal{D}(G_0) - 1).$$

3. Let $G_0 = -G_0$, $\mathcal{D}(G_0) > 2$ and $U = \prod_{i=1}^{k} g_i \in \mathcal{B}(G_0)$ an irreducible block with $\sigma(U) = k = \mathcal{D}(G_0)$. Set $A = -U \cdot U$, $V_i = -g_i \cdot g_i$ for $1 \leq i \leq k$ and $Z = V_1 \ldots V_k$. Then $Z' = -U \cdot U$ is the only factorization of A containing U and thus

$$t(G_0) \geq t(\mathcal{B}(G_0), U) = d(Z, Z') = \mathcal{D}(G_0).$$

Finally, **1.** implies that $t(H) = t(G_0)$. $\square$

The rest of this section is devoted to a detailed study of the tame degree of elementary 2-groups. We shall prove the following result.

Theorem 3.9. $t(C_2) = 0$ *and for every* $r \geq 2$ *we have*

$$1 + \frac{r(r-1)}{2} \leq t(C_2^r) \leq 1 + \frac{r^2}{2}.$$

Furthermore, equality holds on the right hand side for all even r.

We do the proof in a series of lemmata. Throughout, we consider C_2^r as a vector space over the field $\mathbb{F}_2$ with two elements. Recall that $\mathcal{D}(C_2^r) = r + 1$.

Lemma 3.10. *Let* $G = C_2^r$ *with* $r \geq 1$ *and* $S \in \mathcal{F}(G)$. *Then* S *is an irreducible block with* $\sigma(S) = \mathcal{D}(G)$ *if and only if* $S = \prod_{i=0}^r e_i$ *where* $\{e_1, \ldots, e_r\}$ *is a basis of* G *and* $e_0 = \sum_{i=1}^r e_i$.

Proof. If $\{e_1, \ldots, e_r\}$ is a basis and $e_0 = \sum_{i=1}^r e_i$, then of course $S = \prod_{i=0}^r e_i$ is an irreducible block with $\sigma(S) = r + 1 = \mathcal{D}(G)$.

Conversely, suppose $S = \prod_{i=0}^r e_i$ is an irreducible block. Then by Lemma 2.12 $\{e_1, \ldots, e_r\}$ generate G as a $\mathbb{Z}$-module and hence it is a generating system over $\mathbb{F}_2$. Since G has rank r, $\{e_1, \ldots, e_r\}$ is an $\mathbb{F}_2$-basis of G. $\square$

A sequence

$$S = \prod_{i=1}^r e_i \in \mathcal{F}(C_2^r) \quad \text{with} \quad r \geq 1$$

is called a *basis sequence*, if $\{e_1, \ldots, e_r\}$ is a basis of C_2^r.

Lemma 3.11. $t(C_2^r) \geq 1 + \frac{r^2}{2}$ *for every even* $r \geq 4$.

Proof. Let $r \geq 4$ even and $G = C_2^r$. We construct a block $A \in \mathcal{B}(G)$, a factorization $Z \in \mathcal{Z}(A)$ and an irreducible block U with $U \mid A$. Then we show that for every factorization $Z' \in \mathcal{Z}(A)$ with $U \mid Z'$

$$d(Z, Z') \geq 1 + \frac{r^2}{2}$$

holds. This implies the assertion.

Let $e_1, \ldots, e_r$ be a basis of G and set $e_0 = \sum_{i=1}^r e_i$. Define

$$U = e_0 \prod_{i=1}^r (e_0 + e_i), \quad V_0 = \prod_{j=0}^r e_j$$

and

$$V_i = (e_0 + e_i) \prod_{\substack{j=1 \\ j \neq i}}^{r} e_j \quad \text{for} \quad 1 \leq i \leq r.$$

Clearly, these are irreducible blocks. Then

$$Z = \prod_{j=0}^{r} V_j$$

and

$$Z' = U \prod_{i=1}^{r} (e_i \cdot e_i)^{r/2}$$

are factorizations of some block $A \in \mathcal{B}(G)$. Obviously, Z' is the only factorization of A containing U and

$$d(Z, Z') = \max \left\{ r+1, \; 1 + r \cdot \frac{r}{2} \right\} = 1 + \frac{r^2}{2}. \quad \square$$

Lemma 3.12. $t(C_2^r) \geq 1 + \frac{r(r-1)}{2}$ *for every odd* $r \geq 3$.

Proof. Let $r \geq 3$ odd, $G = C_2^r$, $e_1, \ldots, e_r$ a basis of G and $e_0 = \sum_{i=1}^{r} e_i$. We proceed in the same way as in the previous lemma. Define

$$U = \prod_{i=1}^{r} (e_0 + e_i)$$

and

$$V_i = (e_0 + e_i) \prod_{\substack{j=1 \\ j \neq i}}^{r} e_j \quad \text{for} \quad 1 \leq i \leq r.$$

These are irreducible blocks and

$$Z = \prod_{j=1}^{r} V_j \,,$$

$$Z' = U \prod_{i=1}^{r} (e_i \cdot e_i)^{\frac{r-1}{2}}$$

are factorizations of some block $A \in \mathcal{B}(G)$. Again, Z' is the only factorization of A containing U and

$$d(Z, Z') = \max\{r, 1 + \frac{r(r-1)}{2}\} = 1 + \frac{r(r-1)}{2}. \quad \square$$

Let G be a finite abelian group. A sequence $S \in \mathcal{F}(G)$ is called *suitable*, if either

> there is some irreducible block $U \in \mathcal{B}(G)$
> with $\sigma(U) \geq 4$ and $U|S$

or

> there are irreducible blocks $U_1, U_2 \in \mathcal{B}(G)$
> with $\sigma(U_1) = \sigma(U_2) = 3$ and $U_1 U_2 | S$.

Lemma 3.13. *Let* $G = C_2^r$ *with* $r \geq 3$ *and* A, B *two distinct basis sequences. Then either* AB *is suitable or* $A = a_1 \cdot (a_1 + b_1) \cdot T$, $B = b_1 \cdot (a_1 + b_1) \cdot T$ *with* $a_1, b_1 \in G$ *and* $T \in \mathcal{F}(G)$.

Proof. Let $S = \gcd(A, B)$ and $k = r - \sigma(S)$. So $k \in \{1, \ldots, r\}$, $A = \prod_{i=1}^{k} a_i \cdot S$, $B = \prod_{j=1}^{k} b_j \cdot S$ with $a_i, b_j \in G$ and $\{a_1, \ldots, a_k\} \cap \{b_1, \ldots, b_k\} = \emptyset$.

If $k = 1$ and $a_1 + b_1 | S$, then the second condition holds. Hence it remains to consider the following two cases.

Case 1. There exist $i, j \in \{1, \ldots, k\}$ such that $a_i + b_j \nmid S$.

We may assume without restriction that $i = j = 1$ and that $a_1 + b_1 \nmid A$. Since $A = \prod_{i=1}^{r} a_i$ is a basis sequence, there exists a set $I \subseteq \{1, \ldots, r\}$ such that

$$a_1 + b_1 = \sum_{i \in I} a_i$$

and hence

$$U = a_1 \cdot b_1 \cdot \prod_{i \in I} a_i \in \mathcal{B}(G).$$

Since $a_1 + b_1 \nmid A$, it follows that $\#I \geq 2$. If $1 \notin I$, then U is irreducible with $\sigma(U) \geq 4$ and $U | AB$.

Suppose $1 \in I$ and set $J = I \setminus \{1\}$. Then

$$V = b_1 \cdot \prod_{j \in J} a_j \in \mathcal{B}(G).$$

Since $b_1 \nmid A$, V is irreducible and $\#J \geq 2$. If $\#J \geq 3$, we are done.

Hence we suppose $J = \{i, j\} \subseteq \{2, \ldots, r\}$ which implies $b_1 = a_i + a_j$. Since B is a basis sequence and $a_1 \nmid B$, there is a subset $\Omega \subseteq \{1, \ldots, r\}$ such that

$$W = a_1 \cdot \prod_{\nu \in \Omega} b_\nu$$

is an irreducible block. If $\#\Omega \geq 3$, we are done. If $\#\Omega = 2$ and $1 \notin \Omega$, then

$$U_1 = b_1 \cdot a_i \cdot a_j \quad \text{and} \quad U_2 = a_1 \cdot \prod_{\nu \in \Omega} b_\nu$$

are two irreducible blocks with size 3 such that $U_1 U_2 \mid AB$.

Finally consider the case that $\Omega = \{1, l\}$. Then

$$a_1 = b_1 + b_l = a_i + a_j + b_l,$$

and thus

$$S = b_l \cdot a_1 \cdot a_i \cdot a_j$$

is an irreducible block dividing AB and with $\sigma(S) = 4$.

Case 2. $k \geq 2$ and for all $i, j \in \{1, \ldots, k\}$ we have $a_i + b_j \mid S$.

If $a_1 + b_1 \neq a_2 + b_2$, then

$$U_1 = a_1 \cdot b_1 \cdot (a_1 + b_1) \quad \text{and} \quad U_2 = a_2 \cdot b_2 \cdot (a_2 + b_2)$$

are two irreducible blocks of size 3 with $U_1 U_2 \mid AB$.
If $a_1 + b_1 = a_2 + b_2$, set

$$U = a_1 \cdot a_2 \cdot b_1 \cdot b_2.$$

Then U is an irreducible block dividing AB with $\sigma(U) = 4$. $\square$

Lemma 3.14. *Let* $G = C_2^r$ *with* $r \geq 3$ *and* A, B, C *pairwise distinct basis sequences. Then either*

$$A = a_1 \cdot (a_1 + b_1) \cdot T, \; B = b_1 \cdot (a_1 + b_1) \cdot T, \; C = a_1 \cdot b_1 \cdot T$$
with $a_1, b_1 \in G, \; T \in \mathcal{F}(G)$ *and* ABC *is suitable*

or

    one of the sequences AB, AC, BC *is suitable.*

Furthermore, if A_1, A_2, A_3, A_4 *are pairwise distinct basis sequences, then* $A_i A_j$ *is suitable for some* $1 \leq i < j \leq 4$.

Proof. Suppose that none of the sequences AB, AC and BC is suitable. Then by Lemma 3.13, $A = a_1 S, B = b_1 S$ with $S = \gcd(A, B) = (a_1 + b_1)T$ and $A = a \cdot \gcd(A, C), C = c \cdot \gcd(A, C)$ where $a_1, b_1, a, c \in G$ and $T \in \mathcal{F}(G)$.

Assume to the contrary that $a = a_1$. This implies that $S = \gcd(A, B) = \gcd(A, C)$ and hence

$$A = aS, \; C = cS, \; B = b_1 S.$$

By Lemma 3.13 we obtain that $a + c \,|\, C$. Since $B \neq C$, it follows that $c \neq b_1$. Since C is a basis sequence with $\sigma(C) = r \geq 3$, we infer that $c \neq a + b_1$. Thus $c, a + b_1, a + c$ are pairwise distinct and hence

$$c \cdot (a + b_1) \cdot (a + c) \,|\, C.$$

Therefore

$$U = b_1 \cdot c \cdot (a + b_1) \cdot (a + c)$$

is an irreducible block dividing BC, a contradiction to BC not suitable.

This shows that $a \neq a_1$ holds. Hence for some $V \in \mathcal{F}(G)$ we have

$$A = aa_1 V, \; B = ab_1 V \quad \text{and} \quad C = ca_1 V.$$

Since $\gcd(B, C) = \gcd(ab_1, \, ca_1)V$ and, again by Lemma 3.13, $\sigma(\gcd(B, C)) = r - 1$, it follows that $\gcd(ab_1, \, ca_1) \in \{a, b_1\} \cap \{c, a_1\}$ and hence

$$\gcd(ab_1, \, ca_1) = b_1 = c.$$

Since C is a basis sequence, $a_1 + b_1 \nmid V$. However, $a_1 + b_1 \mid A$ and thus $a = a_1 + b_1$. This implies that

$$V = T \quad \text{and} \quad C = a_1 \cdot b_1 \cdot T.$$

Setting

$$U_1 = U_2 = a_1 \cdot b_1 \cdot (a_1 + b_1)$$

we have that $U_1 U_2 \mid ABC$.

The statement on the four basis sequences is an immediate consequence of the previous assertion. □

Lemma 3.15. *Let $G = C_2^r$ with $r \geq 3$ and $A_1, \ldots, A_l \in \mathcal{F}(G)$ pairwise distinct basis sequences with $1 \leq l \leq r+1$. Then there exist $k \leq l-1$ irreducible blocks $U_1, \ldots, U_k$ with $U_1 \ldots U_k \mid A_1 \ldots A_l$ and $\sigma(U_1 \ldots U_k) \geq 3(l-1)$.*

Proof. We argue by induction on l. Obviously, the assertion is true for $1 \leq l \leq 2$.

Suppose $l = 3$. By Lemma 3.14 $A_1 A_2 A_3$ is suitable. So the only case which remains to be considered is that there is one irreducible block U dividing $A_1 A_2 A_3$ with $4 \leq \sigma(U) \leq 5$. However, since

$$\sigma(A_1 A_2 A_3) - \sigma(U) \geq 3r - 5 \geq r + 1,$$

there is an irreducible block V such that $UV \mid A_1 A_2 A_3$ and clearly $\sigma(UV) \geq 6$.

Suppose $l = 4$. We have to find $k \leq 3$ irreducible blocks $U_1, \ldots, U_k$ with $\sigma(U_1 \ldots U_k) \geq 9$ such that $U_1 \ldots U_k \mid A_1 \ldots A_l$. By Lemma 3.14 there exist two basis sequences, say A_1 and A_2, such that $A_1 A_2$ is suitable. Clearly, there is at least one irreducible block with size greater than or equal to three dividing $A_3 A_4$. So it remains to consider the case that there are irreducible blocks U, V with $U \mid A_1 A_2$, $V \mid A_3 A_4$ and $(\sigma(U), \sigma(V)) \in \{(4,3), (4,4), (5,3)\}$. Again, since

$$\sigma(A_1 A_2 A_3 A_4) - \sigma(U) - \sigma(V) \geq 4r - 8 \geq r + 1,$$

there is an irreducible block $W \in \mathcal{B}(G)$ such that $UVW \mid A_1 \ldots A_4$ and $\sigma(UVW) \geq 9$.

Let $l \geq 5$ and assume that the assertion is true for all smaller l. As in the case $l = 4$ we may suppose that $A_1 A_2$ is suitable. By induction hypothesis there are $U_3, \ldots, U_k$, $k \leq l - 1$, with $U_3 \ldots U_k \mid A_3 \ldots A_l$ and $\sigma(U_3 \ldots U_k) = 3(l-3) + \lambda$ for some $\lambda \geq 0$. If there exist two irreducible blocks U_1, U_2 with $\sigma(U_1) = \sigma(U_2) = 3$ and $U_1 U_2 \mid A_1 A_2$, or an irreducible block U with $\sigma(U) \geq 6$ and $U \mid A_1 A_2$, or if $\lambda \geq 2$, then the assertion follows.

Thus it remains to consider the case that $4 \leq \sigma(U) \leq 5$ and $0 \leq \lambda \leq 1$. Then we have

$$\sigma(A_1 \ldots A_l) - \sigma(U) - \sigma(U_3 \ldots U_k) \geq$$
$$lr - 5 - 3(l-3) - 1 = l(r-3) + 3 \geq$$
$$5r - 12 \geq r + 1,$$

since $l \geq 5$ implies that $r \geq 4$. Hence there is an irreducible block $V \in \mathcal{B}(G)$ such that $UVU_3 \ldots U_k \mid A_1 \ldots A_l$ and

$$\sigma(UVU_3 \ldots U_k) \geq 6 + 3(l-3) = 3(l-1). \quad \square$$

Lemma 3.16. *Let* $G = C_2^r$ *with* $r \geq 3$. *Then* $t(C_2^r) \leq 1 + \frac{r^2}{2}$.

Proof. Let $A \in \mathcal{B}(G)$,

$$Z = V_1 \ldots V_m \in \mathcal{Z}(A)$$

a factorization of A and $U \in \mathcal{B}(G)$ irreducible with $U \mid A$. We have to find a factorization $Z' \in \mathcal{Z}(A)$ with $U \mid Z'$ and

$$d(Z, Z') \leq 1 + \frac{r^2}{2}.$$

Without restriction we may suppose that

$$U \mid \prod_{i=1}^{n} V_i \quad \text{with} \quad n \leq \sigma(U)$$

and

$$U \nmid \prod_{\substack{i=1 \\ i \neq j}}^{n} V_i \quad \text{for any} \quad j \in \{1, \ldots, n\}.$$

Then

$$\prod_{i=1}^{n} V_i = UW$$

and we have to find a factorization $Z'' \in \mathcal{Z}(W)$ with $\sigma(Z'') \leq \frac{r^2}{2}$. Then

$$Z' = U \cdot Z''V_{n+1} \ldots V_m \in \mathcal{Z}(A)$$

and

$$d(Z, Z') \leq \max\{n, \, 1 + \sigma(Z'')\} \leq 1 + \frac{r^2}{2}.$$

Clearly,

$$\sigma(W) = \sum_{i=1}^{n} \sigma(V_i) - \sigma(U) \leq \sum_{i=1}^{n} (\sigma(V_i) - 1) \leq n \cdot r$$

If $\sigma(W) \leq r^2$, then we are done. So suppose that $\sigma(W) > r^2$. Then $r + 1 = n = \sigma(U)$ and there is some $i \in \{1, \ldots r + 1\}$ with $\sigma(V_i) = r + 1$. Set $U = \prod_{i=1}^{r+1} g_i$ and suppose that $g_i \,|\, V_i$ for $1 \leq i \leq r + 1$. Furthermore, suppose that

$$V_i = g_i A_i \quad \text{for} \quad 1 \leq i \leq l$$

with basis sequences A_i (here Lemma 3.10 comes in) and that

$$\sigma(V_i) \leq r \quad \text{for} \quad l + 1 \leq i \leq r + 1.$$

By Lemma 3.15 there exist $k \leq l - 1$ irreducible blocks $U_1, \ldots, U_k$ with $\sigma(U_1 \ldots U_k) \geq 3(l - 1)$ and $U_1 \ldots U_k | A_1 \ldots A_l$. Thus

$$W = U_1 \ldots U_k W'$$

for some block $W' \in \mathcal{B}(G)$. Then

$$\begin{aligned}
\sigma(W') &= \sigma(W) - \sigma(U_1 \ldots U_k) \\
&= \sigma(V_1 \ldots V_{r+1}) - \sigma(U) - \sigma(U_1 \ldots U_k) \\
&\leq lr + (r + 1 - l)(r - 1) - 3(l - 1) \\
&= r^2 - 2(l - 1).
\end{aligned}$$

Therefore W has a factorization Z'' with

$$\sigma(Z'') \leq k + \frac{1}{2}\sigma(W') \leq \frac{r^2}{2},$$

which completes the proof. $\square$

Proof of Theorem 3.9. Since $\mathcal{B}(C_2)$ is factorial, $t(C_2) = 0$. Next consider the case $r = 2$. By Theorem 3.8 it follows that

$$3 = \mathcal{D}(C_2^2) \leq t(C_2^2) \leq 1 + \frac{\mathcal{D}(C_2^2)(\mathcal{D}(C_2^2) - 1)}{2} = 4.$$

It is easy to check that indeed $t(C_2^2) = 3$ holds. For $r \geq 3$ the assertion follows from Lemma 3.11, Lemma 3.12 and Lemma 3.16. $\square$

4. FACTORIZATIONS OF LARGE ELEMENTS

There is the following structure theorem for sets of lengths in weakly Krull domains:

let R be a weakly Krull domain of finite type with finite t-class group. Then there exists some $M \in \mathbb{N}$ such that for all $a \in R^\bullet$ $L(a)$ is an almost arithmetical progression bounded by M (cf. [HK2; Theorem 6.4]).

Indeed, an arithmetical sharper result holds true involving chains of factorizations (cf. [Ge7; Theorem 6.2]). However, there are sets of lengths in the above domains which are no arithmetical progressions. As usual, we say a subset $L \subseteq \mathbb{Z}$ is an *arithmetical progression* (with distance $d \in \mathbb{N}$), if $L = \{m, m+d, m+2d, \ldots, m+\kappa d\}$ for some $m \in \mathbb{Z}$ and some $\kappa \in \mathbb{N}_0$.

In this section we study factorizations of large elements i.e., multiples of some fixed element. The arithmetical significance of such results will become more clear in the next section, where we deal with density theorems for factorizations (see Theorems 5.3 and 5.4). A general prototype of results we want to achieve is the following Theorem 4.1 whose proof might be found in [Ge7; Theorem 5.1].

Theorem 4.1. *Let H be a reduced atomic monoid which is locally tame and has finite catenary degree.*

Then there exists some element $a^* \in H$ and some constant $C \in \mathbb{N}$ such that for all $a \in H$ with $a^* | a$ the following holds: for all

z, $z' \in \mathcal{Z}(a)$ there is a C-chain of factorizations $(z_i)_{i=-1}^{k+1}$ from $z = z_{-1}$ to $z_{k+1} = z'$ such that $|\sigma(z_i) - \sigma(z_{i-1})| \in \{0, \min\Delta(H)\}$ for $1 \leq i \leq k$.

The above result has the following consequence for the structure of sets of lengths. Let $a \in H$ be a multiple of a^* and let $z, z' \in \mathcal{Z}(a)$ with $\min L(a) = \sigma(z)$, $\max L(a) = \sigma(z')$ and let $(z_i)_{i=-1}^{k+1}$ have the above properties. Then $\{\sigma(z_i) \,|\, 0 \leq i \leq k\}$ is an arithmetical progression with distance $d = \min\Delta(H)$ and

$$L(a) = \{l_1, \ldots, l_\alpha, m, m + d, \ldots, m + \kappa d, n_1, \ldots, n_\beta\}$$

with $l_1 < \cdots < l_\alpha < m \leq m + \kappa d < n_1 < \cdots < n_\beta$, $m = \min\{\sigma(z_0), \sigma(z_k)\}$, $m + \kappa d = \max\{\sigma(z_0), \sigma(z_k)\}$, $\alpha \leq C - 2$ and $\beta \leq C - 2$ (cf. [Ge7; Corollary 5.2).

Under such general assumptions as in Theorem 4.1 no sharper result can be achieved. The next result for Krull monoids, proved in [Ge6; Theorem 3.1], is as sharp as possible. It implies in particular that sets of lengths of large elements are arithmetical progressions with distance 1 (cf. Lemma 2.4).

Theorem 4.2. *Let H be a reduced Krull monoid with divisor theory $H \hookrightarrow \mathcal{F}(P)$, finite divisor class group G and let $\beta : \mathcal{F}(P) \to \mathcal{F}(G)$ denote the associated block homomorphism. Suppose that each class contains a prime divisor.*

Then there exists some element $A^* \in \mathcal{B}(G)$ such that for all $a \in H$ with $A^* \,|\, \beta(a)$ the following holds: for all $z, z' \in \mathcal{Z}(a)$ there is a 3-chain of factorizations $(z_i)_{i=0}^{k}$ from $z = z_0$ to $z_k = z'$.

In this section we shall prove the following two theorems.

Theorem 4.3. *Let $H \subseteq D = \mathcal{F}(P) \times T$ be a saturated atomic submonoid of a reduced atomic monoid D with finite class group G and let $\beta : \mathcal{F}(P) \times T \to \mathcal{F}(G) \times T$ denote the associated block homomorphism. Suppose that $\#G \geq 3$, $g \cap P \neq \emptyset$ for every $g \in G$ and that H is locally tame with finite catenary degree.*

Then there exist some element $A^* \in \mathcal{B}(G)$ and some constant $C \in \mathbf{N}$ such that for all $a \in H$ with $A^* \,|\, \beta(a)$ the following holds: for all $z \in \mathcal{Z}(a)$ there exist some $z' \in \mathcal{Z}(a)$ with $\sigma(z') = \max L(a)$ and a C-chain of factorizations $(z_i)_{i=-1}^{k}$ from $z = z_{-1}$ to $z_k = z'$ such that $|\sigma(z_i) - \sigma(z_{i-1})| \leq 1$ for $1 \leq i \leq k$.

Theorem 4.4. *Let $H \subseteq D = \mathcal{F}(P) \times T$ be a saturated atomic submonoid of a reduced atomic monoid D with finite class group G and let $\beta : \mathcal{F}(P) \times T \to \mathcal{F}(G) \times T$ denote the associated block homomorphism. Suppose that $\#G \geq 3$, $g \cap P \neq \emptyset$ for every $g \in G$ and that T is a finite product of finitely primary monoids where each monoid has rank greater than 1.*

Then there exist some element $A^* \in \mathcal{B}(G)$ and some constant $C \in \mathbf{N}$ such that for all $a \in H$ with $A^* | \beta(a)$ the following holds: for all $z \in \mathcal{Z}(a)$ with $\sigma(z) = \min L(a)$ there exist some $z' \in \mathcal{Z}(a)$ with $\sigma(z') = \max L(a)$ and a C-chain of factorizations $(z_i)_{i=0}^k$ from $z = z_0$ to $z_k = z'$ such that $|\sigma(z_i) - \sigma(z_{i-1})| \leq 1$ for $1 \leq i \leq k$.

We discuss the significance of these two results and their relationship. Let all assumptions be as in Theorem 4.3 and let $a \in H$ be such that $A^* | \beta(a)$. Arguing as above, we infer that

$$L(a) = \{l_1, \ldots, l_\alpha, m, m+1, \ldots, m+\kappa\}$$

with $l_1 < \cdots < l_\alpha < m$, $\kappa \in \mathbf{N}_0$ and $\alpha \leq C - 2$. Theorem 4.3 has its model in [Ge3] where examples are given showing the necessity of the assumptions. In particular, there are examples - one will be given below - showing that in general $L(a)$ is not an arithmetical progression with distance 1.

Next, suppose the situation of Theorem 4.4. Then, by Theorem 2.7 H has finite catenary degree and by Theorem 3.6 H is locally tame. Hence the assumptions of 4.3 are satisfied. However, Theorem 4.4 implies that for all $a \in H$ with $A^* | \beta(a)$ the sets of lengths $L(a)$ are arithmetical progressions with distance 1. Clearly, the assertion of 4.4 is much stronger than just a statement on sets of lengths.

We give an example of a simple monoid H satisfying the assumptions of 4.3 and all assumptions of 4.4 except the assumption concerning the ranks of the finitely primary monoids. We show that the assertion of 4.4 does not hold in H.

Example. Let $G = \mathbf{Z}/3\mathbf{Z} = \{\bar{0}, \bar{1}, \bar{2}\}$ and $T = [3, 10, 11] \subseteq (\mathbf{N}_0, +)$ be the numerical monoid generated by 3, 10 and 11. Clearly, T is a reduced finitely primary monoid of rank 1. Define a content homomorphism

$$\iota : \mathcal{F}(G) \times T \to G$$

by $\iota(g) = g$ for all $g \in G$, $\iota(3) = \bar{0}$, $\iota(10) = \bar{2}$ and $\iota(11) = \bar{1}$. We study the monoid

$$H = \mathrm{Ker}(\iota) \subseteq \mathcal{F}(G) \times T.$$

Then $H \subseteq \mathcal{F}(G) \times T$ is a saturated submonoid with class group isomorphic to G and each class contains a prime divisor. Clearly, $\mathcal{B}(G) \subseteq H$ is a divisor closed submonoid and $1 = \min\Delta(H)$.

H is reduced and finitely generated with $\mathcal{A}(H) = \{u_i \mid 0 \leq i \leq 8\}$ where

$$u_0 = \bar{0}, \ u_1 = \bar{1}^3, \ u_2 = \bar{2}^3, \ u_3 = \bar{1} \cdot \bar{2}, \ u_4 = \bar{1} \cdot 10,$$
$$u_5 = \bar{1}^2 \cdot 11, \ u_6 = \bar{2} \cdot 11, \ u_7 = \bar{2}^2 \cdot 10, \ u_8 = 3.$$

Clearly, u_0 is the only prime element.

We assert that there exists no $a^* \in H$ such that for every $a \in H$ with $a^* \mid a$ the set of lengths $L(a)$ is an arithmetical progression with distance 1. For this we define for $k \geq 2$ an element

$$a_k = u_4^k \, u_6^k \in H$$

and show that

$$\min L(a_k) = 2k \quad \text{but} \quad 2k + 1 \notin L(a_k).$$

This implies the assertion as follows. Let $a^* \in H$ be given. Then $a^* = u_0^r b$ for some $b \in H$ and some $r \in \mathbf{N}_0$. Obviously, there exists some $k \geq 2$ such that $a^* \mid u_0^r a_k$ and $L(u_0^r a_k) = r + L(a_k)$ fails to be an arithmetical progression.

Let $k \geq 2$. Then

$$L(a_k) = \left\{ \sum_{i=1}^{8} l_i \in \mathbf{N} \,\middle|\, a_k = \prod_{i=1}^{8} u_i^{l_i} \right\}.$$

Clearly, $a_k = \prod_{i=1}^{8} u_i^{l_i}$ for some $1 = (l_i)_{i=1}^{8} \in \mathbf{N}_0^8$ if and only if 1 satisfies the following system of equations:

$$\begin{array}{ll} \text{I} & 3l_1 + l_3 + l_4 + 2l_5 = k \\ \text{II} & 3l_2 + l_3 + l_6 + 2l_7 = k \\ \text{III} & 10l_4 + 11l_5 + 11l_6 + 10l_7 + 3l_8 = 21k \end{array}$$

Let $1 \in \mathbb{N}_0^8$ be a solution with $\sum_{i=1}^8 l_i \le 2k+1$. It is sufficient to verify that $\sum_{i=1}^8 l_i = 2k$.

Since

$$l_4 + 2l_5 \le k, \quad l_6 + 2l_7 \le k$$

and

$$5(l_4 + 2l_5) + 5(l_6 + 2l_7) + 5l_4 + l_5 + 6l_6 + 3l_8 = 21k,$$

it follows that

$$5l_4 + l_5 + 6l_6 + 3l_8 \ge 11k.$$

Using $l_6 = k - 3l_2 - l_3 - 2l_7$ we infer that

$$-3l_2 - l_3 + 5l_4 + l_5 + 5l_6 - 2l_7 + 3l_8 \ge 10k.$$

Therefore,

$$10k + 5 \ge 5 \sum_{i=1}^8 l_i \ge -3l_2 - l_3 + 5l_4 + l_5 + 5l_6 - 2l_7 + 3l_8 \ge 10k$$

and hence

$$5 \ge 5l_1 + 8l_2 + 6l_3 + 4l_5 + 7l_7 + 2l_8. \tag{$*$}$$

This implies that $l_1 + l_5 \le 1$ and $l_2 = l_3 = l_7 = 0$. From (II) it follows that $l_6 = k$. Assume to the contrary, that $l_1 + l_5 = 1$. Then $(*)$ implies that $l_8 = 0$. By (I) we infer that $l_4 \in \{k-3, k-2\}$ and thus (III) gives

$$21k = 10l_4 + 11l_5 + 11k \le (10k - 20) + 11 + 11k,$$

a contradiction. Therefore we obtain $l_1 = l_5 = 0$. Thus $l_4 = k$, $l_8 = 0$ and $\sum_{i=1}^8 l_i = 2k$.

We start with the preparations for the proof of the Theorems 4.3 and 4.4. Let $H \subseteq \mathcal{F}(P) \times T$ be as in Theorem 4.3, G the class group and $\mathcal{B} \subseteq \mathcal{F}(G) \times T$ the associated block monoid. By Proposition 4.2 in [Ge5] it is sufficient to prove the assertions for the block monoid $\mathcal{B}$ instead of H. By Propositions 2.8 and 3.7 $\mathcal{B}$ is locally tame and has finite catenary degree $c(\mathcal{B}) = \mathfrak{c}$.

Lemma 4.5. *For every* $l \in \mathbf{N}$ *there exists a block* $A_l \in \mathcal{B}(G)$ *with* $\max L(A_l) = 3l$ *and factorizations* $x_0, x_1, \ldots, x_l \in \mathcal{Z}(A_l)$ *with* $\sigma(x_\nu) = 2l + \nu$ *for* $0 \leq \nu \leq l$.

Proof. Let $l \in \mathbf{N}$. Since $\#G \geq 3$, we infer that $\mathcal{D}(G) \geq 3$. Hence there exists an irreducible block $U = g_1 g_2 g_3 \in \mathcal{B}(G)$. We set $A_l = (-U \cdot U)^l$ and

$$x_\nu = (-U \cdot U)^{l-\nu} \prod_{i=1}^{3} (-g_i \cdot g_i)^\nu$$

for $0 \leq \nu \leq l$. $\square$

Throughout the rest of this section, let

$$A_{\mathfrak{c}} = \prod_{i=1}^{3} (-g_i \cdot g_i)^{\mathfrak{c}} \in \mathcal{B}(G)$$

be as above and let $x_0, \ldots, x_{\mathfrak{c}} \in \mathcal{Z}(A_{\mathfrak{c}})$ have the properties of Lemma 4.5.

Lemma 4.6. *If* $a \in \mathcal{B}$ *with* $\prod_{i \in I} (-g_i \cdot g_i) \mid a$ *for a finite set* I *and elements* $g_i \in G$, *then there exists a factorization* $z \in \mathcal{Z}(a)$ *with* $\sigma(z) = \max L(a)$ *and* $\prod_{i \in I} (-g_i \cdot g_i) \mid z$ *in* $\mathcal{Z}(\mathcal{B})$.

Proof. We proceed by induction on $\#I$. If $I = \emptyset$, then there is nothing to show. Let $a \in \mathcal{B}$ with $\prod_{i \in I} (-g_i \cdot g_i) \mid a$,

$$z = \prod_{j \in J} (-g_j \cdot g_j) u_1 \ldots u_k \in \mathcal{Z}(a)$$

with $u_\nu \in \mathcal{A}(\mathcal{B})$, $J \subseteq I$ and $\sigma(z) = \max L(a)$. If $I \setminus J \neq \emptyset$, take some $i \in I \setminus J$. Then $-g_i \cdot g_i \mid u_1 \ldots u_k$ and after a suitable renumeration we have

$$u_1 = -g_i \cdot v_1 \ldots v_l$$
$$u_2 = g_i \cdot w_1 \ldots w_m$$

with $v_\nu, w_\mu \in \mathcal{A}(\mathcal{F}(G) \times T)$. Then $u_1 u_2 = (-g_i \cdot g_i) x$ with $x = v_1 \ldots v_l w_1 \ldots w_m$. Since $\sigma(z) = \max L(a)$, it follows that $x \in \mathcal{A}(\mathcal{B})$. Thus

$$z' = \prod_{j \in J} (-g_j \cdot g_j) \cdot (-g_i \cdot g_i) \cdot x u_3 \ldots u_k \in \mathcal{Z}(a),$$

and the assertion follows by induction hypothesis. $\square$

Lemma 4.7. *Let $a \in \mathcal{B}$ and $z = x_\mathfrak{c} \cdot y \in \mathcal{Z}(a)$ with $y \in \mathcal{Z}(b)$ for some $b \in \mathcal{B}$. Then there exists some $z' \in \mathcal{Z}(a)$ with $\sigma(z') = maxL(a)$ and a $4\mathfrak{c}$-chain of factorizations $(z_i)_{i=0}^k$ from $z = z_0$ to $z_k = z'$ such that $|\sigma(z_i) - \sigma(z_{i-1})| \leq 1$ for $1 \leq i \leq k$.*

Proof. By Lemma 4.6 there exists a factorization $z' \in \mathcal{Z}(a)$ with $\sigma(z') = maxL(a)$ and $z' = x_\mathfrak{c} \cdot y'$ for some $y' \in \mathcal{Z}(b)$. By definition of the catenary degree, there exists a chain of factorizations $(y_i)_{i=0}^k$ from $y = y_0$ to $y_k = y'$ with $d(y_{i-1}, y_i) \leq \mathfrak{c}$ for all $1 \leq i \leq k$.

Hence it remains to find a chain of factorizations from $x_\mathfrak{c} y_{i-1}$ to $x_\mathfrak{c} y_i$ for $1 \leq i \leq k$ which satisfies the wanted properties. Let $1 \leq i \leq k$ and set $w = x_\mathfrak{c} y_{i-1}$ and $w' = x_\mathfrak{c} y_i$. Using Lemma 2.4 and interchanging y_i, y_{i-1} if necessary, we suppose that

$$\sigma(y_i) - \sigma(y_{i-1}) = \mathfrak{c} - m \leq \mathfrak{c} - 2$$

for some $m \in \{2, \ldots, \mathfrak{c}\}$.

We define $w_\nu = x_\nu y_i$ for $m \leq \nu \leq \mathfrak{c}$. By definition, $w_\mathfrak{c} = w'$. Then, for $m < \nu \leq \mathfrak{c}$,

$$d(w_{\nu-1}, w_\nu) = d(x_{\nu-1}, x_\nu) \leq maxL(A_\mathfrak{c}) \leq 3\mathfrak{c}$$

and

$$|\sigma(w_{\nu-1}) - \sigma(w_\nu)| = |\sigma(x_{\nu-1}) - \sigma(x_\nu)| = 1.$$

Finally, we infer that

$$d(w, w_m) = d(x_\mathfrak{c} y_{i-1}, x_m y_i) \leq d(x_\mathfrak{c}, x_m) + d(y_{i-1}, y_i)$$
$$\leq maxL(A_\mathfrak{c}) + \mathfrak{c} = 4\mathfrak{c}$$

and

$$\sigma(w) - \sigma(w_m) = \sigma(x_\mathfrak{c}) + \sigma(y_{i-1}) - \sigma(y_i) - \sigma(x_m) = 0. \quad \square$$

Proof of Theorem 4.3. We set

$$C = \max\{4\mathfrak{c}, t_\mathcal{B}(\mathcal{B}, x_\mathfrak{c})\} \quad \text{and} \quad A^* = A_\mathfrak{c}.$$

Let $a, b \in \mathcal{B}$ with $a = A^* b$ and let $z \in \mathcal{Z}(a)$ be given. There exists some $y \in \mathcal{Z}(b)$ such that $x_\mathfrak{c} y \in \mathcal{Z}(a)$ and

$$d(z, x_\mathfrak{c} y) \leq t_\mathcal{B}(\mathcal{B}, x_\mathfrak{c}) \leq C.$$

Lemma 4.7 gives us a C-chain $(z_i)_{i=0}^k$ from $z_0 = x_c y$ to some $z_k = z' \in Z(a)$ with $\sigma(z') = \max L(a)$ such that $(z_i)_{i=-1}^k$ with $z_{-1} = z$ has the required properties. $\square$

We tackle the proof of Theorem 4.4. Hence, we further suppose that $T = \prod_{i=1}^\lambda D_i$ where D_i are finitely primary monoids of exponent α and of rank greater than 1.

The assumption on the ranks is needed for the following lemma.

Lemma 4.8. *Every $a \in B \cap T$ has a factorization $z \in Z(a)$ with $\sigma(z) < C_1 = (4\alpha \exp(G) + \alpha)\lambda$.*

Proof. This follows from Lemma 5.6 in [Ge5]. $\square$

Lemma 4.9. *Let $a \in B$ and $z = u_1 \ldots u_m \in Z(a)$ where for $1 \le i \le m$ $u_i = a_i S_i \in A(B)$ with $1 \ne a_i \in T$, $S_i \in F(G)$ with $0 \le \sigma(S_i) \le 1$ and where $m \ge C_2 = 2(C_1 + D(G))$. Then a has a factorization $z' \in Z(a)$ with $\sigma(z') < \sigma(z)$.*

Proof. By definition of Davenport's constant, there exists a set $I \subseteq \{1, \ldots, m\}$ such that $c = \prod_{i \in I} a_i \in B \cap T$ and $m - \#I < D(G)$. Without restriction suppose that $I = \{1, \ldots, n\}$. Then $B = \prod_{i=1}^n S_i \in B(G)$. By Lemma 4.8 c has a factorization x with $\sigma(x) < C_1$. Clearly, B has a factorization y with $\sigma(y) \le \frac{n}{2}$. Therefore,

$$z' = xyu_{n+1} \ldots u_m \in Z(a)$$

is a factorization of a with

$$\sigma(z') < C_1 + \frac{n}{2} + D(G) \le m = \sigma(z) . \quad \square$$

Lemma 4.10. *Let $C_3 = C_2(D(G) - 2)$,*

$$A = \prod_{g \in G} g^{3C_3 D(G)} \in F(G)$$

and $a \in B$ with $A \mid a$. Then for every $z \in Z(a)$ with $\sigma(z) = \min L(a)$ there exists a 3-chain of factorizations $(z_i)_{i=0}^k$ from $z = z_0$

to some $z_k = z'$ with $\sigma(z') = \sigma(z) + 1$ such that $k \leq C_3$ and $d(z, z') \leq 3C_3$.

Proof. Let

$$z = u_1 \ldots u_n \in \mathcal{Z}(a)$$

with $\sigma(z) = \min L(a)$, $u_i \in \mathcal{A}(\mathcal{B}) \setminus \mathcal{F}(G)$ for $1 \leq i \leq m$ and $u_i \in \mathcal{A}(\mathcal{B}(G))$ for $m+1 \leq i \leq n$. Suppose there is a 3-chain $(z_i)_{i=0}^k$ with $k \leq C_3$ and $\sigma(z_k) = \sigma(z) + 1$. Then by the triangle inequality in Lemma 2.1 we infer that

$$d(z, z_k) \leq \sum_{i=0}^{k-1} d(z_i, z_{i+1}) \leq 3C_3 .$$

To find such a chain we distinguish two cases.

Case 1. $m \geq C_2$. By Lemma 4.9 no factorization $y \in \mathcal{Z}(a)$ with $\sigma(y) = \min L(a)$ is divisible by $\prod_{i=1}^{C_2} v_i$ with $v_i = a_i S_i$ where $1 \neq a_i \in T$ and $S_i \in \mathcal{F}(G)$ with $0 \leq \sigma(S_i) \leq 1$.

We consider $u_1, \ldots, u_{c_2}$. By the above observation there is some $i \in \{1, \ldots, C_2\}$ such that $u_i = a_i S_i$ with $\sigma(S_i) \geq 2$. Say $i = 1$, and set $S_1 = h_1 \ldots h_r$ with $r \geq 2$.

Since $\prod_{i=1}^{C_2} S_i \mid \prod_{g \in G} g^{C_2(\mathcal{D}(G)-1)}$ and $A \mid a$, there is some $i \in \{C_2 + 1, \ldots, n\}$ such that $u_i = (h_1 + h_2)v$ with $v \in \mathcal{F}(G) \times T$. Then $w_1' = a_1(h_1 + h_2)h_3 \ldots h_r = a_1 S_1' \in \mathcal{A}(\mathcal{B})$ with $\sigma(S_1') < \sigma(S_1)$ and $w_1'' = h_1 h_2 v \in \mathcal{B}$ is a product of at most two irreducible elements of $\mathcal{B}$. This gives rise to a factorization z_1 of a with $d(z, z_1) \leq 3$ and $\sigma(z) \leq \sigma(z_1) \leq \sigma(z) + 1$.

If $\sigma(z) = \sigma(z_1)$, then we proceed with z_1 and w_1' in the same way as with z and u_1. After $\sigma(S_1) - 1 \leq \mathcal{D}(G) - 2$ steps at most we arrive at a factorization $\bar{z}$ with $\bar{u}_1 = \bar{a}_1 \bar{S}_1$, $\sigma(\bar{S}_1) = 1$ or with $\sigma(\bar{z}) = \sigma(z) + 1$.

If it is necessary, we repeat this process with $u_2, u_3, \ldots, u_{C_2}$. Since no factorization $y \in \mathcal{Z}(a)$ with $\sigma(y) = \min L(a)$ is divisible by a product $\prod_{i=1}^{C_2} a_i S_i$, with $1 \neq a_i \in T$ and $S_i \in \mathcal{F}(G)$ with $0 \leq \sigma(S_i) \leq 1$, we arrive after $k \leq C_2(\mathcal{D}(G) - 2)$ steps at a factorization z_k with $\sigma(z_k) = \sigma(z) + 1$.

Case 2. $m < C_2$. Let $U = g_1 g_2 g_3 \in \mathcal{B}(G)$ be an irreducible block. Then $\prod_{i=1}^{3}(-g_i \cdot g_i)$ has a factorization of length 2. Hence no factorization $y \in \mathcal{Z}(a)$ with $\sigma(y) = \min L(a)$ is divisible by $\prod_{i=1}^{3}(-g_i \cdot g_i)$ in $\mathcal{Z}(\mathcal{B})$.

Suppose that $-g_1 \cdot g_1$ does not divide z. Then there is some block $w \in \{u_{m+1}, \ldots, u_n\}$ with $g_1 \mid w$ and $\sigma(w) \geq 3$. Say $u_{m+1} = g_1 h_1 \ldots h_r$ with $r \geq 2$. There is some $i \in \{m+2, \ldots, n\}$ such that $u_i = (h_1 + h_2)S$ with $S \in \mathcal{F}(G)$. Then $w_1 = g_1(h_1 + h_2)h_3 \ldots h_r \in \mathcal{A}(\mathcal{B}(G))$, $g_1 \mid w_1$, $\sigma(w_1) < \sigma(w)$ and $w_1' = h_1 h_2 S \in \mathcal{B}(G)$ is a product of at most two irreducible elements of $\mathcal{B}(G)$. This gives rise to a factorization z_1 of a with $d(z, z_1) \leq 3$ and $\sigma(z) \leq \sigma(z_1) \leq \sigma(z) + 1$.

If $\sigma(z) = \sigma(z_1)$, then we proceed with z_1 and w_1 in the same way as with z and w. After $\sigma(w) - 2 \leq \mathcal{D}(G) - 2$ steps at most we arrive at a factorization $\bar{z}$ which is either divisible by $-g_1 \cdot g_1$ or for which $\sigma(\bar{z}) = \sigma(z) + 1$ holds.

If it is necessary, we repeat this process with g_2 and g_3. Since no factorization $y \in \mathcal{Z}(a)$ with $\sigma(y) = \min L(a)$ is divisible by $\prod_{i=1}^{3}(-g_i \cdot g_i)$, we arrive after $k \leq 3(\mathcal{D}(G)-2)$ steps at a factorization z_k with $\sigma(z_k) = \sigma(z) + 1$. $\square$

Lemma 4.11. *Let* $M \in \mathbf{N}$, A *as in Lemma 4.10 and* $a \in \mathcal{B}$ *with* $A^M \mid a$. *Then for every* $z \in \mathcal{Z}(a)$ *with* $\sigma(z) = \min L(a)$ *there exists a* $3C_3$*-chain of factorizations* $(z_i)_{i=0}^{M}$ *with* $z = z_0$ *and* $\sigma(z_i) = \min L(a) + i$ *for* $0 \leq i \leq M$.

Proof. We argue by induction on M. For $M = 1$ the assertion follows from Lemma 4.10. Let $M \geq 2$, $a \in \mathcal{B}$ with $A^M \mid a$ and $z \in \mathcal{Z}(a)$ with $\sigma(z) = \min L(a)$. By Lemma 4.10 there is some $z' \in \mathcal{Z}(a)$ such that $z = xy$, $z' = x'y$ with $y \in \mathcal{Z}(b)$ for some $b \in \mathcal{B}$, $\sigma(z') = 1 + \sigma(z)$ and $d(z, z') = d(x, x') = \sigma(x') \leq 3C_3$. Since for every $g \in G$

$$v_g(A) = 3C_3\mathcal{D}(G) \geq \sigma(x')\mathcal{D}(G),$$

it follows that $A^{M-1} \mid b$. Clearly, $\sigma(y) = \min L(b)$ and hence induction hypothesis implies the assertion. $\square$

Lemma 4.12. *Set* $M = t_\mathcal{B}(\mathcal{B}, x_c)$, $C_4 = M(3C_3+1)$ *and let* $a \in \mathcal{B}$ *such that* $A^M \mid a$. *Then for every* $z \in \mathcal{Z}(a)$ *with* $\sigma(z) = \min L(a)$ *there exists a* C_4*-chain of factorizations* $(z_i)_{i=0}^{k}$ *from* $z = z_0$ *to* $z_k = x_c y \in \mathcal{Z}(a)$ *such that* $|\sigma(z_i) - \sigma(z_{i-1})| \leq 1$ *for* $1 \leq i \leq k$.

Proof. Let $z \in \mathcal{Z}(a)$ with $\sigma(z) = \min L(a)$. Clearly, $A_c \mid A^M \mid a$. Hence, by definition of tameness there exists a factorization $z' \in \mathcal{Z}(a)$

with $z' = x_c y$ and $d(z, z') \leq M$. By Lemma 2.4 it follows that $\sigma(z') - \sigma(z) \leq M - 2$. Set $k = \sigma(z') - \sigma(z) + 1$. Then by Lemma 4.11 there is a $3C_3$-chain of factorizations $(z_i)_{i=0}^{k-1}$ with $z = z_0$ and $\sigma(z_{k-1}) = \sigma(z) + k - 1 = \sigma(z')$. We set $z_k = z'$ and infer that

$$d(z_{k-1}, z') \leq d(z_{k-1}, z) + d(z, z')$$
$$\leq \sum_{i=0}^{k-2} d(z_i, z_{i+1}) + d(z, z')$$
$$\leq (k-1)3C_3 + M \leq M(3C_3 + 1) . \quad \square$$

Proof of Theorem 4.4. We set

$$C = \max\{4c, C_4\} \quad \text{and} \quad A^* = A^M$$

with A as in Lemma 4.10 and $M = t_B(\mathcal{B}, x_c)$.

Let $a \in \mathcal{B}$ with $A^* | a$ and let $z \in \mathcal{Z}(a)$ with $\sigma(z) = \min L(a)$ be given. By Lemma 4.12 there is a C-chain of factorizations $(z_i)_{i=0}^{k_1}$ from $z = z_0$ to some $z' = x_c y$ such that $|\sigma(z_i) - \sigma(z_{i-1})| \leq 1$ for $1 \leq i \leq k_1$. By Lemma 4.7 there exists some $z'' \in \mathcal{Z}(a)$ with $\sigma(z'') = \max L(a)$ and a C-chain $(z_i)_{i=k_1}^{k}$ from $z' = z_{k_1}$ to $z_k = z''$ such that $|\sigma(z_i) - \sigma(z_{i-1})| \leq 1$ for $k_1 \leq i \leq k$. Thus $(z_i)_{i=0}^{k}$ is the required chain. $\quad \square$

5. QUANTITATIVE ASPECTS OF FACTORIZATIONS

Quantitative theory of non-unique factorizations was initiated by W. Narkiewicz in the sixties. The reader is referred to his book [Na] and to the short survey given in [Ch-G]. For a long time investigations were restricted to rings of integers (i.e., principal orders) in algebraic number fields. Only quite recently first results were carried over from principal orders to arbitrary orders (cf. [G-HK-K].

Let R be a ring of integers and let $Z \subseteq \mathcal{H}(R)$ denote the set of principal ideals aR with $L(a)$ being an arithmetical progression with distance 1. Then Z has density 1 (see [Ge1; Satz1] and [Ge6] for a refinement). It is the aim of this section to prove similar results for non-principal orders. For this we need the results from section 4 and the analytical machinery from [G-HK-K].

Definition 5.1. 1. A *norm (function)* on a reduced monoid H is a monoid homomorphism $|\cdot|: H \to (\mathbf{N}, \cdot)$ satisfying $|a| = 1$ if and only if $a = 1$.

2. An *arithmetical order formation* $[\mathcal{F}(P), T, H, |\cdot|]$ consists of a free abelian monoid $\mathcal{F}(P)$, reduced monoids T, H and of a norm on $\mathcal{F}(P) \times T$ such that the following properties are satisfied:

i) $H \subseteq \mathcal{F}(P) \times T$ is a saturated submonoid and $G = \mathcal{F}(P) \times T/H$ is a finite abelian group (called the class group of the formation).

ii) for every $g \in G$

$$\sum_{p \in P \cap g} |p|^{-s} = \frac{1}{\#G} \log \frac{1}{s-1} + h_g(s)$$

holds in the half-plane $\mathrm{Re}(s) > 1$, where $h_g(s)$ is regular in the half-plane $\mathrm{Re}(s) > 1$ and in some neighborhood of $s = 1$.

iii) there is some $r \in \mathbf{N}$ such that

$$\#\{t \in T | \ |t| \leq x\} \ll (\log x)^r$$

for every $x \geq 2$.

Let $[\mathcal{F}(P), H | \cdot |]$ be an arithmetical formation in the sense of [Ch-G; section 4]. Then, by definition $[\mathcal{F}(P), T, H, | \cdot |]$ is an arithmetical order formation with $T = \{1\}$. Hence arithmetical order formations are generalizations of arithmetical formations and they are modelled on orders in global fields. We discuss this main example. For further examples the reader is referred to [G-HK-K].

Let K be a global field (i. e., either an algebraic number field or an algebraic function field in one variable over a finite field). Let $S(K)$ denote the set of all non-archimedean places and for $v \in S(K)$ let R_v be the corresponding valuation ring. For a finite subset $S \subset S(K)$, $S \neq \emptyset$ in the function field case,

$$R_S = \bigcap_{v \in S(K) \backslash S} R_v \subseteq K,$$

is called the *holomorphy ring* of K associated with S. R_S is a Dedekind domain with quotient field K.

Let $R \subseteq K$ be a holomorphy ring and $\mathfrak{o} \subseteq R$ an order (i. e., $\mathfrak{o}$ has quotient field K and $R = \bar{\mathfrak{o}}$ is a finite $\mathfrak{o}$-module). In particular, o is one-dimensional noetherian, hence weakly Krull, and has finite Picard group. The monoid $\mathcal{I}(o)$ of integral invertible ideals equals $\mathcal{I}_t(o)$ and thus $\mathcal{I}(R) = \mathcal{F}(P) \times T$ with P and T as explained in section 1. For an ideal $I \in \mathcal{I}(o)$ set $|I| = \#(o/I)$ and let $\mathcal{H}(o) \subseteq \mathcal{I}(o)$ denote the submonoid of principal ideals. Then $[\mathcal{F}(P), T, \mathcal{H}(o), |\cdot|]$ is an arithmetical order formation ([G-HK-K; Proposition 3]).

Let $[\mathcal{F}(P), T, H, |\cdot|]$ be an arithmetical order formation. For a subset $Z \subseteq \mathcal{F}(P) \times T$ let

$$Z(x) = \#\{a \in Z|\ |a| \leq x\}$$

denote the associated counting function. The following sets have been studied for every $k \in \mathbf{N}$

$$M_k = \{a \in H \mid \operatorname{max}L(a) \leq k\},$$
$$G_k = \{a \in H \mid \#L(a) \leq k\}$$

and

$$F_k = \{a \in H \mid \#\mathcal{Z}(H) \leq k\}\ .$$

Theorem 5.2. *Let* $[\mathcal{F}(P), T, H, |\cdot|]$ *be an arithmetical order formation and let* Z *be either* M_k, G_k *or* F_k *for some* $k \in \mathbf{N}$. *Then, for* x *tending to infinity,*

$$Z(x) \asymp x(\log x)^{-A}(\log\log x)^B$$

with $A \in [0, 1]$ *and* $B \in \mathbf{N_0}$.

Proof. see [G-HK-K; Theorems 1, 2 and 3] $\square$

Let $[\mathcal{F}(P), T, H, |\cdot|]$ be an arithmetical order formation and $Z \subseteq H$. We say that Z has *(natural) density* δ, if

$$\lim_{x\to\infty} \frac{Z(x)}{H(x)} = \delta\ .$$

Theorem 5.3. *Let* $[\mathcal{F}(P), T, H, |\cdot|]$ *be an arithmetical order formation with class group* G *and* $\beta : \mathcal{F}(P) \times T \to \mathcal{F}(G) \times T$ *the associated block homomorphism. Let* $Z \subseteq H$ *be a subset containing* $\{a \in H \mid A^* \mid \beta(a)\}$ *for some* $A^* \in \mathcal{F}(G)$. *Then* Z *has density* 1.

Proof. Let $A^* \in \mathcal{F}(G)$. Obviously, it is sufficient to show that

$$E = \{a \in H \mid A^* \mid \beta(a)\}$$

has density 1. Since for every $x \geq 1$

$$\frac{H(x)}{E(x)} = 1 + \frac{(H \setminus E)(x)}{E(x)} \; ,$$

it suffices to verify that, for x tending to infinity, $E(x) \gg x$ and that $(H \setminus E)(x) \ll x(\log x)^{-k}$ for some $k > 0$.

We begin with $H \setminus E$. Clearly,

$$H \setminus E = \{a \in H \mid A^* \nmid \beta(a)\} \subseteq \bigcup_{g \in G} \bigcup_{i=0}^{v_g(A^*)-1} \Omega_{g,i}$$

with

$$\Omega_{g,i} = \{a \in H \mid v_g(\beta(a)) = i\} \; .$$

Let $g \in G$ with $v_g(A^*) \geq 1$ and $i \in \{0, \ldots, v_g(A^*) - 1\}$. For every $t \in T$ we consider

$$(\Omega_{g,i})_t = \{b \in \mathcal{F}(P) \mid b \in -[t] \text{ and } v_g(\beta(b)) = i\} \; .$$

By Proposition 8 in [G-HK-K] it follows that

$$(\Omega_{g,i})_t(x) = C_{[t]}(x) x(\log x)^{-\frac{1}{\#G}} (\log \log x)^i$$

with $C_{[t]}(x) \asymp 1$ as $x \to \infty$. Thus the assumptions of Proposition 5 in [G-HK-K] are satisfied and we obtain that

$$\Omega_{g,i}(x) \asymp x(\log x)^{-\frac{1}{\#G}} (\log \log x)^i \; .$$

Hence the assertion on $(H \setminus E)(x)$ follows.

Obviously,

$$E = \{a \in \mathcal{F}(P) \times T \mid a \in H \text{ and } A^* \mid \beta(a)\}$$
$$\supseteq \{a \in \mathcal{F}(P) \mid a \in H \text{ and } A^* \mid \beta(a)\} = E' \ .$$

Proposition 8 in [G-HK-K] implies that $E'(x) \asymp x$ and thus $E(x) \gg x$ for $x \to \infty$. $\square$

Building on Theorem 5.3, Theorem 4.3 and Theorem 4.4 imply density results for orders in holomorphy rings of global fields. We formulate this explicitly only for Theorem 4.4.

For an atomic monoid H and some $C \in \mathbf{N}$ let $Z(C) \subseteq H$ consist of those elements $a \in H$ having the following property:

for all $z \in \mathcal{Z}(a)$ with $\sigma(z) = \min L(a)$ there exists some $z' \in \mathcal{Z}(a)$ with $\sigma(z') = \max L(a)$ and a C-chain of factorizations $(z_i)_{i=0}^{k}$ from $z = z_0$ to $z_k = z'$ such that $|\sigma(z_i) - \sigma(z_{i-1})| \leq 1$ for $1 \leq i \leq k$. In particular,

$$L(a) = \{\sigma(z_i) \mid 1 \leq i \leq k\}$$

is an arithmetical progression with distance 1.

Theorem 5.4. *Let o be an order in a holomorphy ring R of a global field K with Picard group $G = Pic(o)$ and conductor $\mathfrak{f} = Ann_o(R/o)$. Suppose that $\#G \geq 3$ and that for every $\mathfrak{p} \in spec(o)$ with $\mathfrak{p} \supseteq \mathfrak{f}$ there are at least two prime ideals P, P' of R such that $P \cap o = P' \cap o = \mathfrak{p}$.*
Then there exists some constant $C \in \mathbf{N}$ such that $Z(C) \subseteq \mathcal{H}(o)$ has density 1.

Proof. This is an immediate consequence of Theorem 4.4 and Theorem 5.3. $\square$

REFERENCES

[A-A-Z1]   D.D. Anderson, D.F. Anderson and M. Zafrullah, *Rings between* $D[X]$ *and* $K[X]$, Houston J. Math. **17**: 109-129 (1991).

[A-A-Z2]   D.D. Anderson, D.F. Anderson and M. Zafrullah, *Factorization in integral domains*, J. Pure Applied Algebra **69**: 1-19 (1990).

[A-M-Z]   D.D. Anderson, J.L. Mott and M. Zafrullah, *Finite character representations for integral domains*, Boll. U.M.I. **6-B** : 613-630 (1992).

[An]   D. F. Anderson, *Elasticity of factorization in integral domains, a survey*, these Proceedings.

[Ch-G]   S. Chapman and A. Geroldinger, *Krull domains and monoids, their sets of lengths and associated combinatorial problems*, these Proceedings.

[Ge1]   A. Geroldinger, *Über nicht-eindeutige Zerlegungen in irreduzible Elemente*, Math. Z. **197**: 505–529 (1988).

[Ge2]   A. Geroldinger, *On the arithmetic of certain not integrally closed noetherian integral domains*, Comm. Algebra **19**: 685–698 (1991).

[Ge3]   A. Geroldinger, *T-block monoids and their arithmetical applications to certain integral domains*, Comm. Algebra **22**: 1603–1615 (1994).

[Ge4]   A. Geroldinger, *On the structure and arithmetic of finitely primary monoids*, Czech. Math. J. : (1996).

[Ge5]   A. Geroldinger, *Chains of factorizations in weakly Krull domains*, Colloq. Math. : (1996).

[Ge6]   A. Geroldinger, *Chains of factorizations in orders of global fields*, Colloq. Math. : (1996).

[Ge7]   A. Geroldinger, *Chains of factorizations and sets of lengths*, J. Algebra.

[G-HK-K]   A. Geroldinger, F. Halter-Koch and J. Kaczorowski, *Non-unique factorizations in orders of global fields*, J. reine angew. Math. **459**: 89–118 (1995).

[G-L]   A. Geroldinger and G. Lettl, *Factorization problems in semigroups*, Semigroup Forum **40**: 23–38 (1990).

[HK1]   F. Halter-Koch, *Elasticity of factorizations in atomic monoids and integral domains*, J. Théorie des Nombres Bordeaux **7**: 367–385 (1995).

[HK2]   F. Halter-Koch, *Finitely generated monoids, finitely primary monoids and factorization properties of integral domains*, these Proceedings.

[HK3]   F. Halter-Koch, *Divisor theories with primary elements and weakly Krull domains*, Boll. U. M. I. **9-B**: 417–441 (1995).

[Na]   W. Narkiewicz, *Elementary and Analytic Theory of Algebraic Numbers*, Springer (1990).

The Theory of Divisibility

JOE L. MOTT, Department of Mathematics 3027, Florida State University, Tallahassee, Florida 32306-3027

1. INTRODUCTION

The study of divisibility and factorization of elements in a (commutative) integral domain (with identity) has a long, rich, and varied history. As is often the case in scientific history, we see a series of minds each building upon the advances of a predecessor. This lineage includes contributions by Kronecker, Dedekind, Lorenzen, Prüfer, van der Waerden, E. Artin, Krull, Dieudonné, Nakayamma, Cohn, Samuel, Kaplansky and many others. These contributions have connections to a wide array of topics not the least of which include the Fundamental Theorem for Finitely Generated Abelian Groups and extends to Andrew Wiles's proof of the famous Last Theorem of Fermat. And since the quest is not yet over, the list is still being completed.

Dedekind observed that the study of divisibility is essentially a subtopic of the study of partially ordered groups. The two areas of study are connected via the concept of the group of divisibility $G(D)$.

One way to view $G(D)$ is as the multiplicative group of principal fractional ideals of D partially ordered by reverse containment. Then the divisibility and factorization properties of elements of D usually have their corresponding order interpretation in $G(D)$. Indeed, some classes of domains are completely characterized by the order properties of $G(D)$; for example, D is a valuation domain, GCD-domain, or pseudo-principal ideal domain $\Leftrightarrow G(D)$ is, respectively, totally ordered [Kr], lattice-ordered, or a complete lattice-ordered group [Mot₃]. The domain D is a UFD (unique factorization domain or factorial domain) $\Leftrightarrow G(D)$ is the cardinal sum of groups G_λ, where each G_λ is order isomorphic to the totally ordered group $\mathbb{Z}$ of integers where there is a one-to-one correspondence between the set of nonassociate primes of D and the set of groups G_λ.

But these are not the only ring-theoretic properties that can be studied profitably by interpreting these properties in the group of divisibility. Let me mention that the complete integral closure of a Bezout domain D has a nice characterization in terms of $G(D)$. Recall that a Bezout domain is a domain in which each finitely generated (integral) ideal is principal. In particular, a Bezout domain is a GCD-domain but not conversely. An element $b \geq 0$ in any lattice ordered abelian group G is *bounded* if there exists $g \in G$ such that $nb \leq g$ for all positive integers n. If D is a Bezout domain with group of divisibility $G(D)$, then the complete integral closure of D has $G(D)/B(G)$ for its group of divisibility where $B(G) = \{x \mid x = b - c$ where b and c are bounded in $G(D)\}$. Therefore, a Bezout domain D is completely integrally closed $\Leftrightarrow G(D)$ contains no nonzero bounded elements.

The early history of the development of the study of divisibility via partially ordered groups has been traced several places in the literature, in particular, in two articles of my own: "The Group of Divisibility and its Applications" [Mot₃]; and "Groups of Divisibility: A Unifying Concept for Integral Domains and Partially Ordered Groups" [Mot₄], in J. Močkoř's book: Groups of Divisibility [Moč₁] and in each of the books by J. Alajbegovic and J. Močkoř [AM] and M. Anderson and T. Feil [AF].

The study of divisibility of elements in an integral domain D and the study of factorization of elements in D are related but not identical. In the study of divisibility, one considers all pairs $(a, b) \in D^* \times D^*$ and asks whether a divides b (denoted by $a|b$) or b divides a. On the other hand, in the study of factorization, one has a particular subset $\mathcal{C}$ of elements of D (for example, $\mathcal{C}$ could be the set of atoms, primary elements, basic elements, etc. of D) and asks whether each $a \in D^*$ can be written as a product of finitely many elements from $\mathcal{C}$, and if so, whether there is a bound on the lengths of the different factorizations of a, and so forth.

After noting this distinction let us narrow the focus of the present article to problems that relate to the group of divisibility in general and in particular to the contributions of several authors since my 1989 survey article appeared. More specifically, the focus will be on finiteness conditions on certain submonoids of $F(D)$, the monoid of fractional ideals of an integral domain D. Dan Anderson has completely solved the problem Glastad and Mott posed in [GM] by characterizing those domains D for which $G(D)$ is finitely generated. Moreover, D.D. Anderson and Mott [AMo] have finished a work started by I.S. Cohen and I. Kaplansky in [CK] by giving several characterizations of atomic domains containing only finitely many atoms. Finally, I review the characterization by D.D. Anderson, Mott and Park [AMP] of those integral domains D for which $F(D)$ is finitely generated.

I regret that time and space prevent me from reviewing the paper [MS] on exact sequences of semi-value groups and the contributions to the study of divisibility by Močkoř, Bastos and others. I include references to some of their papers so that the interested reader can locate them.

2. PRELIMINARIES

Terminology and notation will follow that of Kaplansky [Kap] or Nagata [N_1]. For instance, a local ring or semilocal ring is always Noetherian while a quasilocal or semiquasilocal ring may not be Noetherian. For an integral domain D, K will usually denote its quotient field and $\overline{D}$ the integral closure of D in K. The group of units of D is denoted by $U(D)$, D^* denotes the nonzero elements of D, and $U(K) = K^*$. A *fractional ideal* of D is a D-submodule A of K for which there exists a nonzero element $d \in D$ such that $dA \subseteq D$. An *integral ideal* of D is just a fractional ideal contained in D. Whether the unmodified term ideal means fractional ideal or integral ideal should be clear from the context, but often the modifying term is added for emphasis. For $a \in K$, the *principal fractional ideal generated by a* is just the set $(a) = aD$.

The set $F(D)$ of nonzero fractional ideals of D forms a commutative monoid under multiplication with identity D. Moreover, $F(D)$ is partially ordered by $A \leq B \Leftrightarrow B \subseteq A$. In particular, if $B \in F(D)$, then $D \leq B \Leftrightarrow B \subseteq D \Leftrightarrow B$ is an integral ideal of D. Thus, the positive cone of $F(D)$ is $F_+(D) = \{B \in F(D) | D \leq B\}$, the submonoid of nonzero integral ideals of D. The set $F^*(D) = \{A \in F(D) | A \text{ is finitely generated}\}$ is a partially ordered submonoid of $F(D)$ with positive cone

$F_+^*(D) = F^*(D) \cap F_+(D)$, the submonoid of nonzero finitely generated integral ideals of D.

The group of units of $F(D)$, and also of $F^*(D)$, is $Inv(D)$, the group of invertible ideals of D. Hence the positive cone $Inv_+(D)$ of $Inv(D)$ is the monoid of integral invertible ideals. The group $P(D)$ of the nonzero principal ideals of D is a subgroup of $Inv(D)$. The group of divisibility of D is usually defined to be the group $G(D) = K^*/U(D)$, partially ordered by $aU(D) \leq bU(D) \Leftrightarrow \frac{b}{a} \in D \Leftrightarrow bD \subseteq aD$. Then the map $P(D) \to G(D)$ given by $aD \to aU(D)$ is clearly an order isomorphism between $P(D)$ and $G(D)$ mapping the positive cone $P_+(D)$ of integral principal ideals to $G_+(D) = \{aU(D)|a \in D\}$. In some contexts, it is important to make the following distinction: if L is a field containing the quotient field K of D, the group $L^*/U(D)$ with positive cone $D^*/U(D)$ is denoted by $G_L(D)$ and is called a *semi-value group*. But for the special case of the quotient field K of D, the subscript is suppressed unless it is needed for clarity. If $G = \prod_\lambda G_\lambda$ (or $\sum_\lambda G$) is the direct product (sum) of partially ordered groups G_λ, then G is called the *cardinal product* (cardinal sum) of the $G_\lambda \Leftrightarrow$ the partial order on G is defined by $(a_\lambda) \leq (b_\lambda) \Leftrightarrow a_\lambda \leq b_\lambda$ in G_λ for each λ.

If $a, b \in D^*$ say a *divides* b in D and write $a|b$ (or, in other words, a is a *divisor* of b, b is *divisible* by a, and b is a *multiple* of a) $\Leftrightarrow \frac{b}{a} \in D \Leftrightarrow bD \subseteq aD$. The two elements $a, b \in D^*$ are *associates* $\Leftrightarrow a|b$ and $b|a \Leftrightarrow aD = bD \Leftrightarrow a = bu$ for some unit $u \in U(D)$. Finally, $a, b \in D^*$ have a *greatest common divisor* (GCD) $\Leftrightarrow$ there is an element $d \in D$ that divides both a and b and is divisible by any $d' \in D$ that divides both a and b.

If d and d_1 are both GCDs of a and b, then d and d_1 are associates and we understand the meaning of the equation $d = GCD(a, b)$ in that light. Moreover, the element d is a GCD of a and $b \Leftrightarrow dD$ equals the intersection of all principal integral ideals of D containing a and b. Thus, it is clear that two elements of D need not possess a GCD because the intersection of a collection of principal integral ideals of D may not itself be a principal ideal.

Least common multiples can be defined analogously. An element $m \in D$ is a *least common multiple* (LCM) of a and b in $D^* \Leftrightarrow m$ is a multiple of both a and b and divides any other multiple of a and b. Any two LCMs of a and b must be associates. With that observation, we can write $m = LCM(a, b)$. Moreover, $m = LCM(a, b) \Leftrightarrow aD \cap bD = mD$. It may be that two elements of D have a greatest common divisor (GCD) but no least common multiple (LCM). For example, in the ring $k[x^2, x^3]$ where k is a field and x is an indeterminate, $1 = \mathrm{GCD}(x^2, x^3)$ but x^2 and x^3 have no LCM. However, in general, the existence of an LCM implies the existence of a GCD. In fact, if $\mathrm{LCM}(a, b) = m$, then $\mathrm{GCD}(a, b) = \frac{ab}{m}$.

If I is a fractional ideal of D, one way of defining I_v, the v-ideal of D, is to set $I_v =$ the intersection of all principal fractional ideals of D containing I. If $a, b \in D^*$, $(a, b)_v = dD$ for some $d \in D \Leftrightarrow \omega(d)$ is the greatest lower bound of $\{\omega(a), \omega(b)\}$, where ω is the natural semivaluation map from $K^* \to G(D)$. Moreover, if $(a, b)_v = dD$, then dD is the intersection of all principal integral ideals of D containing a and b, that is, $d = \mathrm{GCD}(a, b)$.

3. FINITELY GENERATED GROUPS OF DIVISIBILITY

In some situations, quite a bit of information about the integral domain D can be inferred strictly from the algebraic properties of the group $G(D)$ alone. For instance, if $G(D)$ is isomorphic to a subgroup of the additive group of rational numbers — that is, $G(D)$ is torsion-free of rank one — then D must be a valuation ring of Krull dimension one and the order on $G(D)$ must have been a total order [GM, Corollary 2.2].

In [GM] Bruce Glastad and I published our work on finitely generated groups of divisibility begun in Bruce's dissertation at Florida State University. As witnessed by the following theorem, the assumption that $G(D)$ is finitely generated places considerable constraints on the ring theoretic structure of D. Our objective in this section is to give a complete characterization of such domains.

THEOREM 3.1. (Glastad and Mott [GM]) *Let D be an integral domain such that $G(D)$ is finitely generated. Let $\overline{D}$ be the integral closure of D in K. Then $\overline{D}$ is a semiquasilocal Bezout domain containing only finitely many prime ideals and at most rank $G(D)$ maximal ideals.*

Proof. For the details of the fairly simply proof see [GM, Theorem 2.1]. □

The next theorem enables us to make the reduction to the case of a quasilocal domain with finitely generated group of divisibility.

THEOREM 3.2. (Glastad and Mott [GM]) *(1) For an integral domain*

D with quotient field K, $G(D)$ is finitely generated if and only if D is semi-quasilocal and $G(D_M)$ is finitely generated for each maximal ideal M of D.

(2) For a one-dimensional domain D with maximal ideals $M_1, M_2, \ldots, M_n$, $G(D)$ is order isomorphic to the cardinal product of the groups $G(D_{M_i})$ for $i = 1, 2, \ldots, n$. Hence rank $G(D) \geq n$.

Proof. Suppose $G(D)$ is finitely generated. Then by Theorem 3.1, $\overline{D}$ is semiquasilocal and, therefore, D is also by the Lying Over Theorem [Kap, Theorem 4.4]. For each maximal ideal M of D, the group $G(D_M)$, being a homomorphic image of $G(D)$, is finitely generated.

Conversely, suppose D is semiquasilocal and $G(D_M)$ is finitely generated for each maximal ideal M of D. Let $M_1, M_2, \ldots, M_n$ be the maximal ideals of D. Then observe that the natural map $G(D) \to G(D_{M_1}) \oplus \cdots \oplus G(D_{M_n})$ is injective. Hence, $G(D)$ is a subgroup of a direct sum of finitely generated abelian groups and so $G(D)$ is finitely generated.

(2) The proof that the natural map is surjective when D is one-dimensional is straightforward; see [AMo, Theorem 3.2]. $\qquad\qquad\qquad\qquad\qquad$ $\square$

Suppose now that D is an integral domain with a unique maximal ideal M such that $G(D)$ is finitely generated. Then with this simplifying reduction, it is possible to separate the problem into two cases determined by whether D/M is finite or infinite. In this context Brandis's Theorem [Br] is very useful.

THEOREM 3.3. (Brandis) *Let K be an infinite field and L an extension field. Moreover, let K^* and L^* denote their respective groups of units. If L^*/K^* is finitely generated, then $L = K$.*

Let me summarize some of the main conclusions of the paper by Glastad and Mott [GM].

THEOREM 3.4. (Glastad and Mott [GM]) *Let D be an integral domain with quotient field K and $G(D)$ finitely generated. Let $\overline{D}$ be the integral closure of D. Then:*

(1) $\overline{D}$ is a finitely generated D-module.

(2) $G(D) \simeq G(\overline{D}) \oplus U(\overline{D})/U(D)$ where $G(\overline{D})$ is free and $U(\overline{D})/U(D)$ is finite.

(3) If P is a prime ideal of D with D/P infinite, then $D_P = \overline{D}_P$ is a valuation domain.

Proof. (1) [GM, Theorem 3.4].

(2) Consider the exact sequence

$$\{1\} \longrightarrow U(\overline{D})/U(D) \longrightarrow G(D) \longrightarrow G(\overline{D}) \longrightarrow \{1\}.$$

Then $G(\overline{D})$ is finitely generated and torsion-free since $\overline{D}$ is integrally closed. Hence $G(\overline{D})$ is free so that the exact sequence splits. By [GM, Theorem 3.9], $U(\overline{D})/U(D)$ is finite.

(3) [GM, Corollary 3.6].   □

EXAMPLES. Examples of domains with finitely generated groups of divisibility abound.

(1) Let $G = \mathbb{Z}_1 \times \mathbb{Z}_2 \times \cdots \times \mathbb{Z}_n$, for $n \geq 1$, where each $\mathbb{Z}_i$ is a copy of $\mathbb{Z}$. Totally order G by lexicographically ordering the n-tuples. Then the Krull-Jaffard-Ohm (Existence) Theorem [Mot₃] implies that G is the group of divisibility of a (discrete) valuation domain D with Krull dimension of D equal to rank G.

(2) Let $G = \mathbb{Z}_1 \times \mathbb{Z}_2 \times \cdots \times \mathbb{Z}_n$, for $n \geq 1$, where each $\mathbb{Z}_i$ is a copy of $\mathbb{Z}$. Then G can be lattice ordered by the cardinal ordering: $G_+ = \{(a_1, \ldots, a_n) | a_i \geq 0$ for each $i\}$. Then the Existence Theorem implies that there is a domain D with group of divisibility G. In fact, D is a principal ideal domain with exactly n maximal ideals.

Each of these classes of examples arises from the Existence Theorem. Of course there are domains whose existence does not depend on the Existence Theorem of Krull, *et al.* The following class of examples was discovered in [GM].

THEOREM 3.5.   (Glastad and Mott [GM]) *Let D be a semiquasilocal domain whose quotient field K is a global field. Then $G(D)$ is finitely generated.*

Proof.   See [GM, Corollary 3.15].   □

In fact, such a domain D in Theorem 3.5 is one-dimensional and Noetherian by the Krull-Akizuki Theorem [N₁, p. 115]. A complete characterization of Noetherian domains with finitely generated groups of divisibility will be given in the next section.

This is where the state of the knowledge about domains with finitely generated groups of divisibility stayed for about ten years. But while working on our joint paper on Cohen-Kaplansky domains, Dan Anderson uncovered the remaining piece of the puzzle to give a complete characterization. The missing piece is the subject of the following theorem. Here $[D : \overline{D}] = \{d \in D | d\overline{D} \subseteq D\}$ is the *conductor* of D in $\overline{D}$, the largest common ideal of D and $\overline{D}$.

THEOREM 3.6. (D.D. Anderson [A$_2$]) *Let D be an integral domain with $G(D)$ finitely generated. Then $\overline{D}/[D : \overline{D}]$ is finite.*

Proof. [A$_2$, Theorem 1].   □

THEOREM 3.7. (D.D. Anderson [A$_2$]) *Let D be an integral domain. Then the following are equivalent:*
(1) $F^*(D)$ is finitely generated.
(2) Inv(D) is finitely generated.
(3) $G(D)$ is finitely generated.
(4) $G(\overline{D})$ is finitely generated and $\overline{D}/[D : \overline{D}]$ is finite.
Moreover, if $G(D)$ is finitely generated, then $F^*(D) \simeq G(\overline{D}) \times B^*(D)$, where $B^*(D) = \{J \in F^*(D) | J\overline{D} = \overline{D}\}$, and $B^*(D)$ is finite.

Proof. (1) $\Rightarrow$ (2) because Inv(D) is generated by those generators of $F^*(D)$ that are invertible.

(2) $\Rightarrow$ (3) since $G(D) \simeq P(D)$, the subgroup of all principal fractional ideals of D, and subgroups of finitely generated abelian groups are finitely generated.

(3) $\Rightarrow$ (4) because if $G(D)$ is finitely generated, then so is $G(\overline{D})$, being a homomorphic image of $G(D)$. Theorem 3.6 implies that $\overline{D}/[D : \overline{D}]$ is finite.

(4) $\Rightarrow$ (1) By Theorem 3.1, $\overline{D}$ is a semiquasilocal Bezout domain. Now $G(\overline{D})$ is free so the exact sequence

$$\{1\} \longrightarrow U(\overline{D})/U(D) \longrightarrow G(D) \longrightarrow G(\overline{D}) \longrightarrow \{1\}$$

splits.

Let $g : G(\overline{D}) \to G(D)$ be a splitting map for the above exact sequence. View g as a map from $P(\overline{D})$ to $P(D)$. Then, in this context, the fact that

g is a splitting map is tantamount to asserting $g(\overline{D}x)\overline{D} = \overline{D}x$ for each $x \in K^*$.

Using the map g, define another map $\theta : F^*(D) \to P(\overline{D}) \times B^*(D)$ by $\theta(J) = (J\overline{D}, g(J\overline{D})^{-1}J)$.

Next show that θ is a well-defined monoid isomorphism. Because $J\overline{D} = g(J\overline{D})\overline{D}$, $g(J\overline{D})^{-1}J\overline{D} = \overline{D}$, θ is well-defined. That θ is a monoid homomorphism follows from the fact that g is a homomorphism. Next, θ is injective because if $\theta(M) = \theta(N)$, then $M\overline{D} = N\overline{D}$, so $g(M\overline{D}) = g(N\overline{D})$. But then $g(M\overline{D})^{-1}M = g(N\overline{D})^{-1}N$ implies $M = N$.

Moreover, to show that θ is surjective, let $(\overline{D}x, J) \in P(\overline{D}) \times B^*(D)$. Thus, $J\overline{D} = \overline{D}$ and

$$\theta(g(\overline{D}x)J) = (g(\overline{D}x)J\overline{D}, (g(g(\overline{D}x)J\overline{D}))^{-1}g(\overline{D}x)J)$$

$$= (g(\overline{D}x)\overline{D}, (g(g(\overline{D}x)\overline{D}))^{-1}g(\overline{D}x)J)$$

$$= (\overline{D}x, (g(\overline{D}x))^{-1}g(\overline{D}x)J)$$

$$= (\overline{D}x, DJ)$$

$$= (\overline{D}x, J).$$

Finally, to show that $F^*(D)$ is finitely generated we need only observe that $B^*(D)$ is finitely generated. Let $J \in B^*(D)$. Then $J\overline{D} = \overline{D}$ and $[D : \overline{D}] = [D : \overline{D}]\overline{D} = [D : \overline{D}](J\overline{D}) = [D : \overline{D}]J \subseteq J \subseteq J\overline{D} = \overline{D}$. So $[D : \overline{D}] \subseteq J \subseteq \overline{D}$. By hypothesis, $\overline{D}/[D : \overline{D}]$ is finite so, in fact, $B^*(D)$ is finite.   □

COROLLARY 3.8. (D.D. Anderson [A$_2$]) *Let D be an integral domain with $G(D)$ finitely generated. Then there exists a finite set $\{A_1, \ldots, A_n\}$ of finitely generated integral ideals of D such that every nonzero finitely generated fractional ideal A of D has the form $A = xA_i$, where $x \in K^*$ and $1 \leq i \leq n$. Hence there exists a natural number k such that each finitely generated fractional ideal of D can be generated by k elements.*

This corollary motivated the following definition.

DEFINITION. Say that a submonoid S of $F(D)$ is *strongly finitely generated* if $G(D)$ is finitely generated and there exist $A_1, \ldots, A_n \in S$ such that every

$A \in S$ can be written in the form $A = xA_i$ for some $x \in K^*$ and some i. Then $A_1, \ldots, A_n$ are said to *strongly generate* S. Of course, the A_i can be chosen as integral ideals of D.

Corollary 3.8 says that the submonoid $F^*(D)$ of all finitely generated fractional ideals of D is strongly finitely generated if $G(D)$ is finitely generated. In Section 4, we will address the issue of $F(D)$ being finitely generated and show that it is equivalent to $F(D)$ being strongly finitely generated.

The next theorem, also due to Dan Anderson [A$_2$], shows that all integral domains with finitely generated groups of divisibility can be realized by a simple pull-back construction.

THEOREM 3.9. (D.D. Anderson [A$_2$]) *Let D be an integral domain with $G(D)$ finitely generated. Then the integral closure $\overline{D}$ of D in K is a semiquasilocal domain with $G(\overline{D})$ finitely generated and $\overline{D}/[D:\overline{D}]$ is finite. Let π be the natural map $\pi : \overline{D} \to \overline{D}/[D:\overline{D}]$. Then $D = \pi^{-1}(D/[D:\overline{D}])$.*

Conversely, suppose that T is an integral domain such that $G(T)$ is finitely generated. Suppose I is a nonzero ideal of T with T/I finite. Let S be a subring of T/I and let π be the natural map $\pi : T \to T/I$. Then $D = \pi^{-1}(S)$ is an integral domain with quotient field K, $I \subseteq [D:T] = \pi^{-1}([S:T/I])$, $\overline{D} = \overline{T}$, and $G(D)$ is finitely generated.

Proof. The first part of this theorem is contained in Theorem 3.7. Suppose then that T is such that $G(T)$ is finitely generated. Then it is obvious that $\pi^{-1}(S) = D$ is a subring of T containing I, $I \subseteq [D:T]$ and D and T have the same quotient field. Since T/I is a finitely generated S-module, T is a finitely generated D-module and $\overline{D} = \overline{T}$.

The group $U(T)/U(D) \simeq U(T/I)/U(S)$ by [A$_2$, Lemma 2] so $U(T)/U(D)$ is finite. From the exact sequence

$$\{1\} \longrightarrow U(T)/U(D) \longrightarrow G(D) \longrightarrow G(T) \longrightarrow \{1\},$$

we conclude that $G(D)$ is finitely generated since the two groups $U(T)/U(D)$ and $G(T)$ are finitely generated.   □

4. COHEN-KAPLANSKY DOMAINS

Recall that in an integral domain D a nonzero, nonunit $d \in D$ is said to be *irreducible* (or an *atom*) if for any factorization $d = st$, where $s, t \in D$, then either s or t is a unit of D. In Cohn's terminology $[C_1]$, the domain D is *atomic* if each nonzero, nonunit of D is a finite product of atoms. Let us also say that if $d \in D$ is a finite product of atoms, then such a factorization is an *atomic factorization* of d. An atom of D corresponds to a minimal positive element in the group divisibility $G(D)$.

In this section, I want to focus on atomic integral domains that contain only a finite number of (nonassociate) atoms. I.S. Cohen and I. Kaplansky first studied these domains in a 1946 paper [CK] so it is appropriate that they be dubbed *Cohen-Kaplansky domains* (or *CK-domains* or CKD) for short. In that paper, Cohen and Kaplansky used the word *prime* for what is now traditionally called an atom.

Four papers are fundamental for our discussion of CK-domains:

(1) Cohen and Kaplansky's 1946 paper [CK],

(2) the 1978 paper by D.D. Anderson $[A_1]$, which gave the first characterization of CK-domains,

(3) the 1982 paper by Glastad and Mott [GM] on finitely generated groups of divisibility reviewed in the previous section, and

(4) the 1992 joint paper by D.D. Anderson and J.L. Mott [AMo].

Cohen and Kaplansky promised a sequel to [CK], but it never appeared. That omission led me to begin a study of CK-domains with the intention of filling what I thought to be a gap in the knowledge (being unaware that D.D. Anderson had finished the characterization of CK-domains in $[A_1]$).

I reached a sticking-point in my labors and casually mentioned it to D.D. Anderson at a meeting in Lawrence, Kansas in the late 1980s. It turned out he had answered my problem in $[A_1]$. That began a collaboration that resulted in our 1992 paper on CK-domains and another paper in 1993 to be discussed in the next section.

The point is this: Had I known about $[A_1]$ likely I would not have started my investigation and perhaps Dan Anderson and I would not have had an occasion for our collaboration. Ignorance sometimes benefits research.

Now let us review some of the results from the four fundamental papers, starting with the Cohen and Kaplansky paper [CK]. Most of the proofs they give were elementary in the sense that few results were used from what is now standard commutative ring theory.

In the following discussion Max(D) will denote the set of maximal ideals D, and for a semilocal ring R, $\widehat{R}$ denotes the J-adic completion of R where J is the Jacobson radical of R, the intersection of all maximal ideals of R. Moreover, to complete our understanding of CK-domains, we need the important concept of analytically irreducible local rings. For (R, M) local, R is *analytically irreducible* (respectively, *analytically unramified*) if $\widehat{R}$ is an integral domain (respectively, is reduced, that is, it contains no nilpotent elements).

THEOREM 4.1.   (Cohen and Kaplansky [CK])

(1) A CK-domain D is a one-dimensional semilocal domain with the property that for each nonprincipal maximal ideal M, D/M is finite and D_M is analytically irreducible.

(2) If D is a CK-domain and M is a maximal ideal of D, then D_M is a CK-domain and the atoms of D_M are the atoms of D that are contained in M, or equivalently, are M-primary.

(3) An integral domain D is a CK-domain $\Leftrightarrow$ D is semiquasilocal and D_M is a CK-domain for each maximal ideal M of D.

(4) If (D, M) is a local CK-domain, then $\widehat{D}$, the M-adic completion of D, is a CK-domain and the atoms of $\widehat{D}$ are precisely the atoms of D.

Proof.   The proof can be obtained by combining Theorems 4, 6, 7 and 9 of [CK]. An alternate proof using more standard results of commutative ring theory can be found in [AMo].

Let me list a simple proof that a CK-domain is a one-dimensional semilocal domain. Let $\{p_1, p_2, \ldots, p_n\}$ be the complete list of nonassociative atoms of D. Then if P is a nonzero prime ideal of D, P has a basis that is a subset of $\{p_1, p_2, \ldots, p_n\}$. Therefore, each prime ideal of D is finitely generated. Cohen's Theorem [N_1, p. 8] or [Kap, Theorem 8] implies D is Noetherian. Moreover, D contains only finitely many prime ideals so D must be a one-dimensional semilocal domain [Kap, Theorem 144].   □

REMARK.   It follows from part (2) of Theorem 4.1 that atoms of a CK-domain are primary. This observation will play a role in Section 5.

Clearly, an integral domain D is a CK-domain $\Leftrightarrow$ the monoid of nonzero integral principal ideals of D is finitely generated. In $[A_1]$, D.D. Anderson investigated commutative rings with identity (possible including zero divisors) with this property and, as a by-product, proved the converse of Theorem 4.1(1). Anderson's terminology in $[A_1]$ was that a ring R is a *finite product ring* (or an *f.p. ring*) if the monoid of integral ideals of R is finitely generated, that is, if there exists a finite set $\{A_1, \ldots, A_s\}$ of integral ideals of R with the property that every integral ideal of R is a product of powers of ideals from the set $\{A_1, \ldots, A_s\}$. In the course of his investigation, Anderson observed that the f.p. condition was equivalent to two other finiteness conditions on a commutative ring R:

(1) R has only finitely many maximal ideals and the prime ideals of R are finite unions of principal ideals.

(2) R has only a finite number of irreducible principal ideals and every principal ideal of R is a finite product of these irreducible principal ideals.

Recall that an ideals $A \neq R$ is *irreducible* if A cannot be written as a product of two ideals both of which properly contain A. Anderson said that an irreducible principal ideal is an *atom* of R and he defined a ring R to be an *atomic ring* if every principal ideal is a finite product of atoms. The element $a \in R$ is called an atom $\Leftrightarrow$ the principal ideal aR is an atom. As you might suspect, Anderson's definition of atom coincides with the usual definition of atom in an integral domain.

The next theorem lists several of Anderson's conclusions about f.p. rings.

THEOREM 4.2. (D.D. Anderson $[A_1]$) *For a commutative ring R, the following are equivalent:*

(1) R is an f.p. ring (that is, the monoid of integral ideals of R is finitely generated).

(2) The monoid of integral principal ideals of R is finitely generated.

(3) R is an atomic ring containing only finitely many atoms.

(4) R is a finite direct sum of finite local rings, special principal ideal rings, and one-dimensional semilocal domains D_i with the property that for each nonprincipal maximal ideal M of D_i, D_i/M is finite and $(D_i)_M$ is analytically irreducible.

In the present context it would be appropriate to call an f.p. ring by the name of *CK-ring*.

After this short detour to consider rings with zero divisors, let us return to the discussion of integral domains in general and CK-domains in particular. The following well-known exercise from Nagata [N$_1$, Exercise 1, p. 122] (or [Kat, Corollary 5]) is useful in giving an alternate characterization of CK-domains.

PROPOSITION 4.3. (Nagata) *Let D be a one-dimensional semilocal domain. Then D is analytically unramified $\Leftrightarrow \overline{D}$ is a finitely generated D-module. In this case, the number of minimal prime ideals of the completion $\widehat{D}$ of D is equal to the number of maximal ideals of $\overline{D}$. In particular, $\widehat{D}$ is a direct product of integral domains $\Leftrightarrow |\operatorname{Max}(D)| = |\operatorname{Max}(\overline{D})|$. Hence for (D, M) local, D is analytically irreducible $\Leftrightarrow \overline{D}$ is a finitely generated D-module and $\overline{D}$ is a discrete valuation domain.*

It is well-known that $\overline{D}$ is a finitely generated D-module $\Leftrightarrow$ the conductor $[D : \overline{D}] = \{x \in K \mid x\overline{D} \subseteq D\}$ is nonzero. Moreover, combining Theorem 4.2 and Proposition 4.3, we get the following characterization of CK-domains.

THEOREM 4.4. (Anderson and Mott [AMo]) *For an integral domain D, the following statements are equivalent:*

(1) D is a CK-domain.

(2) D is a one-dimensional semilocal domain and for each nonprincipal maximal ideal M of D, D/M is finite, and D_M is analytically irreducible.

(3) D is a one-dimensional semilocal domain with D/M finite for each nonprincipal maximal ideal M of D, $\overline{D}$ is a finitely generated D-module (equivalently $[D : \overline{D}] \neq 0$), and $|\operatorname{Max}(D)| = |\operatorname{Max}(\overline{D})|$.

(4) $\overline{D}$ is a semilocal PID, $|\operatorname{Max}(D)| = |\operatorname{Max}(\overline{D})|$, $\overline{D}$ is a finitely generated D-module, and for each nonprincipal maximal ideal M of D, D/M is finite.

Proof. (1) $\Leftrightarrow$ (2) is contained in Theorem 4.2 simplified to the domain case.

(2) $\Leftrightarrow$ (3) follows from Proposition 4.3. Hence the new contribution of Theorem 3.4 is the equivalence of (3) and (4).

(3) $\Rightarrow$ (4) because the integral closure $\overline{D}$ of a one-dimensional semilocal domain D is a semilocal PID. This follows because each valuation overring is a DVR and $\overline{D}$ is Noetherian by Krull-Akizuki [N$_1$, p. 115], $\overline{D}$ is a Prüfer

domain by Nagata's Independence of Valuation Theorem [N_1, p. 38] or [Kap, Theorem 107]. Therefore, $\overline{D}$ is a Dedekind domain with only finitely many maximal ideals; each maximal ideal being invertible is therefore principal [Kap, Theorem 60].

(4) $\Rightarrow$ (3) since $\overline{D}$ is a finitely generated D-module and $\overline{D}$ Noetherian implies D is Noetherian by Eakin's Theorem [E]. □

An integral domain D is a CK-domain $\Leftrightarrow$ the monoid $P_+(D)$ of integral principal ideals of D is finitely generated, and because $P_+(D)$ is isomorphic to the positive cone of $G(D)$, it follows that for a CK-domain D, $G(D)$ is finitely generated. The converse is false even for Noetherian domains. To see how close the one concept is to the other for Noetherian domains, I list a characterization of Noetherian domains with finitely generated groups of divisibility. This characterization follows easily from results in Section 3 and results proved in [AMo].

THEOREM 4.5. (Anderson and Mott [AMo]) *Suppose that D is a Noetherian integral domain. Then the following statements are equivalent:*

(1) $G(D)$ is finitely generated.

(2) D is semilocal and $G(D_M)$ is finitely generated for each maximal ideal M of D.

(3) D is a one-dimensional semilocal domain and for each nonprincipal maximal ideal M of D, D/M is finite and D_M is analytically unramified.

(4) D is a one-dimensional semilocal domain, $\overline{D}$ is a finitely generated D-module (equivalently $[D : \overline{D}] \neq 0$) and for each nonprincipal maximal ideal M of D, D/M is finite.

Proof. (1) $\Leftrightarrow$ (2) Theorem 3.2.

(3) $\Leftrightarrow$ (4) Proposition 4.3.

To prove (2) $\Leftrightarrow$ (3) it suffices to show that if (D, M) is a local domain then $G(D)$ is finitely generated $\Leftrightarrow$ either D is a DVR or D is one-dimensional, analytically unramified, and D/M is finite. By [GM, Theorem 3.5] if D/M is infinite, then $G(D)$ is finitely generated $\Leftrightarrow$ D is a DVR. For the case where D/M is finite the equivalence follows from [GM, Theorem 4.1]. □

Now compare Theorem 4.4(2) with Theorem 4.5(3). The only difference is that for a CK-domain, the localization at a nonprincipal maximal ideal is analytically irreducible, while in a Noetherian domain with finitely generated group of divisibility such a localization is only analytically unramified. Thus, if (D, M) is one-dimensional local domain with D/M finite and the completion $\widehat{D}$ is reduced, but not a domain (equivalently, $\overline{D}$ is a finitely generated D-module with $|\operatorname{Max}(\overline{D})| > 1$), $G(D)$ is finitely generated but D is not a CK-domain. See [GM, Example 3.12] for such a domain.

Nevertheless, a CK-domain can be characterized, in terms of its group of divisibility.

THEOREM 4.6. (Anderson and Mott [AMo]) *Let D be an integral domain. Then the following statements are equivalent:*

(1) D is a CK-domain with $n = |\operatorname{Max}(D)|$.

(2) $G(D) \simeq \mathbb{Z}^n \oplus F$ where F is finite and $n = |\operatorname{Max}(D)| = |\operatorname{Max}(\overline{D})|$.

(3) D is Noetherian, $G(D)$ is finitely generated, and $n = |\operatorname{Max}(D)| = |\operatorname{Max}(\overline{D})|$.

Proof. (1) $\Rightarrow$ (2) By Theorem 4.4, $\overline{D}$ is a semilocal PID with n maximal ideals, so $G(\overline{D}) \simeq \mathbb{Z}^n$. Hence $G(D) \simeq G(\overline{D}) \oplus U(\overline{D})/U(D)$ is isomorphic to $\mathbb{Z}^n \oplus F$ where $F = U(\overline{D})/U(D)$ is finite by Theorem 3.4.

(2) $\Rightarrow$ (3) By Theorems 3.1 and 3.4, $\overline{D}$ is Bezout domain with $G(\overline{D}) \simeq \mathbb{Z}^n$. Also $n = |\operatorname{Max}(D)| \leq |\operatorname{Max}(\overline{D})| \leq \operatorname{rank} G(\overline{D}) = n$, so $n = |\operatorname{Max}(D)| = |\operatorname{Max}(\overline{D})|$. Let $M_1, \ldots, M_n$ be the maximal ideals of $\overline{D}$. By induction on n, $\overline{D}$ is a PID with n maximal ideals. By Eakin's Theorem, D is Noetherian since $\overline{D}$ is a finitely generated D-module.

(3) $\Rightarrow$ (1) follows by combining Theorems 4.4 and 4.5. $\square$

REMARKS. As immediate corollaries of Theorem 4.6, we observe:

(1) An integral domain D is a PID with n maximal ideals $\Leftrightarrow G(D) \simeq \mathbb{Z}^n$ and D has at least n maximal ideals.

(2) An integral domain D is a local CK-domain $\Leftrightarrow G(D) \simeq \mathbb{Z} \oplus F$ where F is finite.

The following theorem summarizes the contents of Theorems 4.2, 4.4 and 4.6.

THEOREM 4.7. *For an integral domain D the following statements are equivalent:*

(1) D is a CK-domain, that is, D is an atomic domain containing only a finite number of atoms.

(2) The monoid $F_+(D)$ of integral ideals of D is finitely generated.

(3) The monoid $F_+^*(D)$ of finitely generated integral ideals of D is finitely generated.

(4) The monoid $\mathrm{Inv}_+(D)$ of invertible integral ideals of D is finitely generated.

(5) The monoid $P_+(D) \simeq G_+(D)$ of integral principal ideals of D is finitely generated.

(6) D is a one-dimensional semilocal domain and for each nonprincipal maximal ideal M of D, D/M is finite, and D_M is analytically irreducible.

(7) D is a one-dimensional semilocal domain, D/M is finite for each nonprincipal maximal ideal M of D, $\overline{D}$ is a finitely generated D-module, and $|\mathrm{Max}(D)| = |\mathrm{Max}(\overline{D})|$.

(8) $\overline{D}$ is a semilocal PID with $|\mathrm{Max}(D)| = |\mathrm{Max}(\overline{D})|$, $\overline{D}$ is a finitely generated D-module, and D/M is finite for each nonprincipal maximal ideal M of D.

(9) $\overline{D}$ is a semilocal PID with $\overline{D}/[D : \overline{D}]$ finite and $|\mathrm{Max}(D)| = |\mathrm{Max}(\overline{D})|$.

(10) $G(D)$ is finitely generated and rank $G(D) = |\mathrm{Max}(D)|$.

(11) D is a Noetherian domain with $G(D)$ finitely generated and $|\mathrm{Max}(D)| = |\mathrm{Max}(\overline{D})|$.

Proof.   Anderson's Theorem 4.2 proves (1), (2), (5) and (6) are equivalent. Theorem 4.4 shows that (1), (6), (7) and (8) are equivalent. (2) $\Rightarrow$ (3) $\Rightarrow$ (4) $\Rightarrow$ (5) is elementary. Theorem 3.6 demonstrates the equivalence of (1), (10) and (11). Finally, (8) $\Leftrightarrow$ (9) follows essentially by an application of Eakin's theorem. $\Box$

REMARK.   In [AMo], four other equivalences are established by using the fact that $\overline{D}/D$ is torsion.

5. SPECIAL ATOMIC FACTORIZATION DOMAINS

In [Mot$_4$] I traced the history of several attempts to replace unique atomic factorization by some weaker concept. These attempts had been influenced by Kummer's success in proving a special case of Fermat's Last Theorem.

In this section I would like to review some recent work on atomic domains for which the unique atomic factorization assumption is weakened.

Recall that an integral domain D satisfies the *ascending chain condition on principal ideals* (ACCP) if there does not exist an infinite strictly ascending chain of principal integral ideals of D. Any UFD, Noetherian domain, Krull domain, or domain with ACCP is atomic, but not conversely. Grams [Gr] gave the first example of an atomic domain without ACCP and Zaks [Z$_3$] added several other examples.

The difference between an integral domain D being atomic and satisfying ACCP is best seen in terms of the group of divisibility $G(D) \simeq P(D)$. The domain D satisfies ACCP $\Leftrightarrow$ *each* chain in $P(D)$ is finite; while D is atomic precisely when for each $x \in R^*$, *some* maximal chain starting at xD is finite. (This follows since there are no principal ideals between $yD \subset zD \Leftrightarrow z/y$ is irreducible.) Thus D is atomic $\Leftrightarrow$ each positive element of $G(D)$ is a finite sum of minimal positive elements, and D satisfies ACCP $\Leftrightarrow$ there does not exist an infinite strictly decreasing sequence of positive elements in $G(D)$.

Many classes of domains can be studied for their various atomic factorizations. Of course, the classical situation is when each nonzero, nonunit element of an atomic domain has an atomic factorization that is unique up to order of factors and associates. Such an integral domain is, as we all know, a *unique factorization domain* (UFD) or *factorial domain*.

In a sequence of at least five papers [AAZ$_1$–AAZ$_5$], D.D. Anderson, D.F. Anderson and M. Zafrullah study various factorization properties of integral domains that are weaker than UFD. For instance, they define in [AAZ$_2$] the domain D to be a *bounded factorization domain* (BFD) if D is atomic and for each nonzero, nonunit $d \in D$ there is a bound on the number of atoms in all atomic factorizations of d. Zaks [Z$_1$] called the domain D a *half-factorial domain* (HFD) if D is atomic and for each nonzero, nonunit $d \in D$ all atomic factorizations of d have the same length. While Zaks first introduced the terminology, Carlitz [C] was the first to study such domains when he showed that the ring D of algebraic integers in a finite field extension of the rationals is an HFD $\Leftrightarrow |C(D)| \leq 2$, where $C(D)$ is the class group for D.

More generally, a Krull domain whose divisor class group has order 2 is an HFD [Z_2].

Grams and Warner [GW] introduced the concept of idf-domain. An integral domain D is an idf-domain (for *irreducible divisor-finite-domain*) if each nonzero, nonunit of D has at most a finite number of nonassociate irreducible divisors. In [AAZ_2] an atomic idf-domain is called a *finite factorization domain* (FFD) since each nonzero, nonunit will have only a finite number of nonassociate divisors (and hence, only a finite number of atomic factorizations up to order and associates).

Each of the above factorization properties for a domain have their corresponding interpretation in $G(D)$. In [AAZ_2], the Andersons and Zafrullah observe the implication relationships depicted in Figure 1 and give examples to show that no other implications are possible.

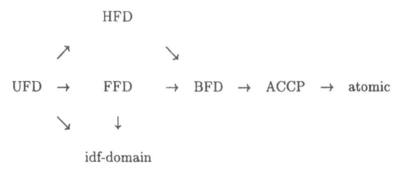

Figure 1. Special Factorization Implications.

Many other research papers [AA_1, AA_2, AAZ_1–AAZ_5, ACS, AP, GH-K, Ru] study atomic domains with various factorization properties that are weaker than UFD. Some papers explore such factorization properties in the more general setting of monoids [H-K_1] and, in particular, monoids with divisor theory [G_1, G_2, H-K_2, Sk_1, Sk_2]. But unique factorization continues to be of interest [Be, H, L].

In the remainder of this section I want to discuss how certain special atomic factorizations relate to the CK-domains discussed in Section 4. The investigation will consider three types of atomic factorization domains: the finite factorization domains (FFD) defined above, generalized CK-domains, and weakly factorial domains.

In the paper [AAZ_5], the Anderson brothers and Zafrullah consider *generalized CK-domains*, atomic domains for which all but a finite number of atoms are prime. An integral domain D is *weakly factorial* of each nonunit of D is a finite product of primary elements. Weakly factorial domains are studied in [AZ] and [AAZ$_5$].

EXAMPLE 5.1. [AAZ$_5$, Example 9] Let K be a finite field with $k \subseteq K$ a subfield. Let $n \geq 1$ be an integer. Then $D = k + x^n K[x]$ is a generalized CK-domain but not a CK-domain. Also $Z[2i]$ is a generalized CK-domain but not a CKD.

In the remark following Theorem 4.1, the observation was made that CKD $\Rightarrow$ weakly factorial. Anderson and Mott [AMo] give necessary and sufficient conditions for a CKD to be an HFD. More can be said. I begin by reviewing some results in [AMu] on finite factorization domains (FFD).

THEOREM 5.1. *For an integral domain D, the following conditions are equivalent:*

(1) D is an FFD.

(2) Every nonzero (principal) ideal of D is contained in only finitely many principal ideals.

(3) For each $x \in G(D)$ with $x \geq 0$, the interval $[0, x]$ is finite.

(4) For any collection of distinct principal ideals $\{(r_\alpha)\}$ of D, $\bigcap_\alpha (r_\alpha) = (0)$.

(5) Every nonzero element of D has only a finite number of factorizations, up to associates.

(6) D is an atomic idf-domain.

Proof. Clearly (1)–(5) are equivalent and (1) $\Rightarrow$ (6). That (6) $\Rightarrow$ (1) was proved in [AAZ$_2$, Theorem 5.1]. $\square$

EXAMPLE 5.2. [AMu]. Let D be a domain with $D/(a)$ finite for each $0 \neq a \in D$. Then D is an FFD. In particular, if D is a subring of the ring of integers in a finite extension of Q, D is an FFD. Rings with the property that each proper homomorphic image is finite were studied by K. Levitz and J.L. Mott [LM]. There we observed that an integral domain D has the property that D/I is finite for each proper (principal) ideal $\Leftrightarrow$ D is a field or D is a one-dimensional Noetherian domain with each residue field finite.

THEOREM 5.2. (Anderson and Mullins [AMu]) *Let $D = \bigcap_\alpha D_\alpha$ be a*

*locally finite intersection of domains D_α each of which is an FFD. Then D
is an FFD.*

Proof. First note that $G(D)$ is isomorphic to a subgroup of the cardinal sum of the groups $G(D_\alpha)$. Then apply Theorem 5.1. □

A special case of the following theorem was proved first in [H-K₁] in the context of semigroups but then generalized to the present form by Anderson and Mullins.

THEOREM 5.3. (Anderson and Mullins [AMu]) *For a Noetherian domain D, the following conditions are equivalent.*

(1) D is an FFD.

(2) If D' is an overring of D with D' a finitely generated D-module, then $U(D')/U(D)$ is finite.

(3) There is an FFD overring D' of D that is integral over D such that if S is an overring of D with $D \subseteq S \subseteq D'$ where S is a finitely generated D-module, then $U(S)/U(D)$ is finite.

Proof. See [AMu, Theorem 6]. □

COROLLARY 5.4. (Anderson and Mullins [AMu]) *Let D be a Noetherian domain with $\overline{D}$ a finitely generated D-module. Then D is an FFD $\Leftrightarrow$ $U(\overline{D})/U(D)$ is finite.*

THEOREM 5.5. *If D is a CK-domain, then D is an FFD.*

Proof. If D is a CKD, then D is Noetherian and $G(D)$ is finitely generated by Theorem 4.7(11). Then Theorem 3.4 $\Rightarrow$ $\overline{D}$ is a finitely generated D-module and $U(\overline{D})/U(D)$ is finite. Therefore, Corollary 5.4 $\Rightarrow$ D is an FFD. □

EXAMPLE 5.3. The converse of Theorem 5.5 is false. Example 11 of [AMu] is an example of a local FFD (R, M) where $\overline{R}$ is not finitely generated so R is not a CK-domain. The ring R is obtained as follows: Let k be a finite field of characteristic $p \neq 0$ and let $Y \in Xk[[X]]$ with X and Y algebraically independent over k. Let $R = V[Y]$ where $V = k[[X]] \cap k(X, Y') \subset W =$

$k[[X]] \cap k(X, Y)$. The authors show $W = \overline{R}$ and that $\overline{R}$ is not a finitely generated R-module.

THEOREM 5.6. (Anderson and Mullins [AMu]) *Let (D, M) be a quasilocal FFD with D/M infinite. Then D is integrally closed. Thus a local domain (D, M) with D/M infinite is an FFD $\Leftrightarrow$ D is integrally closed.*

Proof. Combine Theorems 4 and 7 of [AMu].   □

Theorem 5.5 can also be realized as a corollary to the following theorem.

THEOREM 5.7. (Anderson and Mullins [AMu]) *A one-dimensional semilocal domain D is an FFD $\Leftrightarrow$ for each maximal ideal M of D with D/M infinite, D_M is a DVR.*

Proof. Let $M_1, \ldots, M_n$ be the maximal ideals of D. since $G(D)$ is order-isomorphic to the cardinal sum $\sum_{i=1}^{n} G(D_{M_i})$ by Theorem 3.2(2), D is FFD $\Leftrightarrow$ each D_{M_i} is an FFD. The quotient ring D_{M_i} has residue field D/M_i. So D_{M_i} is an FFD $\Leftrightarrow$ either D/M_i is finite (Example 1) or D/M_i is infinite and, therefore, D_{M_i} integrally closed by Theorem 5.6 and, hence, a DVR.□

Now let us turn our attention to the relationship between generalized CK-domains and FFDs.

THEOREM 5.8. *A generalized CK-domain is an FFD.*

Proof. Let D be a generalized CK-domain and let S be the multiplicative system generated by all the prime elements of D. Then consider the exact sequence:

$$\{1\} \rightarrow \langle S \rangle \rightarrow G(D) \rightarrow G(D_S) \rightarrow \{1\}.$$

By Corollary 1.7 of [AAZ$_1$] any multiplicative system generated by prime elements in an atomic domain is a *splitting* multiplicative system. This means that $G(D)$ is order isomorphism to the cardinal sum $\langle S \rangle \oplus G(D_S)$ by [AAZ$_1$, Theorem 2.2] and [MS, Proposition 4.1]. By [AAZ$_1$, Corollary 1.4] the finite number of atoms of D that were not prime become all the atoms of D_S. Moreover, D_S is atomic by [AAZ$_1$, Theorem 3.1]. Therefore, we conclude D_S is a CK-domain, and, in turn, an FFD by Theorem 5.5. Theorem 3.1 of [AAZ$_1$] implies that D is an FFD.   □

EXAMPLE 5.4.   Let $k_1 \subset k_2$ be fields where k_1 is infinite. Let $D = k_1 + Xk_2[X]$. Then Anderson and Zafrullah observe in [AZ] that D is a weakly factorial domain. Hence, D is not an FFD by [AAZ$_2$, Proposition 5.2] or [AMu, Example 5]. For another example, let (D, M) be a one-dimensional non-integrally closed local domain with D/M infinite. Then D is a weakly factorial domain but not an FFD by Theorem 5.6 or 5.7.

THEOREM 5.9.   (Anderson, Anderson, and Zafrullah [AAZ$_5$]) *Let D be an atomic domain in which all but a finite number of atoms are primary. Then every atom is primary and D is weakly factorial.*

Proof.   See [AAZ$_5$, Theorem 4].   □

COROLLARY 5.10.   (Anderson, Anderson, and Zafrullah [AAZ$_5$]) *A generalized CK-domain is weakly factorial.*

We now have the following diagram of the relationship between CKD, FFD generalized CK-domains, and weakly factorial domains; no implication can be reversed.

$$\text{FFD}$$

$$\nearrow \qquad \uparrow$$

$$\text{CKD} \;\rightarrow\; \text{generalized CK-domain} \;\rightarrow\; \text{weakly factorial domain}$$

Figure 2. Additional Special Factorization Implications.

The main theorem of [AAZ$_5$] is the following.

THEOREM 5.11.   (Anderson, Anderson, and Zafrullah [AAZ$_5$]) *An integral domain D is a generalized CK-domain if and only if*
(1) $\overline{D}$ is a UFD,
(2) $D \subseteq \overline{D}$ is a root extension (that is, for each $x \in \overline{D}$, there is a positive integer $n = n(x)$ with $x^n \in D$),

(3) the conductor $[D : \overline{D}]$ is principal as an ideal of $\overline{D}$, and

(4) $\overline{D}/[D : \overline{D}]$ is finite.

Moreover, in the proof of their main theorem, the authors observe that a generalized CK-domain is in a natural way an intersection of a UFD and a CK-domain. In addition, they observe that every generalized CK-domain D may be realized as a pull-back of the following type:

$$
\begin{array}{ccc}
D & \hookrightarrow & T \\
\downarrow & & \downarrow \\
S & \hookrightarrow & T/I
\end{array}
$$

where T is a UFD, I is a principal ideal of T with T/I finite, and S is a subring of T/I such that $S \subseteq T/I$ is a root extension.

I close this section with the following characterization of weakly factorial domains from [AZ]. We need the concept of t-class group. For a nonzero fractional ideal I of D, define $I_t = \cup\{(a_1, \ldots, a_n)_v | 0 \neq (a_1, \ldots, a_n) \subseteq I\}$ where $I_v = (I^{-1})^{-1} = [D : [D : I]]$. A fractional ideal I is said to be a t-ideal if $I_t = I$ and I is t-invertible if there is a fractional ideal J of D such that $(IJ)_t = D$. For any integral domain D, the set of t-invertible t-ideals of D forms a group under the t-product $A * B = (AB)_t$. The t-class group of D, denoted by $C\ell_t(D)$, is the group of t-invertible t-ideals modulo the subgroup of principal fractional ideals $P(D)$. Let $X^{(1)} = \{$ height-one prime ideals of $D\}$.

THEOREM 5.12. (Anderson and Zafrullah [AZ]) *Let D be an integral domain. Then the following statements are equivalent:*

(1) Every convex directed subgroup of $G(D)$ is a cardinal summand of $G(D)$.

(2) For each saturated multiplicatively closed subset S of D, $\langle S \rangle$ is a cardinal summand of $G(D)$.

(3) For each prime ideal P of D, $\langle D\backslash P \rangle$ is a cardinal summand of $G(D)$.

(4) D is weakly factorial.

(5) $D = \bigcap_{P \in X^{(1)}} D_P$ is a locally finite intersection and the natural map $G(D)$ into the cardinal sum $\sum_{P \in X^{(1)}} G(D_P)$ is an order isomorphism.

(6) If P is a prime ideal of D minimal over a proper principal ideal (x), then ht $P = 1$ and $(x)D_P \cap D$ is principal.

(7) $\bigcap_{P \in X^{(1)}} D_P$ is a locally finite intersection and $C\ell_t(D) = 0$.

Proof. See [AZ].   □

EXAMPLE 5.5. Let D be a Dedekind domain that is not a PID. Then D is FFD since any Krull domain is FFD [AAZ_2, Theorem 5.1]. But D is not weakly factorial since by Theorem 5.12(7), the class group of D is nontrivial.

6. DOMAINS FOR WHICH $F(D)$ IS FINITELY GENERATED

The purpose of this section is to determine when the monoid $F(D)$ of nonzero fractional ideals of an integral domain D is finitely generated. Of course, to say that $F(D)$ is finitely generated means that there exist $I_1, \ldots, I_n \in F(D)$ such that every $I \in F(D)$ may be written as the product $I = I_1^{s_1} \cdots I_n^{s_n}$ where the s_i are nonnegative integers. Theorem 4.7 includes the following equivalences:

(1) D is CK-domain, that is, D is atomic and contains only finitely many atoms,

(2) $F_+(D)$ is finitely generated,

(3) $F_+^*(D)$ is finitely generated,

(4) $Inv_+(D)$ is finitely generated, and

(5) $P_+(D) \simeq G_+(D)$ is finitely generated.

It is easy to see that $F_+(D)$ finitely generated implies that $F(D)$ is finitely generated. More can be said.

THEOREM 6.1.   (Anderson, Mott, and Park [AMP]) *Let D be an integral domain. Then $F(D)$ finitely generated implies that $G(D)$ is finitely generated.*

Proof. Suppose that $F(D)$ is generated by $I_1, \ldots, I_n$. Suppose $I \in Inv(D)$ and $I = I_1^{s_1} \cdots I_n^{s_n}$ where the exponents $s_i \geq 0$ for each i. For each i where $s_i \geq 1$, then I_i is invertible since a factor of an invertible ideal is invertible. Thus the invertible elements of $\{I_1, \ldots, I_n\}$ generate $Inv(D)$. So $F(D)$ finitely generated $\Rightarrow$ the group $Inv(D)$ is finitely generated $\Rightarrow$ the subgroup $P(D) \simeq G(D)$ is finitely generated.   □

The converse of Theorem 6.1 is false. One can see that if V is a rank one *nondiscrete* valuation domain with $G(V) \simeq \mathbb{Z}^n$, where $n \geq 2$, then $F(V)$ is not finitely generated.

If $F(D)$ is finitely generated, then $G(D)$ is finitely generated and hence by Theorem 3.4, $\overline{D}$ is a finitely generated D-module. Therefore, if $A \in F(\overline{D})$, then there is $0 \neq r \in D$ such that $rA \subseteq D$. But $\overline{D}$ a finitely generated D-module implies there is $0 \neq d \in D$ such that $drA \subseteq d\overline{D} \subseteq D$. Thus, $A \in F(D)$ and $F(\overline{D})$ is submonoid of $F(D)$. We know that a subgroup of finitely generated abelian group is finitely generated, but in general it may be that a submonoid of a finitely generated monoid is not finitely generated. Nevertheless, $F(D)$ finitely generated does imply $F(\overline{D})$ is finitely generated.

THEOREM 6.2. (Anderson, Mott, and Park [AMP]) *Let D be an integral domain such that $F(D)$ is generated by $I_1, \ldots, I_n$.*

(1) Let D_1 be an overring of D with $D \subseteq D_1 \subseteq \overline{D}$. Then $I_1 D_1, \ldots, I_n D_1$ generate $F(D_1)$. Thus $F(D)$ is finitely generated $\Rightarrow F(D_1)$ is also finitely generated.

(2) Let S be a multiplicatively closed subset of D. Then $I_1 D_S, \ldots, I_n D_S$ generate $F(D_S)$.

Proof. (1) Let $B \in F(D_1)$. As above, there exists $0 \neq r \in D$ such that $rB \subseteq D_1 \subseteq \overline{D}$. Since $\overline{D}$ is a finitely generated D-module, there exists $0 \neq d \in D$ such that $drB \subseteq D$; so $B \in F(D)$. Thus, $B = I_1^{s_1} \cdots I_n^{s_n}$ where each $s_i \geq 0$. But then $B = BD_1 = (I_1 D_1)^{s_1} \cdots (I_n D_1)^{s_n}$ and $I_1 D_1, \ldots, I_n D_1$ generate $F(D_1)$.

(2) Let $B \in F(D_S)$. Then $B = AD_s$ for some $A \in F(D)$. Since $A = I_1^{s_1} \cdots I_n^{s_n}$, it follows that $B = AD_S = (I_1 D_S)^{s_1} \cdots (I_n D_S)^{s_n}$ and $I_1 D_S, \ldots, I_n D_S$ generate $F(D_S)$. $\square$

Corollary 3.8 motivated the concept of a strongly finitely generated submonoid of $F(D)$. In particular, $F(D)$ strongly finitely generated means that $G(D)$ is finitely generated and there exist $A_1, \ldots, A_n \in F(D)$ such that every $A \in F(D)$ can be written in the form $A = xA_i$ for some $x \in K^*$ and for some $1 \leq i \leq n$. The following conclusion is immediate.

PROPOSITION 6.3. *Let D be an integral domain. Then $F(D)$ is strongly finitely generated $\Leftrightarrow G(D)$ is finitely generated and $F(D)/G(D)$ is finite.*

Certainly $F(D)$ strongly finitely generated $\Rightarrow F(D)$ is finitely generated. The converse is also true.

THEOREM 6.4. (Anderson, Mott, and Park [AMP]) *For an integral domain the following conditions are equivalent:*

(1) $F(D)$ is strongly finitely generated.

(2) $F(D)$ is finitely generated.

(3) $F(\overline{D})$ is finitely generated and $\overline{D}/[D:\overline{D}]$ is finite.

(4) $F(\overline{D})$ is strongly finitely generated and $\overline{D}/[D:\overline{D}]$ is finite.

(5) $\overline{D}$ is a semiquasilocal Bezout domain such that $\overline{D}_M$ is a finite rank discrete valuation domain for each maximal ideal M of $\overline{D}$ and $\overline{D}/[D:\overline{D}]$ is finite.

Proof. As I have said, (1) $\Rightarrow$ (2) is clear. Moreover, (2) $\Rightarrow$ (3) because if $F(D)$ finitely generated, then Theorem 6.2 implies that $F(\overline{D})$ is finitely generated. Also, $F(D)$ finitely generated implies that $G(D)$ is finitely generated by Theorem 6.1 and hence that $\overline{D}/[D:\overline{D}]$ is finite by Theorem 3.6 or 3.7. Clearly (4) $\Rightarrow$ (3) since strongly finitely generated implies finitely generated.

To conclude (3) $\Rightarrow$ (5) first note that $F(\overline{D})$ finitely generated implies that $G(\overline{D})$ is finitely generated by Theorem 6.1. But then we conclude from Theorem 3.1 that $\overline{D}$ is a Bezout domain with only finitely many prime ideals. Let M be a maximal ideal of $\overline{D}$. Then for the valuation domain $\overline{D}_M$, Theorem 6.2 implies that $F(\overline{D}_M)$ is finitely generated and $G(\overline{D}_M)$ is finitely generated so $\overline{D}_M$ has finite Krull dimension. Thus for each pair of adjacent primes $P \subset Q$ in $\overline{D}_M$, $F(\overline{D}_Q)$ is finitely generated so $F_+(\overline{D}_Q)$ is countable and $F_+(\overline{D}_Q/P\overline{D}_Q)$ is countable. Lemma 3.1 of [AMP] shows that $\overline{D}_Q/P\overline{D}_Q$ is a DVR. Thus, $\overline{D}_M$ is a discrete valuation domain of finite rank.

(5) $\Rightarrow$ (4) First observe that $G(\overline{D})$ is finitely generated. Let $M_1, \ldots, M_n$ be the maximal ideals of $\overline{D}$. Then each $\overline{D}_{M_i}$ is a discrete valuation domain of finite rank, so $G(\overline{D}_{M_i})$ is a free abelian group of rank equal to $\dim \overline{D}_{M_i}$. Since $G(\overline{D})$ is isomorphic to a subgroup of the cardinal sum $G(\overline{D}_{M_1}) \oplus \cdots \oplus G(\overline{D}_{M_n})$, $G(\overline{D})$ is also finitely generated.

The remainder of the proof that (5) $\Rightarrow$ (4) requires Lemmas 4.1, 4.2, 4.3 and 4.4 from [AMP]. Thus (3) $\Leftrightarrow$ (4) $\Leftrightarrow$ (5).

Finally we conclude the final implication (4) $\Rightarrow$ (1) from the following theorem.   □

THEOREM 6.5. (Anderson, Mott, and Park [AMP]) *Let D be an integral domain. Then $F(D)$ is strongly finitely generated $\Leftrightarrow$ $F(\overline{D})$ is strongly finitely generated and $\overline{D}/[D:\overline{D}]$ is finite.*

Proof. ($\Rightarrow$) Suppose $F(D)$ is strongly finitely generated. Then $G(D)$ is finitely generated by definition and hence $G(\overline{D})$ is finitely generated and $\overline{D}/[D:\overline{D}]$ is finite by Theorem 3.7. Since $F(D)$ is finitely generated, the homomorphism $F(D) \to F(\overline{D})$ given by $A \to A\overline{D}$ is surjective. We know from Theorem 6.2 that $F(\overline{D})$ is finitely generated but we can say more here. Suppose $A_1, \ldots A_n$ are strong generators for $F(D)$. Then if $A \in F(D)$, $A = xA_i$ for some $x \in K^*$ and some $1 \leq i \leq n$. But then $A\overline{D} = xA_i\overline{D}$, and it follows that $A_i\overline{D}, \ldots, A_n\overline{D}$ are strong generators for $F(\overline{D})$.

($\Leftarrow$) First observe that $F(\overline{D})$ strongly finitely generated implies that $G(\overline{D})$ is finitely generated. Since $\overline{D}/[D:\overline{D}]$ is finite, Theorem 3.7 implies that $G(D)$ is finitely generated. Let $A_1, \ldots, A_n$ be a set of strong generators for $F(\overline{D})$. For each i, the set $B_D(A_i) = \{C \in F(D) | C\overline{D} = A_i\}$ is finite by Lemma 5.1 of [AMP]. Let $B_D(A_i) = \{B_{i_1}, \ldots B_{i_{n_i}}\}$. Then for $A \in F(D)$, $A\overline{D} = xA_i$ for some $x \in K^*$ and some $1 \leq i \leq n$. Therefore $x^{-1}A\overline{D} = A_i$ and $x^{-1}A = B_{ij}$ for some $1 \leq j \leq n_i$. Thus $A = xB_{ij}$. We conclude that $\bigcup_{i=1}^{n} B_D(A_i)$ is a set of strong generators for $F(D)$.   □

REMARK. Any integral domain D with $F(D)$ finitely generated can be obtained as a pull-back construction (see [AMP, Corollary 5.4]).

REFERENCES

[A] M. Anderson, *Lattice-ordered groups of divisibility: an expository intro-duction*, Ordered Algebraic Structures (Curacao, 1988), Kluwer Academic Publ., Dordrecht-Boston-London (1989), 3–9.

[A₁] D.D. Anderson, *Some finiteness conditions on a commutative ring*, Houston J. Math. **4** (1978), 289–299.

[A₂] D.D. Anderson, *Integral domains with finitely generated groups of divisibility*, Proc. Amer. Math. Soc. **112** (1991), 613–618.

[AA$_1$] D.D. Anderson and D.F. Anderson, *Elasticity of factorizations in integral domains II*, Houston J. Math. **20** (1994), 1–15.

[AA$_2$] D.D. Anderson and D.F. Anderson, *Elasticity of factorizations in integral domains*, J. Pure Appl. Algebra **80** (1992), 217–235.

[AAZ$_1$] D.D. Anderson, D.F. Anderson and M. Zafrullah, *Factorization in integral domains II*, J. Algebra **152** (1992), 78–93.

[AAZ$_2$] D.D. Anderson, D.F. Anderson and M. Zafrullah, *Factorization in integral domains*, J. Pure Appl. Algebra **69** (1990), 1–19.

[AAZ$_3$] D.D. Anderson, D.F. Anderson and M. Zafrullah, *Splitting the t-class group*, J. Pure Appl. Algebra **74** (1991), 17–37.

[AAZ$_4$] D.D. Anderson, D.F. Anderson and M. Zafrullah, *Rings between $D[x]$ and $K[x]$*, Houston J. Math. **17** (1991), 109–129.

[AAZ$_5$] D.D. Anderson, D.F. Anderson and M. Zafrullah, *Atomic domains in which almost all atoms are prime*, Comm. Algebra **20**, (1992), 1447–1462.

[ACS] D.F. Anderson, S.T. Chapman and W.W. Smith, *Factorization sets and half-factorial sets in integral domains*, J. Algebra **178** (1995), 92–121.

[AF] M. Anderson and T. Feil, *Lattice-Ordered Groups: An Introduction*, D. Reidel Publishing Co., Dordrecht, Holland (1988).

[AM] J. Alajbegovic and J. Močkoř, *Approximation theorems in commutative algebra: Classical and categorical methods*, Mathematics and its Applications (East European Series), Vol. 59, Kluwer Academic Publ., Dordrecht-Boston-London (1992).

[AMa] D.D. Anderson and L.A. Mahaney, *On primary factorizations*, J. Pure Appl. Algebra **54** (1988), 141–154.

[AMu] D.D. Anderson and B. Mullins, *Finite factorization domains*, Proc. Amer. Math. Soc. **124** (1996), 389–396.

[AMo] D.D. Anderson and J.L. Mott, *Cohen-Kaplansky domains: integral domains with a finite number of irreducible elements*, J. Algebra **148** (1992), 17–41.

[AMP] D.D. Anderson, J.L. Mott and J. Park, *Finitely generated monoids of fractional ideals*, Comm. Algebra **21** (1993), 615–634.

[AP] D.F. Anderson and P. Pruis, *Length functions on integral domains*, Proc. Amer. Math. Soc. **113** (1991), 933–937.

[AZ] D.D. Anderson and M. Zafrullah, *Weakly factorial domains and groups of divisibility*, Proc. Amer. Math. Soc. **109** (1990), 907-913.

[Ba$_1$] G.G. Bastos, *Quasi-semi-valuations and groups of divisibility*, Math.

Student **50** (1982), 257–263 (1987).

[Ba₂] G.G. Bastos, *A new class of ordered abelian groups which are not groups of divisibility*, C.R. Acad. Sci. Paris Ser. I Math. **306** (1988), 17–20.

[Be] R.A. Beauregard, *When is F[x, y] a unique factorization domain?*, Proc. Amer. Math. Soc. **117** (1993), 67–70.

[Br] A. Brandis, *Uber die multiplikative Struktur von Korpererweiterungen*, Math. Z. **87** (1965), 71–73.

[C] L. Carlitz, *A characterization of algebraic number fields with class number two*, Proc. Amer. Math. Soc. **11** (1960), 391–392.

[C₁] P.M. Cohn, *Bezout rings and their subrings*, Proc. Cambridge Philos. Soc. **64** (1968), 251–264.

[C₂] P.M. Cohn, *Unique factorization domains*, Amer. Math. Monthly **80** (1973), 1–18.

[CK] I.S. Cohen and I. Kaplansky, *Rings with a finite number of primes, I*, Trans. Amer. Math. Soc. **60** (1946), 468–477.

[CS₁] S.T. Chapman and W.W. Smith, *An analysis using the Zaks-Skula constant of element factorization in Dedekind domains*, J. Algebra **159** (1993), 176–190.

[CS₂] S.T. Chapman and W.W. Smith, *Factorization in Dedekind domains with finite class group*, Israel J. Math. **71** (1990), 65–95.

[CS₃] S.T. Chapman and W.W. Smith, *On the HFD, CHFD and k-HFD properties in Dedekind domains*, Comm. Algebra **20** (1992), 1955–1987.

[D] J. Dieudonné, *Sur la théorie de la divisibilité*, Bull. Soc. Math. France **49** (1941), 1–12.

[E] P.M. Eakin, *The converse to a well-known theorem on Noetherian rings*, Math. Ann. **177** (1968), 278–282.

[G₁] A. Geroldinger, *Arithmetical characterizations of divisor class groups*, Arch. Math. (Basel) **54** (1990), 455–464.

[G₂] A. Geroldinger, *Arithmetical characterizations of divisor class groups, II*, Acta Math. Univ. Comenian. (N.S.) **61** (1992), 193–208.

[GH-K] A. Geroldinger and F. Halter-Koch, *On the asymptotic behavior of lengths of factorizations*, J. Pure Appl. Algebra **77** (1992), 239–252.

[Gl] A.M.W. Glass, *A directed d-group that is not a group of divisibility*, Czechoslovak Math. J. **34** (109) (1984), no. 3, 475–476.

[GM] B. Glastad and J.L. Mott, *Finitely generated groups of divisibility*, Ordered fields and real algebraic geometry (San Francisco, Calif., 1981), 231–247. Contemp. Math. 8, Amer. Math. Soc., Providence, R.I. (1982).

[Gr] A. Grams, *Atomic rings and the ascending chain condition for principal ideals*, Proc. Cambridge Philos. Soc. **75** (1974), 321–329.

[GW] A. Grams and H. Warner, *Irreducible divisors in domains of finite character*, Duke Math. J. **42** (1975), 271–284.

[H-K₁] F. Halter-Koch, *Finiteness theorems for factorizations*, Semigroup Forum **44** (1992), 112–117.

[H-K₂] F. Halter-Koch, *Ein Approximationssatz für Halbgruppen mit Divisorentheorie*, Resultate Math. **19** (1991), 74-82.

[H] R.C. Heitmann, *Characterization of completions of unique factorization domains*, Trans. Amer. Math. Soc. **337** (1993), 379–387.

[He] W. Heinzer, *Some remarks on complete integral closure*, J. Austral. Math. Soc. **9** (1969), 310–314.

[HL] W. Heinzer and D. Lantz, *ACCP in polynomial rings: a counterexample*, Proc. Amer. Math. Soc. **121** (1994), 975–977.

[J] P. Jaffard, *Contribution à l'étude des groupes ordonnés*, J. Math. Pures Appl. (9) **32** (1953), 203–280.

[Kap] I. Kaplansky, *Commutative Rings*, Revised Edition, The University of Chicago Press, Chicago (1974).

[Kat] D. Katz, *On the number of minimal prime ideals in the completion of a local domain*, Rocky Mountain J. Math. **16** (1986), 575–578.

[Kr] W. Krull, *Allgemeine Bewertungstheorie*, J. Reine Angew. Math. **167** (1931), 160–196.

[L] J. Lang, *Purely inseparable extensions of unique factorization domains*, Publ. Res. Inst. Math. Sci. **26** (1990), 453–471.

[LM] K.B. Levitz and J.L. Mott, *Rings with finite norm property*, Canad. J. Math. **24** (1972), 557–565.

[Ma] J. Martinez, *Some pathology involving pseudo ℓ-groups as groups of divisibility*, Proc. Amer. Math. Soc. **40** (1973), 333–340.

[Moč₁] J. Močkoř, *Groups of divisibility*, Mathematics and its Applications (East European Series), D. Reidel Publishing Co., Dordrecht-Boston, Mass. (1983).

[Moč₂] J. Močkoř, *On o-ideals of groups of divisibility*, Czechoslovak Math. J. **31 (106)** (1981), 390–403.

[Moč₃] J. Močkoř, *Topological groups of divisibility*, Colloq. Math. **39** (1978), 301–311.

[Moč₄] J. Močkoř, *Topological groups of divisibility*, General topology and its relations to modern analysis and algebra, IV (Proc. Fourth Prague Topological Sympos., Prague, 1976), Part B, Soc. Czechoslovak Mathematicians and Physicists, Prague (1977), 291–295.

[Moč₅] J. Močkoř, *A realization of groups of divisibility*, Comment. Math. Univ. St. Paul. **26** (1977), 61–74.

[Moč₆] J. Močkoř, *Semi-valuations and d-groups*, Czechoslovak Math. J. **32** **(107)** (1982), 77–89.

[Moč₇] J. Močkoř, *The group of divisibility of* $\widehat{Z}$, Arch. Math. (Brno) **20** (1984), 31–38.

[Mot₁] J.L. Mott, *The group of divisibility of Rees rings*, Math. Japon. **20** (1975), 85–87.

[Mot₂] J.L. Mott, *Convex directed subgroups of a group of divisibility*, Canad. J. Math. **26** (1974), 532–542.

[Mot₃] J.L. Mott, *The group of divisibility and its applications*, Conference on Commutative Algebra (Univ. Kansas, Lawrence, Kan., 1972), Lecture Notes in Math., Vol. 311, Springer, Berlin (1973), 194–208.

[Mot₄] J.L. Mott, *Groups of divisibility: a unifying concept for integral domains and partially ordered groups*, Lattice-ordered groups, Math. Appl., Vol. 48, Kluwer Academic Publ., Dordrecht-Boston-London (1989), 80–104.

[Mot₅] J.L. Mott, *Nonsplitting sequences of value groups*, Proc. Amer. Math. Soc. **44** (1974), 39–42.

[MS] J.L. Mott and M. Schexnayder, *Exact sequences of semi-value groups*, J. Reine Angew. Math. **283/284** (1976), 388–401.

[N₁] M. Nagata, *Local Rings*, Interscience, New York (1962).

[N₂] M. Nagata, *A remark on the unique factorization theorem*, J. Math. Soc. Japan **9** (1957), 143–145.

[O] J. Ohm, *Semi-valuations and groups of divisibility*, Can. J. Math. **21** (1969), 576–591.

[P] H. Prüfer, *Untersuchungen über Teilbarkeitseigenschaften in Körpern*, J. Reine Angew. Math. **168** (1932), 1–36.

[Ru] D. Rush, *An arithmetic characterization of algebraic number fields with a given class group*, Math. Proc. Cambridge Philos. Soc. **94** (1983), 23–28.

[Sa₁] P. Samuel, *Unique factorization*, Amer. Math. Monthly **75** (1968), 945–952.

[Sa₂] P. Samuel, *On unique factorization domains*, Illinois J. Math. **5** (1961), 1–17.

[Sa₃] P. Samuel, *About Euclidean rings*, J. Algebra **19** (1971), 282–301.

[Sh] P. Sheldon, *Two counterexamples involving complete integral closure infinite dimensional Prüfer domains*, J. Algebra **27** (1973), 462–474.

[Sk$_1$] L. Skula, *Divisorentheorie einer Halbgruppe*, Math. Z. **114** (1970), 113-120.

[Sk$_2$] L. Skula, *On c-semigroups*, Acta Arith. **31** (1976), 247–257.

[Z$_1$] A. Zaks, *Half factorial domains*, Bull. Amer. Math. Soc. **82** (1976) 721–723.

[Z$_2$] A. Zaks, *Half-factorial-domains*, Israel J. Math. **37** (1980), 281–302.

[Z$_3$] A. Zaks, *Atomic rings without A.C.C. on principal ideals*, J. Algebra **74** (1982), 223–231.

Some Generalizations of GCD-Domains

D. D. ANDERSON Department of Mathematics, The University of Iowa, Iowa City, IA 52242

R. O. QUINTERO Departamento de Física y Matemáticas, Universidad de Los Andes-NURR, Trujillo-Venezuela

1. INTRODUCTION

In this paper we introduce several new generalizations of GCD-domains and study the relationships between them and the domains having the GL-property or the PSP-property introduced by Arnold and Sheldon [2], and the domains having the IP-property or the CP-property introduced by D.F. Anderson [1].

Throughout R will be an integral domain with quotient field K. We let $R^* = R - \{0\}$ and $U(R)$ be its group of units. We say that R is *atomic* if every nonzero nonunit of R is a product of irreducible elements. For a nonzero fractional ideal I of R, let $I^{-1} = \{x \in K \mid xI \subseteq R\}$. We will denote $(I^{-1})^{-1}$ by I_v. For a polynomial f in $R[X]$, the *content of* f, denoted by A_f, is the ideal of R generated by the coefficients of f. The polynomial f is said to be *primitive* over R if its coefficients have no common factors and to be *superprimitive* [7] in case $A_f^{-1} = R$, or equivalently $(A_f)_v = R$. Given $a_1, \ldots, a_n$ in R, we will denote their *greatest common divisor in R* and *least common multiple in R*, provided they exist, by $[a_1, \ldots, a_n]$ and $\mathrm{lcm}(a_1, \ldots, a_n)$, respectively. Following I. Kaplansky [5], we say that a domain R is a *GCD-domain* if each pair of nonzero elements of R has a greatest common divisor in R. Throughout, $\mathbb{Z}$, $\mathbb{Q}$, and $\mathbb{R}$ will denote the ring of integers, rational numbers, and real numbers, respectively. In other aspects of commutative ring theory, our notation and terminology follow [4] and [5].

In the second section of this paper we define the various new generalizations of GCD-domains and of domains satisfying *Gauss's Lemma* (i.e., if f and g are two primitive polynomials over R, then the same is true of fg), and give some basic relationships among them and the other classes of domains investigated in [1] and [2]. In the last section we present new characterizations of the GL-property and the PSP-property and give several counterexamples.

2. NEW GENERALIZATIONS AND BASIC RESULTS

Let R be an integral domain. We begin this section by recalling some definitions and introducing the new generalizations mentioned above.

DEFINITION 2.1. A domain R has the *AP-property* if every atom (irreducible element) of R is prime.

DEFINITION 2.2. A domain R has the *PP-property* if whenever a, b, and c are elements of R such that $[a, b] = [a, c] = 1$, then $[a, bc] = 1$.

DEFINITION 2.3. A domain R has the *GL-property* if the product of two primitive polynomials is primitive (i.e., R satisfies Gauss's Lemma) [2].

DEFINITION 2.4. A domain R has the *GL2-property* if R has the GL-property for linear polynomials (i.e., f, g linear primitive $\Rightarrow fg$ is primitive).

DEFINITION 2.5. A domain R has the *PSP-property* if every primitive polynomial is superprimitive [2].

It is clear that

$$\text{GL-property} \Rightarrow \text{GL2-property}.$$

We have the following implications (see [2] and [7]):

$$\text{GCD-domain} \Rightarrow \text{PSP-property} \Rightarrow \text{GL-property} \Rightarrow \text{AP-property}. \qquad (2.1)$$

Observe that the classes of domains introduced generalize the class of GCD-domains, but not all of the classes satisfy Gauss's Lemma (as Example 3.13 shows).

LEMMA 2.6. *For an integral domain R we have the following implications:*

$$GL2\text{-property} \Rightarrow PP\text{-property} \Rightarrow AP\text{-property}.$$

PROOF. GL2 $\Rightarrow$ PP. Assume that $[a, b] = [a, c] = 1$. Then the linear polynomials $f = a + bX$ and $g = a + cX$ are primitive over R. By the GL2-property, $fg = a^2 + a(b + c)X + bcX^2$ is also primitive. So we have that $[a^2, a(b + c), bc] = 1$. But this clearly implies $[a, bc] = 1$.

PP $\Rightarrow$ AP. Let a be an atom of R, and suppose $a|bc$. If $a \nmid b$ and $a \nmid c$, then $[a, b] = [a, c] = 1$. So $[a, bc] = 1$, but this is a contradiction. Therefore, either $a|b$ or $a|c$, and so a is prime. ∎

So diagram 2.1 can be expanded to the following

$$
\begin{array}{ccccc}
\text{GCD-domain} & \Rightarrow & \text{PSP-property} & \Rightarrow & \text{GL-property} \\
& & & & \Downarrow \\
& & & & \text{GL2-property} \qquad (2.2) \\
& & & & \Downarrow \\
& & & & \text{PP-property} \Rightarrow \text{AP-property}
\end{array}
$$

We will now recall some other known definitions and introduce more generalizations, but in this case, of domains that satisfy Gauss's Lemma (as diagram 3.1 shows).

DEFINITION 2.7. A domain R has the *IP-property* if for every nonzero integral ideal I of R, I_v is the intersection of all the principal integral ideals of R which contain I [1].

DEFINITION 2.8. A domain R has the *FIP-property* if R has the IP-property for nonzero finitely generated ideals (i.e., $I_v = \bigcap \{(d) \mid I \subseteq (d) \subseteq R\}$ for I a nonzero finitely generated ideal).

DEFINITION 2.9. A domain R has the *CP-property* if each proper integral v-ideal of R is contained in a proper principal integral ideal of R [1].

DEFINITION 2.10. A domain R has the *FCP-property* if R has the CP-property for nonzero finitely generated ideals (i.e., for I a nonzero finitely generated ideal, $I_v \subsetneq R \Rightarrow I_v \subseteq (d)$ for some $d \notin U(R)$).

DEFINITION 2.11. A domain R has the *PSP2-property* if R has the PSP-property for linear polynomials (i.e., f linear primitive $\Rightarrow f$ superprimitive).

DEFINITION 2.12. A domain R has the *D-property* if whenever a, b, and c are elements of R such that $[a, b] = 1$ and $a|bc$, then $a|c$.

We have the following implications [1, Lemma 2.1]:

$$\text{GCD-domain} \Rightarrow \text{IP-property} \Rightarrow \text{CP-property}. \qquad (2.3)$$

It is clear from the definitions that

$$\begin{array}{rcl} \text{IP-property} & \Rightarrow & \text{FIP-property,} \\ \text{CP-property} & \Rightarrow & \text{FCP-property, and} \\ \text{PSP-property} & \Rightarrow & \text{PSP2-property.} \end{array}$$

On the other hand, by [1, Proposition 3.1], we have the following equivalence:

$$\text{FCP-property} \Leftrightarrow \text{PSP-property}.$$

LEMMA 2.13. *For an integral domain R we have the following implications:*

$$\text{PSP2-property} \Rightarrow D\text{-property} \Rightarrow PP\text{-property}.$$

PROOF. PSP2 $\Rightarrow$ D. Suppose $[a, b] = 1$ and $a|bc$. Then $bc = ar$ for some $r \in R$. By hypothesis, $(a, b)^{-1} = R$. Observe that

$$\frac{c}{a}(a, b) \subseteq (c, r) \subseteq R.$$

So $\frac{c}{a} \in R$. Thus $a|c$, and R has the D-property.

  D $\Rightarrow$ PP. Suppose $[a, b] = [a, c] = 1$. Let d be a common divisor of a and bc. We claim that $[d, b] = 1$. In fact, if $e|d, b$, then $e|a, b$, so $e \in U(R)$. Since $[d, b] = 1$ and $d|bc$, by the D-property, $d|c$. Then $d|a, c$. Thus $d \in U(R)$. So we have proved that $[a, bc] = 1$. Thus R has the PP-property. ∎

So combining diagrams 2.2, 2.3, and the last implications we get the following diagram:

$$\begin{array}{ccccccccc} \text{GCD} & \Rightarrow & \text{IP} & \Rightarrow & \text{CP} & & & & \\ & & \Downarrow & & \Downarrow & & & & \\ & & \text{FIP} & \Rightarrow & \text{FCP} & \Leftrightarrow & \text{PSP} & \Rightarrow & \text{GL} \\ & & & & & & \Downarrow & & \Downarrow \\ & & & & & & \text{PSP2} & \Rightarrow & \text{GL2} \\ & & & & & & \Downarrow & & \Downarrow \\ & & & & & & \text{D} & \Rightarrow & \text{PP} & \Rightarrow & \text{AP} \end{array} \qquad (2.4)$$

3. MAIN RESULTS AND COUNTEREXAMPLES

In this section we present the main results of this paper, as well as give some examples which show that the reverse implications of some of the introduced implications given above in diagram 2.4 are false. We start by giving a new characterization of the GL-property. The proof is a modification of the usual proof of Gauss's Lemma.

THEOREM 3.1. *For an integral domain R, the following conditions are equivalent:*

(1) *R has the GL-property,*
(2) *R has the GL2-property, and*
(3) *R has the PP-property.*

PROOF. It suffices to prove (3) $\Rightarrow$ (1). Suppose that the polynomials $f = a_0 + a_1X + \ldots + a_sX^s$ and $g = b_0 + b_1X + \ldots + b_tX^t$ are primitive, but $fg = c_0 + c_1X + \ldots + c_{s+t}X^{s+t}$ is not primitive. Let r be a nonunit divisor of the c_l's. Suppose that $r|a_0, \ldots, r|a_{i-1}$, but $r\nmid a_i$. If $[r, a_i] \neq 1$, we can replace r by a nonunit divisor r' of r and a_i. Then r' divides all the c_l's and $r'|a_0, \ldots, a_i$. Continuing this process we can find some nonunit factor r^* of r' which divides all the c_l's and $a_0, \ldots, a_{j-1}$, but $[r^*, a_j] = 1$. By applying the same procedure to the coefficients of g, there exists a nonunit $y \in R$ such that $y|c_0, \ldots, c_{s+t}, a_0, \ldots, a_{j-1}, b_0, \ldots, b_{k-1}$ and $[y, a_j] = [y, b_k] = 1$. Then $y|c_{j+k} = \sum_{p+q=j+k} a_p b_q$ and $y|a_p b_q$ for $p \neq j$ and $q \neq k$. Hence $y|a_j b_k$, a contradiction since $[y, a_j b_k] = 1$ and y is a nonunit element of R. ∎

COROLLARY 3.2. *If R has the D-property, then R has the GL-property.*

So diagram 2.4 transforms into the following one:

$$
\begin{array}{ccccc}
\text{GCD} & \Rightarrow & \text{IP} & \Rightarrow & \text{CP} \\
& & \Downarrow & & \Downarrow \\
& & \text{FIP} & \Rightarrow & \text{FCP} \equiv \text{PSP} \\
& & & & \Downarrow \\
& & & & \text{PSP2} \\
& & & & \Downarrow \\
& & & & \text{D} \\
& & & & \Downarrow \\
& & \text{GL} \equiv \text{GL2} \equiv \text{PP} & \Rightarrow & \text{AP}
\end{array} \qquad (3.1)
$$

Another characterization of the PSP-property is given below.

THEOREM 3.3. *For an integral domain R the following conditions are equivalent:*

(1) *R has the PSP-property, and*
(2) *For $a_1, \ldots, a_n \in R$, $[a_1, \ldots, a_n]$ exists $\Rightarrow (a_1, \ldots, a_n)_v$ is a principal ideal.*

PROOF. (1) $\Rightarrow$ (2). Assume that $[a_1, \ldots, a_n] = d \in R$. Then $\left[\frac{a_1}{d}, \ldots, \frac{a_n}{d}\right] = 1$, and by the PSP-property, $\left(\frac{a_1}{d}, \ldots, \frac{a_n}{d}\right)_v = R$, or equivalently, $(a_1, \ldots, a_n)_v = dR$.

(2) $\Rightarrow$ (1). Now assume that $[a_1, \ldots, a_n] = 1$, so $(a_1, \ldots, a_n)_v = dR$ for some $d \in R$. Since $(a_1, \ldots, a_n) \subseteq (a_1, \ldots, a_n)_v$, $d|a_1, \ldots, a_n$. So $d \in U(R)$. Hence R has the PSP-property. ∎

The next theorem gives several characterizations of an integral domain that satisfies the PSP2-property.

THEOREM 3.4. *For an integral domain R, the following are equivalent:*

(1) *for all $a, b \in R^*$, $(a) \cap (b) = (ab)$, or a and b have a nonunit common factor,*
(2) *for $a, b \in R^*$, $[a, b] = 1 \Rightarrow (a) \cap (b) = (ab)$,*
(3) *for $a, b \in R^*$, $[a, b] = 1 \Rightarrow (a, b)_v = R$,*
(4) *R has the PSP2-property,*
(5) *for $a, b \in R^*$, $[a, b]$ exists $\Rightarrow$ lcm(a, b) exists,*
(6) *every irreducible linear polynomial over R is prime,*
(7) *R has the D-property, and*
(8) *for $a, b, c \in R$, $a|c$, $b|c$ and $[a, b] = 1 \Rightarrow ab|c$.*

PROOF. Clearly (1) $\Leftrightarrow$ (2) and (3) $\Leftrightarrow$ (4).

(2) $\Leftrightarrow$ (3) follows from

$$(a) \cap (b) = (ab) \Leftrightarrow (a^{-1}) \cap (b^{-1}) = R \Leftrightarrow (a, b)^{-1} = R \Leftrightarrow (a, b)_v = R.$$

To prove (5) $\Rightarrow$ (2), assume that $[a, b] = 1$. Then by (5), lcm(a, b) exists, and by [4, Exercise 1(e), page 76], $1 = \frac{ab}{\text{lcm}(a,b)}$. So lcm$(a, b) = ab$, and hence $(a) \cap (b) = (ab)$.

For (3) $\Rightarrow$ (5), suppose $[a, b] = d$. Then $a = dx$, and $b = dy$ where $[x, y] = 1$. So by (3), $(x, y)_v = R$. Then $(a, b)_v = d(x, y)_v = (d)$. Thus

$$(a) \cap (b) = (ab)[(b^{-1}) \cap (a^{-1})] = (ab)(a, b)^{-1} = (ab)(d)^{-1} = \left(\frac{ab}{d} \right),$$

which clearly implies that lcm$(a, b) = \frac{ab}{d}$.

To prove (3) $\Leftrightarrow$ (6) first observe that $f = a + bX$ is irreducible over $R \Leftrightarrow [a, b] = 1$, and by [5, Exercises 1, 3, and 4, page 102] $f = a + bX$ is prime $\Leftrightarrow (A_f)_v = R$.

To show (2) $\Rightarrow$ (8) assume that $a|c$, $b|c$, and $[a, b] = 1$. Then by (2), $(a) \cap (b) = (ab)$, but $c \in (a) \cap (b)$, hence $ab|c$.

For (8) $\Rightarrow$ (2), assume that $[a, b] = 1$. It suffices to prove the inclusion $(a) \cap (b) \subseteq (ab)$. Let $c \in (a) \cap (b)$, then $a|c$, $b|c$ and by (8), $ab|c$. So $c \in (ab)$.

Since (4) $\Rightarrow$ (7) (Lemma 2.13) it only remains to prove (7) $\Rightarrow$ (4). Suppose $[a, b] = 1$ and $0 \neq t \in (a, b)^{-1}$, say $t = \frac{u}{v}$ $(u, v \in R)$. So $ta = r$ and $tb = s$ for some $r, s \in R$. Then $tas = rs = tbr$, so $as = br$. So $a|br$. Then, by the D-property, $a|r$, say $r = aw$ where $w \in R$. Hence, $t = w$, which proves that $(a, b)^{-1} = R$, and so R has the PSP2-property. ∎

Finally, diagram 3.1 transforms into the following one:

$$
\begin{array}{ccccc}
\text{GCD} & \overset{(i)}{\Rightarrow} & \text{IP} & \overset{(ii)}{\Rightarrow} & \text{CP} \\
& \Downarrow (iii) & \overset{(v)}{} & & \Downarrow (iv) \\
& \text{FIP} & \overset{(v)}{\Rightarrow} & \text{FCP} \equiv \text{PSP} & \\
& & & \Downarrow (vi) & \\
& & & \text{PSP2} \equiv \text{D} & \\
& & & \Downarrow (vii) & \\
& & \text{GL} \equiv \text{GL2} \equiv \text{PP} & \overset{(viii)}{\Rightarrow} & \text{AP}
\end{array}
\tag{3.2}
$$

REMARK 3.5. We have the following equivalences:

(1) If R is completely integrally closed, IP-property $\Leftrightarrow$ CP-property [1, Proposition 2.6].
(2) If R is a domain in which every finitely generated proper ideal is in the radical of some principal ideal, PSP-property $\Leftrightarrow$ GL-property [2, Proposition 2.1].
(3) $R[X]$ has the IP-property $\Leftrightarrow R[X]$ is a GCD-domain [6, Lemma 16] and $R[X]$ has the PSP-property $\Leftrightarrow R[X]$ has the AP-property [2].

COROLLARY 3.6. *If R is atomic, then all the conditions given in diagram 3.2 are equivalent.*

PROOF. Let R be an atomic domain satisfying the AP-property. Then every nonzero nonunit of R is a product of principal primes and hence R is a UFD. ∎

We will now give a counterexample to the reverse implication of each implication given in diagram 3.2 except (vi).

EXAMPLE 3.7. The counterexample for the reverse implication of (i) is $R = \mathbf{Z}_{(2)} + M$, where $M = XV$ and $V = \mathbb{R}[[X]] = \mathbb{R} + M$ [1, Example 2.4].

EXAMPLE 3.8. The counterexample for the reverse implication of (ii) is $R = \mathbf{Z}_{(2)} + M$, where $M = (X^2, X^3)D$ is the unique maximal ideal of $D = \mathbb{Q}[[X^2, X^3]] = \mathbb{Q} + M$ [1, Example 2.5].

EXAMPLE 3.9. The counterexample for the reverse implication of (iv) is $R = F + M$, where F is a proper subfield of K, and $V = K + M$ is a rank-one nondiscrete valuation ring (see [1, page 171]).

EXAMPLE 3.10. The counterexample for the reverse implication of (iii) is the same ring given in the previous example. Thus $R = F + M$. Clearly R does not have the IP-property. We now show that R satisfies the FIP-property. Let I be a nonzero finitely generated ideal of R; we may assume that I is not principal. Let $I = (m_1, \ldots, m_n)R$ where $m_1, \ldots, m_n \in M$ with $Vm_1 \supseteq \cdots \supseteq Vm_n$. If $Vm_1 \supsetneq Vm_i$, then $m_i \in Rm_1$, so we may assume that $Vm_1 = \cdots = Vm_n$. So $IV = Vm_1$. For $i > 1$, $m_i = u_i m_1$ where $u_i \in U(V)$, so $u_i = k_i + n_i$ where $k_i \in K$ and $n_i \in M$. Now $k_i m_1 + n_i m_1$ and $n_i m_1 \in Rm_1$, so $I = (m_1, k_2 m_1, \ldots, k_n m_1)R$ where the k_i's $\in K - F$. Hence $I = Wm_1 + Mm_1$ where $W = F + Fk_2 + \cdots + Fk_n$. By [3, Theorem 4.3], $I_v = Vm_1$ (since W is not a fractional ideal of F). Suppose that $I \subseteq Rv$ where $v \in R$. Then $Vm_1 \subsetneq Vv$ for if $Vm_1 = Vv$, then $v = \lambda m_1$ where $\lambda \in U(V)$ and hence $Rm_1 \subseteq R\lambda m_1 \subseteq R$ gives $\lambda \in U(R)$. But then $Rm_2 \subseteq Rv = Rm_1$, a contradiction. And if $Vm_1 \subsetneq Vv$ where $v \in M$, then $Rm_1 \subseteq Rv$. So $I_p = \bigcap\{Rv \mid I \subseteq Rv \subseteq R\} \subseteq \bigcap\{Vv \mid Vv \supsetneq Vm_1\} = Vm_1$.

EXAMPLE 3.11. The counterexample for the reverse implication of (v) is the ring R given in Example 3.8. We show that R does not satisfy the FIP-property (i.e. CP $\not\Rightarrow$ FIP) which clearly implies that FCP $\not\Rightarrow$ FIP. In fact, take $I = (X^2, X^3)R = \mathbf{Z}_{(2)}X^2 + \mathbf{Z}_{(2)}X^3 + \mathbb{Q}[[X]]X^4$. If $I \subseteq yR$ where $y = a_0 + a_2X^2 + \cdots \in R$, then $X^3 \in I$ gives $a_0 \neq 0$. So either $a_0 \in 2\mathbf{Z}_{(2)}$ and hence $y \in 2R$, or $a_0 \in U(\mathbf{Z}_{(2)})$ and hence $y \in U(R)$. So $I_p = 2R$. But $I_v \subseteq R \cap \frac{1}{X}R = \frac{1}{X}(R \cap XR) = X^2\mathbb{Q}[[X]] \subsetneq 2R$.

EXAMPLE 3.12. The counterexample for the reverse implication of (vii) is the ring

$$R = F[\{X^\alpha \mid \alpha \in \mathbb{Q}_0^+\}, \{Y^\alpha \mid \alpha \in \mathbb{Q}_0^+\}, \{X^\alpha Z^\beta \mid \alpha, \beta \in \mathbb{Q}^+\}, \{Y^\alpha Z^\beta \mid \alpha, \beta \in \mathbb{Q}^+\}]$$

where F denotes the field of two elements, $\mathbb{Q}^+$ denotes the set of positive rational numbers, and $\mathbb{Q}_0^+ = \mathbb{Q}^+ \cup \{0\}$ (see [2, Example 2.5] and proof of [2, Proposition 2.9]).

EXAMPLE 3.13. The counterexample for the reverse implication of (viii) is the ring

$$R = F[\{X_i^\alpha Y_j^\beta Z^{-\delta} \mid \alpha, \beta, \delta \in \mathbb{Q}^+; i, j = 1, 2\}]$$

with F and $\mathbb{Q}^+$ as in Example 3.12 [2, Example 2.10].

To conclude we will give a domain that does not have the **AP-property**.

EXAMPLE 3.14. Let $R = \mathbf{Z} + (2X, X^2)\mathbf{Z}[X]$. Thus R is the subring of $\mathbf{Z}[X]$ formed by all the polynomials with the coefficient of X in $2\mathbf{Z}$. First observe that 2 is an atom in R. But $2 | 4X^2 = (2X)^2$, and $2 \nmid 2X$. So 2 is not prime. Thus R does not have the **AP-property**.

REFERENCES

[1] D.F. Anderson, Integral V-ideals, *Glasgow Math. J.* **22** (1981), 167–172.

[2] J.T. Arnold and P.B. Sheldon, Integral domains that satisfy Gauss's lemma, *Michigan Math. J.* **22** (1975), 39–51.

[3] E. Bastida and R. Gilmer, Overrings and divisorial ideals of the form $D + M$, *Michigan Math. J.* **20** (1973), 79–95.

[4] R. Gilmer, *Multiplicative Ideal Theory*, Queen's Papers in Pure and Appl. Math., vol. 90, Queen's Univ., Kingston, Ontario, 1992 (originally published by Marcel Dekker, Inc., New York, 1972).

[5] I. Kaplansky, *Commutative Rings*, Polygonal Publishing House, Washington, New Jersey, 1994 (previous editions published by University of Chicago Press, 1974, and Allyn and Bacon, Boston, 1970).

[6] R. Matsuda, Note on integral domains that satisfy Gauss's lemma, *Math. Japon.* **41** (1995), 625–630.

[7] H.T. Tang, Gauss' lemma, *Proc. Amer. Math. Soc.* **35** (1972), 372–376.

Factorization in Commutative Rings with Zero Divisors, II

D.D. ANDERSON   Department of Mathematics, The University of Iowa, Iowa City, IA 52242

SILVIA VALDES-LEON   Department of Mathematics, University of Southern Maine, Portland, ME 04103

Dedicated to James A. Huckaba on the occasion of his sixtieth birthday.

0. INTRODUCTION

As the title of this conference proceedings indicates, the central theme of the conference was factorization in integral domains. Much of classical multiplicative ideal theory for integral domains can be extended to commutative rings with zero divisors. Two excellent references for this are [18] and [22]. Likewise, much of the theory of factorization and divisibility in integral domains can be extended to commutative rings with zero divisors. For this, the reader is referred to [8], which contains an extensive bibliography.

The purpose of this paper is threefold. First, we briefly review some of the results from [8] showing how certain results on factorization in integral domains can be extended to commutative rings with zero divisors. This is covered in Section 1. The main theme here is the many different ways to define irreducible elements. Also, several new results are given.

Second, we study factorization in modules. This is started in Section 2, which considers various definitions of associates and primitive elements in modules. One reason to study factorization in an R-module M is to study factorization in the idealization $R(M)$ and symmetric algebra $S(M)$. As we showed in [8], idealization is a useful construction to get commutative rings with bad factorization properties. Section 3 begins a study of the relationship between factorization in R and $R(M)$ and factorization in R and $S(M)$. Section 4 studies factorization in torsion-free modules. Particular emphasis is given to unique factorization modules (factorial modules) and factorable modules.

Third, we study the factorization properties of the regular elements (i.e., nonzero divisors) of a commutative ring. One way to do this is to observe that the regular elements reg(R) of a commutative ring R form a commutative cancellative monoid and then to exploit the highly developed theory of factorization in commutative cancellative monoids. We pay especial attention to Krull rings and the corresponding Krull monoids.

Throughout this paper R denotes a commutative ring with identity with total quotient ring $T(R)$. We denote the group of units of R by $U(R)$, the set of zero divisors of R by $Z(R)$, and the set of regular elements or nonzero divisors of R by reg(R). An ideal I is regular if it contains a regular element. Our references for results from multiplicative ideal theory and for commutative rings with zero divisors are [15] and [18], respectively.

1. FACTORIZATION IN COMMUTATIVE RINGS WITH ZERO DIVISORS

The purpose of this section is to survey the results of [8] and to give further extensions of the theory of factorization in integral domains to commutative rings with zero divisors. In later sections, factorization in modules will be investigated.

Let R be an integral domain. A nonzero, nonunit element $a \in R$ is *irreducible*, or is an *atom*, if for any factorization $a = bc$ in R, either b or c is a unit. Equivalently, a nonzero, nonunit element $a \in R$ is an atom $\Leftrightarrow a = bc$ implies either b or c is an associate of $a \Leftrightarrow (a)$ is a maximal proper principal ideal of R. The integral domain R is said to be *atomic* if every nonzero, nonunit element of R is a finite product of atoms. If an integral domain R satisfies the ascending chain condition on principal ideals (ACCP), then R is atomic, but the converse is false. Of course, R is a UFD precisely when the factorization into atoms is unique up to order and associates: if $a_1 \cdots a_n = b_1 \cdots b_m$ where each a_i, b_j is an atom, then $n = m$ and after reordering, if necessary, a_i and b_i are associates. Two recurrent themes in the study of factorization are (1) how close can factorization into atoms be to being unique without the domain being a UFD and (2) how bad can factorization into atoms be in an atomic domain that is not a UFD.

A survey of various conditions weaker than unique factorization can be found in [2]. We repeat several of these definitions here. Let R be an atomic domain. Then R is *half-factorial* if $a_1 \cdots a_n = b_1 \cdots b_m$, where each a_i, b_j is an atom, then $n = m$. Also, R is a *finite factorization domain* (FFD) if one of the following three equivalent conditions holds: (1) every nonzero element of R has only a finite number of factorizations up to order and associates, (2) each nonzero element of R has only finitely many nonassociate factors, (3) R is atomic and each nonzero element of R has at most a finite number of nonassociate irreducible divisors. And R is a *bounded factorization domain* (BFD) if for each nonzero, nonunit element $a \in R$, there exists a natural number $N(a)$, so that if $a = a_1 \cdots a_n$ where each a_i is a nonunit, then $n \leq N(a)$. Now for an integral domain R, we have UFD $\Rightarrow$ FFD $\Rightarrow$ BFD $\Rightarrow$ ACCP and UFD $\Rightarrow$ half-factorial $\Rightarrow$ BFD, but none of these implications can be reversed. A Noetherian integral domain is a BFD, but need not be an FFD. A Krull domain is an FFD. These results can all be found in [2]. For results on FFD's, also see [5].

One of the measures of different lengths of factorizations into atoms is elasticity. This is discussed in detail in the survey article [9] by David F. Anderson in these proceedings. Let R be an atomic domain and let $a \in R$ be a nonzero nonunit. Define the *elasticity* $\rho(a)$ of a by $\rho(a) = \sup\{m/n \mid a_1 \cdots a_n = b_1 \cdots b_m$ where each a_i, b_j is an atom$\}$. So either $\rho(a)$ is a rational number greater than or equal to 1, or $\rho(a) = \infty$. The *elasticity of* R is $\rho(R) = \sup\{\rho(a) \mid a$ is a nonzero, nonunit element of $R\}$. Clearly, R is half-factorial if and only if $\rho(R) = 1$. So $\rho(R)$ measures now far R is from being half-factorial. For each real number $r \geq 1$ or $r = \infty$, there is a Dedekind domain R with torsion class group such that $\rho(R) = r$. Moreover, if r is rational, we may choose $Cl(R)$ to be finite. However, for a Krull domain R such that only a finite number of divisor classes contain prime ideals (which is the case if $Cl(R)$ is finite), $\rho(R) = \rho(a)$ for some $a \in R$ and hence is rational [1].

We are now ready to survey the results of [8] concerning factorization in commutative rings with zero divisors. The point of view of [8] was to define various notions of irreducible elements via different notions of associates. Briefly, if $\backsimeq$ was an "associate" relation on R, then a nonunit $a \in R$ was said to be "$\backsimeq$-irreducible" if $a = bc \Rightarrow a \backsimeq b$ or $a \backsimeq c$. In the case of an integral domain, all these notions agree with the usual one.

Let R be a commutative ring with 1 and group of units $U(R)$. Let $a, b \in R$. Then a and b are *associates*, denoted $a \sim b$, if $a|b$ and $b|a$, that is, $aR = bR$. If $a = ub$ for some $u \in U(R)$, we say a and b are *strong associates*, denoted $a \approx b$. Finally, we say that a and b are *very strong associates*, denoted $a \cong b$, if (i) a and b are associates and (ii) either $a = b = 0$ or $a \neq 0$ and $a = rb \Rightarrow r \in U(R)$. Clearly, $a \cong b \Rightarrow a \approx b \Rightarrow a \sim b$, but examples given in [8] show that none of these implications can be reversed. Now $\approx$ and $\sim$ are equivalence relations, even congruences on the monoid $(R, \cdot)$. However, while $\cong$ is symmetric and transitive, it need not be reflexive. For $a \cong a \Leftrightarrow a = 0$ or $a = ra \Rightarrow r \in U(R)$. A ring satisfying this last condition for all $a \in R$ is called *présimplifiable*. Thus $\cong$ is an equivalence relation (even a congruence on $(R, \cdot)$) $\Leftrightarrow \cong$ is reflexive $\Leftrightarrow R$ is présimplifiable. It is not hard to prove that R is présimplifiable $\Leftrightarrow Z(R) \subseteq 1 - U(R) = \{1 - u \mid u \in U(R)\}$ $\Leftrightarrow Z(R) \subseteq \operatorname{rad}(R)$ where $Z(R)$ and $\operatorname{rad}(R)$ are the set of zero divisors and Jacobson radical of R, respectively. If R is an integral domain or is quasilocal, then R is présimplifiable. If R is présimplifiable, then R is indecomposable.

We next define three different forms of "irreducible elements". Let a be a nonunit of R. Then a is *irreducible* (respectively, *strongly irreducible*, *very strongly irreducible*) if $a = bc \Rightarrow a \sim b$ or $a \sim c$ (respectively, $a \approx b$ or $a \approx c$, $a \cong b$ or $a \cong c$). So a very strongly irreducible $\Rightarrow a$ is strongly irreducible $\Rightarrow a$ is irreducible, but examples given in [8] show that none of these implications can be reversed. All three forms agree if R is présimplifiable. Note that we do allow $a = 0$. In this case, 0 is (very strongly) irreducible if and only if R is an integral domain.

The following theorems whose proofs may be found in [8] give characterizations of the various forms of irreducible elements.

THEOREM 1.1. *Let R be a commutative ring. For a nonunit $a \in R$, the following conditions are equivalent.*

(1) a *is irreducible, that is,* $a = bc \Rightarrow a \sim b$ *or* $a \sim c$.
(2) $a = a_1 \cdots a_n \Rightarrow a \sim a_i$ *for some* i.
(3) $a \sim bc \Rightarrow a \sim b$ *or* $a \sim c$.
(4) $a \sim a_1 \cdots a_n \Rightarrow a \sim a_i$ *for some* i.
(5) *There exists a prime ideal P of R with $a \in P$ so that if $(a) \subseteq (b) \subseteq P$, then $(a) = (b)$.*

THEOREM 1.2. *Let R be a commutative ring. For a nonunit $a \in R$, the following conditions are equivalent.*

(1) *a is strongly irreducible, that is, $a = bc \Rightarrow a \approx b$ or $a \approx c$.*
(2) *$a = a_1 \cdots a_n \Rightarrow a \approx a_i$ for some i.*
(3) *$a \approx bc \Rightarrow a \approx b$ or $a \approx c$.*
(4) *$a \approx a_1 \cdots a_n \Rightarrow a \approx a_i$ for some i.*

THEOREM 1.3. *Let R be a commutative ring. For a nonzero nonunit $a \in R$, the following conditions are equivalent.*

(1) *a is very strongly irreducible, that is, $a = bc \Rightarrow a \cong b$ or $a \cong c$.*
(2) *$a = a_1 \cdots a_n \Rightarrow a \cong a_i$ for some i.*
(3) *$a = bc \Rightarrow b$ or c is a unit.*
(4) *$a = a_1 \cdots a_n \Rightarrow$ all a_i except one is a unit.*
(5) *$a \cong a$ and $a \cong bc \Rightarrow a \cong b$ or $a \cong c$.*
(6) *$a \cong a$ and $a \cong a_1 \cdots a_n \Rightarrow a \cong a_i$ for some i.*

A fourth type of irreducible element was also studied in [8]. A nonunit $a \in R$, R a commutative ring, is *m-irreducible* if (a) is a maximal element in the set of proper principal ideals of R. The next theorem relates m-irreducible elements to the other forms of irreducible elements.

THEOREM 1.4. *Let R be a commutative ring and let $a \in R$ be a nonunit.*

(1) *a is m-irreducible if and only if $a = bc \Rightarrow b \in U(R)$ or $a \sim b$.*
(2) *If $a \neq 0$ is very strongly irreducible, then a is m-irreducible.*
(3) *If a is m-irreducible, then a is strongly irreducible.*
(4) *If a is irreducible, $S = \{b \in R \mid (a) \subset (b)\}$ is multiplicatively closed and $a/1$ is m-irreducible in R_S.*

Each form of irreducibility leads to a form of atomicity. A commutative ring R is *atomic* (respectively, *strongly atomic, very strongly atomic, m-atomic, p-atomic*) if every nonzero, nonunit element of R is a finite product of irreducible (respectively, strongly irreducible, very strongly irreducible, m-irreducible, prime) elements. Now very strongly atomic $\Rightarrow$ m-atomic $\Rightarrow$ strongly atomic $\Rightarrow$ atomic, but none of these implications can be reversed. As expected, if R satisfies ACCP, then R is atomic. Also, if R satisfies ACCP (respectively, is atomic, strongly atomic, m-atomic, very strongly atomic, p-atomic), then R is a finite direct product of indecomposable rings satisfying ACCP (respectively, which are atomic, strongly atomic, m-atomic, very strongly atomic, p-atomic).

There are a number of ways to generalize the notion of a unique factorization domain to commutative rings with zero divisors. A. Bouvier, C. Fletcher, S. Galovich, and S. Mori have all given definitions. For a thorough discussion, see [8, Section 4]. We give three generalizations. Let R be a commutative ring and $a \in R$ a nonunit. Two factorizations of a into nonunits $a = a_1 \cdots a_n = b_1 \cdots b_m$ are said to be *isomorphic* (respectively, *strongly isomorphic, very strongly isomorphic*) if $n = m$ and there exists a permutation $\sigma \in S_n$ such that $a_i \sim b_{\sigma(i)}$ (respectively, $a_i \approx b_{\sigma(i)}$, $a_i \cong b_{\sigma(i)}$). Two factorizations of a into nonunits $a = a_1 \cdots a_n = b_1 \cdots b_m$ are said to be *homomorphic* (respectively, *strongly homomorphic, very strongly homomorphic*) if for each $i \in \{1, \ldots, n\}$ there exists a $j \in \{1, \ldots, m\}$ with $a_i \sim b_j$ (respectively, $a_i \approx b_j$, $a_i \cong b_j$) and for each $i \in \{1, \ldots, m\}$, there exists a $j \in \{1, \ldots, n\}$ with $b_i \sim a_j$ (respectively, $b_i \approx a_j$, $b_i \cong a_j$). Let $\alpha \in \{$atomic, strongly atomic, very strongly atomic, m-atomic} and $\beta \in$ {isomorphic, strongly isomorphic, very strongly isomorphic}. Then R is an (α, β)-*unique factorization ring* if (1) R is α and (2) any two factorizations of a nonzero, nonunit element into irreducible elements of the type used to define α are β. For any choice of α and β, an (α, β)-unique factorization ring is présimplifiable. Hence the various types of (α, β)-unique factorization coincide. So we drop the prefix (α, β) and just say *unique factorization ring*.

THEOREM 1.5.

(1) (Bouvier, Galovich) R *is a unique factorization ring if and only if R is either* (a) *a UFD,* (b) *an SPIR, or* (c) *a quasilocal ring* (R, M) *with* $M^2 = 0$.

(2) R *is atomic and any two factorizations of a nonzero nonunit into irreducibles are homomorphic if and only if R is either* (a) *a UFD,* (b) *a finite direct product of SPIR's and fields, or* (c) *a quasilocal ring* (R, M) *with* $M^2 = 0$.

(3) (S. Mori) R *is p-atomic if and only if R is a finite direct product of UFD's and SPIR's.*

The notions of HFD and BFD also have natural generalizations to commutative rings with zero divisors. We define R to be a *half-factorial ring* (HFR) if (1) R is atomic and (2) if $0 \neq a_1 \cdots a_n = b_1 \cdots b_m$ where a_i, b_j are atoms of R, then $n = m$. And R is a *bounded factorization ring* (BFR) if for each nonzero nonunit $a \in R$, there exists a natural number $N(a)$ so that if $a = a_1 \cdots a_n$, where each a_i is a nonunit, then $n \leq N(a)$. It is easily seen that R a unique factorization ring $\Rightarrow R$ is an HFR $\Rightarrow R$ is a BFR $\Rightarrow R$ is présimplifiable and R satisfies ACCP. In [8] we showed that a locally Noetherian ring R is a BFR $\Leftrightarrow R$ is présimplifiable $\Leftrightarrow$ $\bigcap_{n=0}^{\infty} I^n = 0$ for each proper (principal) ideal of R.

Each of the three equivalent conditions given to define an FFD leads to a different generalization to commutative rings. In [8], we defined R to be a *finite factorization ring* (FFR) if every nonzero nonunit of R has only a finite number of factorizations up to order and associates; R was called a *weak finite factorization ring* (WFFR) if every nonzero nonunit of R has only a finite number of nonassociate divisors; and R was called an *atomic idf-ring* if R is atomic and each nonzero element of R has at most a finite number of nonassociate irreducible divisors. Clearly if R is an FFR then R is a WFFR and if R is a WFFR then R is an atomic idf-ring. But $\mathbf{Z}_2 \times \mathbf{Z}_2$ is a WFFR that is not an FFR (consider $(0, 1) = (0, 1)^n$) and $\mathbf{Z}_{(2)} \times \mathbf{Z}_{(2)}$ is an atomic idf-ring that is not a WFFR (consider $(0, 1) = (0, 1)(2, 1)^n$). In [8, Proposition 6.6] it is shown that the following conditions on a commutative ring R are equivalent: (1) R is an FFR, (2) R is a BFR and WFFR, (3) R is présimplifiable and a WFFR, (4) R is a BFR and an atomic idf-ring, and (5) R is présimplifiable and an atomic idf-ring.

We next consider the question of when $R[X]$ or $R[[X]]$ is an FFR.

PROPOSITION 1.6. *Let R be a commutative ring and let $n_1, n_2 \in R$.*

(1) $n_1 + X \sim n_2 + X$ *in* $R[X] \Leftrightarrow n_1 = n_2$.

(2) *Suppose that* $\bigcap_{k=1}^{\infty} (n_2)^k = 0$. *Then* $n_1 + X \sim n_2 + X$ *in* $R[[X]] \Leftrightarrow n_1 = n_2$.

Proof. (1) Suppose that $n_1 + X = (n_2 + X)(a_0 + a_1 X + \cdots + a_k X^k)$ where $a_k \neq 0$. If $k \geq 1$, then $a_k X^{k+1} = 0$, a contradiction. So $n_1 + X = a_0(n_2 + X) = a_0 n_2 + a_0 X$ $\Rightarrow a_0 = 1$.

(2) Suppose that $n_1 + X = (n_2 + X) \left(\sum_{i=0}^{\infty} a_i X^i \right)$. So $n_1 = a_0 n_2$, $1 = a_1 n_2 + a_0$, $0 = a_2 n_2 + a_1$, $\ldots$, $0 = a_{k+1} n_2 + a_k$, $\ldots$. If $a_1 = 0$, then $a_0 = 1$, so $n_1 = n_2$. Suppose that $a_1 \neq 0$. Then $a_1 = -a_2 n_2 = a_3 n_2^2 = \cdots = (-1)^k a_{k+1} n_2^k = \ldots$. So $0 \neq a_1 \in \bigcap_{k=1}^{\infty} (n_2)^k$, a contradiction.

THEOREM 1.7. *If $R[X]$ is an atomic idf-ring or $R[[X]]$ is a WFFR, then either R is an integral domain or R is a finite local ring.*

Proof. We first show that R must be indecomposable. Suppose that R is not indecomposable, say $R = R_1 \times R_2$, so $R[X] = R_1[X] \times R_2[X]$ and $R[[X]] = R_1[[X]] \times R_2[[X]]$. Now $R_1[X]$ has infinitely many nonassociate irreducible elements and for each irreducible element $f \in R_1[X]$, $(f, 1)$ is an irreducible element of $R[X] = R_1[X] \times R_2[X]$ that is a factor of $(0, 1)$ since $(0, 1) = (0, 1)(f, 1)$. But this contradicts the hypothesis that $R[X]$ is an atomic idf-ring. Now $R_1[[X]]$ has infinitely many nonassociate irreducible elements unless R_1 is a field in which case $R_1[[X]]$ is a DVR. If $R_1[[X]]$ has infinitely many nonassociate irreducible elements, we may proceed as in the polynomial case. Thus we may assume that $R_1[[X]]$ is a DVR. But then X is irreducible in $R_1[[X]]$ so the factorization $(0, 1) = (0, 1)(X, 1)^n$ shows that $(0, 1)$ has infinitely many nonassociate factors. But this contradicts the hypothesis that $R[[X]]$ is a WFFR.

Suppose that R is not an integral domain. Let $0 \neq m \in R$ with $\text{ann}(m) \neq 0$. Let $n_1, n_2 \in \text{ann}(m)$. Then $0 \neq mX = m(n_1 + X) = m(n_2 + X)$ and $n_1 + X \nsim n_2 + X$ if $n_1 \neq n_2$ in both $R[X]$ and $R[[X]]$ by Proposition 1.6. (For the power series case, observe that n_1 and n_2 cannot be units and if $\bigcap_{k=1}^{\infty} (n_2)^k \neq 0$, then $(n_2)^l = (n_2)^{l+1} = \ldots$ for large l since R is a WFFR. Then $(n_2^l) = (n_2^l)^2 \neq 0$ forces R to be decomposable, a contradiction.) Hence $\text{ann}(m)$ must be finite. So $\text{ann}(m)$ contains a simple ideal Rt which is also finite. Then $Rt \cong R/M$ for some maximal ideal M and R/M is finite. But $M = \text{ann}(t)$ is also finite. Hence R itself is finite. So R is a finite direct product of finite local rings. Since R is indecomposable, R is a finite local ring.

COROLLARY 1.8. *The polynomial ring $R[X, Y]$ is an atomic idf-ring if and only if R is an FFD. If the power series ring $R[[X, Y]]$ is a WFFR, then R is a completely integrally closed FFD. Hence for R a Noetherian ring, $R[[X, Y]]$ is a WFFR if and only if R is a Krull domain.*

Proof. If $R[X, Y] = R[X][Y]$ is an atomic idf-ring, then $R[X]$ being infinite must be an integral domain. But for R an integral domain, R is an FFD if and only if $R[X, Y]$ is an FFD.

If $R[[X, Y]] = R[[X]][[Y]]$, then $R[[X]]$ being infinite must be an integral domain. Then $R[[X]]$ is an FFD and hence R must be a completely integrally closed FFD [5]. If R is a Noetherian integral domain, then $R[[X, Y]]$ is an FFD if and only if R is a Krull domain.

For R a finite local ring, $R[X]$ and $R[[X]]$ are both BFR's. So $R[X]$ (respectively, $R[[X]]$) is an FFR if and only if $R[X]$ (respectively, $R[[X]]$) is an atomic idf-ring. We end this section with the following question.

QUESTION 1.9. For R a finite local ring, when is $R[X]$ or $R[[X]]$ an FFR?

2. ASSOCIATES AND PRIMITIVES IN MODULES

Let R be a commutative ring with total quotient ring $T(R)$ and group of units $U(R)$. In Section 1 we defined three different associate relations on R. We briefly review these before extending them to modules. Let $a, b \in R$. Then a and b were called *associates* if $a|b$ and $b|a$, that is, $Ra = Rb$. If $a = ub$ for some $u \in U(R)$, we said that a and b were *strong associates*. Finally, we defined a and b to be *very strong associates* if (i) a and b are associates and (ii) either $a = b = 0$ or $a \neq 0$ and $a = rb \Rightarrow r \in U(R)$. We denoted these relations by $a \sim b$, $a \approx b$, and $a \cong b$, respectively. Clearly $a \cong b \Rightarrow a \approx b \Rightarrow a \sim b$ and $\sim$ and $\approx$ are equivalence relations on R. But $\cong$ need not be an equivalence relation. In fact, $\cong$ is an equivalence relation $\Leftrightarrow R$ is *présimplifiable*, that is, $x = xy \Rightarrow x = 0$ or $y \in U(R)$. Moreover, R is présimplifiable $\Leftrightarrow Z(R) \subseteq 1 - U(R) \Leftrightarrow Z(R) \subseteq \mathrm{rad}(R)$ where $Z(R)$ is the set of zero divisors of R and $\mathrm{rad}(R)$ is the Jacobson radical of R. We now extend these definitions and results to R-modules.

DEFINITION 2.1. Let S be an R-algebra and M an S-module. Let $m, n \in M$. Then m and n are R-*associates*, denoted $m \sim_R n$, $\Leftrightarrow Rm = Rn$. If $m = un$ for some $u \in U(R)$, then m and n are *strong R-associates*, denoted $m \approx_R n$. Finally, we say that m and n are *very strong R-associates*, denoted $m \cong_R n$, if $m \sim_R n$ and either $m = n = 0$ or $m \neq 0$ and $m = rn$ implies $r \in U(R)$.

When $R = S$, we will usually just write $m \sim n$ instead of $m \sim_R n$. Our new definitions of the various types of associates agree with the original ones in the case $R = S = M$. Clearly $m \cong_R n \Rightarrow m \approx_R n \Rightarrow m \sim_R n$ and examples given in [8] show that even in the ring case, none of these implications can be reversed. For an easy module example, we may take $R = \mathbb{Z}$ and $M = \mathbf{Z}_p$, where $p \geq 5$ is prime. Then $\bar{1} \sim \bar{2}$ but $\bar{1} \not\approx \bar{2}$ and $\bar{1} \approx \bar{1}$ but $\bar{1} \not\cong \bar{1}$. However, if R is Artinian, then $\sim_R$ and $\approx_R$ coincide for any R-module M [27].

As in the ring case, $\sim_R$ and $\approx_R$ are equivalence relations on M (even "R-congruences" on M in the sense that $m \sim_R n \Rightarrow rm \sim_R rn$ and $m \approx_R n \Rightarrow rm \approx_R rn$ for $r \in R$). But $\cong_R$ is an equivalence relation on M if and only if M is R-*présimplifiable*, that is, $m = rm$ implies $m = 0$ or $r \in U(R)$. Theorem 2.2 [8] carries over to modules. We get that $\cong_R$ is an equivalence relation (even an R-congruence) on M $\Leftrightarrow \cong_R$ is reflexive $\Leftrightarrow \sim_R, \approx_R, \cong_R$ coincide on $M \Leftrightarrow M$ is R-présimplifiable. It is easily checked that M is R-présimplifiable $\Leftrightarrow Z(M) \subseteq 1 - U(R) = \{1 - u \mid u \in U(R)\}$ $\Leftrightarrow Z(M) \subseteq \mathrm{rad}(R)$. Here $Z(M) = \{r \in R \mid rm = 0 \text{ for some } 0 \neq m \in M\}$. From the last implication we see that for a ring R, every R-module is R-présimplifiable if and only if R is quasilocal. Also, a direct sum $\bigoplus M_i$ or direct product $\prod M_i$ of R-modules $\{M_i\}$ is R-présimplifiable if and only if each M_i is R-présimplifiable. Certainly a torsion-free R-module (R an integral domain) is R-présimplifiable.

Unlike the case for rings (cf. [8, Theorem 2.15]), the various associate relations are not preserved by direct products. If $(m_i), (n_i) \in \prod M_i$ with $(m_i) \sim (n_i)$, then certainly each $m_i \sim n_i$. But the converse is false. Let F be a field with at least two distinct nonzero elements x and y. Then $1 \sim 1$ and $x \sim y$ but $(1, x) \not\sim (1, y)$ in $F \times F$.

We then used the three different associate relations on R to define three different types of irreducible elements. We defined [8, Def. 2.4] a nonunit element $a \in R$ to be *irreducible* (respectively, *strongly irreducible*, *very strongly irreducible*) if $a = bc$ $\Rightarrow a \sim b$ or $a \sim c$ (respectively, $a \approx b$ or $a \approx c$, $a \cong b$ or $a \cong c$). Here for $a \neq 0$

a very strongly irreducible $\Rightarrow$ a is m-irreducible (that is, Ra is a maximal proper principal ideal of R) $\Rightarrow$ a is strongly irreducible $\Rightarrow$ a is irreducible, but none of the implications can be reversed. Note that an irreducible element can not be a unit, but it is allowed to be 0, in which case R is an integral domain. A nonzero, nonunit element $a \in R$ is very strongly irreducible $\Leftrightarrow$ $a = bc$ implies b or c is a unit. We next extend these definitions to R-modules. To avoid confusion with ring irreducible elements, we use the word "primitive" instead of "irreducible".

DEFINITION 2.2. Let S be an R-algebra, M be an S-module, and $0 \neq m \in M$. Then m is R-primitive (respectively, strongly R-primitive, very strongly R-primitive) if for $a \in R$ and $n \in M$, $m = an$ implies $m \sim_R n$ (respectively, $m \approx_R n$, $m \cong_R n$). And m is R-superprimitive if $bm = an$ for $a, b \in R$ implies $a|b$ in R.

Again, when $R = S$, we will usually drop the prefix "R-". Note that if M is an R-module and $n, m \in M$ with $\text{ann}_R(m) = 0$, then $n \sim m \Leftrightarrow n \approx m \Leftrightarrow n \cong m$. Thus if either M is R-présimplifiable or $\text{ann}_R(m) = 0$, then m is R-primitive $\Leftrightarrow$ m is strongly R-primitive $\Leftrightarrow$ m is very strongly R-primitive.

The next proposition gives some properties of the various notions of primitive.

PROPOSITION 2.3. Let S be an R-algebra, M be a nonzero S-module, and let $0 \neq m \in M$.

(1) m is R-primitive $\Leftrightarrow$ Rm is a maximal cyclic R-submodule of M.

(2) m is strongly R-primitive $\Leftrightarrow$ $m = an$ for $a \in R$ and $n \in M$ implies $m = un$ for some $u \in U(R)$.

(3) m is very strongly R-primitive $\Leftrightarrow$ $m = an$ for $a \in R$ and $n \in M$ implies $a \in U(R)$.

(4) m R-superprimitive $\Rightarrow$ m is very strongly R-primitive $\Rightarrow$ m is strongly R-primitive $\Rightarrow$ m is R-primitive. If $\text{ann}_R(m) = 0$, m R-primitive $\Rightarrow$ m is very strongly R-primitive.

(5) m R-superprimitive $\Rightarrow$ $\text{ann}_R(m) = 0$.

Proof. (1) ($\Rightarrow$) Suppose $Rm \subseteq Rn$. Then $m = an$, so $m \sim_R n$, that is, $Rm = Rn$. ($\Leftarrow$) Suppose $m = an$. Then $Rm \subseteq Rn \Rightarrow Rm = Rn$, so $m \sim_R n$.

(2) and (3) are just restatements of the definitions.

(4) Suppose that m is R-superprimitive. Let $m = an$ where $a \in R$. Then $1m = m = an$, so $a|1$ in R and hence $a \in U(R)$. So m is very strongly R-primitive. The remaining implications of the first sentence of (4) are obvious. Suppose that m is R-primitive with $\text{ann}_R(m) = 0$. Then $m = an$ for $a \in R$ gives $n = bm$ for some $b \in R$, so $m = abm$ and hence $ab = 1$ so $a \in U(R)$.

(5) Suppose $bm = 0$ where $b \in R$. Then $bm = 0m$ gives $0|b$ so $b = 0$.

EXAMPLE 2.4.

(1) Let $M = R = S$. Then $0 \neq m \in R$ primitive $\Rightarrow$ Rm is a maximal cyclic submodule of R, that is, $Rm = R$, so $m \in U(R)$. And if $m \in U(R)$, then m is superprimitive: if $bm = an$, then $b = m^{-1}an$ so $a|b$. So for $M = R = S$, all the forms of primitive coincide and each is equivalent to being a unit.

(2) Let $R = \mathbf{Z}$ and $M = \mathbf{Z}_n$, $n \geq 2$. Let $0 \leq m < n$. Then $\overline{m}$ is primitive $\Leftrightarrow$ $\mathbf{Z}\overline{m} = \mathbf{Z}_n \Leftrightarrow (n, m) = 1$. Now $\overline{m}$ is strongly primitive $\Leftrightarrow (n, m) = 1$ and for $1 \leq i \leq n - 1$ with $(i, n) = 1$, we have $\overline{i} = \pm \overline{m}$. This forces $\phi(n) \leq 2$. So $n = 2, 3, 4$, or 6. So $\overline{m}$ is strongly primitive in $\mathbf{Z}_n$ only for $n = 2$, $m = 1$;

$n = 3$, $m = 1, 2$; $n = 4$, $m = 1, 3$; and $n = 6$, $m = 1, 5$. Now $\overline{m}$ is never very strongly primitive or superprimitive. So in $\mathbf{Z}_5$, $\overline{1}$ is primitive but not strongly primitive and in $\mathbf{Z}_2$, $\overline{1}$ is strongly primitive, but not very strongly primitive.

(3) Suppose that R is an integral domain and $M = R[X]$. Let $0 \neq f \in R[X]$. Let A_f be the ideal of R generated by the coefficients of f. As remarked in Proposition 2.3, here the notions of primitive, strongly primitive, and very strongly primitive coincide. Now f is primitive $\Leftrightarrow A_f \subseteq Rr \Rightarrow r \in U(R)$ $\Leftrightarrow$ the gcd of the coefficients is 1. So this is just the classical definition of a primitive polynomial. And f is superprimitive $\Leftrightarrow (A_f)_v = R$. (Here v denotes the v-operation, so $(A_f)_v = ((A_f)^{-1})^{-1}$.) For if $(A_f)_v = R$ and $bf = ag$, then $Rb = b(A_f)_v = (A_{bf})_v = (A_{ag})_v = a(A_g)_v$ so $a|b$ and hence f is superprimitive. If f is superprimitive, then $A_f \subseteq \frac{a}{b}R \Rightarrow bf = ag$ for some $g \in R[X] \Rightarrow a|b \Rightarrow R \subseteq \frac{a}{b}R$. So $(A_f)_v = R$. Tang [26] had previously defined $f \in R[X]$ to be superprimitive if $(A_f)_v = R$. Note that here f primitive $\nRightarrow f$ superprimitive. For if $R = K[t^2, t^3]$, K a field and t an indeterminate over K, and $f = t^2 + t^3 X$, then f is (very strongly) primitive, but f is not superprimitive since $(A_f)_v = (t^2, t^3)R \neq R$.

Our definition of primitive does not agree with that of A.-M. Nicolas [23, 24]. For R an integral domain and M a torsion-free R-module, she defined $0 \neq m \in M$ to be "irreducible" if $m = an \Rightarrow a \in U(R)$, which is what we defined to be very strongly primitive. And she defined $0 \neq m \in M$ to be "primitive" if for $0 \neq a \in R$ and $n \in N$, $an \in Rm \Rightarrow n \in Rm$. In the case for M torsion-free, the next proposition shows that this is equivalent to m being superprimitive. The equivalence of (2) and (3) is essentially given in [23].

PROPOSITION 2.5. *Suppose that R is an integral domain with quotient field K and M is a torsion-free R-module. For $0 \neq m \in M$, the following conditions are equivalent:*

(1) *m is superprimitive;*
(2) *$Km \cap M = Rm$;*
(3) *for $0 \neq a \in R$ and $n \in M$, $an \in Rm \Rightarrow n \in Rm$.*

Proof. (1) $\Rightarrow$ (2). Let $\frac{b}{a}m = n \in M$, so $bm = an$. Then $a|b$ so $n \in Rm$. (2) $\Rightarrow$ (3). Suppose $an \in Rm$, say $an = bm$. Then $n = \frac{b}{a}m \in Km \cap M = Rm$. (3) $\Rightarrow$ (1). Suppose $bm = an$. So $an \in Rm$ gives $n \in Rm$, say $n = rm$. So $bm = an = ram \Rightarrow b = ra$.

However, in general neither our superprimitive nor Nicolas's primitive implies the other. Let us call $0 \neq m \in M$ N-*primitive* if for $a \in R$ and $n \in M$, $an \in Rm - \{0\} \Rightarrow n \in Rm$. Note that m N-primitive implies m is primitive (our definition). For $n > 1$, $\overline{1}$ is an N-primitive element of the $\mathbf{Z}$-module $\mathbf{Z}_n$, but we have already observed that no element of $\mathbf{Z}_n$ is superprimitive (in fact, $\overline{1}$ need not even be strongly primitive). Thus if $M = Rm$ is cyclic, m is N-primitive, but m need not be superprimitive nor even strongly primitive. Next, put $R = \mathbf{Z}_{12}$ and $M = \mathbf{Z}_{12}[X]$. Then $f = \overline{2} + \overline{3}X$ is superprimitive, but not N-primitive. For suppose $b(\overline{2} + \overline{3}X) = ag$. Then $\mathbf{Z}_{12}b = A_{b(\overline{2}+\overline{3}X)} = A_{ag} \subseteq \mathbf{Z}_{12}a$. So $f = \overline{2} + \overline{3}X$ is superprimitive. However, $\overline{3}(\overline{2} + \overline{3}X + \overline{4}X^2) = \overline{3}(\overline{2} + \overline{3}X) \in \mathbf{Z}_{12}f - \{0\}$, but $\overline{2} + \overline{3}X + \overline{4}X^2 \notin \mathbf{Z}_{12}f$.

We next define various types of atomicity for modules.

DEFINITION 2.6. Let R be a commutative ring and M an R-module. Let $I = \{$irreducible, strongly irreducible, m-irreducible, very strongly irreducible, prime$\}$ and $J = \{R$-primitive, strongly R-primitive, very strongly R-primitive, R-super-primitive$\}$. Let $m \in M$. A factorization $m = an$ where $a \in R$ and $n \in M$ is called a β-*factorization* if n is β for $\beta \in J$. And we call $m = a_1 \cdots a_s n$ (where $a_i \in R$) an (α, β)-*factorization* if each a_i is α ($\alpha \in I$) and n is β ($\beta \in J$). (We allow $s = 0$ in which case $m = \mu n$ where $\mu \in U(R)$.) We say that M is β-*atomic* (respectively, (α, β)-*atomic*) if every $0 \neq m \in M$ has a β-factorization (respectively, (α, β)-factorization).

Thus M is primitive-atomic $\Leftrightarrow$ every element of M is contained in a maximal cyclic submodule of M. Note that M is (α, β)-atomic if R is α-atomic and M is β-atomic. Now certainly if M is (α, β)-atomic, then M is β-atomic, but R need not be α-atomic. For example, if S is a simple R-module, then S is (α, β)-atomic for any α and for $\beta =$ primitive, but R itself may be chosen not to be α-atomic.

DEFINITION 2.7. Let R be a commutative ring and M an R-module, and let α and β be as in Definition 2.6. Let $m \in M$. Two β-factorizations $an_1 = m = bn_2$ of m are R-*isomorphic* (respectively, *strongly R-isomorphic, very strongly R-isomorphic*) if $n_1 \sim_R n_2$ (respectively, $n_1 \approx_R n_2$, $n_1 \cong_R n_2$). And two (α, β)-factorizations $a_1 \cdots a_k n_1 = m = b_1 \cdots b_l n_2$ of m are R-*isomorphic* (respectively, *strongly R-isomorphic, very strongly R-isomorphic*) if (1) $n_1 \sim_R n_2$ (respectively, $n_1 \approx_R n_2$, $n_1 \cong_R n_2$) and (2) $k = l$ and there exists a permutation $\sigma \in S_k$ such that $a_i \sim b_{\sigma(i)}$ (respectively, $a_i \approx b_{\sigma(i)}$, $a_i \cong b_{\sigma(i)}$).

Let α and β be as in Definition 2.6 and let $\gamma \in \{R$-isomorphic, strongly R-isomorphic, very strongly R-isomorphic$\}$. We say that a nonzero R-module M is an $((\alpha, \beta), \gamma)$-*unique factorization module* (respectively, (β, γ)-*unique factorization module*) if (1) M is (α, β)-atomic (respectively, β-atomic) and (2) any two (α, β)-factorizations (respectively, β-factorizations) of a nonzero element of M are γ. We say that M is an (α, β)-*half-factorial module* if M is (α, β)-atomic and for $0 \neq m \in M$ any two (α, β)-factorizations of m have the same length, that is, if $a_1 \cdots a_s n = m = b_1 \cdots b_t n'$ where each a_i, b_j is α and n, n' are β, then $s = t$. And M is a *bounded factorization module* if for each $0 \neq m \in M$, there exists a natural number $N(m)$ so that if $m = a_1 \cdots a_k n$ where a_i is a nonunit of R and $n \in M$, then $k \leq N(m)$.

While we will not do so here, one can extend the notion of elasticity and the various forms of homomorphic factorizations from rings to modules.

Clearly an $((\alpha, \beta), \gamma)$-unique factorization module is an (α, β)-half-factorial module and a bounded factorization module is présimplifiable. However, unlike the case for rings, an $((\alpha, \beta), \gamma)$-unique factorization module need not be présimplifiable. For let R be a ring with no nonzero elements that are α (so R is not a unique factorization ring unless it is a field) and take $M = R/\mathcal{M}$ where $\mathcal{M}$ is a maximal ideal of R. Then M is an $((\alpha, \beta), \gamma)$-unique factorization module where $\beta =$ primitive (and examples can be constructed for $\beta =$ strongly primitive) and $\gamma \in \{R$-isomorphic, strongly R-isomorphic$\}$ since each $0 \neq m \in M$ is primitive and an (α, β)-factorization of m is just $m = \mu n$ where $\mu \in U(R)$. But M is R-présimplifiable $\Leftrightarrow R$ is quasilocal. Thus if R is not quasilocal, M is not R-présimplifiable and hence is not a bounded factorization module. However, on the positive side we have the following theorem.

THEOREM 2.8. *Let R be a commutative ring and M a nonzero R-module.*

(1) *If M is a faithful bounded factorization module, then R is a bounded factorization ring.*

(2) *Let M be an $((\alpha, \beta), \gamma)$-half-factorial module where each β element m of M has $\text{ann}(m) = 0$. Then R is half-factorial and M is présimplifiable.*

(3) *Let M be an $((\alpha, \beta), \gamma)$-unique factorization module where each β element m of M has $\text{ann}(m) = 0$. Then R is a unique factorization ring and M is présimplifiable. (And if α = prime, R is a p-atomic unique factorization ring.)*

Proof. (1) Let $0 \neq a \in R$ be a nonunit. Since M is faithful, there is an $m \in M$ with $am \neq 0$. Suppose that $a = a_1 \cdots a_l$ where each a_i is a nonunit. Then $0 \neq am = a_1 \cdots a_l m$, so $l \leq N(am)$. Hence R is a bounded factorization ring.

(2), (3) We first show that M is a bounded factorization module. Suppose $m = a_1 \cdots a_t n \neq 0$ where each a_i is a nonunit. Let $m = b_1 \cdots b_s m'$ be an (α, β)-factorization of m. Let $a_{11} \cdots a_{1k_1} n_1$ be an (α, β)-factorization of $a_1 n$. Note that $k_1 \geq 1$. For if $a_1 n$ is β, then $Ra_1 n$ is a maximal cyclic submodule. Hence $a_1 Rn = Rn$ and so there is an $r \in R$ with $n = ra_1 n$ and hence $(1 - ra_1)n = 0$. But since $\text{ann}(n) \subseteq \text{ann}(a_1 n) = 0$, $1 - ra_1 = 0$ and hence a_1 is a unit, a contradiction. Continuing, let $a_{i1} \cdots a_{ik_i} n_i$ be an (α, β)-factorization of $a_i n_{i-1}$. And as before $k_i \geq 1$. So $m = a_t \cdots a_1 n = a_{11} \cdots a_{1k_1} a_{21} \cdots a_{2k_2} \cdots a_{t1} \cdots a_{tk_t} n_t$ is an (α, β)-factorization of m of length $k_1 + \cdots + k_t \geq t$. If M is either an (α, β)-half-factorial module or an (α, β)-unique factorization module, $s = k_1 + \cdots + k_t \geq t$. Hence M is a bounded factorization module. By (1), R is a bounded factorization ring. Hence R is atomic and présimplifiable. Suppose that $0 \neq a_1 \cdots a_s = b_1 \cdots b_t$ where a_i, b_j are atoms (since R is présimplifiable, all the forms of atomicity coincide). Let $m \in M$ be β. Then $a_1 \cdots a_s m = b_1 \cdots b_t m$. So if M is $((\alpha, \beta), \gamma)$-half-factorial $s = t$ and thus R is half-factorial. If M is an $((\alpha, \beta), \gamma)$-unique factorization module, then $s = t$ and after reordering $a_i \sim b_i$. So R is a unique factorization ring. $\blacksquare$

Of course, an important case in Theorem 2.8 in which each β element m of M has $\text{ann}(m) = 0$ is when β = superprimitive or M is torsion-free. Torsion-free unique factorization modules will be considered in Section 4.

Recall (Theorem 1.5) that a unique factorization ring R is either (a) a UFD, (b) an SPIR, or (c) a quasilocal ring $(R, \mathcal{M})$ with $\mathcal{M}^2 = 0$. The next theorem characterizes unique factorization modules over the last two types of unique factorization rings. Since every nonzero vector space over a field is a unique factorization module, we exclude the trivial case where R is a field.

THEOREM 2.9.

(1) *Let $(R, (p))$ be an SPIR that is not a field and let M be a nonzero R-module. Then M is a (β, γ)-unique factorization module or an $((\alpha, \beta), \gamma)$-unique factorization module for any α and γ and $\beta \in \{$primitive, strongly primitive, very strongly primitive$\}$ (respectively, β = superprimitive) if and only if $pM = 0$ or M is cyclic (respectively, $M \approx R$).*

(2) *Let $(R, \mathcal{M})$ be a quasilocal ring with $\mathcal{M}^2 = 0$ that is not a field and let M be a nonzero R-module. M is a (β, γ)-unique factorization module for $\beta \in \{$primitive, strongly primitive, very strongly primitive$\}$ (respectively, β = superprimitive) and any γ if and only if $\mathcal{M}M = 0$ or M is cyclic*

(*respectively*, $M \approx R$). Let $\alpha \in$ {atomic, strongly atomic, very strongly atomic, m-atomic}. M *is an* $((\alpha, \beta), \gamma)$-*unique factorization module for* $\beta \in$ {primitive, strongly primitive, very strongly primitive} (*respectively*, $\beta =$ superprimitive) *if and only if* $MM = 0$ *or* $M \approx R$ (*respectively*, $M \approx R$). *For* $\alpha = p$-*atomic*, M *is an* $((\alpha, \beta), \gamma)$-*unique factorization module for* $\beta \in$ {primitive, strongly primitive, very strongly primitive} (*respectively*, $\beta =$ superprimitive) *if and only if either* $MM = 0$ *or* R *is an SPIR and* M *is cyclic* (*respectively*, R *is an SPIR and* $M \approx R$).

Proof. First note that since R is quasilocal, every R-module is présimplifiable. Thus the notions of primitive, strongly primitive, and very strongly primitive coincide as do the various forms of R-isomorphic. If M is a nonzero R-module with $MM = 0$ (M the maximal ideal of R), then every nonzero element of M is primitive and M is a (β, γ)-unique factorization module and an $((\alpha, \beta), \gamma)$-unique factorization module for any α, β, and γ except for $\beta =$ superprimitive. The case where $\beta =$ superprimitive follows from the fact (Proposition 2.3) that if m is superprimitive, then $\mathrm{ann}(m) = 0$.

(1) Suppose that $p^n = 0$, but $p^{n-1} \neq 0$ where $n \geq 2$. Now over an SPIR, any module is isomorphic to a direct sum of ideals of R, say $M \approx \bigoplus_{\alpha \in \Lambda} p^{i_\alpha} Re_\alpha$ where $0 \leq i_\alpha \leq n - 1$ and $Re_\alpha \approx R$. Note that $m \in M$ is primitive $\Leftrightarrow m = \sum p^{i_\alpha} r_\alpha e_\alpha$ where almost all $r_\alpha = 0$, but some r_α is a unit. Suppose that some $i_{\alpha_0} \leq n - 2$ (equivalently, $pM \neq 0$). Then $p^{i_{\alpha_0}} e_{\alpha_0}$ is primitive and $p\left(p^{i_{\alpha_0}} e_{\alpha_0}\right) \neq 0$. Now for any $n \in \sum_{\alpha \in \Lambda - \{\alpha_0\}} p^{i_\alpha} Re_\alpha$ with $pn = 0$, $p\left(p^{i_{\alpha_0}} e_{\alpha_0}\right) = p\left(p^{i_{\alpha_0}} e_{\alpha_0} + n\right)$ where $p^{i_{\alpha_0}} e_{\alpha_0} + n$ is also primitive. But $p^{i_{\alpha_0}} e_{\alpha_0} \sim p^{i_{\alpha_0}} e_{\alpha_0} + n \Leftrightarrow n = 0$. The result follows.

(2) Assume that $MM \neq 0$. Let $n \in M$ with $Mn \neq 0$. Then Rn is a maximal cyclic submodule of M and hence n is primitive. Let $s \in \mathrm{Soc}(M)$, the socle of M. For $m \in M$ with $mn \neq 0$, $mn = m(n + s)$ where n and $n + s$ are both primitive. If $n \sim n + s$, then $s \in Rn$. But then $\mathrm{Soc}(M) \subseteq Mn$ and hence $\mathrm{Soc}(M) = Mn$. If n' is another primitive element of M, then $n' \notin Mn = \mathrm{Soc}(M)$, so $mn' \neq 0$ for some $m \in M$. Then $mn' \in \mathrm{Soc}(M)$ gives $0 \neq mn' = rn$ for some $r \in R$. Thus unique factorization gives $n' \sim n$ and hence $M = Rn$ is cyclic. Let $m \in M$ with $mn \neq 0$ and let $0 \neq r \in \mathrm{ann}(n)$. Then $0 \neq mn = (m + r)n$. Now $m \sim m + r$ gives $m + r = sm$ for some $s \in R$, so $r = (s - 1)m$. Since $r \neq 0$, $s - 1$ must be a unit, so $m = (s - 1)^{-1} r \in \mathrm{ann}(n)$, a contradiction. Thus if M is an $((\alpha, \beta), \gamma)$-unique factorization module, we must have $\mathrm{ann}(n) = 0$ and hence $M = Rn \approx R$.

3. RINGIFICATION

One way to study factorization in an R-module M is to embed M into a ring containing isomorphic copies of R and M that preserves the R-module structure of M. Two natural such rings are the symmetric algebra $S_R(M)$ of M and the idealization $R(M)$ of M. Here $S_R(M) = \bigoplus_{n \geq 0} S_R^n(M)$ is a graded R-algebra where $S_R^n(M) = M \otimes \cdots \otimes M / b_n$ where b_n is the R-submodule of $M^{\otimes n} = M \otimes \cdots \otimes M$ (n copies) generated by all $m_1 \otimes \cdots \otimes m_n - m_{\sigma(1)} \otimes \cdots \otimes m_{\sigma(n)}$ where $\sigma \in S_n$. We identity $S_R^0(M) = R$ and $S_R^1(M) = M$. So in $S_R(M)$ elements of M are multiplied via tensor product with the appropriate relations to make $S_R(M)$ a commutative R-algebra and in some sense $S_R(M)$ is the freest R-algebra generated by R and M. The idealization $R(M) = R \oplus M$ has multiplication defined by $(r, m)(s, n) =$

$(rs, rn + sm)$. Here R may be identified with the subring $\{(r, 0) \mid r \in R\}$ of $R(M)$ and M may be identified with the ideal $\{(0, m) \mid m \in M\}$. So $(r, 0)(0, m) = (0, rm)$ and the product of two elements of M is 0. Now $R(M)$ may be thought of as the R-algebra generated by R and M with the most relations. Note that $R(M)$ is naturally isomorphic to $S_R(M) / \bigoplus_{n \geq 2} S_R^n(M)$.

As expected the relations $\sim_R$, $\approx_R$, and $\cong_R$ on M are related to the relations $\sim$, $\approx$, and $\cong$ on $R(M)$ and $S_R(M)$. For $R(M)$ things go as expected and the proof of the following proposition is left to the reader. (One only needs to observe that $U(R(M)) = \{(r, m) \mid r \in U(R), m \in M\}$.)

PROPOSITION 3.1. *Let M be an R-module and $R(M)$ be the idealization of M. Let $m, n \in M$.*

(1) $m \sim_R n \Leftrightarrow (0, m) \sim (0, n)$ *in $R(M)$.*
(2) $m \approx_R n \Leftrightarrow (0, m) \approx (0, n)$ *in $R(M)$.*
(3) $m \cong_R n \Leftrightarrow (0, m) \cong (0, n)$ *in $R(M)$.*
(4) $R(M)$ *is présimplifiable $\Leftrightarrow R$ is présimplifiable and M is R-présimplifiable.*

PROPOSITION 3.2. *Let M be an R-module and $S_R(M)$ be the symmetric algebra of M. Let $m, n \in M$.*

(1) $m \sim_R n \Leftrightarrow m \sim n$ *in $S_R(M)$.*
(2) $m \approx_R n \Leftrightarrow m \approx n$ *in $S_R(M)$.*
(3) $m \cong_R n \Leftarrow m \cong n$ *in $S_R(M)$. But $\Rightarrow$ need not hold.*
(4) *If $S_R(M)$ is présimplifiable, then R is présimplifiable and M is R-présimplifiable, but the converse need not hold.*

Proof. (1) $m \sim_R n \Leftrightarrow Rm = Rn \Leftrightarrow S_R(M)m = S_R(M)n \Leftrightarrow m \sim n$.

(2) Now $m \approx_R n \Leftrightarrow m = un$ for some $u \in U(R)$. Now $u \in U(R) \Rightarrow u \in U(S_R(M))$, so $m \approx_R n \Rightarrow m \approx n$. Conversely, suppose $m \approx n$. So $m = vn$ where $v \in U(S_R(M))$. But then $v = v_0 + v_1 + \cdots$ where $v_i \in S_R^i(M)$ and $v_0 \in U(R)$. So $m = v_0 n$ and hence $m \approx_R n$.

(3) Suppose that $m \cong n$ in $S_R(M)$. Then $m \sim n$ in $S_R(M)$, so that $m \sim_R n$. Suppose that $m \neq 0$. If $m = rn$ in M, then $m = rn$ in $S_R(M)$; so $r \in U(S_R(M))$. Hence $r \in U(R)$, so $m \cong_R n$. However, $\Rightarrow$ need not hold. Suppose $R = \mathbf{Z}_{(2)}(\mathbf{Z}_4)$ and let $a = (0, 1)$. Now from [8, Example 6.1] we have $a \cong_R a$ but $a \not\cong_{R[X]} a$. Take $M = R$, so $S_R(M) = S_R(R) \cong R[X]$. Then $a \cong_R a$, but $a \not\cong a$ in $S_R(M)$.

(4) Suppose $S_R(M)$ is présimplifiable. Then $\cong$ is an equivalence relation on $S_R(M)$. Hence $\cong$ is an equivalence relation on R and $\cong_R$ is an equivalence relation on M. Thus R is présimplifiable and M is R-présimplifiable. The ring $R = \mathbf{Z}_{(2)}(\mathbf{Z}_4)$ given in (3) is présimplifiable, being quasilocal. However, as (3) shows, $S_R(M) = S_R(R) \cong R[X]$ is not présimplifiable.

We would like m to be primitive in M if and only if $(0, m)$ is irreducible in $R(M)$. As we next show, this is the case when R is an integral domain.

THEOREM 3.3. *Let R be an integral domain and let M be a nonzero R-module. Let $0 \neq m \in M$.*

(1) m *is primitive $\Leftrightarrow (0, m)$ is irreducible in $R(M)$.*
(2) m *is strongly primitive $\Leftrightarrow (0, m)$ is strongly irreducible in $R(M)$.*
(3) m *is very strongly primitive $\Leftrightarrow (0, m)$ is very strongly irreducible in $R(M)$.*

Proof. This is Proposition 5.1 [8].

Thus in the case where R is an integral domain, our definitions of primitive, strongly primitive, and very strongly primitive seem to be appropriate. However, as the next theorem shows, in general there does not appear to be a reasonable definition of primitive so that m is primitive in M if and only if $(0, m)$ is irreducible in $R(M)$.

THEOREM 3.4.

(1) *Suppose that R has a nontrivial idempotent and let M be a nonzero R-module. Then no element $(0, m)$ of $0 \oplus M$ is irreducible in the idealization $R(M)$.*

(2) *However, if R is indecomposable and $0 \neq m \in M$ is superprimitive, then $(0, m)$ is a very strongly irreducible element of $R(M)$.*

Proof. (1) Let $e \neq 0, 1$ be idempotent. Then for $m \in M$,

$$(0, m) = (e, m)(1 - e, m),$$

but $(0, m) \not\sim (e, m)$ and $(0, m) \not\sim (1 - e, m)$. So $(0, m)$ is not irreducible.

(2) Suppose that $(0, m) = (a_1, n_1)(a_2, n_2)$, so $a_1 a_2 = 0$ and $a_1 n_2 + a_2 n_1 = m$. First, suppose $a_2 = 0$. Then $m = a_1 n_2$. Since m is superprimitive, $a_1 \in U(R)$. Thus $(a_1, n_1) \in U(R(M))$. Next, suppose $a_2 \neq 0$. Then $a_2 m = a_2^2 n_1$. Since m is superprimitive, $a_2^2 | a_2$. So $a_2 = \lambda a_2^2$ for some $\lambda \in R$. Then λa_2 is idempotent. Hence $\lambda a_2 = 1$ since R is indecomposable and $a_2 \neq 0$. So $(a_2, n_2) \in U(R(M))$. Thus $(0, m)$ is very strongly irreducible.

Thus for $R = \mathbf{Z}_2 \times \mathbf{Z}_2$, 1_R is superprimitive, but $(0, 1)$ is not irreducible in $R(R)$. Also, m may be superprimitive as an element of M, but not irreducible as an element of $S_R(M)$. For example, X is irreducible in $R[X]$ if and only if R is indecomposable. Thus for $R = M$, since $S_R(R) \cong R[X]$, if R is not indecomposable, 1 is superprimitive in M but not irreducible in $S_R(R)$.

4. FACTORIZATION IN TORSION-FREE MODULES

In this section we study factorization in torsion-free modules with especial emphasis given to various forms of unique factorization. We also discuss the previous work of Nicolas [23, 24] and Costa [11]. Let R be an integral domain with quotient field K, and M a nonzero torsion-free R-module. Since R is an integral domain, the various forms of irreducible elements all agree, so we will just use the term irreducible. However, as is customary for integral domains, we will usually assume an irreducible element is nonzero. Since a torsion-free module is présimplifiable, the notions of primitive, strongly primitive, and very strongly primitive coincide, so we will just use the term primitive. So $0 \neq m \in M$ is primitive precisely when Rm is a maximal cyclic submodule of M. Recall from Section 2 that $0 \neq m \in M$ is superprimitive if $bm = an$ for $a, b \in R$ and $n \in M$ implies $a | b$. By Proposition 2.5, m is superprimitive $\Leftrightarrow Km \cap M = Rm \Leftrightarrow$ for $0 \neq a \in R$ and $n \in M$, $an \in Rm \Rightarrow n \in Rm$. Also, the various forms of "isomorphic β-factorizations" coincide as do the various forms of "isomorphic (α, β)-factorizations" given in Definition 2.7. Moreover, since M is torsion-free, the factorizations $an_1 = bn_2 \neq 0$ are R-isomorphic if either $a \sim b$ or $n_1 \sim n_2$.

Thus M is primitive-atomic (respectively, superprimitive-atomic) if for each $0 \neq m \in M$, $m = an$ for some $a \in R$ and primitive (respectively, superprimitive) $n \in M$. And we will simply say that M is *atomic* (respectively, *super-atomic*) if M is (irreducible, primitive)-atomic (respectively, (irreducible, superprimitive)-atomic). The following theorem gives some simple conditions for M to be atomic.

THEOREM 4.1. *Let R be an integral domain and M a nonzero torsion-free R-module. If M satisfies ACC on cyclic submodules, then M is atomic. More generally, if R is atomic and each cyclic submodule of M is contained in a maximal cyclic submodule, then M is atomic.*

Proof. The second statement, which is stated by Nicolas [23, Théorème 1.7], is obvious. As to the first statement, if M satisfies ACC on cyclic submodules, so does R since R is isomorphic to a cyclic submodule of M. Since R satisfies ACCP, R is atomic. Moreover, every submodule of M is contained in a maximal cyclic submodule.

The next theorem gives some equivalent conditions for M to be superprimitive-atomic. Of course, M is primitive-atomic $\Leftrightarrow$ each cyclic submodule of M is contained in a maximal cyclic submodule.

THEOREM 4.2. *Let R be an integral domain and M a nonzero torsion-free R-module. Then the following conditions are equivalent.*

(1) *M is a (primitive, R-isomorphic)-unique factorization module.*
(2) *M is a (superprimitive, R-isomorphic)-unique factorization module.*
(3) *M is primitive-atomic and every primitive element of M is superprimitive.*
(4) *M is superprimitive-atomic.*

Proof. (1) $\Rightarrow$ (2). It is enough to show that if $m \in M$ is primitive, then m is superprimitive. Suppose that $bm = an \neq 0$. Then $n = cn'$ where n' is primitive, so $bm = acn'$. Then $b \sim ac$ so $a|b$. Thus m is superprimitive. (2) $\Rightarrow$ (3). Clearly M is primitive-atomic. Let m be primitive. Then $m = cm'$ where m' is superprimitive. Since Rm is a maximal cyclic submodule of M, $Rm = Rm'$ and hence since M is torsion-free, $m = um'$ for some $u \in U(R)$. Thus m is superprimitive. (3) $\Rightarrow$ (4). Clear. (4) $\Rightarrow$ (1). Clearly M is primitive-atomic. Suppose that $0 \neq an = a'n'$ where n and n' are primitive. As in (2) $\Rightarrow$ (3), both n and n' are actually superprimitive. Then $a|a'$ and $a'|a$, so $a \sim a'$. Hence $n \sim n'$ since M is torsion-free.

Following [24, Def. 3.2], we give the next definition. We also define two other types of "unique factorization modules".

DEFINITION 4.3. Let R be an integral domain and M a nonzero torsion-free R-module. Then M is *factorable* if M satisfies any of the equivalent conditions of Theorem 4.2. We call M a *unique factorization module* (respectively, *factorial module*) if M is an ((irreducible, primitive), R-isomorphic)-unique factorization module (respectively, ((irreducible, superprimitive), R-isomorphic)-unique factorization module).

The study of factorization in torsion-free modules was begun in Nicolas [23]. She defined the notions that we have called very strongly primitive and superprimitive. As remarked in the paragraph preceding Proposition 2.5, she used the term "irreducible" for what we have defined to be very strongly primitive and she used the term "primitive" in the sense of our superprimitive (see Proposition 2.5). Let M be a nonzero torsion-free module over the integral domain R. She defined M to be "factorial" if (1) M is atomic, (2) p irreducible in $R \Rightarrow p$ is prime in R, and (3) every primitive element of M is superprimitive. She showed that if M is factorial, then R is a UFD. It is easy to see that her definition and our definition of factorial module agree (see Theorem 4.4). In the second paragraph of page 38 of [23], Nicolas remarked (in our terminology) that if M is a unique factorization module over R and if $a_1 \cdots a_n = b_1 \cdots b_m$ where each a_i, b_j is an atom of R, then $n = m$ and after reordering $a_i \sim b_i$. But she said she was unable to prove that R must be atomic. In Theorem 4.4 we show that R must indeed be atomic. This shows that the notions of factorial module and unique factorization module coincide. In [24], Nicolas defined a factorable module to be a torsion-free module satisfying either of the equivalent conditions (1) or (3) of Theorem 4.2. Later, Costa [11] considered unique factorization in torsion-free modules and showed that for a torsion-free R-module M, the symmetric algebra $S_R(M)$ is a factorial R-module if and only if $S_R(M)$ is a UFD. He also considered other properties on a torsion-free module weaker than being factorial. Studying factorization in M via $S_R(M)$ can best be done when $S_R(M)$ is an integral domain. For a discussion of when $S_R(M)$ is an integral domain, see [12]. The following theorem gives a number of equivalent conditions for M to be factorial. In particular, it shows that the notions of factorial module and unique factorization module coincide.

THEOREM 4.4. *For an integral domain R and nonzero torsion-free R-module M, the following conditions are equivalent.*

(1) *M is a factorial module as defined in Definition 4.3.*
(2) *M is a factorial module as defined by Nicolas.*
(3) *R is a UFD and M is factorable.*
(4) *R is atomic and M is a unique factorization module.*
(5) *R is a unique factorization module.*

Proof. We note that the equivalence of (1)–(4) is essentially given by Nicolas. (1) $\Rightarrow$ (4). It suffices to show that R is atomic. Let $0 \neq a \in R$ be a nonunit and let $m \in M$ be superprimitive. Then $am = a_1 \cdots a_s n$ where each $a_i \in R$ is irreducible and $n \in M$ is superprimitive. Since m and n are both superprimitive, $a \sim a_1 \cdots a_s$. Thus a is a product of atoms and hence R is atomic. (4) $\Rightarrow$ (3). Suppose that $a_1 \cdots a_s = b_1 \cdots b_t$ where $a_i, b_j \in R$ are atoms. Let $n \in M$ be primitive. Then $a_1 \cdots a_s n = b_1 \cdots b_t n \Rightarrow s = t$ and after reordering, if necessary, $a_i \sim b_i$. So R is a UFD. Let $m \in M$ be primitive. It suffices to show that m is superprimitive. Write $am = bn \neq 0$ where $b \in R$ and $n \in M$. Let $a = a_1 \cdots a_l$, $b = b_1 \cdots b_t$, and $n = c_1 \cdots c_s n'$ where a_i, b_j, c_k are all atoms and n' is primitive. Then $a_1 \cdots a_l m = b_1 \cdots b_t c_1 \cdots c_s n'$. So $a_1 \cdots a_l \sim (b_1 \cdots b_t)(c_1 \cdots c_s)$. Thus $b = b_1 \cdots b_t | a_1 \cdots a_l = a$, so m is superprimitive. Note that for (1) $\Rightarrow$ (4) and (4) $\Rightarrow$ (3) we could have just quoted Theorem 2.8(3). (3) $\Rightarrow$ (2). This follows from the definitions and Theorem 4.2. (2) $\Rightarrow$ (1). Clearly M is super-atomic. Suppose that $a_1 \cdots a_s n = b_1 \cdots b_t n'$ where the a_i, b_j's are atoms and n and n' are superprimitive. Then $a_1 \cdots a_s \sim b_1 \cdots b_t$. By hypothesis each a_i, b_j is prime. But it is well known that a factorization of an element into primes is unique up to order and associates. (4) $\Rightarrow$ (5). Clear. (5) $\Rightarrow$ (4). Theorem 2.8(3).

5. FACTORIZATION OF REGULAR ELEMENTS

There are basically two approaches to generalizing results about integral domains to commutative rings with zero divisors. The first is to directly extend a definition concerning elements or ideals of an integral domain to all the elements or ideals of a commutative ring. For example, by this approach the notion of a BFD generalizes to a BFR. Also, a Prüfer domain generalizes to an arithmetical ring (i.e., every finitely generated ideal is locally principal). This is the approach taken in Section 1. The second approach is to extend a definition concerning nonzero elements or ideals of an integral domain to just the regular elements or regular ideals of a commutative ring. For example, Prüfer domains have Prüfer rings as a generalization since Prüfer rings are defined to be commutative rings in which every finitely generated regular ideal is invertible. See [20] and [6] for results on Prüfer rings that typify this approach. One of the earliest papers to use this approach is [25] where it was shown that if a commutative ring R has ascending chain condition on regular principal ideals and if each pair of regular elements has a gcd, then the elements of reg(R) have unique factorization into irreducible elements up to order and associates. Also see [7] where a lattice-theoretic approach is used to study the regular ideals of a general commutative ring. In this section we follow the second approach.

One simplification in dealing with only the regular elements of a ring R is that the three associate relations $\sim$, $\approx$, and $\cong$ all agree for regular elements. Hence for a regular nonunit $a \in R$, the notions of irreducible, strongly irreducible, very strongly irreducible, and m-irreducible all coincide, so we will simply use the term irreducible. Likewise, the various forms of atomicity when restricted to regular elements also coincide. We will say that R is r-*atomic* if every regular, nonunit element of R is a product of irreducible elements. Finally, we say that R satisfies r-*ACCP* if every ascending chain of regular principal ideals stabilizes. Clearly if R satisfies r-ACCP, then R is r-atomic, but the converse is false.

There are natural generalizations of integral domains satisfying certain factorization properties to commutative rings with zero divisors whose regular elements satisfy the corresponding factorization properties. We define R to be *factorial* if (1) R is r-atomic and (2) if $a_1 \cdots a_n = b_1 \cdots b_m$ where each a_i, b_j is a regular irreducible element of R, then $n = m$ and after reordering a_i and b_i are associates. The ring R is an r-*half-factorial ring* (r-HFR) if (1) R is r-atomic and (2) if $a_1 \cdots a_n = b_1 \cdots b_m$ where each a_i, b_j is a regular irreducible element of R, then $n = m$. The *regular elasticity of* R is r-$\rho(R) = \sup \{\rho(a) \mid a \in \text{reg}(R), a \text{ is a nonunit}\}$. Clearly R is an r-HFR if and only if r-$\rho(R) = 1$. A ring R is defined to be an r-*bounded factorization ring* (r-BFR) if for each regular nonunit $a \in R$, there exists a natural number $N(a)$ so that if $a = a_1 \cdots a_n$ where each a_i is a nonunit, then $n \leq N(a)$. We define R to be an r-*finite factorization ring* (r-FFR) if one of the following three equivalent conditions holds: (1) every regular element of R has only a finite number of factorizations up to order and associates, (2) every regular element of R has only finitely many nonassociate factors, (3) R is r-atomic and each regular element of R has at most a finite number of nonassociate irreducible divisors. (That these three conditions are equivalent is identical to the proof of the integral domain case. Also see the remarks below about FF-monoids.) It is clear that factorial $\Rightarrow$ r-HFR $\Rightarrow$ r-BFR $\Rightarrow$ r-ACCP and that factorial $\Rightarrow$ r-FFR $\Rightarrow$ r-BFR. Moreover, none of these implications can be reversed.

Let H be a commutative cancellative monoid that is written multiplicatively. We denote the units of H by $H^\times$ and the quotient group of H by $\langle H \rangle$. The notions $a|b$, associate, irreducible, and prime have obvious definitions in H. (Since H is a cancellative monoid, we don't need to worry about the various forms of associate and irreducible.) An ideal A of H is a nonempty subset A of H with the property that $a \in A$ and $h \in H$ implies that $ah \in A$. Of course, for us here the most important example of a monoid will be reg(R), the set of regular elements of a commutative ring R under multiplication. For $a, b \in$ reg(R), $a|b$ in R (equivalently, $Rb \subseteq Ra$) $\Leftrightarrow$ $a|b$ in reg(R) (equivalently, reg$(R)b \subseteq$ reg$(R)a$), $a \sim b$ in $R \Leftrightarrow a \sim b$ in reg(R), $a \in U(R) \Leftrightarrow a \in$ reg$(R)^\times$, and a is irreducible as an element of $R \Leftrightarrow$ a is irreducible as an element of reg(R). Note that $\langle$reg$(R)\rangle = U(T(R))$. If p is a regular prime of R, then certainly p is prime as an element of reg(R). However, as we will later see, an element $a \in$ reg(R) that is prime as an element of reg(R) need not be a prime element of R.

The various definitions for integral domains satisfying certain factorization properties also carry over to monoids. Let H be a commutative cancellative monoid. Then H is *atomic* if every nonunit of H may be written as a finite product of irreducible elements. The monoid H is *factorial* if H is atomic and for atoms $a_i, b_j \in H$ with $a_1 \cdots a_n = b_1 \cdots b_m$, we have $n = m$ and after reordering $a_i \sim b_i$. It is well known that H is factorial $\Leftrightarrow H \approx G \times (\bigoplus \mathbb{N}_0)$ where G is a group and $\mathbb{N}_0 = \{0, 1, 2, \ldots\}$ under addition $\Leftrightarrow$ every nonunit element of H is a product of prime elements. We say that H is a *half-factorial (HF-) monoid* if H is atomic and for atoms $a_i, b_j \in H$ with $a_1 \cdots a_n = b_1 \cdots b_m$ we have $n = m$. Halter-Koch [16] has already defined H to be a *bounded factorization (BF-) monoid* if for each $a \in H$, the lengths of factorizations are bounded and H to be a *finite factorization (FF-) monoid* if H is atomic and each nonunit $a \in H$ possesses only finitely many different factorizations (into irreducibles) up to order and associates. He showed [16, Theorem 2] that the equivalent definitions of an FFD are also equivalent for a monoid. (Applying this result to reg(R) gives another proof that the different definitions of an r-FFR are indeed equivalent.) For each property α defined above, a commutative ring R is r-α if and only if the monoid reg(R) satisfies α.

For an atomic monoid H, we can define the two "length" functions $l_H : H \to \mathbb{N}_0$ and $L_H : H \to \mathbb{N}_0 \cup \{\infty\}$ by $l_H(a) = L_H(a) = 0$ for $a \in H^\times$ and for $a \in H - H^\times$, $l_H(a) = \inf\{n \mid a = a_1 \cdots a_n$ where each a_i is an atom$\}$ and $L_H(a) = \sup\{n \mid a = a_1 \cdots a_n$ where each a_i is an atom$\}$. So H is an HF-monoid $\Leftrightarrow$ $l_H = L_H \Leftrightarrow l_H$ is a length function and H is a BF-monoid $\Leftrightarrow$ im $L_H \subseteq \mathbb{N}_0$. For $a \in H - H^\times$ define $\rho(a) = L_H(a)/l_H(a)$ and for $a \in H^\times$ put $\rho(a) = 1$. Then the elasticity of H is $\rho(H) = \sup\{\rho(a) \mid a \in H\}$. Hence r-$\rho(R) = \rho(reg(R))$. So H is an HF-monoid $\Leftrightarrow \rho(H) = 1$. For $a \in H$, put $\bar{l}_H(a) = \lim_{n \to \infty} l_H(a^n)/n$ and $\bar{L}_H(a) = \lim_{n \to \infty} L_H(a^n)/n$. In [13] it is shown that both limits exist (with possibly $\bar{L}_H(a) = \infty$) and hence we get functions $\bar{l}_H : H \to [0, \infty)$ and $\bar{L}_H : H \to [0, \infty) \cup \{\infty\}$. Finally, $\bar{\rho}(a) = \lim_{n \to \infty} \rho(a^n) = \bar{L}_H(a)/\bar{l}_H(a)$ exists for each $a \in H - H^\times$. Let R be a commutative ring. Taking $H =$ reg(R) and $a \in$ reg(R), put $l_R(a) = l_{\text{reg}(R)}(a)$, $L_R(a) = L_{\text{reg}(R)}(a)$, $\bar{l}_R(a) = \lim_{n \to \infty} l_R(a^n)/n = \bar{l}_{\text{reg}(R)}(a)$, and $\bar{L}_R(a) = \lim_{n \to \infty} L_R(a^n)/n = \bar{L}_{\text{reg}(R)}(a)$. As previously mentioned, [13] shows that both limits exist with possibly $\bar{L}_R(a) = \infty$. Likewise, $\bar{\rho}(a) = \lim_{n \to \infty} \rho(a^n) = \bar{L}_R(a)/\bar{l}_R(a)$ for each $a \in$ reg$(R) - U(R)$. We later show that for R a Krull ring (defined in the next paragraph) much more can be said.

The notion of a Krull domain has been generalized to both Krull rings and Krull monoids. Kennedy [19] defined a commutative ring R to be a Krull ring if $R = \bigcap_\alpha (V_\alpha, P_\alpha)$ where each (V_α, P_α) is a rank one discrete valuation pair with total quotient ring $T(R)$, each P_α is a regular prime ideal, and if v_α is the associated valuation of (V_α, P_α), then for each $a \in \mathrm{reg}(R)$, $v_\alpha(a) = 0$ for almost all α. While Kennedy required $R \neq T(R)$, we let a total quotient ring be a Krull ring with empty defining family. Equivalently, R is a Krull ring if and only if R is completely integrally closed and R has ACC on integral regular v-ideals (a regular ideal A is a v-ideal if $\left(A^{-1}\right)^{-1} = A$)[19, 21]. Chouinard [10] defined a cancellative monoid S to be a Krull monoid if there exists a family $(v_i)_{i \in I}$ of discrete valuations on $\langle S \rangle$ (that is, each $v_i : \langle S \rangle \to \mathbb{Z}$ is a group homomorphism) such that $S = \bigcap V_i$ where $V_i = \{x \in \langle S \rangle \mid v_i(x) \geq 0\}$ and for every $x \in S$, the set $\{i \in I \mid v_i(x) > 0\}$ is finite. So a group is a Krull monoid with $I = \emptyset$. He showed that S is a Krull monoid if and only if S is completely integrally closed (that is, if $x \in \langle S \rangle$ with $rx^n \in S$ for all $n \geq 1$ where $r \in S$, then $x \in S$) and S has ACC on integral v-ideals (an ideal A of S is a v-ideal $\Leftrightarrow \left(A^{-1}\right)^{-1} = A \Leftrightarrow A = \bigcap \{Sx \mid A \subseteq Sx$ where $x \in \langle S \rangle\}\}$). For results on Krull monoids, also see [14].

Now an integral domain R is a Krull domain if and only if $(R - \{0\}, \cdot)$ is a Krull monoid. Unfortunately, the theorem we would like, namely that R is a Krull ring if and only if $\mathrm{reg}(R)$ is a Krull monoid, is not true. While a Krull ring R has $\mathrm{reg}(R)$ a Krull monoid, $\mathrm{reg}(R)$ can be a Krull monoid, even factorial, without R being a Krull ring. However, if R is a Marot ring and $\mathrm{reg}(R)$ is a Krull monoid, then R is a Krull ring. Recall that a ring R is a Marot ring if every regular ideal of R is generated by regular elements. A Noetherian ring, or more generally a ring with $Z(R)$ a finite union of prime ideals, is a Marot ring. If R is a Marot ring and A is a regular fractional ideal of R, then $\left(A^{-1}\right)^{-1} = \bigcap \{Rx \mid A \subseteq Rx\}$. In fact, in Theorems 5.1, 5.3, 5.5, and 5.6 the Marot hypothesis can be replaced by the condition that $\left(A^{-1}\right)^{-1} = \bigcap \{Rx \mid A \subseteq Rx\}$ for each regular fractional ideal A of R (which was called Property (D) in [4]).

THEOREM 5.1. *Let R be a commutative ring and let $\mathrm{reg}(R)$ be the multiplicative monoid of regular elements of R.*

(1) *If R is a Krull ring, then $\mathrm{reg}(R)$ is a Krull monoid.*

(2) *If R is a Marot ring and $\mathrm{reg}(R)$ is a Krull monoid, then R is a Krull ring.*

Proof. (1) Suppose that R is a Krull ring. Let $R = \bigcap_\alpha (V_\alpha, P_\alpha)$ be the defining family of rank one discrete valuation pairs. If v_α is the associated valuation for V_α, then $v_\alpha^* : \langle \mathrm{reg}(R) \rangle \to \mathbb{Z}$ given by $v_\alpha^* = v_\alpha|_{\mathrm{reg}(R)}$ is a discrete valuation on $\langle \mathrm{reg}(R) \rangle$ with $\mathrm{reg}(R) = \bigcap V_\alpha^*$ where $V_\alpha^* = \{x \in \langle \mathrm{reg}(R) \rangle \mid v_\alpha^*(x) \geq 0\}$ and the intersection is locally finite. Hence $\mathrm{reg}(R)$ is a Krull monoid.

(2) Suppose that R is a Marot ring and $\mathrm{reg}(R)$ is a Krull monoid. Let $\hat{R}$ be the complete integral closure of R. If $x \in \hat{R}$ is regular, then $x \in R$ since $\mathrm{reg}(R)$ is completely integrally closed. So R and $\hat{R}$ have the same regular elements. Since R is a Marot ring, $R = \hat{R}$, so R is completely integrally closed. Let A be a regular ideal of R. Now for regular $x \in T(R)$, $A \subseteq Rx \Leftrightarrow A \cap \mathrm{reg}(R) \subseteq \mathrm{reg}(R)x$. So if A is a v-ideal, $A = \bigcap \{Rx \mid Rx \supseteq A\}$ easily gives that $A \cap \mathrm{reg}(R) = \bigcap \{\mathrm{reg}(R)x \mid \mathrm{reg}(R)x \supseteq A \cap \mathrm{reg}(R)\}$ and hence $A \cap \mathrm{reg}(R)$ is a v-ideal of $\mathrm{reg}(R)$. Moreover, $(A \cap \mathrm{reg}(R))R = A$ since R is a Marot ring. Let

$A_1 \subseteq A_2 \subseteq \cdots$ be an ascending chain of integral regular v-ideals of R. Then $A_1 \cap \operatorname{reg}(R) \subseteq A_2 \cap \operatorname{reg}(R) \subseteq \cdots$ is an ascending chain of v-ideals of $\operatorname{reg}(R)$. Since $\operatorname{reg}(R)$ is a Krull monoid, there is an $n \geq 1$ so that $A_n \cap \operatorname{reg}(R) = A_k \cap \operatorname{reg}(R)$ for all $k \geq n$. But then $A_n = (A_n \cap \operatorname{reg}(R)) R = (A_k \cap \operatorname{reg}(R)) R = A_k$ for all $k \geq n$. Hence R is a Krull ring.

EXAMPLE 5.2. (A ring R with $\operatorname{reg}(R)$ a Krull monoid, even factorial, but R not a Krull ring.) Let $D = K[X,Y]$, K a field and $A = \bigoplus \{D/M \mid M \in \max(D), Y \notin M\}$. Put $D_2 = K[Y, X^2, XY, X^3]$ and $R_2 = D_2 \oplus A$ (idealization). So R_2 is the ring of [7, Example 4.3]. Since $(X^2, 0) \in R_2$, but $(X, 0) \notin R_2$, R_2 is not (completely) integrally closed and hence R_2 is not a Krull ring. However, $\operatorname{reg}(R_2) = \{(\alpha Y^m, a) \mid \alpha \in K - \{0\}, m \geq 0, a \in A\}$. Since $(\alpha Y^m, a) \sim (Y^m, 0)$ [7, page 407, second paragraph], $\operatorname{reg}(R_2) \cong U(R_2) \times \mathbb{N}_0$ is a Krull monoid, even factorial. Note that $(Y, 0)$ is a prime element of $\operatorname{reg}(R)$, but $(Y, 0) R_2 = Y D_2 \oplus A$ is not a prime ideal of R_2 since $(X^2, 0) \cdot (XY, 0) = (X^3 Y, 0) \in (Y, 0) R_2$ but $(X^2, 0) \notin (Y, 0) R_2$ and $(XY, 0) \notin (Y, 0) R_2$.

If R is a Krull ring, then the regular fractional v-ideals $D(R)$ form a group under the v-product $A * B = (AB)_v$ with subgroup $\operatorname{Princ}(R)$ of regular fractional principal ideals. The quotient group $\operatorname{Cl}(R)$ is called the *divisor class group of R* [19]. In a similar fashion, we define $D(H)$, $\operatorname{Princ}(H)$, and the *divisor class group* $\operatorname{Cl}(H)$ for a Krull monoid H [10]. We next show that for R a Krull ring, $\operatorname{Cl}(\operatorname{reg}(R))$ is naturally isomorphic to a subgroup of $\operatorname{Cl}(R)$; moreover, if further R is a Marot ring, then $\operatorname{Cl}(\operatorname{reg}(R))$ and $\operatorname{Cl}(R)$ are naturally isomorphic. But we give an example where $\operatorname{Cl}(\operatorname{reg}(R)) = 0$, but $\operatorname{Cl}(R) \approx \mathbf{Z}$.

THEOREM 5.3. *Let R be a Krull ring. Then the map $\psi : D(\operatorname{reg}(R)) \to D(R)$ given by $\psi(A) = (RA)_v$ is a group monomorphism with $\psi(\operatorname{Princ}(\operatorname{reg}(R))) = \operatorname{Princ}(R)$. Moreover, if $\psi(A)$ is principal, so is A. Thus the induced map $\bar{\psi} : \operatorname{Cl}(\operatorname{reg}(R)) \to \operatorname{Cl}(R)$ given by $\bar{\psi}([A]) = [(RA)_v]$ is a group monomorphism. If further R is a Marot ring, then ψ and $\bar{\psi}$ are both surjective.*

Let A be a v-ideal of $\operatorname{reg}(R)$. If $(RA)_v$ is a prime ideal of R, then A is a prime ideal of $\operatorname{reg}(R)$. If further R is a Marot ring, the converse is true. Thus for R a Marot ring, only a finite number of divisor classes of $\operatorname{Cl}(R)$ contain a prime ideal if and only if only a finite number of divisor classes of $\operatorname{Cl}(\operatorname{reg}(R))$ contain a prime ideal. And each (nonzero) class of $\operatorname{Cl}(R)$ contains a prime ideal if and only if the same is true of $\operatorname{Cl}(\operatorname{reg}(R))$.

Proof. Let A be a fractional v-ideal of $\operatorname{reg}(R)$. Then certainly $(RA)_v$ is a regular fractional v-ideal of R. Clearly ψ is a group homomorphism with $\psi(\operatorname{Princ}(\operatorname{reg}(R))) = \operatorname{Princ}(R)$. Suppose that $\psi(A) = \psi(B)$, that is, $(RA)_v = (RB)_v$. Let $b \in B$ and suppose that $A \subseteq y \operatorname{reg}(R)$ where $y \in \langle \operatorname{reg}(R) \rangle = U(T(R))$. Then $Rb \subseteq (RB)_v = (RA)_v \subseteq Ry$, so $b \operatorname{reg}(R) \subseteq y \operatorname{reg}(R)$. Thus $B \subseteq y \operatorname{reg}(R)$ and hence $B \subseteq \bigcap \{y \operatorname{reg}(R) \mid y \operatorname{reg}(R) \supseteq A, y \in \langle \operatorname{reg}(R) \rangle\} = A_v = A$. Interchanging the roles of A and B gives that $A = B$. Thus ψ is a monomorphism. Moreover, if $\psi(A)$ is principal, then $(RA)_v = Rx = (Rx \operatorname{reg}(R))_v$ where $x \in U(T(R)) = \langle \operatorname{reg}(R) \rangle$ and hence $A = x \operatorname{reg}(R)$. Thus the induced homomorphism $\bar{\psi}$ is also a monomorphism.

Suppose that R is a Marot ring. Let B be a regular fractional v-ideal of R. Since R is a Marot ring, $B = (\{x_\alpha\})$ where $x_\alpha \in U(T(R)) = \langle \operatorname{reg}(R) \rangle$. Put $A = \{x_\alpha\} \operatorname{reg}(R)$. Since B is a fractional ideal of R, A is a fractional ideal of

$\operatorname{reg}(R)$. Now $RA = B$, so $\psi(A) = (RA)_v = B_v = B$. Hence ψ is surjective and thus $\bar{\psi}$ is also surjective.

Suppose that $(RA)_v$ is a prime ideal of R. Let $xy \in A$ where $x, y \in \operatorname{reg}(R)$. So $xy \in (RA)_v$. Now $(RA)_v$ is a prime ideal; suppose that $x \in (RA)_v$. Then $Rx \subseteq (RA)_v$, so by the proof in the first paragraph, $x \operatorname{reg}(R) \subseteq A$. Thus A is a prime ideal of $\operatorname{reg}(R)$. Suppose that R is a Marot ring and that A is a prime v-ideal of $\operatorname{reg}(R)$. Then $(RA)_v$ is a v-ideal of R that is prime with respect to regular elements. By [3, Corollary 3.5] and [4], $(RA)_v$ is a prime ideal. The last two statements are now immediate.

EXAMPLE 5.4. (A Krull ring R with $\operatorname{Cl}(\operatorname{reg}(R)) \subsetneq \operatorname{Cl}(R)$.) Let D be a Dedekind domain with maximal ideal M that is *not* principal, but some power of M is principal, say $m > 1$ is the least positive integer with $M^n = (t)$ principal. Let $A = \bigoplus \{D/Q \mid Q \neq M \text{ is a maximal ideal of } D\}$ and take $R = D \oplus A$ to be the idealization of D and A. Then $\{t^n R\}_{n=0}^{\infty}$ is the set of regular principal ideals of R. So $\operatorname{reg}(R) \approx U(R) \times (\mathbb{N}_0, +)$ and $\operatorname{Cl}(\operatorname{reg}(R)) = 0$. Let $P = M \oplus A$, so P is the unique regular prime ideal of R and $P^n = tR$, so P is invertible but not principal. Note that P is not an intersection of principal fractional ideals of R. So $\operatorname{Cl}(R) \neq 0$; in fact, $\operatorname{Cl}(R) = \langle [P] \rangle \approx \mathbb{Z}$. For details, see [7, Example 3.6]. Also, note that $t \operatorname{reg}(R)$ is a prime ideal of $\operatorname{reg}(R)$, but tR is not a prime ideal of R.

An integral domain R is factorial if and only if R is a Krull domain with $\operatorname{Cl}(R) = 0$. However, Example 5.2 shows that a factorial ring need not be a Krull ring. And Example 5.4 gives an example of a Krull ring R that is factorial but with $\operatorname{Cl}(R) \neq 0$.

THEOREM 5.5. *If R is a Krull ring with $\operatorname{Cl}(R) = 0$, then R is factorial. If R is a Marot ring, then R is factorial $\Leftrightarrow R$ is a Krull ring with $\operatorname{Cl}(R) = 0$*

Proof. Suppose that R is a Krull ring with $\operatorname{Cl}(R) = 0$. By Theorem 5.1, $\operatorname{reg}(R)$ is a Krull monoid. By Theorem 5.3, $\operatorname{Cl}(\operatorname{reg}(R)) \subseteq \operatorname{Cl}(R) = 0$, so $\operatorname{Cl}(\operatorname{reg}(R)) = 0$. Thus $\operatorname{reg}(R)$ is a factorial monoid and hence R is factorial.

Suppose that R is a factorial Marot ring. Then $\operatorname{reg}(R)$ is factorial and hence a Krull monoid. By Theorem 5.1, R is a Krull ring. Since $\operatorname{reg}(R)$ is factorial, $\operatorname{Cl}(\operatorname{reg}(R)) = 0$. By Theorem 5.3, $\operatorname{Cl}(R) = 0$.

If R is a unique factorization ring or a p-atomic ring, then it easily follows from Theorem 1.5 that R is a Marot Krull ring with $\operatorname{Cl}(R) = 0$ and (hence) that R is factorial. However, even a Marot Krull ring R with $\operatorname{Cl}(R) = 0$ need not be a unique factorization ring nor a p-atomic ring. For as remarked by Matsuda [21, 7.11], if D is a UFD and T is a total quotient ring that is not a field, SPIR, or quasilocal ring (T, M) with $M^2 = 0$, then $R = D \times T$ is a Marot Krull ring with $\operatorname{Cl}(R) = \operatorname{Cl}(D) \oplus \operatorname{Cl}(T) = 0$, but R is not a unique factorization ring nor a p-atomic ring.

Our last theorem uses Krull monoids to generalize several rationality results for Krull domains to the regular elements of Krull rings.

THEOREM 5.6. *Let R be a Krull ring and let x be a regular nonunit of R. Then $\bar{l}_R(x)$ and $\bar{L}_R(x)$ are each positive rational numbers. Moreover, there is a natural number m so that $\bar{l}_R(x) = l_R\left(x^{km}\right)/km$ and $\bar{L}_R(x) = L_R\left(x^{km}\right)/km$ for all natural numbers k. Hence $\bar{\rho}(x)$ is a positive rational number and there is a natural number m so that $\bar{\rho}(x) = L_R\left(x^{km}\right)/l_R\left(x^{km}\right)$ for all natural numbers k.*

Let R be a Krull ring such that only a finite number of divisor classes of $\mathrm{Cl}(\mathrm{reg}(R))$
contain a prime ideal of $\mathrm{reg}(R)$ *(e.g., R is a Krull ring with* $\mathrm{Cl}(R)$ *finite or R is a*
Marot Krull ring in which only a finite number of divisor classes contain a prime
ideal). Then r-$\rho(R)$ is rational. Moreover, r-$\rho(R) = \rho(a)$ for some regular $a \in R$.

Proof. Since $\mathrm{reg}(R)$ is a Krull monoid, the results of the first paragraph follow from
[13, Corollary 5].

Let R be a Krull ring such that only a finite number of divisor classes of $\mathrm{Cl}(\mathrm{reg}(R))$
contain a prime ideal of $\mathrm{reg}(R)$. Note that if $\mathrm{Cl}(R)$ is finite, then by Theorem 5.3,
$\mathrm{Cl}(\mathrm{reg}(R))$ is also finite and hence satisfies the hypothesis of the second paragraph.
Likewise, if R is a Marot Krull ring with only a finite number of divisor classes
containing a prime ideal, then the same is true of $\mathrm{Cl}(\mathrm{reg}(R))$ by Theorem 5.3. Then
by the proof of [1, Theorem 10] modified to $\mathrm{reg}(R)$ or using the techniques of [17],
the result follows.

REFERENCES

[1] D.D. Anderson, D.F. Anderson, S.T. Chapman, and W.W. Smith, Rational elasticity of factorizations in Krull domains, *Proc. Amer. Math. Soc.* **117** (1993), 37–43.

[2] D.D. Anderson, D.F. Anderson, and M. Zafrullah, Factorization in integral domains, *J. Pure Appl. Algebra* **69** (1990), 1–19.

[3] D.D. Anderson and R. Markanda, Unique factorization rings with zero divisors, *Houston J. Math.* **11** (1985), 15–30.

[4] D.D. Anderson and R. Markanda, Corrigendum: "Unique factorization rings with zero divisors", *Houston J. Math.* **11** (1985), 423–426.

[5] D.D. Anderson and B. Mullins, Finite factorization domains, *Proc. Amer. Math. Soc.* **124** (1996), 389–396.

[6] D.D. Anderson and J. Pascual, Characterizing Prüfer rings via their regular ideals, *Comm. Algebra* **15** (1987), 1287–1295.

[7] D.D. Anderson and J. Pascual, Regular ideals in commutative rings, sublattices of regular ideals, and Prüfer rings, *J. Algebra* **111** (1987), 404–426.

[8] D.D. Anderson and S. Valdes-Leon, Factorization in commutative rings with zero divisors, *Rocky Mountain J. Math.* **26** (1996), 439–480.

[9] D.F. Anderson, Elasticity of factorizations in integral domains, a survey, these Proceedings.

[10] L.G. Chouinard II, Krull semigroups and divisor class groups, *Canad. J. Math.* **33** (1981), 1459–1468.

[11] D.L. Costa, Unique factorization in modules and symmetric algebras, *Trans. Amer. Math. Soc.* **224** (1976), 267–280.

[12] D.L. Costa, On the torsion-freeness of the symmetric powers of an ideal, *J. Algebra* **80** (1983), 152–158.

[13] A. Geroldinger and F. Halter-Koch, On the asymptotic behaviour of lengths of factorizations, *J. Pure Appl. Algebra* **77** (1992), 239–252.

[14] R. Gilmer, *Commutative semigroup rings*, University of Chicago Press, Chicago and London, 1984.

[15] R. Gilmer, *Multiplicative ideal theory*, Queen's Papers in Pure and Appl. Math., vol. 90, Queen's University, Kingston, Ontario, 1992.

[16] F. Halter-Koch, Finiteness theorems for factorizations, *Semigroup Forum* **44** (1992), 112–117.

[17] F. Halter-Koch, Elasticity of factorizations in atomic monoids and integral domains, *Journal de Théorie des Nombres de Bordeaux* **7** (1995), 367–385.

[18] J.A. Huckaba, *Commutative rings with zero divisors*, Marcel Dekker, Inc., New York, 1988.

[19] R.E. Kennedy, Krull rings, *Pacific J. Math.* **89** (1980), 131–136.

[20] M.D. Larsen and P.J. McCarthy, *Multiplicative theory of ideals*, Academic Press, New York, 1971.

[21] R. Matsuda, On Kennedy's problems, *Comment. Math. Univ. St. Paul.* **31** (1982), 143–145.

[22] R. Matsuda, Generalizations of multiplicative ideal theory to commutative rings with zero divisors, *Bull. Fac. Sci. Ibaraki Univ. Ser. A* (1985), no. 17, 49–101.

[23] A.-M. Nicolas, Modules factoriels, *Bull. Sci. Math. (2)* **95** (1971), 33–52.

[24] A.-M. Nicolas, Extensions factorielles et modules factorables, *Bull. Sci. Math. (2)* **98** (1974), 117–143.

[25] T. Skolem, Eine Bemerkung über gewisse Ringe mit Anwendung auf die Produktzerlegung von Polynomen, *Norsk Mat. Tidsskr.* **21** (1939), 99–107.

[26] H.T. Tang, Gauss' lemma, *Proc. Amer. Math. Soc.* **35** (1972), 372–376.

[27] Jay A. Wood, Characters and codes over finite rings, preprint.

On t-Invertibility, IV

D.D. ANDERSON   Department of Mathematics, The University of Iowa, Iowa City, IA 52242

MUHAMMAD ZAFRULLAH   MTA Teleport, 1440 Briggs Chaney Road, Silver Spring, MD 20905

This paper is the fourth in a series of papers ([2], [8], [11]) investigating t-invertibility. Let D be an integral domain with quotient field K. A *star operation* $*$ on D is a closure operation on the set $F(D)$ of nonzero fractional ideals of D which satisfies $(a)^* = (a)$ and $(aA)^* = aA^*$ for all $(a), A \in F(D)$. A fractional ideal A is called a $*$-*ideal* if $A = A^*$. For $A, B \in F(D)$, the $*$-*product* of A and B is $(AB)^* = (A^*B)^* = (A^*B^*)^*$. The star operation $*$ is said to have *finite character* if $A^* = \bigcup\{B^* \mid 0 \neq B \subseteq A \text{ is finitely generated}\}$. Examples of star operations include (1) the d-operation $A_d = A$, (2) the v-operation $A_v = (A^{-1})^{-1}$, and (3) the t-operation $A_t = \bigcup\{B_v \mid 0 \neq B \subseteq A \text{ is finitely generated}\}$. The d-operation and t-operation have finite character while the v-operation need not have finite character. Thus if A is finitely generated, $A_v = A_t$ and we use v and t interchangeably. A v-ideal is also called a divisorial ideal. For results on star operations, see [4], [6], [9].

A fractional ideal A of D is $*$-*invertible* if there exists a fractional ideal B of D with $(AB)^* = D$ and in this case we can take $B = A^{-1}$. Thus A is t-*invertible* if and only if $(AA^{-1})_t = D$. Suppose that $*$ has finite character. Then the union of a chain of $*$-ideals is again a $*$-ideal, each proper integral $*$-ideal is contained in a maximal $*$-ideal, and a maximal $*$-ideal is prime. Also, if P is a prime $*$-ideal containing A, then P can be shrunk to a prime $*$-ideal $Q \supseteq A$ so that there are no prime $*$-ideals strictly between A and Q. A fractional ideal A is $*$-invertible if and only if (1) A has finite type or is $*$-*finite* (i.e., $A^* = (a_1, \dots, a_n)^*$ for some finite set of elements $\{a_1, \dots, a_n\}$) and (2) A_P is principal for each maximal $*$-ideal P. For details, see [6] or [9]. Thus a t-invertible t-ideal is a divisorial ideal of finite type.

It is well known (for example, see [10]) that D is a Krull domain if and only if every $A \in F(D)$ is t-invertible if and only if every integral t-ideal of D is a finite t-product of prime ideals if and only if every t-ideal of D is a finite t-product of t-invertible prime t-ideals. It is also well known that if I is an integral t-ideal of a

Krull domain, then any ascending chain of divisorial ideals starting at I stabilizes; in fact, there are only finitely many divisorial ideals containing I. The purpose of this paper is to show (Corollary 4) that for any integral domain D and any t-ideal I of D, the following conditions are equivalent: (1) I is a t-product of t-invertible prime t-ideals, (2) every prime t-ideal minimal over I is t-invertible, (3) every (prime) t-ideal containing I is t-invertible, (4) there is a finite set of t-invertible prime t-ideals $\{P_1, \ldots, P_n\}$ so that every t-ideal containing I is a t-product of powers of the P_i's. Moreover, if (1)–(4) hold, every t-ideal containing I is divisorial (of finite type) and the set of t-ideals containing I is finite. The equivalence of (1) and (3) (with "(prime)" deleted) is given in [5, Theorem 1.3].

It is well known that for a Noetherian ring R, there are only finitely many primes minimal over an ideal I and that for a Mori domain D (i.e., D has ACC on divisorial ideals) there are only finitely many minimal primes (necessarily divisorial) over a divisorial ideal I [7, Proposition 2.3]. Also, for any ring R and ideal I, if all the prime ideals minimal over I are finitely generated, then there are only finitely many primes minimal over I [1]. We first generalize this result to finite character star operations. Note that the result stated for Mori domains is an immediate corollary.

THEOREM 1. Let D be an integral domain, $*$ a finite character star operation on D, and I a nonzero ideal of D. Suppose that every prime $*$-ideal minimal over I is $*$-finite. Then there are only finitely many prime $*$-ideals minimal over I.

Proof. We may assume that $I \neq D$ is a $*$-ideal. The proof is a simple modification of [1, Theorem]. Let $S = \{P_1 \cdots P_n \mid \text{each } P_i \text{ is a prime } *\text{-ideal minimal over } I\}$. If for some $C = P_1 \cdots P_n \in S$, we have $C \subseteq I$, then any prime $*$-ideal minimal over I contains some P_i, so $\{P_1, \ldots, P_n\}$ is the set of prime $*$-ideals minimal over I. Hence we may assume $C \not\subseteq I$ for each $C \in S$. Consider the set $T = \{J \mid J \text{ is a } *\text{-ideal of } D \text{ with } J \supseteq I \text{ and } C \not\subseteq J \text{ for each } C \in S\}$ partially ordered by set inclusion. Since $*$ has finite character and each element of S is $*$-finite, T is inductive. Hence by Zorn's Lemma, T has a maximal element Q which is easily seen to be a prime $*$-ideal. But then $Q \supseteq I$ and Q can be shrunk to a prime $*$-ideal P minimal over I. Thus $P \in S$, a contradiction. ∎

COROLLARY 2. Suppose that D is an integral domain, $*$ is a finite character star operation on D, and I is a nonzero ideal of D. Suppose that every prime $*$-ideal minimal over I is $*$-invertible. Then I is contained in only finitely many minimal prime $*$-ideals (each of which is $*$-invertible).

Proof. A $*$-invertible $*$-ideal is $*$-finite. ∎

The hypothesis in Theorem 1 and Corollary 2 that $*$ has finite character is essential. It is known (see, for example, [4, Example 42.6]) that there exists a non-Noetherian almost Dedekind domain $\mathbf{Z}^*$ with exactly one non-invertible maximal ideal M. Since $\mathbf{Z}^*$ is not Dedekind, $\mathbf{Z}^*$ has a nonzero principal ideal (a) that is contained in M and in infinitely many other maximal ideals. Since $\mathbf{Z}^*$ is completely integrally closed, every ideal of $\mathbf{Z}^*$ is v-invertible. Note that $M_v = \mathbf{Z}^*$. Now every prime v-ideal minimal over (a) is v-invertible (in fact, invertible), but there are infinitely many prime v-ideals minimal over (a).

We next give a (finite character) star operation version of the promised result that a t-ideal I is a t-product of t-invertible prime t-ideals if and only if every (prime) t-ideal containing I is t-invertible. Since, unlike the case for t-ideals, a $*$-invertible prime $*$-ideal need not be a maximal $*$-ideal (for example, for $* = d$), we must replace "t-invertible prime t-ideal" by "$*$-invertible maximal $*$-ideal". For example, a non-maximal principal prime ideal (p) of a domain D is a d-product of d-invertible prime d-ideals, but (p) and D are the only d-invertible ideals containing (p). Like Theorem 1 and Corollary 2, in Theorem 3 it is essential that $*$ has finite character. For example, the v-operation on the integral domain $\mathbf{Z}^*$ mentioned in the previous paragraph satisfies (2), (3), and (5) of Theorem 3, but not (1) and (4) for the ideal $I = (a)$. Also, for (V, M) a (rank one) valuation domain with value group $(\mathbb{R}, +)$, every v-ideal of V is principal. For a nonzero nonunit $a \in V$, every v-ideal containing (a) is principal (and hence v-invertible), but (a) is not a v-product of v-invertible prime v-ideals (in fact, V has no prime v-ideals). We note that the equivalence of (1) and (5) of Theorem 3 is given without proof in [5, Theorem 1.3'] and the equivalence of (1) and (5) of Corollary 4 is given in [5, Theorem 1.3].

THEOREM 3. Let D be an integral domain and $*$ a finite character star operation on D. For a nonzero proper $*$-ideal I of D, the following conditions are equivalent.

(1) I is a $*$-product of $*$-invertible maximal $*$-ideals.
(2) Every prime $*$-ideal minimal over I is a $*$-invertible maximal $*$-ideal.
(3) Every prime $*$-ideal containing I is $*$-invertible.
(4) There is a finite set of $*$-invertible prime $*$-ideals $\{P_1, \ldots, P_n\}$ so that every $*$-ideal containing I is a $*$-product of powers of the P_i's (and hence is $*$-invertible).
(5) Every $*$-ideal containing I is $*$-invertible.

Moreover, in the case where (1)–(5) hold, every $*$-ideal J containing I is divisorial of finite type and the set of $*$-ideals containing I is finite.

Proof. (1)$\Rightarrow$(2). Suppose $I = (P_1^{a_1} \cdots P_s^{a_s})^*$ where each P_i is a $*$-invertible maximal $*$-ideal. Let P be a prime $*$-ideal minimal over I. Then $P_1^{a_1} \cdots P_s^{a_s} \subseteq I \subseteq P$, so some $P_i \subseteq P$ and hence $P_i = P$ is a $*$-invertible maximal $*$-ideal. (2)$\Rightarrow$(3). Clear. (3)$\Rightarrow$(4). By Corollary 2, the set of prime $*$-ideals minimal over I is finite; say it is $\{P_1, \ldots, P_n\}$. Moreover, each P_i is a maximal $*$-ideal. For if there is a prime $*$-ideal $Q \supsetneq P_i$, then the hypothesis that both P_i and Q are $*$-invertible leads to the contradiction that $Q = D$. Let $J \supseteq I$ be a proper $*$-ideal. Since the set of prime $*$-ideals containing J is a subset of $\{P_1, \ldots, P_n\}$, it suffices to prove that if I is a $*$-ideal and there are only finitely many prime $*$-ideals $P_1, \ldots, P_n$ containing I, each of which is a $*$-invertible maximal $*$-ideal, then I is a $*$-product of powers of $P_1, \ldots, P_n$. Note that there exists a natural number a_1 so that $I \subseteq (P_1^{a_1})^*$, but $I \not\subseteq (P_1^{a_1+1})^*$. For suppose $I \subseteq (P_1^n)^*$ for all $n \geq 1$. Since P_1^n is $*$-invertible, $(P_1^n)^*_{P_1} = P_1^n P_1$ is principal (see the top of page 1193 [2]). Now $Q = \bigcap_{n=1}^{\infty} P_1^n P_1$ is a prime t-ideal of D_{P_1} with $I_{P_1} \subseteq Q \subsetneq P_1 P_1$. Hence $Q \cap D$ is a prime t-ideal of D with $I \subseteq Q \cap D \subsetneq P_1$. But then $Q \cap D$ is a prime $*$-ideal contradicting the fact that P_1 is minimal over I. Hence we can write $I = (P_1^{a_1} B)^*$ where $B = \left((P_1^{a_1})^{-1} I \right)^* \subseteq D$ is $*$-invertible and $\{P_2, \ldots, P_n\}$ is the set of prime $*$-ideals containing B. By induction, $B^* = (P_2^{a_2} \cdots P_n^{a_n})^*$, so $I = (P_1^{a_1} \cdots P_n^{a_n})^*$.

(4)$\Rightarrow$(1). The proof of (3)$\Rightarrow$(4) showed that each P_i is actually a maximal $*$-ideal. Hence I is a $*$-product of $*$-invertible maximal $*$-ideals. Finally, clearly (4)$\Rightarrow$(5) and (5)$\Rightarrow$(3).

Suppose that (1)–(5) hold for I and let J be a $*$-ideal containing I. Then J is a $*$-product of $*$-invertible prime $*$-ideals and hence J itself is $*$-invertible. But a $*$-invertible $*$-ideal J is divisorial. (For $x \in J_v$ gives $xJ^{-1} \subset D$ and hence $x \in xD = x\left(J^{-1}J\right)^* = \left(xJ^{-1}J\right)^* \subseteq J^* = J$.) Finally, since the set of $*$-ideals containing I is $\left\{(P_1^{b_1} \cdots P_n^{b_n})^* \mid 0 \leq b_i \leq a_i\right\}$ where $I = (P_1^{a_1} \ldots P_n^{a_n})^*$, the set of $*$-ideals containing I is finite. ∎

COROLLARY 4. For an integral domain D and proper t-ideal I of D, the following conditions are equivalent.

(1) I is a t-product of t-invertible prime t-ideals.
(2) Every prime t-ideal minimal over I is t-invertible.
(3) Every prime t-ideal containing I is t-invertible.
(4) There is a finite set of t-invertible prime t-ideals $\{P_1, \ldots, P_n\}$ so that every t-ideal containing I is a t-product of powers of the P_i's (and hence is t-invertible).
(5) Every t-ideal containing I is t-invertible.

Moreover, in the case where (1)–(5) hold, every t-ideal J containing I is divisorial of finite type and the set of t-ideals containing I is finite.

Proof. This follows from Theorem 3 since a t-invertible prime t-ideal is a maximal t-ideal. ∎

COROLLARY 5. Let D be an integral domain. Suppose that an ideal I of D is a finite product of invertible prime ideals. Then every divisorial ideal containing I is invertible.

Proof. An invertible (prime) ideal is a (maximal) t-invertible t-ideal. ∎

The previously mentioned example of a valuation domain with value group $(\mathbb{R}, +)$ shows that the converse of Corollary 5 is false.

REMARK 6. The proof of (3)⇒(4) of Theorem 3 shows that if the $*$-ideal I (where $*$ is a finite character star operation on D) is contained in only a finite number of $*$-invertible prime $*$-ideals $\{P_1, \ldots, P_n\}$ and is contained in no other prime $*$-ideals, then $I = (P_1^{a_1} \cdots P_n^{a_n})^*$ where each $a_i \geq 1$. Moreover, it is easily checked that this factorization is unique up to order. Since any $*$-ideal J containing I has the form $J = (P_1^{b_1} \cdots P_n^{b_n})^*$ where $0 \leq b_i \leq a_i$, the number of $*$-ideals containing I is $(a_1 + 1) \cdots (a_n + 1)$. Thus if $I = (x)$, the number m of nonassociate factors of x satisfies $1 \leq m \leq (a_1 + 1) \cdots (a_n + 1)$ with $m = (a_1 + 1) \cdots (a_n + 1)$ if and only if each P_i is principal. This gives another proof that a Krull domain is a FFD (i.e., D is an atomic domain in which every nonzero element has only a finite number of factorizations up to associates).

A well known result of Kaplansky states that an integral domain D is a UFD if and only if every nonzero prime ideal of D contains a nonzero principal prime. Several other Kaplansky-type results are given in [3] and [5]. Let us call an element $0 \neq x \in D$ a *Krull element* if (x) is a t-product of prime t-ideals. (Here we consider (1) to be an empty t-product of prime t-ideals.) We then have the following theorem.

THEOREM 7. An integral domain D is a Krull domain if and only if every nonzero prime ideal of D contains a Krull element.

Proof. If D is a Krull domain every nonzero nonunit of D is a Krull element. Conversely, suppose that every nonzero prime ideal of D contains a Krull element. Let S be the set of all Krull elements of D. By Remark 6, S is a saturated multiplicatively closed subset of D. If D is not a Krull domain, then some nonzero element x of D is not a Krull element, so $(x) \cap S = \emptyset$. But then enlarging (x) to a prime ideal P with $P \cap S = \emptyset$, we get a contradiction. ∎

REFERENCES

[1] D.D. Anderson, A note on minimal prime ideals, *Proc. Amer. Math. Soc.* **122** (1994), 13–14.

[2] D.D. Anderson and M. Zafrullah, On t-invertibility III, *Comm. Algebra* **21** (1993), 1189–1201.

[3] D.D. Anderson and M. Zafrullah, On a theorem of Kaplansky, *Boll. Un. Mat. Ital. A (7)* **8** (1994), 397–402.

[4] R. Gilmer, *Multiplicative ideal theory*, Queen's Papers in Pure and Appl. Math. **90**, Queen's University, Kingston, Ontario, 1992.

[5] R. Gilmer, J. Mott, and M. Zafrullah, t-invertibility and comparability, *Commutative Ring Theory* (P.-J. Cahen, D.L. Costa, M. Fontana, and S.-E. Kabbaj, eds.), Lecture Notes in Pure and Appl. Math. **153**, Marcel Dekker, New York, 1994, pp. 141–150.

[6] M. Griffin, Some results on v-multiplication rings, *Canad. J. Math.* **19** (1967), 710–722.

[7] E.G. Houston, T.G. Lucas, and T.M. Viswanathan, Primary decomposition of divisorial ideals in Mori domains, *J. Algebra* **117** (1988), 327–342.

[8] E. Houston and M. Zafrullah, On t-invertibility II, *Comm. Algebra* **17** (1989), 1955–1969.

[9] P. Jaffard, *Les systèmes d'idéaux*, Dunod, Paris, 1960.

[10] B.G. Kang, On the converse of a well-known fact about Krull domains, *J. Algebra* **124** (1989), 284–299.

[11] S. Malik, J.L. Mott, and M. Zafrullah, On t-invertibility, *Comm. Algebra* **16** (1988), 149–170.

Factorization in Subrings of $K[X]$ or $K[[X]]$

DAVID F. ANDERSON Department of Mathematics, The University of Tennessee, Knoxville Tennessee 37996-1300.

JAENAM PARK Department of Mathematics, The University of Tennessee, Knoxville Tennessee 37996-1300.

1. INTRODUCTION

Let R be an integral domain. A nonzero nonunit $r \in R$ is said to be *irreducible* (or an *atom*) if whenever $r = ab$ with $a, b \in R$, then either a or b is a unit of R. The domain R is *atomic* if every nonzero nonunit of R is a product of irreducible elements of R. The factorization of an element into the product of irreducibles need not be unique (up to units and order). In fact, the factorization of each element is unique precisely when R is a unique factorization domain (UFD), or equivalently, when each atom is prime. Following Zaks [20], we define an atomic domain R to be a *half-factorial domain* (HFD) if each factorization of a nonzero nonunit of R into a product of irreducible elements has the same length. A UFD is obviously a HFD, but the converse fails since any Krull domain R with divisor class group $Cl(R) = \mathbb{Z}_2$ is a HFD [20], but not a UFD. In order to measure how far an atomic domain R is from being a UFD, we define the *elasticity* of R as $\rho(R) = sup\{\frac{m}{n} \mid x_1 \cdots x_m = y_1 \cdots y_n$, for $x_i, y_j \in R$ irreducible$\}$. Thus $1 \leq \rho(R) \leq \infty$, and $\rho(R) = 1$ if and only if R is a HFD. The elasticity was introduced by Valenza [19] and has been studied extensively in the papers [1], [2], [3], [16], and [18](see [9] for a survey article on elasticity.).

Two of the simplest examples of HFDs which are not UFDs are the domains $F + XK[X]$ and $F+XK[[X]]$, where F is a proper subfield of a field K ([5, Theorem 5.3] and [7, Corollary 7.2], or Remark 4.6(d).). In this paper, we compute $\rho(R)$ for more general subrings of either the polynomial ring $K[X]$ or the power series ring $K[[X]]$ over a field K. This paper is divided into four sections including the introduction. In the second section, we introduce the necessary notation and machinery to prove the results in the final two sections. In the third section, we compute $\rho(R)$ for certain subrings R of $K[[X]]$. In the final section, we do the harder case; we compute $\rho(R)$ for a subring of $K[X]$ of the form $R = K_0 + K_1X + \cdots + K_{n-1}X^{n-1} + X^n K[X]$, where $K_0 \subsetneq K_1 \subsetneq \cdots \subsetneq K_{n-1}$ are proper subfields of a finite field K (Theorem 4.4). These computations are special cases of the following general problem:

PROBLEM. Let k be a subfield of a field K and $\{V_n \mid 0 \leq n < \infty\}$ a family of k-submodules of K containing k such that $V_0 = k$ and $V_iV_j \subseteq V_{i+j}$ for all $i, j \geq 0$. Let $A = \bigoplus_{n=0}^{\infty} V_n X^n \subseteq K[X]$ and $B = \prod_{n=0}^{\infty} V_n X^n \subseteq K[[X]]$. Compute $\rho(A)$ and $\rho(B)$. □

In this paper, we concentrate on the case when the V_n's form an ascending sequence of subrings of K with $K = \bigcup_{n=0}^{\infty} V_n$. Some special cases of these calculations have been done in [1, Theorem 2.7 and Examples 3.5 and 3.6]. Another special case which has received considerable attention is when S is a numerical submonoid of $\mathbb{Z}_+$, and $V_n = K$ if $n \in S$ and $V_n = \{0\}$ if $n \notin S$; so $A = K[S]$ and $B = K[[S]]$ (cf. [10], [11], [12], and [13]). We start with the following motivating theorem which will be proved in the next sections (Theorems 3.1 and 4.1).

THEOREM. Let $R_0 \subseteq R_1 \subseteq R_2 \subseteq \cdots$ be an ascending sequence of subrings of a field K with $K = \bigcup_{n=0}^{\infty} R_n$ and R_0 a field. Let $A = \bigoplus_{n=0}^{\infty} R_n X^n \subseteq K[X]$ and $B = \prod_{n=0}^{\infty} R_n X^n \subseteq K[[X]]$.

(1) If each $R_n \subset K$, then $\rho(A) = \rho(B) = \infty$.

(2) Suppose that $R_{n-1} \subset R_n = K$ for some $n \geq 1$. Then

(a) $\rho(B) = n$.

(b) If $n = 1$, then $\rho(A) = 1$.

(c) If $n \geq 2$, then $\rho(A) < \infty$ if and only if K is finite. $\square$

We will thus assume that $R_{n-1} \subset R_n = K$ for some $n \geq 1$. If K is a finite field, then each R_m is also a field; and thus $B = R_0 + R_1 X + \cdots + R_{n-1} X^{n-1} + X^n K[[X]]$ is a Cohen-Kaplansky domain (CK domain) [7, Theorem 7.1] and $A = R_0 + R_1 X + \cdots + R_{n-1} X^{n-1} + X^n K[X]$ is a generalized CK domain [6, Theorem 6]. (Recall that an atomic domain R is a *CK domain* if it has a finite number of nonassociate atoms and is a *generalized CK domain* if almost all atoms of R are prime (see [7] and [6]).)

Throughout, R will denote an integral domain with R^* its set of nonzero elements and $U(R)$ its group of units, and K will always be a field and X an indeterminate over K. Also, $\mathbb{Z}, \mathbb{Q}$ and $\mathbb{R}$ will denote respectively the sets of integers, rational numbers, and real numbers, and $\mathbb{Z}_+$ and $\mathbb{R}_+$ will denote respectively the sets of nonnegative integers and nonnegative real numbers. Throughout, $\subset$ will always mean proper inclusion. As usual, $a/0 = \infty/0 = \infty/a = \infty + a = \infty - a = \infty$ for any positive $a \in \mathbb{R}$. General references for any undefined factorization terminology are [1] and [4].

2. BASIC RESULTS

In this section, we recall and develop the necessary tools for the rest of the paper. We first recall some facts about semi-length functions. A function $\varphi : R^* \to \mathbb{R}_+$ is called a *semi-length function* on R if (1) $\varphi(xy) = \varphi(x) + \varphi(y)$ for all $x, y \in R^*$, and (2) $\varphi(x) = 0$ if and only if $x \in U(R)$. Semi-length functions were introduced in [1] to determine upper and lower bounds for $\rho(R)$. By [1, Theorem 2.1], if φ is a semi-length function on an atomic domain R, then $1 \leq \varphi(R) \leq M^*/m^*$, where $M^* = M^*(R, \varphi) = sup\{\varphi(x) \mid x \in R$ is irreducible, but not prime$\}$, $m^* = m^*(R, \varphi) = inf\{\varphi(x) \mid x \in R$ is irreducible, but not prime$\}$, and $M^* = m^* = 1$ when R is a UFD. Let D be a subring of an integral domain R and φ a semi-length function defined on R. Then $\varphi \mid_D$ is a semi-length function on D if and only if $U(R) \cap D = U(D)$ [1, Example 1.3]. For $R = K[[X]]$, $\varphi = ord$, where $ord(f) = n$ if $f = a_n x^n + \cdots \in R$ with $a_n \neq 0$, defines a semi-length function on R. We will also denote its restriction to a subring of $K[[X]]$ by ord.

The next concept, the Davenport constant of an abelian group, has proved extremely useful in studying factorization problems (cf. [1], [9], [10], [11], [12], [15] and [16]). Recall that for a finite abelian group G, the *Davenport constant* of G, denoted by $D(G)$, is the smallest positive integer d such that for each sequence $S \subseteq G$ with $|S| = d$, some nonempty subsequence of S has sum 0. For any finite abelian group G, $D(G) \leq |G|$. If G is an infinite abelian group, then $D(G) = \infty$ in the sense that for each positive integer d there is a sequence $S \subseteq G$ with $|S| = d$ and no nonempty subsequence of S has sum 0. In general, there is no known formula for $D(G)$. However, if p is prime, then the p-group $G = (\mathbb{Z}_p)^{n_1} \oplus (\mathbb{Z}_{p^2})^{n_2} \oplus \cdots \oplus (\mathbb{Z}_{p^r})^{n_r}$ has $D(G) = n_1(p-1) + n_2(p^2 - 1) + \cdots + n_r(p^r - 1) + 1$. In particular, for the finite field $K = GF(p^n)$ considered as an abelian group, $K \cong (\mathbb{Z}_p)^n$, and hence $D(G) = n(p-1) + 1$. General references for the Davenport constant are [15] and [17].

Let R be an atomic domain with $\rho(R)$ a rational number. We say that $\rho(R)$ is *realized by a factorization* if there is a factorization $x_1 \cdots x_m = y_1 \cdots y_n$ with each $x_i, y_j \in R$ irreducible such that $\rho(R) = m/n$. If R is either a generalized CK domain or a Krull domain with finite divisor class group, then $\rho(R)$ is realized by a factorization ([3], [1]). However, there are Dedekind domains R with torsion divisor class group such that $\rho(R)$ is rational, but $\rho(R)$ is not realized by a factorization [1]. In Theorem 3.2 and Example 3.3, we will give some elementary examples of domains R such that $\rho(R)$ is rational, but $\rho(R)$ is not realized by a factorization.

We will concentrate on subrings of $K[X]$ or $K[[X]]$ of the form $A = \bigoplus_{n=0}^{\infty} R_n X^n$ or $B = \prod_{n=0}^{\infty} R_n X^n$, respectively, where $\{R_n \mid 0 \leq n < \infty\}$ is an ascending sequence of subrings of K with $K = \bigcup_{n=0}^{\infty} R_n$. For the domains A and B to be atomic, we next show that R_0 must be a field. In this case, the domains A and B actually satisfy ACCP.

THEOREM 2.1. Let $\{R_n \mid 0 \leq n < \infty\}$ be an ascending sequence of subrings of a field K with $K = \bigcup_{n=0}^{\infty} R_n$. Let $A = \bigoplus_{n=0}^{\infty} R_n X^n \subseteq K[X]$ and $B = \prod_{n=0}^{\infty} R_n X^n \subseteq K[[X]]$. Then A and B are each atomic if and only if R_0 is a field. Moreover, if R_0 is a field, then A and B each satisfy ACCP.

Proof. First suppose that R_0 is not a field. Let $r \in R_0$ be a nonzero nonunit. Since $K = \bigcup_{n=0}^{\infty} R_n$, there is a least integer $m \geq 1$ such that $r^{-1} \in R_m$. Then it is easily verified that $r^{-1} X^m$ is not a product of atoms in either A or B. Conversely, suppose that R_0 is a field. In this case, an elementary degree (resp., order) argument shows that A (resp., B) satisfies ACCP, and hence is atomic. □

REMARK 2.2. (a) Let A and B be as in Theorem 2.1 with R_0 a field. Then X is irreducible in both A and B. However, X is not prime in either A or B if some $R_{n-1} \subset R_n$. (For $a \in R_n - R_{n-1}$, $X \mid (aX^n)^2$, but $X \nmid aX^n$.) Also, by comparing coefficients, one easily checks that if $f \in A$ is prime in $K[X]$ with $f(0) \neq 0$, then f is also prime in A (cf. [11, Lemma 2.2(a)] and [12, Lemma 2.2(b)]).

(b) From [14, Corollaire], it follows that $A = \bigoplus_{n=0}^{\infty} R_n X^n$ has Krull dimension one when R_0 and $\bigcup_{n=0}^{\infty} R_n$ are fields. Also, $B = \prod_{n=0}^{\infty} R_n X^n$ is quasilocal when R_0 is a field. □

3. SUBRINGS OF K[[X]]

We first prove the power series ring part of the Theorem from the Introduction for the ring $R = \prod_{n=0}^{\infty} R_n X^n \subseteq K[[X]]$. In this case, $\rho(R)$ does not depend on the field K, but only on when $R_{n-1} \subset R_n$. However, if $R \neq K[[X]]$, then R is a (generalized) CK domain if and only if K is a finite field.

THEOREM 3.1. Let $R_0 \subseteq R_1 \subseteq R_2 \subseteq \cdots$ be an ascending sequence of subrings of a field K with R_0 a field, $K = \bigcup_{n=0}^{\infty} R_n$, and $R = \prod_{n=0}^{\infty} R_n X^n = \{\sum a_n X^n \in K[[X]] \mid a_n \in R_n\}$.

(1) If each $R_n \subset K$, then $\rho(R) = \infty$.

(2) Suppose that $R_{n-1} \subset R_n = K$ for some $n \geq 1$. Then $\rho(R) = n$.

Proof. Let $a \in R_n - R_{n-1}$. Then aX^n is irreducible in R, and the factorization

$(aX^n)^m = (a^m X^n)X^{n(m-1)}$ yields $\rho(R) \geq (n(m-1)+1)/m = n - n/m + 1/m$ for each integer $m \geq 1$. Thus $\rho(R) \geq n$. If each $R_n \subset K$, then for each $N \geq 1$ there is an $n \geq N$ with $R_{n-1} \subset R_n$. In this case, $\rho(R) \geq n \geq N$, and hence $\rho(R) = \infty$. Thus (1) holds. Now suppose $R_{n-1} \subset R_n = K$. By above, $\rho(R) \geq n$. Let $\varphi(f) = ord(f)$ for $f \in R^*$. Clearly $m^* = m^*(R, \varphi) = 1$ since X is irreducible, but not prime, in R by Remark 2.2(a), and $M^* = M^*(R, \varphi) \leq n$ since f is not irreducible in R if $\varphi(f) > n$. Thus $\rho(R) \leq M^*/m^* \leq n$, and hence $\rho(R) = n$. Thus (2) holds. $\square$

We may thus assume that $R = R_0 + R_1 X + \cdots + R_{n-1}X^{n-1} + X^n K[[X]]$, where $K(= R_n)$ is a field, $R_0 \subseteq R_1 \subseteq \cdots \subseteq R_{n-1} \subset K$ is an ascending sequence of subrings of K with R_0 a field, and R_{n-1} is a proper subring of K. If R_{n-1} is a field, then the factorization $(aX^n)(a^{-1}X^n) = X^{2n}$ for $a \in K - R_{n-1}$ shows that $\rho(R)$ is realized by a factorization. However, in general this is not the case. We next determine precisely when $\rho(R)$ is realized by a factorization. Recall that if V is a valuation domain with quotient field K and $ab \in V$ for some $a, b \in K$, then either $a \in V$ or $b \in V$.

THEOREM 3.2. Let $n \geq 1$ and $R = R_0 + R_1 X + \cdots + R_{n-1}X^{n-1} + X^n K[[X]]$, where K is a field, $R_0 \subseteq R_1 \subseteq \cdots \subseteq R_{n-1} \subset K$ is an ascending sequence of subrings of K with R_0 a field, and R_{n-1} is a proper subring of K. Then $\rho(R) = n$, and $\rho(R)$ is realized by a factorization if and only if R_{n-1} is not a valuation domain with quotient field K.

Proof. By Theorem 3.1, $\rho(R) = n$. First, suppose that R_{n-1} is not a valuation domain with quotient field K. We can thus choose $a \in K - R_{n-1}$ such that $a^{-1} \in K - R_{n-1}$. Then aX^n and $a^{-1}X^n$ are each irreducible in R, and the factorization $(aX^n)(a^{-1}X^n) = X^{2n}$ shows that $\rho(R)$ is realized by a factorization.

Next, suppose that R_{n-1} is a valuation domain with (proper) quotient field K (thus $n \geq 2$ since R_0 is a proper subfield of K). Note that each irreducible $g \in R$ has $ord(g) \leq n$. If $\rho(R) = n$ is realized by a factorization, then there is a factorization $f = f_1 \cdots f_r = g_1 \cdots g_{rn}$ for some irreducibles $f_i, g_j \in R$. Let $m = ord(f)$. Then $r \leq m \leq rn$ from the left-hand factorization of f and $rn \leq m \leq rn^2$ from the right-hand factorization. Thus $m = rn$, so each $ord(f_i) = n$ and $ord(g_i) = 1$. Let each $f_i = a_i X^n + \cdots$ with $a_i \in K - R_{n-1}$ and $g_i = b_i X + \cdots$ with $b_i \in R_1 \subseteq R_{n-1}$. Then $a_1 \cdots a_r = b_1 \cdots b_{rn} \in R_{n-1}$, and hence some $a_i \in R_{n-1}$ since R_{n-1} is a valuation domain with quotient field K, a contradiction. Thus $\rho(R)$ is not realized by a factorization. $\square$

EXAMPLE 3.3. Let t be an indeterminate over $\mathbb{Q}$ and let $R = \mathbb{Q} + \mathbb{Q}[t]_{(t)}X + X^2\mathbb{Q}(t)[[X]]$ and $S = \mathbb{Q}+\mathbb{Q}[t]X+X^2\mathbb{Q}(t)[[X]]$ be subrings of $\mathbb{Q}(t)[[X]]$. By Theorem 3.1, $\rho(R) = \rho(S) = 2$. Moreover, by Theorem 3.2, $\rho(R)$ is not realized by a factorization since $\mathbb{Q}[t]_{(t)}$ is a valuation domain with quotient field $\mathbb{Q}(t)$. However, $\rho(S)$ is realized by the factorization $(aX^2)(a^{-1}X^2) = X^4$, where $a = (t+1)/(t-1) \in \mathbb{Q}(t)$. $\square$

Let $F \subseteq K$ be a pair of finite fields and V an F-submodule of K. In [7, Lemma 6.6], it was shown that $R = F + VX + X^2K[[X]]$ is a HFD if $K = V^2(= \{ab \mid a, b \in V\})$. We generalize this to

THEOREM 3.4. Let F be a subfield of a field K, V a proper F-submodule of K containing F, and $R = F + VX + X^2K[[X]]$. Then either $\rho(R) = 1$ or $\rho(R) = 2$. Moreover, $\rho(R) = 1$ if and only if $K = V^2$.

Proof. As in the proof of Theorem 3.1, $\rho(R) \leq 2$. First, suppose that $K = V^2$. Let $f = a_2X^2 + a_3X^3 + \cdots \in R$ with $0 \neq a_2 \in K$. Then $a_2 = bc$ with $0 \neq b, c \in V$, and thus $f = bX(cX + b^{-1}a_3X^2 + \cdots)$ is irreducible in R. Hence $f \in R$ is irreducible if and only if $ord(f) = 1$. Thus $\rho(R) = 1$.

Next, suppose that $V^2 \subset K$. Choose $a \in K - V^2$. Then aX^2 is irreducible in R, and the factorization $(aX^2)^n = (a^nX^2)X^{2n-2}$ yields $\rho(R) \geq (2n-1)/n = 2 - 1/n$ for each integer $n \geq 2$. Hence $\rho(R) = 2$. $\square$

Note that if V is actually a subring of K, then $V^2 = V$; so Theorem 3.4 is a special case of Theorem 3.1. It would be interesting to determine necessary and sufficient conditions on V so that $\rho(R)$ is realized by a factorization.

4. SUBRINGS OF K[X]

The proof of Theorem 3.1(1) also gives Theorem 4.1(1), the analogue for polynomial rings. However, the proofs of the third part of Theorem 4.1 and Theorems 4.3 and 4.4 are much harder and use ideas similar to those developed in [10], [11], and [12] to compute $\rho(K[S])$ for the semigroup ring $K[S]$, where K is a field and S is a numerical submonoid of $\mathbb{Z}_+$. By Theorem 4.1(2), we may assume that $R = K_0 + K_1X + \cdots + K_{n-1}X^{n-1} + X^nK[X]$, where K is a finite field. In Theorem 4.4, we will give an explicit formula for $\rho(R)$ in terms of the Davenport constant of a certain abelian group associated with the finite fields $K_1, ..., K_{n-1}$, and K. Note that in this case, we need not consider the case when the K_m's are not fields since a subring of a finite field is again a finite field.

THEOREM 4.1. Let $R_0 \subseteq R_1 \subseteq R_2 \subseteq \cdots$ be an ascending sequence of subrings of a field K with R_0 a field, $K = \bigcup_{n=0}^{\infty} R_n$, and $R = \bigoplus_{n=0}^{\infty} R_nX^n = \{\sum a_nX^n \in K[X] \mid a_n \in R_n\}$.

(1) If each $R_n \subset K$, then $\rho(R) = \infty$.

(2) If $R_0 \subset R_1 = K$, then $\rho(R) = 1$ (R is a HFD).

(3) Suppose that $R_{n-1} \subset R_n = K$ for some $n \geq 2$. Then $\rho(R) < \infty$ if and only if K is a finite field. Moreover, if $\rho(R) < \infty$, then $\rho(R)$ is realized by a factorization.

Proof. The proof of part (1) is the same as part (1) of Theorem 3.1. So suppose that $R_{n-1} \subset R_n = K$. If $R_0 \subset R_1 = K$, then R is a HFD by [5, Theorem 5.3]. Thus we may assume that $n \geq 2$. If K is a finite field, then R is a generalized CK domain by [6, Theorem 6]. An elementary generalization of [3, Theorem 7] shows that in this case $\rho(R) < \infty$ and that $\rho(R)$ is realized by a factorization. We now show that $\rho(R) = \infty$ if K is an infinite field. First, suppose that K is infinite with $char\,K = p > 0$. Then K/R_{n-1} is also an infinite (additive) abelian group, and hence $D(K/R_{n-1}) = \infty$. Thus for each integer $m \geq 1$, there are $a_1, ..., a_m \in K - R_{n-1}$ such that no subsum is in R_{n-1}. Also, choose $a \in K - R_{n-1}$. Then $f = aX^n(1+a_1X)\cdots(1+a_mX)$ is irreducible in R. Choose an integer k such that $N = p^k \geq n$. Since $char\,K = p$ and each $1+a_i^NX^N \in R$, the factorization $f^N =$

$(a^N X^n) X^{(N-1)n} (1 + a_1^N X^N) \cdots (1 + a_m^N X^N)$ yields $\rho(R) \geq ((N-1)n + 1 + m)/N$ for each $m \geq 1$. Hence $\rho(R) = \infty$. Next, suppose that $char\, K = 0$; thus $\mathbb{Q} \subseteq R_0$. Choose $a \in K - R_{n-1}$. Let $g = 1 - a^n X^n = (g_1 \cdots g_s)(g_{s+1} \cdots g_t) \in R$, where each $g_i \in K[X]$ is irreducible in $K[X]$ with $g_i(0) = 1$, and $g_1, ..., g_s \notin R$ and $g_{s+1}, ..., g_t \in R$. (Note that $s \geq 1$ since $1 - aX \notin R$ is an irreducible factor of g). By comparing coefficients, one easily checks that $g_1 \cdots g_s \in R$ (hence $s \geq 2$) and $g_i^m \notin R$ for $1 \leq i \leq s$ and any integer $m \geq 1$ (here we use that $\mathbb{Q} \subseteq R_0$) (cf. [11, Lemma 2.2(b)] and [12, Lemma 2.2(a)]). Thus for $1 \leq i \leq s$ and any fixed integer $m \geq 1$, $f_i = aX^n g_i^m$ is irreducible in R. The factorization $f_1 \cdots f_s = (a^s X^n) X^{n(s-1)} (g_1 \cdots g_s)^m$ then yields $\rho(R) \geq (n(s-1) + 1 + m)/s$ for each integer $m \geq 1$. Hence $\rho(R) = \infty$. $\square$

REMARK 4.2. Let $R = R_0 + R_1 X + \cdots + R_{n-1} X^{n-1} + X^n K[X]$ be as in Theorem 4.1 above with R_0 a proper subfield of K. Then R is a FFD if and only if K is finite. (Recall that an integral domain R is a *finite factorization domain* (FFD) if each nonzero element of R has only a finite number of nonassociate divisors (see [4] and [8]).) If K is finite, then R is a FFD by [8, Theorem 3]. Conversely, if R is a FFD, then K^*/R_0^* is finite by [8, Theorem 4], and hence K is finite by Brandis' Theorem. In fact, if K is finite, then R is a strong FFD in the sense that each nonzero element of R has only finitely many divisors (cf. [8, Theorem 5]). $\square$

We first do the special case when $R = k + FX + X^2 K[X]$. In this case, for $GF(p^n) = F \subset K = GF(p^{mn})$ we obtain an explicit formula for $\rho(R)$ in terms of m, n and p. The general case will be done in Theorem 4.4.

THEOREM 4.3. Let F be a proper subfield of a field K, k a subfield of F, and $R = k + FX + X^2 K[X]$. Then $\rho(R) = (3 + D(K/F))/2$. Moreover, if $GF(p^n) = F \subset K = GF(p^{mn})$, then $\rho(R) = (n(m-1)(p-1) + 4)/2$.

Proof. Consider K as an (additive) abelian group and F a proper subgroup of K. Thus K/F is nonzero. Note that $D(K/F) < \infty$ if and only if K/F is finite, if and only if K is a finite field. Let $1 \leq m \leq D(K/F) - 1$. We may then choose $a_1, ..., a_m \in K - F$ such that no subsum is in F. Pick $a \in K - F$. Then $f = aX^2(1 + a_1 X) \cdots (1 + a_m X)$ and $g = a^{-1} X^2 (1 - a_1 X) \cdots (1 - a_m X)$ are each irreducible elements of R. Since each $1 - a_i^2 X^2 \in R$, the factorization $fg =$ $X^4(1 - a_1^2 X^2) \cdots (1 - a_m^2 X^2)$ yields $\rho(R) \geq (4 + m)/2$. If K/F is infinite, then $D(K/F) = \infty$, and hence $\rho(R) = \infty$. Thus $\rho(R) \geq (3 + D(K/F))/2$ for any proper subfield F of K. So we may assume that K is finite.

For $f = aX^n f_1 \cdots f_m \in K[X]$ with $0 \neq a \in K$, and each $f_i \in K[X]$ irreducible in $K[X]$ with $f_i(0) = 1$, define $\varphi(f) = 2n + m$. Then φ is a semi-length function on $K[X]$, and φ restricts to a semi-length function on R since $U(K[X]) \cap R =$ $U(R) = k$. Suppose that $f = aX^n f_1 \cdots f_m$ (as above) is irreducible in R with each $f_i = 1 + a_i X + \cdots \in K[X]$. Clearly $n \leq 2$. First, suppose that $n = 2$. Then $m \leq D(K/F) - 1$ since otherwise (after possibly reordering) we would have $a_1 + \cdots + a_i \in F$ for some $1 \leq i \leq m$. Then $f_1 \cdots f_i \in R$, and hence $f =$ $(aX^2 f_{i+1} \cdots f_m)(f_1 \cdots f_i)$ is not irreducible in R, a contradiction. Next, suppose that $n = 0$ or $n = 1$. We show that $m \leq D(K/F)$. If not, then (after possibly reordering) we would have $a_1 + \cdots + a_i \in F$ for some $1 \leq i < m$. Hence $f_1 \cdots f_i \in R$. If $n = 0$, then by comparing coefficients, one checks that $f_{i+1} \cdots f_m \in R$ (cf. [11, Lemma 2.2(b)] and [12, Lemma 2.2(a)]). Hence $f = (af_{i+1} \cdots f_m)(f_1 \cdots f_i)$ is not irreducible in R, a contradiction. If $n = 1$, then $f = (aX f_{i+1} \cdots f_m)(f_1 \cdots f_i)$ is not irreducible in R, a contradiction. So in either case, $M^* = M^*(R, \varphi) \leq$ $4 + D(K/F) - 1 = 3 + D(K/F)$. Also, $m^* = m^*(R, \varphi) = 2$ since X is irreducible, but not prime, in R, while $1 + aX$ is prime in R for any $0 \neq a \in F$ by Remark 2.2(a). Thus $\rho(R) \leq M^*/m^* \leq (3 + D(K/F))/2$; so we have equality. The "moreover" statement follows since $K/F = GF(p^{mn})/GF(p^n) \cong (\mathbb{Z}_p)^{n(m-1)}$ as an abelian group, and hence $D(K/F) = n(m - 1)(p - 1) + 1$. $\square$

We next do the general case for $n \geq 2$. Let $K_0 \subseteq K_1 \subseteq \cdots \subseteq K_{n-1}$ be an ascending sequence of subfields of K with K_{n-1} a proper subfield of K and $R =$ $K_0 + K_1 X + \cdots + K_{n-1} X^{n-1} + X^n K[X]$. As in [11] and [12], we define a surjective map $\pi_1 : 1 + XK[[X]] \rightarrow K^{n-1}$ by $\pi_1(\sum a_i X^i) = (a_1, ..., a_{n-1})$. Then π_1 is a group homomorphism from the multiplicative group $1 + XK[[X]]$ to the group $(K^{n-1}, *)$ with multiplication defined by $(a_1, ..., a_{n-1}) * (b_1, ..., b_{n-1}) = (c_1, ..., c_{n-1})$, where $(\sum a_i X^i)(\sum b_i X^i) = \sum c_i X^i$ and $a_0 = b_0 = c_0 = 1$. Then $H = (K_1 \oplus \cdots \oplus K_{n-1}, *)$ is a subgroup of the abelian group $G = (K^{n-1}, *)$, and hence G/H is an abelian group. Let $\pi_2 : G \to G/H$ be the projection map and $\pi = \pi_2 \circ \pi_1$. Note that for $f \in K[X]$ with $f(0) = 1$, we have $\pi(f) = 0$ if and only if $f \in R$. Also, note that $D(G/H) < \infty$ if and only if G/H is finite, if and only if K is a finite field. Moreover, when K is a finite field, G, and hence G/H, is a finite p-group, where $p = char\, K$.

THEOREM 4.4. For $n \geq 2$, let $K_0 \subseteq K_1 \subseteq \cdots \subseteq K_{n-1}$ be an ascending sequence of subfields of K with K_{n-1} a proper subfield of K, and $R = K_0 + K_1 X + \cdots + K_{n-1} X^{n-1} + X^n K[X]$. Then $\rho(R) = (2n + D(G/H) - 1)/2$, where G/H is the abelian group $(K \oplus \cdots \oplus K)/(K_1 \oplus \cdots \oplus K_{n-1})$ defined above.

Proof. By the above comments and Theorem 4.1(3), the formula gives the desired result $\rho(R) = \infty$ when K is infinite. We may thus assume that K is a finite field, and hence $D(G/H) < \infty$. Pick $a \in K - K_{n-1}$. For $m = D(G/H) - 1$, we can choose $a_i \in G - H$, $1 \leq i \leq m$, such that no subsum (with respect to $*$) is in H. For each a_i, there are irreducible $f_i, g_i \in K[X]$ with $\pi_1(f_i) = a_i$ and $\pi_1(g_i) = -a_i$ (the inverse of a_i in G) [11, Lemma 2.1]. Thus no subproduct of the f_i's or g_i's is in R. Hence $f = aX^n f_1 \cdots f_m$ and $g = a^{-1} X^n g_1 \cdots g_m$ are each irreducible in R. However, each $\pi(f_i g_i) = 0$, so each $f_i g_i \in R$, in fact, each is irreducible in R. The factorization $fg = X^{2n}(f_1 g_1) \cdots (f_m g_m)$ then yields $\rho(R) \geq (2n + D(G/H) - 1)/2$.

For $f = aX^k f_1 \cdots f_m \in K[X]$ with $0 \neq a \in K$, and each $f_i \in K[X]$ irreducible in $K[X]$ with $f_i(0) = 1$, define $\varphi(f) = 2k + m$. Then φ defines a semi-length function on $K[X]$, and φ restricts to a semi-length funtion on R since $U(K[X]) \cap R = U(R) = K_0$. If $f = aX^k f_1 \cdots f_m$ (as above) is irreducible in R, then clearly $0 \leq k \leq n$. First, we show that $m \leq D(G/H) - 1$ when $k = n$. If $D(G/H) \leq m$, then (after possibly reordering) $\pi(f_1 \cdots f_i) = \pi(f_1) * \cdots * \pi(f_i) = 0$ in G/H for some $1 \leq i \leq m$. Thus $f_1 \cdots f_i \in R$, and hence $f = (aX^n f_{i+1} \cdots f_m)(f_1 \cdots f_i)$ is not irreducible in R, a contradiction. Next, we show that $m \leq D(G/H)$ when $0 \leq k < n$. If $D(G/H) < m$, then (after possibly reordering) $\pi(f_1 \cdots f_i) = \pi(f_1) * \cdots * \pi(f_i) = 0$ in G/H for some $1 \leq i < m$. Thus $f_1 \cdots f_i \in R$, and by comparing coefficients, one checks that $aX^k f_{i+1} \cdots f_m \in R$ (cf. [11. Lemma 2.2(b)] and [12, Lemma 2.2(a)]). Hence $f = (aX^k f_{i+1} \cdots f_m)(f_1 \cdots f_i)$ is not irreducible in R, a contradiction. Thus $M^* = M^*(R, \varphi) \leq 2n + D(G/H) - 1$. Also, $m^* = m^*(R, \varphi) = 2$ since X is irreducible, but not prime, in R, and any irreducible $f \in K[X]$ with $f \in R$ and $f(0) = 1$ is prime in R by Remark 2.2(a). Hence $\rho(R) \leq M^*/m^* \leq (2n + D(G/H) - 1)/2$, and we have equality. $\square$

EXAMPLE 4.5. (a) Let $R = K_0 + K_1 X + K_2 X^2 + X^3 K[X]$, where $K_0 \subseteq K_1 = GF(p^a) \subseteq K_2 = GF(p^{ab}) \subset K = GF(p^{abc})$ with $a, b \geq 1$ and $c \geq 2$. Then $G/H = (K \oplus K)/(K_1 \oplus K_2)$ with $|G/H| = p^{a(2bc-b-1)}$. First assume that $p \geq 3$. Then each element of G/H has order $\leq p$, so $D(G/H) = a(2bc - b - 1)(p - 1) + 1$. Hence $\rho(R) = (a(2bc - b - 1)(p - 1) + 6)/2$ by Theorem 4.4. Next, suppose that $p = 2$. Then each nonzero element of G/H has order 2 or 4; so $G/H \cong (\mathbb{Z}_2)^i \oplus (\mathbb{Z}_4)^j$ for $i, j \geq 1$ with $i + 2j = 2abc - ab - a$. Moreover, one can check that $(\bar{a}, \bar{b}) \in G/H$ has order $\leq 2 \Leftrightarrow a^2 \in K_2 \Leftrightarrow a \in K_2$. Thus G/H has 2^{abc-a} elements of order ≤ 2. Hence $i + j = abc - a$, and thus $i = ab - a$ and $j = abc - ab$. Hence $D(G/H) = (ab - a) + 3(abc - ab) + 1 = a(3bc - 2b - 1) + 1$; so $\rho(R) = (a(3bc - 2b - 1) + 6)/2$ by Theorem 4.4.

(b) Let $R = K_0 + K_1 X + \cdots + K_{n-1} X^{n-1} + X^n K[X]$ be as in Theorem 4.4 with $\mathrm{char}\, K = p \geq n$, $|K_1| = a_1, ..., |K_{n-1}| = a_{n-1}$, and $|K| = a_n$. Then $G/H \cong (\mathbb{Z}_p)^m$, where $m = (n-1)a_n - (a_1 + \cdots + a_{n-1}) + 1$; so $\rho(R) = ((p-1)m + 6)/2$ by Theorem 4.4. $\square$

REMARK 4.6. (a) When $n = 1$, we have $R = F + XK[X]$. In this case, we define $G/H = \{0\}$, and thus $D(G/H) = 1$. Hence the formula in Theorem 4.4 gives $\rho(R) = 1$, so R is a HFD. We thus recover the result mentioned in the Introduction and Theorem 4.1 that $R = F + XK[X]$ is a HFD. That $F + XK[[X]]$ is also a HFD follows directly from Theorem 3.1(2). (That $F + XK[X]$ and $F + XK[[X]]$ are HFD's also follows from Remark 4.6(d).)

(b) For a specific case, $D(G/H)$, and hence $\rho(R)$, may be calculated as in Example 4.5 above. However, in general, it seems to be rather difficult to give an explicit formula for $D(G/H)$ since this group does not have the usual coordinate-wise multiplication. For some special cases of similar calculations when $H = \{0\}$, see [11, Section 3]. (Also, cf. Remark 4.6(e).)

(c) Note that the formulas for $\rho(K[S])$ obtained in [10], [11], and [12] are similar to, but not a consequence of, those derived in Theorems 4.3 and 4.4 for $\rho(R)$. For example, let $GF(p^n) = F \subset K = GF(p^{mn})$, $A = K[X^2, X^3] = K + X^2K[X]$, and $B = F + FX + X^2K[X]$. Then $\rho(A) = (mn(p-1)+3)/2$ by [10, Theorem 2.4] and $\rho(B) = (n(m-1)(p-1)+4)/2$ by Theorem 4.3. In particular, $\rho(A) = \rho(B)$ if and only if $n = 1$ and $p = 2$. In this case, $F = \mathbb{Z}_2$, $K = GF(2^m)$, and $\rho(A) = \rho(B) = (m+3)/2$ for $m \geq 2$. (By Remark (d) below, we could also let $A = F + X^2K[X]$ throughout.)

(d) Note that in all our calculations, $\rho(R)$ is independent of the field R_0. This is a special case of the following more general result: Let T be an atomic domain of the form $T = K + M$, where M is a maximal ideal of T and the field K is a subring of T. If k is a subfield of K, then the subring $R = k + M$ is also atomic, and moreover, $\rho(R) = \rho(T)$ (cf. [1, Example 3.7]).

(e) The abelian group G/H in Theorem 4.4 is just $Pic(R)$, the Picard group of $R = K_0 + K_1X + \cdots + K_{n-1}X^{n-1} + X^nK[X]$ (one can see this by using the Mayer-Vietoris exact sequence for (U, Pic)). The domain $A = K_0 + K_1X + \cdots + K_{n-1}X^{n-1} + X^nK[[X]]$ is quasilocal, so $Pic(A) = 0$, and hence $D(Pic(A)) = 1$. The analogue of the formula in Theorem 4.4 for A is then $\rho(A) = (2n + D(Pic(A)) - 1)/2 = (2n + 1 - 1)/2 = n$, as in Theorem 3.1(2).

Let K be a finite field and $S = \langle n_1, ..., n_r \rangle$ a numerical submonoid of $\mathbb{Z}_+$ with $g(S) + n_1 = n_r$ (here $g(S) = max\{n \in \mathbb{Z}_+ \mid n \notin S\}$ is the Frobenius number of S). In [12, Theorem 3.2], we showed that $\rho(K[S]) = (2n_r - n_1 + n_1 D(G(K, S)))/(2n_1)$. As above, $G(K, S)$ is just $Pic(K[S])$. The analogue of this formula for $R = K_0 + K_1X + \cdots + K_{n-1}X^{n-1} + X^nK[X]$ with $n_1 = 1$ and $n_r = n$ is then $\rho(R) = (2n - 1 + D(Pic(R)))/2$, as in Theorem 4.4. $\square$

ACKNOWLEDGMENTS

The second author gratefully acknowledges support received under The Inha University Research Foundation and the hospitality of The University of Tennessee at Knoxville.

REFERENCES

1. D.D. Anderson and D.F. Anderson, Elasticity of factorizations in integral domains, *J. Pure Appl. Algebra 80*: 217-235 (1992).

2. D.D Anderson and D.F. Anderson, Elasticity of factorizations in integral domains, II, *Houston J. Math. 20*: 1-15 (1994).

3. D.D. Anderson, D.F. Anderson, S. Chapman, and W.W. Smith, Rational elasticity of factorizations in Krull domains, *Proc. Amer. Math. Soc. 117*: 37-43 (1993).

4. D.D. Anderson, D.F. Anderson, and M. Zafrullah, Factorization in integral domains, *J. Pure Appl. Algebra 69*: 1-19 (1990).

5. D.D. Anderson, D.F. Anderson, and M. Zafrullah, Rings between $D[X]$ and $K[X]$, *Houston J. Math. 17*: 109-129 (1991).

6. D.D. Anderson, D.F. Anderson, and M. Zafrullah, Atomic domains in which almost all atoms are prime, *Comm. Algebra 20*: 1447-1462 (1992).

7. D.D. Anderson and J.L. Mott, Cohen-Kaplansky domains: integral domains with a finite number of irreducible elements, *J. Algebra 148*: 17-41 (1992).

8. D.D. Anderson and B. Mullins, Finite factorization domains, *Proc. Amer. Math. Soc. 124(2)*: 389-396 (1996).

9. D.F. Anderson, Elasticity of factorizations in integral domains, A survey, these Proceedings.

10. D.F. Anderson, S. Chapman, F. Inman, and W.W. Smith, Factorization in $K[X^2, X^3]$, *Arch. Math. 61*: 521-528 (1993).

11. D.F. Anderson and S. Jenkens, Factorization in $K[X^n, X^{n+1}, ..., X^{2n-1}]$, *Comm. Algebra 23*: 2561-2576 (1995).

12. D.F. Anderson and C. Scherpenisse, Factorization in $K[S]$, to appear in Lecture Notes in Pure and Applied Mathematics, Marcel Dekker, New York, in press, 1996.

13. D.F. Anderson and J. Winner, Factorization in $K[[S]]$, these Proceedings.

14. P-J. Cahen, Spectre d'anneaux de polynômes sur une suite croissante d'anneaux, *Arch. Math. 49*: 281-285 (1987).

15. S. Chapman, On the Davenport constant, the cross number, and their applications in factorization theory, *Lecture Notes in Pure and Applied Mathematics*, Marcel Dekker, New York, *171*: 167-190 (1995).

16. S. Chapman and W.W. Smith, An analysis using the Zaks-Skula constant of element factorizations in Dedekind domains, *J. Algebra 159*: 176-190 (1993).

17. A. Geroldinger and R. Schneider, On Davenport's constant, *J. Combin. Theory Ser. A. 61*: 147-152 (1992).

18. J.L. Steffan, Longuers des décompositions en produits d'éléments irréducibles dans un anneau de Dedekind, *J. Algebra 102*: 229-236 (1986).

19. R.J. Valenza, Elasticity of factorizations in number fields, *J. Number Theory 36*: 212-218 (1990).

20. A. Zaks, Half-factorial domains, *Israel J. Math. 37*: 281-302 (1980).

Factorization in $K[[S]]$

DAVID F. ANDERSON Department of Mathematics, The University of Tennessee, Knoxville, TN 37996-1300, USA

JANICE WINNER Department of Mathematics, The University of Kentucky, Lexington, KY 40506, USA

ABSTRACT: Let K be a field, $S = \langle n_1, \ldots, n_r \rangle$ a numerical semigroup, and $K[[S]]$ the (power series) semigroup ring. In this paper, we study lengths of factorizations in $K[[S]]$. We show that $n_r/n_1 \leq \rho(K[[S]]) \leq (g(S)+n_1)/n_1$, where $g(S)$ is the Frobenius number of S. This inequality is an equality when $g(S) + n_1 = n_r$, but in general $\rho(K[[S]])$ may lie strictly between the two bounds.

1 INTRODUCTION

If D is a UFD, then any two factorizations of a nonzero nonunit into the product of irreducible elements have the same length. However, this need not be true for an arbitrary atomic domain (an integral domain is *atomic* if each nonzero nonunit is a product of irreducible elements (atoms)). Following Zaks [15], we say that an atomic domain D is a *half-factorial domain* (HFD) if whenever $x_1 \cdots x_m = y_1 \cdots y_n$ with each $x_i, y_j \in D$ irreducible, then $m = n$. For any field K, the domains $A = K[X^2, X^3]$ and $B = K[[X^2, X^3]]$ are probably the simplest examples of atomic domains which are not HFDs since X^2 and X^3 are each irreducible elements, and $X^6 = (X^3)^2 = (X^2)^3$. More generally, let $S = \langle n_1, \ldots, n_r \rangle$ be a proper numerical semigroup, i.e., a proper submonoid of $\mathbf{Z}_+$ under addition with

243

$\mathbf{Z}_+ - S$ finite (thus $n_1 \geq 2$ and $r \geq 2$). Then the (polynomial) semigroup ring $A = K[S] = \{\sum a_s X^s \in K[X] \mid a_s \in K, \, s \in S\} = K[X^{n_1}, \ldots, X^{n_r}]$ and (power series) semigroup ring $B = K[[S]] = \{\sum a_s X^s \in K[[X]] \mid a_s \in K, \, s \in S\} = K[[X^{n_1}, \ldots, X^{n_r}]]$ are each a one-dimensional Noetherian (and hence atomic) domain which is not an HFD since X^{n_1} and X^{n_r} are each irreducible elements of both A and B and $X^{n_1 n_r} = (X^{n_1})^{n_r} = (X^{n_r})^{n_1}$.

Recently there has been much activity on HFDs, other factorization properties weaker than unique factorization, and on invariants that measure different lengths of factorizations. For an atomic domain D, define $\rho(D) = \sup\{m/n \mid x_1 \cdots x_m = y_1 \cdots y_n, \text{ for } x_i, y_j \in D \text{ irreducible }\}$. Thus $1 \leq \rho(D) \leq \infty$, and $\rho(D) = 1$ if and only if D is an HFD. $\rho(D)$, called the *elasticity* of D, was introduced by Valenza [14] and has been studied in more detail in [1], [2], [3], [10], and [13] (see [5] for a survey article on elasticity). We say that $\rho(D)$ is *realized by a factorization* if there are irreducible $x_i, y_j \in D$ with $x_1 \cdots x_m = y_1 \cdots y_n$ and $\rho(D) = m/n$.

In [1, Example 3.6], we showed that $\rho(K[[X^n, X^{n+1}, \ldots, X^{2n-1}]]) = (2n-1)/n$ for any field K. In this paper, we study $\rho(R)$ for an arbitrary subring R of $K[[X]]$ generated by monomials over K. Such subrings R are just the semigroup rings $K[[S]]$ for $S = \langle n_1, \ldots, n_r \rangle$ a numerical semigroup. In Theorem 2.3, we show that $n_r/n_1 \leq \rho(K[[S]]) \leq (g(S) + n_1)/n_1$, where $g(S)$ is the Frobenius number of the numerical semigroup S; clearly this inequality is an equality when $g(S) + n_1 = n_r$. However, for $S = \langle 4, 6, 7 \rangle$ and any field K, $\rho(K[[S]]) = 3$, while $n_3/n_1 = 7/4 < 3 < 13/4 = (g(S) + n_1)/n_1$ (Theorem 3.7(c)). In general, $\rho(K[[S]])$ is always finite, but may depend on the field K. For example, $\rho(\mathbf{Z}_2[[X^3, X^5]]) = 10/3$, while $\rho(\mathbb{C}[[X^3, X^5]]) = 3$ (cf. Theorem 3.2).

In a series of papers ([6], [7], and [9]), we computed $\rho(R)$ for R a subring of $K[X]$ generated by monomials over K. In this case, $\rho(R) < \infty$ if and only if K is a finite field [6, Theorem 3.5]. In particular, we showed in [9, Theorem 2.4] that $\rho(K[S]) \leq (n_1 D(G(K, S)) + 2g(S) + n_1)/(2n_1)$, where $D(G(K, S))$ is the Davenport constant of a certain abelian group $G(K, S)$ associated with K and S. Moreover, this is an equality when $g(S) + n_1 = n_r$ [9, Theorem 3.2]. It turns out that $G(K, S) = \text{Pic}(K[S])$, the Picard group of $K[S]$. Since $K[[S]]$ is local, $\text{Pic}(K[[S]]) = 0$, and hence $D(\text{Pic}(K[[S]])) = 1$. Thus the analogue of the $\rho(K[S])$ formula for $\rho(K[[S]])$ simplifies to

$\rho(K[[S]]) \leq (g(S) + n_1)/n_1$. In general, factorization in $K[[S]]$ is much simpler than in $K[S]$. In [8], we have also computed $\rho(R)$ for some other classes of subrings R of either $K[X]$ or $K[[X]]$.

In the second section, we first review some basic facts about numerical semigroups and semi-length functions. Then we give easily computed upper and lower bounds for $\rho(K[[S]])$ and determine a sufficient condition on S for this upperbound to equal $\rho(K[[S]])$. In the third section, we investigate several special cases for S. Computing $\rho(K[[S]])$ usually amounts to determining the solvability of an infinite system of equations over the field K (cf. the proofs of Theorem 3.2 or Theorem 3.7).

Throughout, D will denote an arbitrary atomic domain with D^* its set of nonzero elements and $U(D)$ its group of units, K a field, and X an indeterminate over K. The fields of rational, real, and complex numbers will be denoted by $\mathbf{Q}$, $\mathbf{R}$, and $\mathbf{C}$, respectively. Also, $\mathbf{Z}_+$, $\mathbf{Q}_+$, and $\mathbf{R}_+$ will denote the sets of nonnegative integers, nonnegative rational numbers, and nonnegative real numbers, respectively. As usual, $a/0 = \infty/0 = \infty/a = \infty + a = \infty$ for any positive $a \in \mathbf{R}$. For other references and related results on factorization in integral domains, see [4] or [5].

This research was begun while the second author was an undergraduate at The University of Kentucky in Lexington and participated in an NSF sponsored REU program with the first author at The University of Tennessee during the summer of 1995.

2 CALCULATING $\rho(K[[S]])$

Let K be a field and $S = \langle n_1, \ldots, n_r \rangle$ a numerical semigroup. In this section, we determine upper and lower bounds for $\rho(K[[S]])$. We first recall some facts about numerical semigroups and semi-length functions.

A *numerical semigroup* S is an additive submonoid of $\mathbf{Z}_+$ with $\mathbf{Z}_+ - S$ finite. It is wellknown that $S = \langle n_1, \ldots, n_r \rangle$ for a unique minimal set of generators with $n_1 < \cdots < n_r$ and $\gcd(n_1, \ldots, n_r) = 1$. Conversely, given positive integers $n_1 < \cdots < n_r$ with $\gcd(n_1, \ldots, n_r) = 1$, $S = \langle n_1, \ldots, n_r \rangle$ is a numerical semigroup. When we write $S = \langle n_1, \ldots, n_r \rangle$, we we will always mean that $\{n_1, \ldots, n_r\}$ is such a minimal generating set for S. The *Frobenius number* of a numerical semigroup S is $g(S) = \max\{n \in \mathbf{Z}_+ |\ n \notin S\}$ (let $g(\mathbf{Z}_+) = 0$). Thus $X^n f \in K[[S]]$ for any $n > g(S)$ and $f \in K[[X]]$. Note that any nonzero submonoid T of $\mathbf{Z}_+$ is isomorphic to a numerical semigroup (just divide out by the gcd of the elements of T); thus any subring of $K[[X]]$ generated by monomials over K is isomorphic to $K[[S]]$ for a suitable numerical semigroup S. For $S = \langle n_1, \ldots, n_r \rangle$, $K[[S]] = K[[X^{n_1}, \ldots, X^{n_r}]]$, and $X^{n_1}, \ldots, X^{n_r}$ are the only irreducible monomials in R. We are interested in the case when S is a proper submonoid of $\mathbf{Z}_+$, i.e., when $n_1 \geq 2$ and $r \geq 2$ (equivalently, $K[[S]] \neq K[[X]]$). Note that in this case, $K[[S]]$ has no prime elements. Also, $K[[S]]$ is local with $U(K[[S]]) = \{f \in K[[S]] | f(0) \neq 0\}$ and analytically irreducible (i.e., its M-adic completion is an integral domain (cf. [1, Theorem 2.12])). Moreover, $K[[S]]$ is a CK-domain (i.e., $K[[S]]$ has, up to associates, only a finite number of irreducible elements) when K is finite. General references for numerical semigroups are [11] and [12].

A function $\varphi : D^* \to \mathbf{R}_+$ is a *semi-length function* on D if (i) $\varphi(xy) = \varphi(x) + \varphi(y)$ for all $x, y \in D^*$, and (ii) $\varphi(x) = 0$ if and only if $x \in U(D)$. Semi-length functions were introduced in [1] to determine upper and lower bounds for $\rho(D)$. By [1, Theorem 2.1], if φ is a semi-length function on D, then $1 \leq \rho(D) \leq M/m$, where $M = M(D, \varphi) = \sup\{\varphi(x) | x \in D$ is irreducible, but not prime $\}$, $m = m(D, \varphi) = \inf\{\varphi(x) | x \in D$ is irreducible, but not prime $\}$, and $M = m = 1$ when D is a UFD. Note that the usual order function on $K[[X]]$ (i.e., for $f(X) = a_n X^n + \cdots \in K[[X]]$ with $a_n \neq 0$, $\operatorname{ord}(f) = n$) defines a semi-length function on $K[[X]]$. We will also denote by ord its restriction to a semi-length function on $K[[S]]$. We next show that in our case $\rho(K[[S]]) = M/m$ (cf. [1, Theorem 2.7] and [1, Theorem 2.12]).

LEMMA 2.1   Let K be a field, $S = \langle n_1, \ldots, n_r \rangle$ a numerical semigroup, $R = K[[S]]$, $m = m(R, \operatorname{ord})$, and $M = M(R, \operatorname{ord})$. Then

   (a) $m = n_1$.
   (b) $n_r \leq M \leq g(S) + n_1$.

Proof: (a) is clear. (b) Clearly $M \geq n_r$ since X^{n_r} is irreducible in R. Suppose that $f \in R$ with $\operatorname{ord}(f) = n > g(S) + n_1$. Then $(n - n_1) + k \in S$ for each integer $k \geq 0$ since $n - n_1 > g(S)$. Hence X^{n_1} divides f in R; so f is not irreducible in R. Thus $M \leq g(S) + n_1$.   □

THEOREM 2.2   Let K be a field, $S = \langle n_1, \ldots, n_r \rangle$ a numerical semigroup, $R = K[[S]]$, and $M = M(R, \mathrm{ord})$. Then $\rho(R) = M/n_1$. Moreover, if $\mathrm{char} K = p > 0$, then $\rho(R)$ is realized by a factorization.

Proof: By Lemma 2.1, $m(R, \mathrm{ord}) = n_1$ and $M \leq g(S) + n_1 < \infty$. We always have $\rho(R) \leq M/n_1$ by [1, Theorem 2.1] mentioned above. Choose an irreducible $f = X^M g \in R$ with $g \in K[[X]]$ and $g(0) = 1$. Choose an integer $s \geq 1$ such that $sM \geq g(S) + 1$. Then $X^{Ms} g^n \in R$ for any integer $n \geq 1$. Thus $f^{kn_1+s} = X^{Mkn_1}(X^{Ms} g^{kn_1+s}) = (X^{n_1})^{Mk}(X^{Ms} g^{kn_1+s})$ in $K[[S]]$ for any integer $k \geq 1$. Hence $\rho(R) \geq (Mk+1)/(kn_1+s) = (M+1/k)/(n_1+s/k)$ for all $k \geq 1$. Thus $\rho(R) \geq M/n_1$, and we have equality.

Next, suppose that $\mathrm{char} K = p > 0$, and let f be as above. Then choose an integer $j \geq 1$ so that $t = p^j \geq g(S) + 1$. Then $f^{n_1 t} = X^{Mn_1 t} g^{n_1 t} = (X^{n_1})^{Mt} g^{n_1 t}$ with $g^{n_1 t} = (g^t)^{n_1} \in U(R)$ because $\mathrm{ord}(g^{n_1 t} - 1) \geq g(S) + 1$ since $\mathrm{char} K = p$ and $t \geq g(S) + 1$. Hence $\rho(R) = Mt/n_1 t = M/n_1$, and $\rho(R)$ is realized by a factorization. $\square$

Thus, to calculate $\rho(K[[S]])$, we need only compute $M = M(K[[S]], \mathrm{ord})$, i.e., we need only find an irreducible $f \in K[[S]]$ with largest possible order. We know of no example where $\rho(K[[S]])$ is not realized by a factorization. Combining our two previous results, we have

THEOREM 2.3   Let K be a field and $S = \langle n_1, \ldots, n_r \rangle$ a numerical semigroup. Then $n_r/n_1 \leq \rho(K[[S]]) \leq (g(S)+n_1)/n_1$. In particular, $\rho(K[[S]]) = n_r/n_1$ and $\rho(K[[S]])$ is realized by a factorization when $g(S) + n_1 = n_r$. Moreover, $\rho(K[[S]]) = M/n_1$, where $M = M(K[[S]], \mathrm{ord}) \in S$ is an integer with $n_r \leq M \leq g(S) + n_1$. $\square$

Note that $g(S) + n_1 = n_r$ if and only if $X^{g(S)+n_1}$ is irreducible in $K[[S]]$ (cf. [9, Proposition 3.1]). Later examples (Theorem 3.7) will show that $\rho(K[[S]])$ may lie strictly between the two bounds given in Theorem 2.3. However, if $g(S) + n_1 = n_r$, then Theorem 2.3 yields $\rho(K[[S]]) = n_r/n_1$. (Note that $g(S) + n_1 \geq n_r$ always holds.) This case was investigated in [9, Section 3], where we said that a numerical semigroup $S = \langle n_1, \ldots, n_r \rangle$ satisfies *property (∗)* if $g(S)+n_1 = n_r$. We now review some facts about such semigroups from [9]. First, suppose that $r = 2$, so $S = \langle n_1, n_2 \rangle$ with $n_1 \geq 2$. Then $g(S) = n_1 n_2 - n_1 - n_2$ [11]. Thus $g(S)+n_1 = n_2$ if and only if $n_1 = 2$ ($= r$). It is clear that $n_1 \geq r$ for any $S = \langle n_1, \ldots, n_r \rangle$ which satisfies property

(∗) since $n_1, \ldots, n_r$ have distinct residues modulo n_1 because $\{n_1, \ldots, n_r\}$ is a minimal generating set for S. However, when $r = 3$, it is no longer true that $S = \langle n_1, n_2, n_3 \rangle$ satisfies property (∗) if and only if $n_1 = r = 3$. The easiest such example is $S = \langle 4, 5, 11 \rangle$. Here, $r = 3$, $n_1 = 4$, $n_3 = 11$, and $g(S) = 7$. Thus $g(S) + n_1 = 11 = n_3$; so $S = \langle 4, 5, 11 \rangle$ satisfies property (∗). (Other examples are given in [9, Example 3.5].) However, S does satisfy property (∗) when $n_1 = r$ [9, Proposition 3.6]. This generalizes the $S = \langle n, n+1, \ldots, 2n-1 \rangle$ case from [1, Example 3.6] (also, cf. [7]). Our next corollary follows directly from the above observations and Theorem 2.3.

COROLLARY 2.4   Let K be a field and $n \geq 1$. Then

(a) $\rho(K[[X^2, X^{2n+1}]]) = (2n+1)/2$.

(b) ([1, Example 3.6]) $\rho(K[[X^n, X^{n+1}, \ldots, X^{2n-1}]]) = (2n-1)/n$.

(c) If $S = \langle n_1, \ldots, n_r \rangle$ is a numerical semigroup with $n_1 = r$, then $\rho(K[[S]]) = n_r/n_1$. □

Let $S = \langle n_1, \ldots, n_r \rangle$ be a proper numerical semigroup (so $n_1 \geq 2$ and $r \geq 2$). Note that if $n_1 = r$, then either $S = \langle r, r+1, \ldots, 2r-1 \rangle$ or $n_r > 2r$. Thus either $n_r/n_1 = 2 - 1/r$, or n_r/n_1 is not an integer and $n_r/n_1 > 2$. We next show that all such possible cases for n_r/n_1 may be realized by a suitable numerical semigroup S which satisfies property (∗).

THEOREM 2.5   Let $q \in \mathbb{Q} - \mathbb{Z}$ with $q > 2$. Then there is a numerical semigroup $S = \langle n_1, \ldots, n_r \rangle$ with $n_1 = r$ and $n_r/r = q$. Thus $\rho(K[[S]]) = q$ for any field K, and $\rho(K[[S]])$ is realized by a factorization.

Proof: Let $q = m/n$ for integers $m, n \geq 2$. Then $q = (j+1) + k/n$ for integers j, k with $j \geq 1$ and $1 \leq k \leq n - 1$. Let $r = n_1 = n$. Then the $r - 1$ integers $jn_1 + (k+1), jn_1 + (k+2), \ldots, jn_1 + (n_1 - 1), (j+1)n_1 + 1,$ $\ldots, (j+1)n_1 + k$ have distinct nonzero residues modulo n_1. Thus n_1, $n_2 = jn_1 + (k+1), \ldots, n_r = (j+1)n_1 + k$ define a numerical semigroup $S = \langle n_1, \ldots, n_r \rangle$ which satisfies property (∗) and has $g(S) = jn_1 + k$. Hence $\rho(K[[S]]) = n_r/n_1 = ((j+1)n_1 + k)/n_1 = (j+1) + k/n_1 = q$ by Corollary 2.4(c). Clearly $\rho(K[[S]])$ is realized by a factorization. □

Note that the proof of the above theorem shows that for each such q there are actually infinitely many numerical semigroups S which satisfy property (∗) since $q = m/n = (tm)/(tn)$ for each integer $t \geq 1$. The integer values

≥ 2 for $\rho(R)$ can be realized from another class of subrings of $K[[X]]$ (see [1, Example 3.5] or [8, Theorem 3.1]). Namely, if k is a proper subfield of K and $R = k + kX + \cdots + kX^{n-1} + X^n K[[X]]$, then $\rho(R) = n$.

3 SPECIFIC CALCULATIONS

In this section, we do some specific calculations for $\rho(K[[S]])$. We first consider the case where $S = \langle m, n \rangle$ with $2 \leq m < n$ and $\gcd(m, n) = 1$. We have already done the $m = 2$ case in Corollary 2.4(a). We next show that if $n \equiv 1 \pmod{m}$, then $\rho(K[[X^m, X^n]]) = (m-1)n/m$. In this case, $g(S) + m = mn - m - n + m = (m-1)n$; so the upper bound for $\rho(R)$ in Theorem 2.3 is realized. However, we always have $\rho(R) > n/m$, the lower bound for $\rho(R)$ in Theorem 2.3, when $m \geq 3$.

THEOREM 3.1   Let K be a field and $2 \leq m < n$ be integers with $n \equiv 1 \pmod{m}$. Then $\rho(K[[X^m, X^n]]) = (m-1)n/m$. Moreover, $\rho(K[[X^m, X^n]])$ is realized by a factorization.

Proof: Let $S = \langle m, n \rangle$ and $R = K[[S]]$, and let $n = km + 1$ with $k \geq 1$. Thus $N = g(S) + m = (m-1)n$, and hence $\rho(R) \leq (m-1)n/m$ by Theorem 2.3. We first show that $f = X^N + eX^{N+1} + \cdots \in R$ is irreducible if $e \neq 0$. If f is not irreducible in R, then $f = (X^\alpha + \cdots)(X^\beta + \cdots)$ in R with $\alpha + \beta = N$ and $\alpha, \beta > 0$. Since $\alpha, \beta \in S$, $\alpha = am + bn$ and $\beta = cm + dn$ for some integers $a, b, c, d \geq 0$. Then $\alpha + \beta = (a+c)m + (b+d)n = (m-1)n$, and hence $n \mid (a+c)m$. Thus $n \mid a+c$ since $\gcd(m, n) = 1$. Hence $a + c = 0$ since otherwise $\alpha + \beta \geq mn > N$, and thus $a = c = 0$. Hence $b + d = m - 1$. We next show that $\alpha + 1 \notin S$. If $\alpha + 1 \in S$, then $\alpha + 1 = im + jn$ for some integers $i, j \geq 0$. Then $(\alpha + 1) + \beta = (m-1)(km+1) + 1$; so $im + (d+j)(km+1) = m(km+1) - km$, and hence $m \mid (d+j)$. Thus $d + j = 0$, and hence $d = j = 0$, a contradiction since $\beta > 0$. Similarly $\beta + 1 \notin S$. Thus f must have zero X^{N+1} coefficient if it factors in R; so f is irreducible if $e \neq 0$. Hence $\rho(R) = (m-1)n/m$ by Theorem 2.3. Now let $g = X^{(m-1)n}(1 - X)$ and $h = X^{(m-1)n}(1 + X + \cdots + X^{m-1})$. By above, g and h are each irreducible in R, and $(gh)^m = (X^m)^{2(m-1)n}(1 - X^m)^m$ with $1 - X^m \in U(R)$. Thus $\rho(R) = (m-1)n/m$ is realized by a factorization. $\square$

Thus the first possible "bad" case (i.e, when $\rho(R) < (g(S) + n_1)/n_1$) to consider would be when $m = 3$ and $n = 5$. We next show that in this case,

$\rho(K[[X^3, X^5]])$ does indeed depend on the field K. We will say that a field K is *quadratically closed* if each $f(x) \in K[X]$ of degree 2 has a root in K, i.e., there are no irreducible $f(X) \in K[X]$ of degree 2. In fact, this case is the only "bad" one when $m = 3$.

THEOREM 3.2   Let K be a field. Then $\rho(K[[X^3, X^5]]) = 3$ if K is quadratically closed, and $\rho(K[[X^3, X^5]]) = 10/3$ if K is not quadratically closed. In either case, $\rho(K[[X^3, X^5]])$ is realized by a factorization.

Proof: Let $R = K[[X^3, X^5]]$. Thus $S = \langle 3, 5 \rangle$ with $g(S) = 7$, so $\rho(R) \leq (7+3)/3 = 10/3$ by Theorem 2.3. We first show that $f = X^9 + a_{10} X^{10} + \cdots \in R$ is irreducible if and only if $a_{10} \neq 0$. If $a_{10} = 0$, then X^3 divides f in R. Conversely, suppose $f = (X^3 + b_5 X^5 + \cdots)(X^6 + c_8 X^8 + \cdots)$ in R. Clearly this is impossible if $a_{10} \neq 0$ since the right-hand side factorization of f has zero X^{10} coefficient. Thus $\rho(R) \geq 9/3 = 3$ by Theorem 2.3.

We next show that there is an irreducible $f \in R$ with $\text{ord}(f) = 10$ if and only if K is not quadratically closed. Theorem 2.3 then yields the desired result. Note that $f = X^{10} + c_{11} X^{11} + c_{12} X^{12} + \cdots \in R$ is reducible if and only if there are $a_i, b_j \in K$ such that $f = (X^5 + a_6 X^6 + a_8 X^8 + \cdots)(X^5 + b_6 X^6 + b_8 X^8 + \cdots)$ in R. Comparing coefficients, we obtain $c_{11} = a_6 + b_6$ and $c_{12} = a_6 b_6$. Such a_6 and b_6 exist if and only if $a_6^2 - c_{11} a_6 + c_{12} = 0$. If such a_6 and b_6 exist, then let $a_8 = a_9 = a_{10} = \cdots = 0$, and then solve recursively for $b_8, b_9, \ldots$. Thus an irreducible $f \in R$ with $\text{ord}(f) = 10$ exists if and only K is not quadratically closed.

We next show that $\rho(K[[X^3, X^5]])$ is realized by a factorization. First, suppose that $\rho(R) = 3$ (i.e., K is quadratically closed). Then, by above, $f = X^9(1 - X)$ and $g = X^9(1 + X + X^2)$ are each irreducible in R. The factorization $(fg)^3 = (X^3)^{18}(1 - X^3)^3$ with $1 - X^3 \in U(R)$ shows that $\rho(R) = 3$ is realized by a factorization. Next, suppose $\rho(R) = 10/3$ (i.e., K is not quadratically closed). First, suppose that $\text{char} K \neq 2$. Choose an irreducible $X^2 + \alpha \in K[X]$. Then, by above, $f = X^{10} + \alpha X^{12}$ is irreducible in R, and the factorization $f^3 = [X^{10}(1 + \alpha X^2)]^3 = (X^3 + \alpha X^5)^3 (X^3)^7$ yields that $\rho(R) = 10/3$ is realized by a factorization. Finally, suppose $\text{char} K = 2$ and $X^2 + \alpha X + \beta \in K[X]$ is irreducible. Then, by above, $f = X^{10} + \alpha X^{11} + \beta X^{12}$ is irreducible in R, and the factorization $f^{24} = [X^{10}(1 + \alpha X + \beta X^2)]^{24} = [X^3(1 + \alpha^8 X^8 + \beta^8 X^{16})]^3 (X^3)^{77}$ yields that $\rho(R) =$

10/3 is realized by a factorization. (The char$K = 2$ case also follows from Theorem 2.2.) □

Note that $g(\langle 3,5 \rangle)+3 = 10$, so $\rho(K[[X^3, X^5]])$ realizes the upperbound in Theorem 2.3 when K is not quadratically closed. However, for K quadratically closed, $\rho(K[[X^3, X^5]]) = 3$ lies strictly between the lower bound 5/3 and the upper bound 10/3 from Theorem 2.3. For specific examples, we have $\rho(\mathbb{C}[[X^3, X^5]]) = 3$ and $\rho(K[[X^3, X^5]]) = 10/3$ for any finite field K. The argument used in the proof of Theorem 3.2 is typical; to determine whether there is an irreducible $f \in K[[S]]$ with ord$(f) = n$ amounts to solving an infinite system of equations over the field K. A solution of this system often depends on the field K.

REMARK. 3.3 Clearly any algebraically closed field is quadratically closed and no finite field is quadratically closed. We next give an example of a subfield of the complex numbers $\mathbb{C}$ which is quadratically closed, but not algebraically closed. Let $K_0 = \mathbb{Q}$. For each integer $n \geq 0$, define recursively $K_{n+1} = K_n(\{\alpha \in \mathbb{C} \mid \alpha^2 \in K_n\})$, and then define $K = \bigcup_{n=0}^{\infty} K_n$. It is easy to verify that the field K is quadratically closed, but not algebraically closed. In a similar manner, for any field k, one can construct the "quadratic closure" K of k in its algebraic closure. □

We next show that when $n \equiv 2 \pmod 3$, the only "bad" case is when $n = 5$.

THEOREM 3.4 Let K be a field and $n \geq 2$. Then $\rho(K[[X^3, X^{3n+2}]]) = 2(3n + 2)/3$ and $\rho(K[[X^3, X^{3n+2}]])$ is realized by a factorization.

Proof: Let $S = \langle 3, 3n+2 \rangle$ and $R = K[[S]]$. Then $N = g(S)+3 = 2(3n+2)$, so $\rho(R) \leq 2(3n + 2)/3$ by Theorem 2.3. We show that $f = X^N + cX^{N+5} + \cdots \in R$ is irreducible if $c \neq 0$. Note that if f is reducible in R, then there are $a_i, b_j \in K$ such that $f = (X^{3n+2} + a_1 X^{3n+3} + a_2 X^{3n+5} + a_3 X^{3n+6} + a_4 X^{3n+8} + \cdots)(X^{3n+2} + b_1 X^{3n+3} + b_2 X^{3n+5} + b_3 X^{3n+6} + b_4 X^{3n+8} + \cdots)$ in R (we need $n \geq 2$ so that $3n + 7 \notin S$). Comparing coefficients, we obtain $a_1 + b_1 = 0$ and $a_1 b_1 = 0$. Thus $a_1 = b_1 = 0$. Hence also $c = a_1 b_3 + a_3 b_1 = 0$, a contradiction. Thus no such a_i and b_j exist, so any such f is irreducible. Hence $\rho(K[[X^3, X^{3n+2}]]) = 2(3n + 2)/3$ by Theorem 2.3. The factorization $[X^{2(3n+2)}(1 - X^5)]^3[X^{2(3n+2)}(1 + X^5 + X^{10})]^3 = (X^3)^{4(3n+2)}(1 - X^{15})$ with

$1 - X^{15} \in U(R)$ shows that $\rho(K[[X^3, X^{3n+2}]]) = 2(3n + 2)/3$ is realized by a factorization. $\square$

By combining our previous three results, we may summarize the $m = 3$ case as follows.

THEOREM 3.5 Let K be a field and $n > 3$ an integer with $n \not\equiv 0 \pmod 3$. Then $\rho(K[[X^3, X^n]]) = 3$ if $n = 5$ and K is quadratically closed. Otherwise, $\rho(K[[X^3, X^n]]) = 2n/3$. Moreover, in either case, $\rho(K[[X^3, X^n]])$ is realized by a factorization. $\square$

We next consider the case when $m = 4$. By Theorem 3.1, the first case which might depend on the field K is when $n = 7$. We will say that a field K is *cubically closed* if each $f(X) \in K[X]$ of degree 3 has a root in K, i.e., there are no irreducible $f(X) \in K[X]$ of degree 3.

THEOREM 3.6 Let K be a field. Then $\rho(K[[X^4, X^7]]) = 5$ if K is cubically closed, and $\rho(K[[X^4, X^7]]) = 21/4$ if K is not cubically closed.
Proof: Let $S = \langle 4, 7 \rangle$ and $R = K[[S]]$. Then $g(S) = 17$, so $\rho(R) \leq (17 + 4)/4 = 21/4$ by Theorem 2.3. One may easily verify that $f = X^{20} + c_{21}X^{21} + \cdots \in R$ is irreducible if $c_{21} \neq 0$. Thus $\rho(R) \geq 20/4 = 5$ by Theorem 2.3. We next determine when there is an irreducibe $f \in R$ with $\text{ord}(f) = 21$. Note that $f = X^{21} + c_{22}X^{22} + \cdots$ is reducible in R if and only if there are $a_i, b_j \in K$ with $f = (X^7 + a_8X^8 + a_{11}X^{11} + \cdots)(X^{14} + b_{15}X^{15} + b_{16}X^{16} + b_{18}X^{18} + \cdots)$. Comparing coefficients yields $c_{22} = b_{15} + a_8$, $c_{23} = b_{16} + a_8 b_{15}$, $c_{24} = a_8 b_{16}$, $\ldots$. Note that f is reducible if and only if there are such $a_8, b_{15}, b_{16} \in K$ since we can then let $a_{11} = a_{12} = \cdots = 0$ and, then solve recursively for $b_{18}, b_{19}, \ldots$. If $c_{24} = 0$, let $a_8 = 0$, and then f is divisible by X^7. So suppose that $c_{24} \neq 0$. Then $b_{16} = c_{24}/a_8$; hence $c_{23} = c_{24}/a_8 + a_8(c_{22} - a_8)$, and thus $a_8^3 - c_{22}a_8^2 + c_{23}a_8 - c_{24} = 0$. If such an a_8 exists, then we may let $a_{11} = a_{12} = \cdots = 0$, and then solve recursively for $b_{18}, b_{19}, \ldots$. In summary, there is an irreducible $f \in R$ with $\text{ord}(f) = 21$ if and only if K is not cubically closed. By Theorem 2.3, $\rho(R) = 21/4$ when K is not cubically closed and $\rho(R) = 20/4 = 5$ otherwise. $\square$

Note that $\rho(K[[X^4, X^7]]) = (g(S) + 4)/10 = 21/4$ when K is not cubically closed, so the upper bound for $\rho(R)$ in Theorem 2.3 is realized in this case. We know of no examples for $S = \langle n_1, n_2 \rangle$ where $\rho(K[[S]]) \neq (g(S) + n_1)/n_1$

when K is a finite field. We next briefly consider the case when $r = 3$, so $S = \langle n_1, n_2, n_3 \rangle$. If $n_1 = 3$, then $\rho(K[[S]]) = n_3/n_1$ by Corollary 2.4(c).

THEOREM 3.7 Let K be a field, $S = \langle n_1, n_2, n_3 \rangle$ a numerical semigroup, and $R = K[[S]]$.

(a) If $S = \langle 4, 5, 6 \rangle$, then $\rho(R) = 3/2$ if either char$K \neq 2$ or K is cubically closed, and $\rho(R) = 2$ otherwise.

(b) If $S = \langle 4, 5, 7 \rangle$, then $\rho(R) = 5/2$ for any field K.

(c) If $S = \langle 4, 6, 7 \rangle$, then $\rho(R) = 3$ for any field K.

Moreover, in each case $\rho(R)$ is realized by a factorization.

Proof: (a) In this case $g(S) = 7$, so $\rho(R) \leq 11/4$ by Theorem 2.3. We first show that no $f \in R$ with ord$(f) = 11$ is irreducible. Suppose that $f = X^{11} + c_{12} X^{12} + c_{13} X^{13} + \cdots = (X^5 + a_6 X^6 + a_8 X^8 + \cdots)(X^6 + b_8 X^8 + b_9 X^9 + \cdots)$ in R. By comparing coefficients, $c_{12} = a_6$, $c_{13} = b_8$, $c_{14} = b_9 + a_6 b_8 + a_8, \ldots$. Let $a_6 = c_{12}$, $a_8 = a_9 = a_{10} = \cdots = 0$, $b_8 = c_{13}$, and solve recursively for the other b_i's. In a similar, manner one may show that there is no irreducible $f \in R$ with ord$(f) = 9$ or ord$(f) = 10$. We next determine when there is an irreducible $f \in R$ with ord$(f) = 8$. Suppose $f = X^8 + c_9 X^9 + c_{10} X^{10} + \cdots = (X^4 + a_5 X^5 + a_6 X^6 + \cdots)(X^4 + b_5 X^5 + b_6 X^6 + \cdots)$. By comparing coefficients, $c_9 = a_5 + b_5$, $c_{10} = a_6 + a_5 b_5 + b_6$, $c_{11} = a_5 b_6 + a_6 b_5$, $\ldots$. If such a_5, a_6, b_5, and b_6 exist, then we can set $a_8 = a_9 = \cdots = 0$ and solve for the rest of the b_i's recursively. We consider several cases. (Case 1): If $c_9 \neq 0$, then set $a_5 = 0$ and $b_5 = c_9 \neq 0$. Then $a_6 = c_{11}/c_9$ and $b_6 = c_{10} - c_{11}/c_9$ are the desired coefficients. (Case 2): $c_9 = 0$ and char$K \neq 2$. Let $a_5 = 1$ and $b_5 = -1$. Then $a_6 = (c_{10} - c_{11} + 1)/2$ and $b_6 = (c_{10} + c_{11} + 1)/2$ are the desired coefficients. (Case 3): Suppose $c_9 = 0$ and char$K = 2$. Then $a_5 + b_5 = 0$ implies $a_5 = b_5$. Hence $a_6 + a_5^2 + b_6 = c_{10}$ and $a_5(a_6 + b_6) = c_{11}$, or $a_5^3 + c_{10} a_5 + c_{11} = 0$. Thus f is irreducible if and only if there are $c_{10}, c_{11} \in K$ such that $X^3 + c_{10} X + c_{11}$ is irreducible in $K[X]$. This happens if and only if K is not cubically closed. (Since char$K = 2$, there is an irreducible $g(X) \in K[X]$ of degree 3 if and only if there is an irreducible $h(X) \in K[X]$ of degree 3 with zero X^2 coefficient.) In summary, there is an irreducible $f \in R$ with ord$(f) = 8$ if and only if char$K = 2$ and K is not cubically closed. In this case, $\rho(R) = 8/4 = 2$ and $\rho(R)$ is realized by a factorization by Theorem 2.2. Otherwise, $\rho(R) = 6/4 = 3/2$ since $M(R, \text{ord}) = 6$. The factorization $(X^6)^4 = (X^4)^6$ shows that in this case, $\rho(R)$ is also realized by a factorization.

(b) In this case, $g(S) = 6$, so $\rho(R) \leq 10/4$ by Theorem 2.3. It is easy to check that $X^{10} + c_{11}X^{11} + \cdots \in R$ is irreducible if $c_{11} \neq 0$. Hence $\rho(R) = 10/4 = 5/2$ by Theorem 2.3. Then $f = X^{10}(1 - X)$ and $g = X^{10}(1 + X + X^2 + X^3)$ are each irreducible in R and the factorization $fg = (X^4)^5(1 - X^4)$ with $1 - X^4 \in U(R)$ yields that $\rho(R) = 5/2$ is realized by a factorization.

(c) In this case, $g(S) = 9$, so $\rho(R) \leq 13/4$ by Theorem 2.3. By comparing coefficients, one can easily check that no $f \in R$ with $\text{ord}(f) = 13$ is irreducible, and that $g = X^{12} + c_{13}X^{13} + \cdots \in R$ is irreducible if $c_{13} \neq 0$. Thus $\rho(R) = 12/4 = 3$ by Theorem 2.3. Then $f = X^{12}(1 - X)$ and $g = X^{12}(1 + X + X^2 + X^3)$ are each irreducible in R, and the factorization $fg = (X^4)^6(1 - X^4)$ with $1 - X^4 \in U(R)$ yields that $\rho(R) = 3$ is realized by a factorization. $\square$

Note that in (a) above, the lower bound of 3/2 for $\rho(R)$ in Theorem 2.3 is realized when either $\text{char} K \neq 2$ or K is cubically closed, but $\rho(R)$ is always less than the upper bound 11/4. In (b), $\rho(R)$ is always the upper bound 5/2 from Theorem 2.3, and in (c) $\rho(R)$ is always strictly between the lower bound 7/4 and the upper bound 13/4 from Theorem 2.3.

REFERENCES

1. D.D. Anderson and D.F. Anderson, Elasticity of factorizations in integral domains, *J. Pure Appl. Algebra: 80*, 217-235 (1992).

2. D.D. Anderson and D.F. Anderson, Elasticity of factorizations in integral domains, II, *Houston J. Math.: 20*, 1-15 (1994).

3. D.D. Anderson, D.F. Anderson, S. Chapman, and W.W. Smith, Rational elasticity of factorizations in Krull domains, *Proc. Amer. Math. Soc.: 117*, 37-43 (1993).

4. D.D. Anderson, D.F. Anderson, and M. Zafrullah, Factorization in integral domains, *J. Pure Appl. Algebra: 69*, 1-19 (1990).

5. D.F. Anderson, Elasticity of factorizations in integral domains, A survey, these Proceedings.

6. D.F. Anderson, S. Chapman, F. Inman, and W.W. Smith, Factorization in $K[X^2, X^3]$, *Arch. Math.: 61*, 521-528 (1993).

7. D.F. Anderson and S. Jenkens, Factorization in $K[X^n, X^{n+1}, \ldots, X^{2n-1}]$, *Comm. Algebra:* *234*, 2561-2576 (1995).

8. D.F. Anderson and J. Park, Factorization in subrings of $K[X]$ or $K[[X]]$, these Proceedings.

9. D.F. Anderson and C. Scherpenisse, Factorization in $K[S]$, to appear in *Lecture Notes in Pure and Applied Mathematics*, Marcel Dekker, New York, in press 1996.

10. S. Chapman and W.W. Smith, An analysis using the Zaks-Skula constant of element factorizations in Dedekind domains, *J. Algebra:* *159*, 176-190 (1993).

11. R. Fröberg, G. Gottlieb, and R. Häggkvist, On numerical semigroups, *Semigroup Forum:* *35*, 63-83 (1987).

12. R. Gilmer, Commutative Semigroup Rings, Chicago Lectures in Mathematics, Univ. of Chicago Press, Chicago, 1984.

13. J.L. Steffan, Longuers des décompositions en produits d'éléments irréductibles dans un anneau de Dedekind, *J. Algebra:* *102*, 229-236 (1986).

14. R.J. Valenza, Elasticity of factorizations in number fields, *J. Number Theory:* *36*, 212-218 (1990).

15. A. Zaks, Half-factorial domains, *Israel J. Math.:* *37*, 281-302 (1980).

Invariant Theory in Characteristic p: Hazlett's Symbolic Method for Binary Quantics

Joseph P. Brennan
Department of Mathematics, North Dakota State University, Fargo,
ND 58105-5075

Abstract

An extension of the symbolic method of invariants of binary
n-tics to fields of arbitrary characteristic due to O. C. Hazlett
is placed in a geometric context.

1   Introduction

The action of the projective special linear group on the projective line
has been one of the central themes in geometry since the late eighteenth
century. The algebraic analogue is the action of the special linear group
on the homogeneous forms of degree n in two variables: the theory of
binary n-tics.

This paper gives contemporary proofs of results of O. C. Hazlett on
the theory of binary n-tics over fields of positive characteristic. The
presence of a field of positive characteristic brings a marked change to
the character of the problem. To this end we examine the Aronhold

symbolic method of classical invariant theory from a geometric point of view and establish Hazlett's method from that perspective.

We concentrate on the importance (for an understanding of a symbolic method) of the invariants of systems of N-linear forms (the invariants of $\otimes^N \mathcal{O}_{\mathbf{P}^1}$). It is the ability to explicitly describe the invariants in this situation on which the symbolic method rests. The theory is illustrated with some examples for the expression of invariants with special emphasis as to the importance in distinguishing between the invariants of the action of the $\mathbf{F}_p$-rational points of SL_2 (formal modular invariants in the old terminology) and the invariants of the group scheme SL_2 (absolute (modular) invariants in the old terminology).

I would like to thank many people for encouraging me to publish these results, in particular, A. Fauntleroy, R. M. Fossum, and to thank the organizers of this meeting for providing a forum for these results; which are on a very different topic than the talk I gave at the conference.

2   Symbolic method

As our primary interest is in the classical example of invariants of binary n-tics (that is the representation of the projective special linear group provided by n points in the projective plane), the discussion here is restricted to this case. For other examples consult the references [7, 14, 15]. For further information concerning the viewpoint of this exposition consult [3, 9, 10].

Let PSL_2 be the group scheme of automorphisms of the projective line $\mathbf{P}^1$. This induces an action of the double cover $G = \mathrm{SL}_2$ of PSL_2 on the structure sheaf $\mathcal{O}_{\mathbf{P}^1}$. This action induces an action of G on $\mathcal{O}_{\mathbf{P}^n}(m)$ by the identification of $\mathcal{O}_{\mathbf{P}^n}(1)$ with $\mathcal{O}_{\mathbf{P}^1}(n)$ on which G acts.

This action canonically induces a comultiplication (Taylor series) on the homogeneous coordinate ring $\oplus^m H^0(\mathcal{O}_{\mathbf{P}^n}(m))$

$$\Delta : \oplus^m H^0(\mathcal{O}_{\mathbf{P}^n}(m)) \longrightarrow \oplus^m H^0(\mathcal{O}_{\mathbf{P}^n}) \otimes \oplus^t H^0(\mathcal{O}_G(t)).$$

Then we say that the action of G on $\mathbf{P}^n$ is *linear* as the map Δ is homogeneous of degree zero in the natural grading of the homogeneous coordinate ring. If we denote by:

$$I : \oplus^m H^0(\mathcal{O}_{\mathbb{P}^n}(m)) \longrightarrow \oplus^m H^0(\mathcal{O}_{\mathbb{P}^n}(m)) \otimes \oplus^t H^0(\mathcal{O}_G(t))$$

the comultiplication associated to the trivial action of G, the equalizer of the maps I and Δ is called the *ring of invariants of the binary n-tic.*

Example 1 *The binary quadratic over a field of characteristic zero.*

The coefficients a_0, a_1, a_2 of the binary quadratic

$$f = a_0 x^2 + 2a_1 xy + a_2 y^2$$

are identified with sections of $\mathcal{O}_{\mathbb{P}^2}(1)$. The special linear group acts on $\mathcal{O}_{\mathbb{P}_2}(1)$ contravariant to the natural action of SL_2 on the the coordinates x, y. Thus if we identify coordinate functions $\xi_1, \xi_2, \eta_1, \eta_2$ for SL_2 :

$$g = \begin{bmatrix} \xi_1 & \eta_1 \\ \xi_2 & \eta_2 \end{bmatrix} \qquad \xi_1 \eta_2 - \xi_2 \eta_1 = 1,$$

we obtain the comultiplication map from:

$$\Delta : a_0 \mapsto a_0 \otimes \xi_1^2 + 2a_1 \otimes \xi_1 \xi_2 + a_2 \otimes \xi_2^2$$

$$\Delta : a_1 \mapsto a_0 \otimes \xi \eta_1 + a_1 \otimes (\xi_1 \eta_2 + \xi_2 \eta_1) + a_2 \otimes \xi_2 \eta_2$$

$$\Delta : a_2 \mapsto a_0 \otimes \eta_1^2 + 2a_1 \otimes \eta_1 \eta_2 + a_2 \eta_2^2$$

A *symbolic method* for $\mathbb{P}^n$ is a G-equivariant morphism $\pi : Y \longrightarrow \mathbb{P}^n$ such that the action of G on Y is linear and $\mathcal{O}_Y(1)$ is a product of twists of G-stable line bundles.

Example 2 *Aronhold's symbolic method (characteristic zero)[1, 4, 6]*

We assume that the characteristic of the base field k is zero. With the identifications of the actions as above the Veronese embedding (with coefficients):

$$\rho_n : \mathbb{P}^1 \longrightarrow \mathbb{P}^n$$

given by:

$$\rho\colon (x\colon y) \mapsto (x^n\colon nx^{n-1}y\colon \cdots \colon \binom{n}{k}x^{n-k}y^k\colon \cdots \colon nxy^{n-1}\colon y^n)$$

is a G-equivariant morphism of schemes associated to the map:

$$\rho_n^*\colon \mathcal{O}_{\mathbf{P}^n}(1) \longrightarrow \mathcal{O}_{\mathbf{P}^1}(n)$$

which is the identification indicated above. The ring of invariants of the binary n-tic in degree m is therefore isomorphic to the invariants of the action of G in $\otimes^m \mathcal{O}_{\mathbf{P}^1}(n)$.

3   Invariants of several binary linear forms

The key in utilizing the symbolic method is that the computation of the invariants of systems of linear forms (the invariants of the action of G on $\otimes^N \mathcal{O}_{\mathbf{P}^1}(n)$ is relatively easy, once some terminology is defined.

Let α_0 and α_1 be sections which generate $\mathcal{O}_{\mathbf{P}^1}$, on which we have the natural G-action. So that if:

$$g = \begin{bmatrix} \xi_1 & \eta_1 \\ \xi_2 & \eta_2 \end{bmatrix} \qquad \xi_1\eta_2 - \xi_2\eta_1 = 1$$

then

$$\Delta\colon \alpha_0 \mapsto \alpha_0 \otimes 1 + \alpha_0 \otimes \xi_1 + \alpha_2 \otimes \xi_2 - \alpha_0 \otimes \xi_2\eta_1 + \alpha_0 \otimes \xi_1\eta_2$$

and

$$\Delta\colon \alpha_1 \mapsto \alpha_1 \otimes 1 + \alpha_0 \otimes \eta_1 + \alpha_1 \otimes \eta_2\alpha_1 \otimes \xi_1\eta_2 + \alpha_1\xi_2\eta_1$$

We denote the generating sections that transform in the same manner in each component of $\otimes^N \mathcal{O}_{\mathbf{P}^1}$ by a different letter (e.g. β_0 and β_1, γ_0 and γ_1).

We define the *symbol* $(\alpha\beta)$ in $\mathcal{O}_{\mathbf{P}^1}(1) \otimes \mathcal{O}_{\mathbf{P}^1}(1)$ by:

$$(\alpha\beta) = \alpha_0 \otimes \beta_1 - \alpha_1 \otimes \beta_0,$$

and similarly the symbol for any pair of letters.

Theorem 1 *The invariants of the action of G on $\otimes^N \mathcal{O}_{\mathbf{P}^1}$ are generated by the symbols.*

Proof: The basis of the argument that follows is classical in the case of characteristic zero [6, Art. 37]. A enumerative proof in a non-symbolic notation can be found in [5, Art 256]. Another similar proof using differential operators can be found in [14, Abs. I §9]. A proof utilizing a combinatorial description of the symbolic notation in characteristic zero is in [13].

The argument is by induction on the number N. In the case of $N = 1$, we take two generating sections $\{\alpha_0, \alpha_1\}$ of $\mathcal{O}_{\mathbf{P}^1}$ which transform as above. Consider an invariant homogeneous form in the α's of degree r:

$$H = C_0 \alpha_0^r + \cdots C_k \alpha_0^{r-k} \alpha_1^k + \cdots C_r \alpha_1^r.$$

Consider the one-parameter subgroup scheme of G associated to the locus $\xi_1 = \eta_2 = 1$ and $\eta_1 = 0$. Then the restriction of the comultiplication Δ to the action of the subgroup scheme, evaluated on H is H, as H is an invariant. On the other hand a direct evaluation of the restriction of Δ at H is given by:

$$C_0(\alpha_0 \otimes 1 + \alpha_1 \otimes \xi_2)^r + \cdots + C_k(\alpha_0 \otimes 1 + \alpha_1 \otimes \xi_2)^{r-k}(\alpha_1^k \otimes 1) + \cdots + C_r \alpha_1^r \otimes 1$$

This leads to a system of equations:

$$
\begin{aligned}
0 &= \binom{r}{1} C_0 \otimes \xi_2 \\
0 &= \binom{r}{2} C_0 \otimes \xi_2^2 + \binom{r-1}{1} C_1 \otimes \xi_2 \\
&\vdots \\
0 &= \binom{r}{r} C_0 \otimes \xi_2^r + \binom{r-k}{r-k} C_k \otimes \xi_2^{r-k} + \cdots + \binom{1}{1} C_{r-1} \otimes \xi_2
\end{aligned}
$$

The last equation is:

$$\sum_{k=0}^{r-1} C_k \otimes \xi_2^{r-k} = 0.$$

For infinite fields this yields: $C_0 = C_1 = \cdots = C_{r-1} = 0$, and the same argument with the one parameter subgroup $\xi_1 = \eta_2 = 1$ and $\xi_2 = 0$ yields $C_r = 0$ and hence that there are no non-constant invariants.

In the case that $N = 2$, the argument is by induction on the degree of the invariant. If the degree of the invariant is zero, the function is a constant and there is nothing to prove. If the degree of the invariant is positive: we take sections $\{\alpha_0, \alpha_1\}$ and $\{\beta_0, \beta_1\}$. Then the polynomials in the symbol $(\alpha\beta)$ are invariant. Suppose that $H(\alpha_0, \alpha_1, \beta_0, \beta_1)$ is an invariant. Then H must be an invariant of a single linear form modulo the relations $\alpha_0 = \beta_0, \alpha_1 = \beta_1$, hence H is an invariant H' of strictly smaller degree times the symbol $(\alpha\beta)$. By the induction hypothesis H' and therefore H is a polynomial in the symbol $(\alpha\beta)$

We next assume that the result holds for values smaller than $N > 2$. Again we argue by induction on the degree of the invariant. If

$$H(\alpha_0, \alpha_1, \beta_0, \beta_1, \cdots, \omega_0, \omega_1)$$

is an invariant of N-linear forms then the specialization

$$H(\beta_0, \beta_1, \beta_0, \beta_1, \cdots, \omega_0, \omega_1)$$

is an invariant of $N - 1$-linear forms, hence is a polynomial in symbols involving the letters $\beta \cdots \omega$, but then H must be a polynomial in the symbols in $\alpha \cdots \omega$ plus $(\alpha\beta)$ times a function H'. The polynomial in the symbols is an invariant; thus H' is an invariant of strictly smaller degree. Hence by the induction hypothesis H' and therefore H is a polynomial in the symbols. $\square$

As a corollary we obtain Aronhold's famous result [1] on the representation of invariants by "symbols" over a field of characteristic zero.

Corollary 2 *Let k be a field of characteristic zero, then the map*

$$\rho_n^*: \mathcal{O}_{\mathbf{P}^n}(1) \longrightarrow \mathcal{O}_{\mathbf{P}^1}(n)$$

induces a map:

$$\bigotimes \rho_n^*: \mathcal{O}_{\mathbf{P}^n}(m) \longrightarrow \overset{m}{\bigotimes} \mathcal{O}_{\mathbf{P}^1}(n)$$

which when restricted to the invariants of the binary n-tic associates a polynomial in the symbols.

Example 3 *The case of the binary quadratic in characteristic zero.*

Again we identify $\mathcal{O}_{\mathbf{P}^2}(1)$ with the coefficients of the binary quadratic

$$f = a_0 x^2 + 2a_1 xy + a_2 y^2.$$

The ring of invariants is given by the ring of polynomials in the discriminant $a_0 a_2 - a_1^2$. It is associated with the symbol $(\alpha\beta)^2$ in $\mathcal{O}_{\mathbf{P}^1}(2) \otimes \mathcal{O}_{\mathbf{P}^1}(2)$.

The above result does not hold if we look at the group scheme of $\mathbb{F}_q$-rational points of G for a finite field. A slight modification will however suffice to extend the result to the case of a finite field. To this end we define the r-th *Frobenius twist* of a letter α for the field $\mathbb{F}_q$ ($q = p^n$) to be the pair of sections $\{\alpha_0^{p^r} \alpha_1^{p^r}\}$. The symbol $(\alpha^{p^r}\beta) \in \mathcal{O}_{\mathbf{P}^1}(p^r) \otimes \mathcal{O}_{\mathbf{P}^1}(1)$ is the element:

$$(\alpha^{p^r}\beta) = \alpha_0^{p^r} \otimes \beta_1 - \alpha_1^{p^r} \otimes \beta_0.$$

Note that:

$$(\alpha^{p^r}\alpha) = \alpha_0^{p^r}\alpha_1 - \alpha_1^{p^r}\alpha_0 \in \mathcal{O}_{\mathbf{P}^1}(p^r + 1).$$

Theorem 3 *The invariants of the action of the $\mathbb{F}_q$-rational points of G on $\otimes^N \mathcal{O}_{\mathbf{P}^1}$ is given by: rational functions of the symbols in the letters and the Frobenius twists of the letters.*

Proof: It is necessary to examine the case $N = 1$. Let $\{\alpha_0, \alpha_1\}$ be sections that transform as above. Consider a non-zero homogeneous invariant of degree r:

$$H(\alpha_0, \alpha_1) = C_0\alpha_0^r + \cdots + C_k\alpha_0^{r-k}\alpha_1^k + \cdots + C_r\alpha_1^r.$$

There is an associated polynomial:

$$H(1, t) = C_0 + \cdots + C_k t^k + \cdots + C_r t^r \in \mathbb{F}_q[t].$$

The polynomial $H(1, t)$ factors into linear terms in an extension field of $\mathbb{F}_q$. So $H(1, t) = C_r \prod_u (t - u)$. As $H(\alpha_0, \alpha_1)$ is an invariant, the set of roots of $H(1, t)$ is a $\mathbb{F}_q$-affine space. Therefore, by the standard inclusion-exclusion argument, $H(1, t)$ is a sum of products of terms of the form:

$$\frac{t^{p^s} - t}{t^{p^k} - t}.$$

The homogenization of this decomposition gives the required expression of $H(\alpha_0, \alpha_1)$ in rational functions of symbols.

The specialization arguments from the proof of the infinite field case will also work here with the modification that for the invariant:

$$H(\alpha_0, \alpha_1, \beta_0, \beta_1, \cdots, \omega_0, \omega)$$

it is necessary to use the specialization:

$$H(\beta_0^{p^k}, \beta_1^{p^k}, \beta_0, \beta_1 \cdots \omega_0, \omega_1)$$

for sufficiently large k. $\square$

4 Hazlett symbols

For positive characteristic, the Aronhold symbolic method is no longer available. Nevertheless, the utility of the language motivated Hazlett [8] to attempt to express the invariants in terms of symbols.

Classically it was known [5, Art. 74] that invariants of binary forms have expression in terms of the roots of the associated polynomials–the locus of points in the projective plane defined by the form.

Let us identify the coefficients of the binary n-tic:

$$a_0 x^n + a_1 x^{n-1} y + \cdots + a_k x^{n-k} y^k + \cdots + a_{n-1} x y^{n-1} + a_n y^n$$

Consider then the morphism:

$$\epsilon \colon \mathbb{P}^1 \times \cdots \times \mathbb{P}^1 \longrightarrow \mathbb{P}^n$$

induced by:

$$\epsilon \colon [(\alpha_0 \colon \alpha_1), \cdots, (\omega_0 \colon \omega_1)] \mapsto (z \colon z e_1 \colon z e_2 \colon \cdots \colon z e_n)$$

where $z = \alpha_0 \cdots \omega_0$ and the e_i are the elementary symmetric functions of the ratios $\frac{\alpha_1}{\alpha_0}, \frac{\beta_1}{\beta_0}, \cdots \frac{\omega_1}{\omega_0}$.

Then for all characteristics, ϵ is a G-equivariant morphism. To it is associated the map:

$$\epsilon^* \colon \mathcal{O}_{\mathbb{P}^n}(1) \longrightarrow \overset{n}{\bigotimes} \mathcal{O}_{\mathbb{P}^1}(1).$$

The observations of the previous section when applied to the morphism ϵ give us the following theorems of Hazlett [8]:

Theorem 4 *Let k be an infinite field, then the map:*

$$\epsilon_m^*: \mathcal{O}_{\mathbf{P}^n}(m) \longrightarrow \overset{n}{\bigotimes} \mathcal{O}_{\mathbf{P}^1}(m)$$

when restricted to the invariants of the binary n-tic associates a polynomial in the symbols.

The case of a finite field is more complicated not only because of the complication in the description of the invariants of systems of linear forms.

Theorem 5 *Let k be a finite field, then the map*

$$\epsilon_m^*: \mathcal{O}_{\mathbf{P}^n}(m) \longrightarrow \overset{n}{\bigotimes} \mathcal{O}_{\mathbf{P}^1}(m)$$

when restricted to the invariants of the k-rational points associates to a power of an invariant a rational function in the symbols in the letters and their Frobenius twists.

Proof: Consider a closed point of $\mathbf{P}_k^n$. There is a finite field extension K/k such that the map ϵ^* is an isomorphism. Hence for every invariant of the K-rational points of G there is a symbolic representation. However, a power of a invariant of the k-rational points is an invariant of the K-rational points, so a power of the invariant will have a symbolic representation. □

The distinction in the infinite case is the fact that the invariants over the algebraic closure are the same as over any infinite subfield.

5 Examples

The following examples are presented as an illustration of the theory. We take the liberty of writing the association of an invariant with a symbol as equality. The examples of the the quadratic over $\mathbb{F}_3$ and the cubic over $\mathbb{F}_2$ are taken from [8].

Example 4 *The binary quadratic over* $\mathbb{F}_2$ *[2, Lecture III,§6]*

Dickson shows that over the field $\mathbb{F}_2$, the invariants of the action of the group of $\mathbb{F}_2$-rational points of G acting on the binary quadratic form:

$$f = a_0 x^2 + a_1 xy + a_2 y^2$$

are generated by a_1, $a_0 a_1 (a_0 + a_1 + a_2)$ and $a_0^2 + a_0 a_1 + a_0 a_2 + a_1 a_2 + a_2^2$.

Examining these invariants one sees that:

$$a_1 = (\alpha\beta) \quad a_0 a_2 (a_0 + a_1 + a_2) = (\alpha^2 \alpha)(\beta^2 \beta)$$

and that

$$a_0^2 + a_0 a_1 + a_0 a_2 + a_1 a_2 + a_2^2$$
$$= [(\alpha^4 \beta^4) + (\alpha^2 \beta)(\alpha\beta)(\beta^2 \alpha) + (\alpha^2 \alpha)(\beta^4 \alpha)(\beta^2 \beta)(\alpha^4 \beta)]/(\alpha\beta)^2.$$

The ring of G-invariants is generated by the section a_1.

Example 5 *The binary quadratic over* $\mathbb{F}_3$ *[2, Lecture III,§8]*

Dickson shows that over the field $\mathbb{F}_3$, the invariants of the action of the group of $\mathbb{F}_3$-rational points of G on the binary quadratic form:

$$f = a_0 x^2 + 2a_1 xy + a_2 y^2$$

are generated by:

$$D = a_0 a_2 - a_1^2, \quad J = a_0 a_2 (a_0 + a_1 + a_2)(a_0 + 2a_1 + a_2),$$
$$B = a_1 (a_1^2 - a_0^2)(a_2 - a_0)(a_2^2 - a_1^2),$$
$$\Gamma = (a_0 + a_2)(2a_0 + 2a_1 + a_2)(2a_0 + a_1 + a_2).$$

As Hazlett[8, §6] shows:

$$D = (\alpha\beta)^2, \quad J = (\alpha^3 \alpha),$$
$$B = (\alpha^3 \beta)^2 (\beta^3 \beta) + (\beta^3 \alpha)^2 (\alpha^3 \alpha)$$

and:

$$\Gamma = [(\beta^3 \alpha)^2 (\alpha^3 \alpha) - (\alpha^3 \beta)^2 (\beta^3 \beta)]/(\alpha\beta)^3.$$

The ring of G-invariants is generated by the section D

Example 6 *The binary cubic over $\mathbb{F}_2$ [8]*

Dickson shows that over the field $\mathbb{F}_2$, the invariants of the action of the group of $\mathbb{F}_2$-rational points of G on the binary cubic form:

$$f = a_0 x^3 + a_1 x^2 y + a_2 x y^2 + a_3 y^3$$

are generated by:

$$D = a_0 a_3 + a_1 a_2, \quad K = a_1 + a_2, \quad k = a_0 a_3 (a_0 + a_1 + a_2 + a_3)$$

$$J = a_0^2 + a_1^2 + a_2^2 + a_3^2 + a_0 a_1 + a_0 a_3 + a_2 a_3$$

and

$$g = a_0^2 a_1^2 + a_0^2 a_1 a_3 + a_0 a_1^2 a_2 + a_0 a_1^2 a_3 + a_0 a_1^3 + a_0 a_1 a_2 a_3 + a_0 a_2^2 a_3 + a_0 a_2 a_3^2$$

$$+ a_1^3 a_2 + a_1^2 a_2^2 + a_1 a_2^3 + a_1 a_2^2 a_3 + a_2^3 a_3 + a_2^2 a_3^2$$

As Hazlett [8, §24] shows:

$$D = (\alpha\beta)(\beta\gamma)(\gamma\alpha) \quad k = (\alpha^2\alpha)(\beta^2\beta)(\gamma^2\gamma)$$

$$J = \frac{(\alpha^4\alpha)(\beta^4\beta)(\gamma^4\gamma)}{(\alpha^2\alpha)(\beta^2\beta)(\gamma^2\gamma)}$$

and

$$K^2 - D = (\alpha^2\gamma)(\beta^2\gamma) + (\alpha^2\beta)(\gamma^2\beta) + (\beta^2\alpha)(\gamma^2\alpha)$$

The ring of G-invariants is generated by the section D.

Neither K nor g is realizable directly as a rational function in the symbols but rather K^2 and g^2 are. The representation of the invariant g^2 is rather involved and is omitted for considerations of space. One might note however that for the computation of a symbolic representation of g^2 one should start with the invariant $(g - kK - JK^2 - K^4)^2$ which has the advantage of having smaller degree in the sections with subscript one.

6 Remarks

The reader will note the fact that Hazlett's method works in all characteristics and had expression classically [5, Art. 74]. The exposition in [9] was the inspiration for framing Hazlett's symbols in this manner.

There is also a corresponding theory [8] for providing a symbolic expression for covariants of binary forms. Computation of examples is considerably more tedious.

Recent work examining the symbolic method [11, 12, 13] has directed attention to a combinatorial understanding of the symbol. The linkage of this combinatorial exploration to the geometric expression explored here will be explored in another work.

References

[1] S. ARONHOLD, *Ueber eine fundamentale Begründung der Invariantentheorie*, J. Reine Angew. Math., 63 (1862), pp. 281–345.

[2] L. E. DICKSON, *On invariants and the theory of numbers*, in The Madison Colloquium, 1913, no. 4 in Colloquium Publications, American Mathematical Society, 1914, pp. 1–110.

[3] J. A. DIEUDONNE, *Introduction to the theory of formal groups*, Marcel Dekker Inc., New York, 1973.

[4] J. A. DIEUDONNE AND J. B. CARRELL, *Invariant theory, old and new*, Adv. in Math., 4 (1970), pp. 1–85.

[5] E. B. ELLIOTT, *An introduction to the algebra of quantics*, Oxford University Press, 1913.

[6] J. H. GRACE AND A. YOUNG, *The algebra of invariants*, Cambridge University Press, 1903.

[7] G. B. GUREVICH, *Foundations of the theory of algebraic invariants*, P. Noordhoff Ltd., Groningen, 1964.

[8] O. C. HAZLETT, *A symbolic theory of formal modular invariants*, Trans. Amer. Math. Soc., 24 (1922), pp. 286–311.

[9] D. MUMFORD AND K. SUOMINEN, *Introduction to the theory of moduli*, in Algebraic Geometry, Oslo 1970, F. Oort, ed., Groningen, 1972, Nordic Summer School in Mathematics (5*th*: Oslo), Wolters-Noordhoff, pp. 171–222.

[10] P. E. NEWSTEAD, *Lectures on introduction to moduli problems and orbit spaces*, no. 51 in Lectures on Mathematics and Physics, Tata Institute of Fundamental Research, Bombay, 1978.

[11] G.-C. ROTA AND J. P. S. KUNG, *Invariant theory of binary forms*, Bull. Amer. Math. Soc. (N.S.), 10 (1984), pp. 27–85.

[12] G.-C. ROTA AND J. A. STEIN, *Symbolic method in invariant theory*, Proc. Nat. Acad. Sci. U.S.A., 83 (1986), pp. 844–847.

[13] G.-C. ROTA AND B. STURMFELS, *Introduction to invariant theory in superalgebras*, in Invariant Theory and Tableaux, D. Stanton, ed., vol. 19 of The IMA Volumes in Mathematics and its Applications, Institute for Mathematics and its Applications, Springer-Verlag, 1990, pp. 1–35.

[14] R. WEITZENBÖCK, *Invariantentheorie*, P. Noordhoff, Groningen, 1923.

[15] E. J. WILCZYNSKI, *Projective differential geometry of curves and ruled surfaces*, B. G. Teubner, Leipzig, 1906.

A Basis for the Ring of Polynomials Integer-Valued on Prime Numbers

Jean-Luc Chabert Université de Picardie, Saint Quentin, France

Scott T. Chapman Trinity University, San Antonio, Texas

William W. Smith The University of North Carolina at Chapel Hill, Chapel Hill, North Carolina

1 INTRODUCTION

Let Z represent the integers, N the nonnegative integers, N^+ the positive integers, Q the rationals, and P the set of prime integers in Z. Since P is a recursively enumerable set, there exist polynomials f in $Z[X_1, \ldots, X_k]$ such that $f(N^k) \cap N^+$ is exactly P (cf. [5]). Here we are interested in the converse problem: to determine polynomials f such that $f(P^k) \subset Z$. Of course, the difficulty is not to find such polynomials (each polynomial in $Z[X_1, \ldots, X_k]$ has this property) but to find all of them. In fact, we may restrict ourselves to the case $k = 1$ by using the following recursive process: if $\{C_n(X)\}_{n=0}^{\infty}$ is a basis of the Z-module $\mathrm{Int}(P, Z)$ and $\{J_m(X_1, \ldots, X_k)\}_{m=0}^{\infty}$ is a basis of the Z-module $\mathrm{Int}(P^k, Z)$ (for some $k \geq 1$), then $\{C_n(X)J_m(X_1, \ldots, X_k)\}_{n \geq 0, m \geq 0}$ is a basis of $\mathrm{Int}(P^{k+1}, Z)$.

Hence, if D is an integral domain with quotient field K and $S \subseteq D$, set

$$\mathrm{Int}(S, D) = \{f(X) \in K[X] | f(s) \in D \text{ for all } s \in S\}.$$

We shall refer to $\mathrm{Int}(S, D)$ as the *ring of polynomials integer-valued on S*

over D. When $S = D$ we shall use the notation $\text{Int}(D)$ (this is the well known ring of integer-valued polynomials on D). In [4], Gilmer has shown for $D = \mathbf{Z}$ that $\text{Int}(S, \mathbf{Z}) = \text{Int}(\mathbf{Z})$ if and only if the set S is prime power complete (i.e., S contains a complete set of residues modulo p^k for each prime $p \in \mathbf{P}$ and $k \in \mathbf{N}$). This result was later extended to the general Dedekind case by McQuillan [7]. Thus, it is clear that the ring $\text{Int}(\mathbf{P}, \mathbf{Z})$ properly contains the ring $\text{Int}(\mathbf{Z})$. In this paper, we will show that $\text{Int}(\mathbf{P}, \mathbf{Z})$ is a free $\mathbf{Z}$-module and will describe an algorithm which constructs for it a free basis. In Section 4 we present this algorithm in a theoretical form (The First Algorithmic Construction 4.2), and then in an alternate form (The Second Algorithmic Construction 4.5) which allows for simplier calculations. Here, for example, are the first seven elements of one such basis:

$$1, \quad X - 1, \quad \frac{(X-1)(X-2)}{2}, \quad \frac{(X-1)(X-2)(X-3)}{2^3 3}, \quad \frac{(X-1)(X-2)(X-3)(X-5)}{2^4 3},$$

$$\frac{(X-1)(X-2)(X-3)(X-5)(X+41)}{2^7 3^2 5}, \quad \frac{(X-1)(X-2)(X-3)(X-5)(X-7)(X+41)}{2^8 3^2 5}.$$

The techniques we use come from commutative algebra and have been useful in the study of the ring of integer-valued polynomials on an arbitrary domain (see [2]). Many of our proofs result from adaptations of previous ones. In reducing our work to a number theoretic setting, we will find Dirichlet's Theorem on prime numbers in an arithmetic progression to be valuable: if r and s are two relatively prime integers, then the arithmetic progression $\{kr + s\}_{k \geq 0}$ contains infinitely many prime numbers [6]. This is the second in a series of papers concerning the ring $\text{Int}(\mathbf{P}, \mathbf{Z})$. In [3], Chabert characterizes polynomials $f(X) \in \text{Int}(\mathbf{P}, \mathbf{Z})$ of degree n by their values on the integers from 1 to 2n-1.

2 PRELIMINARY RESULTS

The following proposition remains valid if $\mathbf{P}$ is replaced by any infinite subset $S \subseteq \mathbf{Z}$.

Proposition 2.1 *Int*$(\mathbf{P}, \mathbf{Z})$ *is a free* $\mathbf{Z}$*-module with a regular basis. That is to say, there exists a basis* $\{f_n(X)\}_{n \in \mathbf{N}}$ *of Int*$(\mathbf{P}, \mathbf{Z})$ *such that* $f_n(X)$ *has degree* n.

Proof: Let $\mathcal{I}_n$ be the set of the leading coefficients of polynomials in $\text{Int}(\mathbf{P}, \mathbf{Z})$ with degree less than or equal to n. It is clear that $\{\mathcal{I}_n\}_{n \in \mathbf{N}}$ is an increasing sequence of $\mathbf{Z}$-submodules of $\mathbf{Q}$. Moreover, each $\mathcal{I}_n$ is a fractional ideal of $\mathbf{Z}$. To see this, let p_i be the ith prime number and let d be the product $\prod_{1 \leq i < j \leq n+1}(p_j - p_i)$. Then, $d\mathcal{I}_n \subset \mathbf{Z}$. Indeed, let $f(X)$ be in $\text{Int}(\mathbf{P}, \mathbf{Z})$ with $\deg(f(X)) = n$. Since $f(p_i)$ belongs to $\mathbf{Z}$ for each

$i = 1, \ldots, n+1$, a Cramer's system with a Vandermonde determinant shows that the coefficients of $f(X)$ belong to $\frac{1}{d}\mathbf{Z}$.

For each n, the fractional ideal $\mathcal{I}_n$ is principal. Hence, for each $n \in \mathbf{N}$, set $\mathcal{I}_n = \frac{1}{d_n}\mathbf{Z}$. Let $\mathcal{B} = \{f_n(X)\}_{n=0}^{\infty}$ be a set of polynomials in $\mathrm{Int}(\mathbf{P}, \mathbf{Z})$ such that each $f_n(X)$ is of degree n with leading coefficient $\frac{1}{d_n}$. We argue that $\mathcal{B}$ is the basis we seek. Obviously $\mathcal{I}_0 = \mathcal{I}_1 = \mathbf{Z}$, and hence any $f(X) \in \mathrm{Int}(\mathbf{P}, \mathbf{Z})$ of degree 1 can be written in the form $f(X) = a_0 f_0(X) + a_1 f_1(X)$, where a_0 and a_1 are integers. If $f(X) \in \mathrm{Int}(\mathbf{P}, \mathbf{Z})$ has degree k and leading coefficient $\frac{a}{d_k}$ for $a \in \mathbf{Z}$, then $h(X) = f(X) - a f_k(X)$ has degree strictly less than $f(X)$. The result now easily follows by induction on k.$\square$

To further illustrate the structure of the ideals $\mathcal{I}_n$, we prove that $\mathcal{I}_2 = \frac{1}{2}\mathbf{Z}$. Let $f(X) = a(X-3)(X-2) + b(X-2) + c$ be in $\mathrm{Int}(\mathbf{P}, \mathbf{Z})$. Then $c = f(2)$ and $b = f(3) - c$ are integers. Also, $g(X) = f(X) - b(X-2) - c = a(X-3)(X-2)$ is in $\mathrm{Int}(\mathbf{P}, \mathbf{Z})$. Since $g(5) = 6a$ and $g(7) = 20a$ are integers, then $2a$ is an integer. Conversely, if $2a$ is an integer, then $f(X)$ is in $\mathrm{Int}(\mathbf{P}, \mathbf{Z})$ since any prime number which is not 2 is odd.

Recall that the sequence of binomial polynomials $\left\{ \begin{pmatrix} X \\ n \end{pmatrix} \right\}_{n=0}^{\infty}$, defined by $\begin{pmatrix} X \\ 0 \end{pmatrix} = 1$ and $\begin{pmatrix} X \\ n \end{pmatrix} = \frac{X(X-1)\cdots(X-n+1)}{n!}$ for $n \geq 1$, is a regular basis of $\mathrm{Int}(\mathbf{Z})$. Since $\mathrm{Int}(\mathbf{Z}) \subset \mathrm{Int}(\mathbf{P}, \mathbf{Z})$, one has $\frac{1}{n!}\mathbf{Z} \subset \mathcal{I}_n$ and the inclusion is an equality for $n = 0, 1, 2$. For integers $n > 2$, Dirichlet's theorem says that there are many prime numbers, sufficiently many to get the following proposition.

Proposition 2.2 *Let n be a nonnegative integer. If $a \begin{pmatrix} X \\ n \end{pmatrix}$ belongs to $\mathrm{Int}(\mathbf{P}, \mathbf{Z})$ then a belongs to $\mathbf{Z}$.*

Proof: Let l be any prime number and let $v = 1 + \sup_{1 \leq m \leq n} v_l(m)$. In the arithmetic sequence $\{kl^v - 1\}_{k \geq 0}$ there is at least one prime $p = k_0 l^v - 1$. We then have

$$\begin{pmatrix} p \\ n \end{pmatrix} = \frac{(k_0 l^v - 1)(k_0 l^v - 2) \ldots (k_0 l^v - n)}{1 \cdot 2 \cdots n}$$

and hence

$$v_l\left(\begin{pmatrix} p \\ n \end{pmatrix}\right) = \sum_{1 \leq m \leq n} (v_l(kl^v - m) - v_l(m)) = 0.$$

Thus $v_l(a) \geq 0$ for each l (i.e., a belongs to $\mathbf{Z}$).$\square$

Despite the previous proposition, it is easy to see using Gilmer's result [4] that $\mathrm{Int}(\mathbf{Z})$ is strictly contained in $\mathrm{Int}(\mathbf{P}, \mathbf{Z})$. For instance, $f(X) =$

$\frac{(X-1)(X-2)(X-3)}{24}$ belongs to Int($\mathbf{P}, \mathbf{Z}$) (the sequence $(X-1)(X-2)(X-3)$ is divisible by at least 2^3 and 3 for any odd prime) but $f(0) = -1/4$. Here is an extension of Proposition 2.2.

Proposition 2.3 *Let n be a positive integer, $m_1, \ldots, m_n$ be any integers, and a be a rational number such that $a(X - m_1) \cdots (X - m_n)$ belongs to Int($\mathbf{P}, \mathbf{Z}$). Then for each prime number $l > n$, $v_l(a) \geq 0$.*

Proof: Let l be a prime number strictly greater than n. If, for each $i = 1, \ldots, n$ we have that $v_l(m_i) = 0$, then $v_l(a) = v_l(a(l - m_1) \cdots (l - m_n)) \geq 0$. Suppose there is an integer m_j such that $v_l(m_j) \geq 1$. Since $m_1, \cdots, m_n$ cannot be a complete set of representatives of $\mathbf{Z}/l\mathbf{Z}$, there exists an integer s such that $v_l(s - m_i) = 0$ for each $i = 1, \ldots, n$. If $v_l(s) > 0$ then $v_l(s - m_j) \geq 1$. Hence, $v_l(s) = 0$. Dirichlet's theorem says that there is a prime number p in the arithmetic sequence $\{kl + s\}_{k \geq 0}$. For such a prime number $p = k_0 l + s$, one has that $v_l(a(p - m_1) \cdots (p - m_n)) = v_l(a(k_0 l + s - m_1) \cdots (k_0 l + s - m_n))$. Thus $v_l(a) \geq 0$.□

Here is now the basic result.

Proposition 2.4 [3, Proposition 2] *Let $f(X)$ be in Int($\mathbf{P}, \mathbf{Z}$), p be a prime in $\mathbf{Z}$, and h be an integer. If p does not divide h, then $f(h)$ belongs to $\mathbf{Z}_{(p)}$.*

Proof: Let d be a non-zero integer such that $df(X)$ belongs to $\mathbf{Z}[X]$. If $v = v_p(d)$, then $d = p^v e$ where p does not divide e. If p does not divide h, then there is at least one prime number in the arithmetic sequence $\{kp^v + h\}_{k \geq 0}$. Let $q = k_0 p^v + h$ be such a prime number. Then $d(f(q) - f(h))$ belongs to $(q - h)\mathbf{Z}$ (i.e., $ep^v(f(q) - f(h))$ belongs to $k_0 p^v \mathbf{Z}$). Since $e \notin p\mathbf{Z}$, $f(q) - f(h)$ belongs to $\mathbf{Z}_{(p)} = \{\frac{a}{b} | a \in \mathbf{Z}, b \in \mathbf{Z} \backslash p\mathbf{Z}\}$. Since $f(q)$ belongs to $\mathbf{Z}$, $f(h)$ belongs to $\mathbf{Z}_{(p)}$.□

Notation 2.5 *For each prime integer p in $\mathbf{Z}$, set*

$$X_p = (\mathbf{Z} \backslash p\mathbf{Z}) \bigcup \{p\}.$$

Hence, X_p is the set of integers consisting of p and all non-multiples of p.

Using our earlier notation,

$$\text{Int}(X_p, \mathbf{Z}_{(p)}) = \{f(X) \in \mathbf{Q}[X] | f(h) \in \mathbf{Z}_{(p)} \text{ for each } h \in X_p\}.$$

As a consequence of Proposition 2.4, we get a characterization of the localization of Int($\mathbf{P}, \mathbf{Z}$) with respect to $\mathbf{Z} \backslash p\mathbf{Z}$, which we denote by Int($\mathbf{P}, \mathbf{Z}$)$_{(p)}$.

Proposition 2.6 [3, Corollaire 3] *For any prime number p, Int($\mathbf{P}, \mathbf{Z}$)$_{(p)}$ = Int($X_p, \mathbf{Z}_{(p)}$).*

Proof: Let p be a prime number. Proposition 2.4 shows that Int($\mathbf{P}, \mathbf{Z}$) is contained in Int($X_p, \mathbf{Z}_{(p)}$). Hence, Int($\mathbf{P}, \mathbf{Z}$)$_{(p)} \subset$ Int($X_p, \mathbf{Z}_{(p)}$)$_{(p)}$ = Int($X_p, \mathbf{Z}_{(p)}$). Conversely, since $\mathbf{P} \subset X_p$, Int($X_p, \mathbf{Z}_{(p)}$) $\subset$ Int($\mathbf{P}, \mathbf{Z}_{(p)}$).□

3   A BASIS OF THE $Z_{(p)}$-MODULE $\text{Int}(P, Z)_{(p)}$

Let p be a fixed prime number. We now construct a basis of the $Z_{(p)}$-module $\text{Int}(P, Z)_{(p)} = \text{Int}(X_p, Z_{(p)})$. First, consider the map: $k \mapsto k + \left[\frac{k-1}{p-1}\right]$. This map is an increasing bijection from Z^+ onto the subset of Z^+ consisting of elements which are not multiples of p.

Notation 3.1 *Let p be a prime integer. Define the sequence $\{u_{p,k}\}_{k=0}^{\infty}$ by $u_{p,0} = p$ and $u_{p,k} = k + \left[\frac{k-1}{p-1}\right]$ for each $k > 0$. For each $n \geq 1$, set*

$$h_{p,n}(X) = \prod_{0 \leq k < n} (X - u_{p,k}).$$

For each prime p and each positive integer n, one has $h_{p,n}(X) \in Z[X]$, $\deg(h_{p,n}(X)) = n$, and $h_{p,n}(u_{p,k}) = 0$ for $k = 0, 1, \ldots, n - 1$.

Lemma 3.2 *For each prime number p and each integer $n > 0$,*

$$v_p(h_{p,n}(u_{p,n})) = \sum_{k \geq 0} \left[\frac{n-1}{p^k(p-1)}\right].$$

Moreover, for each integer m which is not a multiple of p, $v_p(h_{p,n}(m)) \geq v_p(h_{p,n}(u_{p,n}))$.

Proof: Since $v_p(u_{p,n} - u_{p,0}) = 0$, we can disregard this term in our computation of $v_p(h_{p,n}(u_{p,n}))$. The sequence $\{u_{p,k}\}_{1 \leq k \leq p-1}$ is a complete set of residues modulo p of the nonmultiples of p, as is each sequence

$$\{u_{p,k}\}_{1+r(p-1) \leq k \leq (r+1)(p-1)}.$$

Then, for $1 \leq k \leq n - 1$, $v_p(u_{p,n} - u_{p,k}) \geq 1$ exactly $\left[\frac{n-1}{p-1}\right]$ times (notice that $v_p(u_{p,n} - u_{p,k}) = 0$ for $\left[\frac{n-1}{p-1}\right](p-1) < k < n$). In the same manner, the sequence $\{u_{p,k}\}_{1 \leq k \leq p(p-1)}$ is a complete set of residues modulo p^2 of nonmultiples of p and each sequence

$$\{u_{p,k}\}_{1+rp(p-1) \leq k \leq (r+1)p(p-1)}$$

is such a complete set of residues. Then, for $1 \leq k \leq n-1$, $v_p(u_{p,n}-u_{p,k}) \geq 2$ exactly $\left[\frac{n-1}{p(p-1)}\right]$ times. Similarly, for each $1 \leq k \leq n - 1$, we have that $v_p(u_{p,n} - u_{p,k}) \geq m + 1$ exactly $\left[\frac{n-1}{p^m(p-1)}\right]$ times. Hence,

$$v_p(h_{p,n}(u_{p,n})) = \sum_{k \geq 0} \left[\frac{n-1}{p^k(p-1)}\right].$$

Repeating this argument for any integer m with $v_p(m) = 0$ yields that $v_p(h_{p,n}(m)) \geq v_p(h_{p,n}(u_{p,n}))$. $\square$

Notation 3.3 *For each prime p and each positive integer n, define the integer $\omega_p(n)$ by*

$$\omega_p(n) = \sum_{k \geq 0} \left[\frac{n-1}{p^k(p-1)} \right]$$

and the polynomials $C_n^p(X)$ and $\Gamma_n^p(X)$ by

$$C_n^p(X) = \frac{h_{p,n}(X)}{p^{\omega_p(n)}} = \prod_{0 \leq k < n} \frac{(X - u_{p,k})}{p^{\omega_p(n)}}$$

and

$$\Gamma_n^p(X) = \frac{h_{p,n}(X)}{h_{p,n}(u_{p,n})} = \prod_{0 \leq k \leq n-1} \frac{(X - u_{p,k})}{u_{p,n} - u_{p,k}}.$$

For $n = 0$ we have that

$$\omega_p(0) = 0 \text{ and } C_0^p(X) = \Gamma_0^p(X) = 1.$$

Proposition 3.4 *The sequence $\{C_n^p(X)\}_{n \geq 0}$ (respectively $\{\Gamma_n^p(X)\}_{n \geq 0}$) is a basis of the $\mathbf{Z}_{(p)}$-module $Int(\mathbf{P}, \mathbf{Z})_{(p)}$.*

Proof: Lemma 3.2 shows that $C_n^p(X)$ and $\Gamma_n^p(X)$ are in $Int(X_p, \mathbf{Z}_{(p)})$ and in fact in $Int(\mathbf{P}, \mathbf{Z})$. Any polynomial $f(X)$ in $\mathbf{Q}[X]$ with degree d may be written

$$f(X) = \sum_{0 \leq k \leq d} a_k C_k^p(X) = \sum_{0 \leq k \leq d} \alpha_k \Gamma_k^p(X)$$

with coefficients a_k and α_k in $\mathbf{Q}$. Clearly, if the a_k and α_k belong to $\mathbf{Z}_{(p)}$, then $f(X)$ belongs to $Int(\mathbf{P}, \mathbf{Z}_{(p)})$. Conversely, we note that for each $k \geq 1$ and each $m = 0, \ldots, k - 1$, $C_k^p(u_{p,m}) = \Gamma_k^p(u_{p,m}) = 0$ and, for each $k \geq 0$, $\Gamma_k^p(u_{p,k}) = 1$ and $C_k^p(u_{p,k})$ is invertible in $\mathbf{Z}_{(p)}$. An argument by induction on the degree d of $f(X)$ shows that if $f(X_p)$ is contained in $\mathbf{Z}_{(p)}$, then the coefficients a_k and α_k are in $\mathbf{Z}_{(p)}$.$\square$

The preceeding inductive proof leads us to the following corollary.

Corollary 3.5 [3, Lemme 6] *Let $f(X)$ be a polynomial with rational coefficients and degree n. Then $f(X)$ belongs to $Int(\mathbf{P}, \mathbf{Z}_{(p)})$ if and only if $f(u_{p,k})$ belongs to $\mathbf{Z}_{(p)}$ for each $k = 0, \ldots, n$.*

Example 3.6 Here are the first elements of the bases of $\text{Int}(X_p, \mathbf{Z}_{(p)})$ determined by the $C_n^p(X)$ for $0 \le n \le 6$ and $p = 2, 3,$ and 5:

$p = 2$

1

$X - 2$

$\dfrac{(X-2)(X-1)}{2}$

$\dfrac{(X-2)(X-1)(X-3)}{2^3}$

$\dfrac{(X-2)(X-1)(X-3)(X-5)}{2^4}$

$\dfrac{(X-2)(X-1)(X-3)(X-5)(X-7)}{2^7}$

$\dfrac{(X-2)(X-1)(X-3)(X-5)(X-7)(X-9)}{2^8}$

$p = 3$

1

$X - 3$

$\dfrac{(X-3)(X-1)}{3}$

$\dfrac{(X-3)(X-1)(X-2)}{3}$

$\dfrac{(X-3)(X-1)(X-2)(X-4)}{3^2}$

$\dfrac{(X-3)(X-1)(X-2)(X-4)(X-5)}{3^2}$

$\dfrac{(X-3)(X-1)(X-2)(X-4)(X-5)(X-7)}{3^2}$

$p = 5$

1

$X - 5$

$(X - 5)(X - 1)$

$(X - 5)(X - 1)(X - 2)$

$(X - 5)(X - 1)(X - 2)(X - 4)$

$\dfrac{(X-5)(X-1)(X-2)(X-3)(X-4)}{5}$

$\dfrac{(X-5)(X-1)(X-2)(X-3)(X-4)(X-6)}{5}.$

The following lemma will be useful for globalization. It shows how some of the integers $u_{p,k}$ in the polynomials $C_n^p(X)$ may be changed so that the new polynomials still form a basis. Clearly, for a fixed prime number p and a fixed degree n, we may replace an integer $u_{p,k}$ by an integer m such that $v_p(u_{p,k} - m) \ge r = \left\lceil \frac{\log u_{p,n}}{\log p} \right\rceil = \sup\{k | p^k < u_{p,n}\}$. It turns out that a weaker condition is sufficient.

Lemma 3.7 *Let p be a prime number and let n be a positive integer. For any integer k such that $0 < k < n$, let $s = s(p, n, k)$ be the greatest integer i such that*

$$tp^i < u_{p,k} \le (t+1)p^i < u_{p,n}$$

for some integer t. If, in the polynomial $C_n^p(X)$, we replace some of the integers $u_{p,k}$ by integers $m_{p,k}$ such that $v_p(u_{p,k} - m_{p,k}) \ge s(p, n, k)$, then the new polynomial we obtain is an element of degree n in a regular basis of the $\mathbf{Z}_{(p)}$-module $\text{Int}(\mathbf{P}, \mathbf{Z})_{(p)}$.

Proof: Since we construct here a new sequence which still satifies the properties of the sequence $\{u_{p,k}\}_{k \ge 0}$ used in Lemma 3.2 and Proposition 3.4, the result immediately follows.□

Example 3.8 We demonstrate this substitution with the polynomial $C_5^p(X)$ for $p = 2, 3,$ and 5.

1. If $p = 2$, then $u_{2,4} = 7$ and $u_{2,5} = 9$. Since $0 \cdot 2^3 < 7 \leq 1 \cdot 2^3 < 9$, $u_{2,4} = 7$ can be replaced by any m such that $m \equiv 7 \pmod{2^3}$.

2. If $p = 3$, then $u_{3,3} = 4$ and $u_{3,5} = 7$. Since $1 \cdot 3 < 4 \leq 2 \cdot 3 < 7$, $u_{3,3} = 4$ can be replaced by any m such that $m \equiv 4 \pmod 3$.

3. If $p = 5$, then $u_{5,4} = 4$ and $u_{5,5} = 6$. Since $0 \cdot 5 < 4 \leq 1 \cdot 5 < 6$, $u_{5,4} = 4$ can be replaced by any m such that $m \equiv 4 \pmod 5$.

4   GLOBALIZATION: BASES OF THE Z-MODULE Int(P, Z)

Proposition 4.1 *For each $n \geq 0$, the set of leading coefficients of polynomials in $\mathrm{Int}(\mathbf{P}, \mathbf{Z})$ with degree less than or equal to n is the fractional ideal*

$$\mathcal{I}_n = \left(\prod_{p \in \mathbf{P}, p \leq n} p^{-\omega_p(n)} \right) \mathbf{Z},$$

where $\omega_p(n) = \sum_{k \geq 0} \left[\frac{n-1}{p^k(p-1)} \right]$.

Proof: For each prime number p, $(\mathcal{I}_n)_{(p)} = p^{-\omega_p(n)} \mathbf{Z}$ since $p^{-\omega_p(n)}$ is the leading coefficient of the element $f_n(X)$ of a regular basis of $\mathrm{Int}(\mathbf{P}, \mathbf{Z})_{(p)}$. As $\omega_p(n) = 0$ if $p > n$, the product is finite and we see again that $\mathcal{I}_n$ is a fractional ideal (this was observed earlier in Proposition 2.1). $\square$

For each nonnegative integer n, set

$$\prod_{p \in \mathbf{P}, p \leq n} p^{-\omega_p(n)} = \frac{1}{d_n}.$$

To construct a basis for $\mathrm{Int}(\mathbf{P}, \mathbf{Z})$, we must find for each n, a polynomial $C_n(X)$ of degree n belonging to $\mathrm{Int}(\mathbf{P}, \mathbf{Z})$ with leading coefficient $\frac{1}{d_n}$. For example

$$C_0(X) = 1, \quad C_1(X) = X, \quad C_2(X) = \frac{(X-1)(X-2)}{2},$$

$$C_3(X) = \frac{(X-1)(X-2)(X-3)}{2^3 \cdot 3}, \quad C_4(X) = \frac{(X-1)(X-2)(X-3)(X-5)}{2^4 3}$$

are polynomials which routine checks show satisfy these conditions for $0 \leq n \leq 4$. We now outline a more precise algorithmic method for constructing these basis elements based on the polynomials $C_n^p(X)$.

The First Algorithmic Construction 4.2 Since, for a given n, we need only to consider the prime numbers p such that $p \leq n$, and since the leading coefficients of the polynomials $C_p^n(X)$ are the prime powers $p^{-\omega_p(n)}$, the Bezout identity shows that there are integers a_n^p such that

$$\sum_{p \in \mathbf{P}, p \leq n} a_n^p \left(\prod_{l \in \mathbf{P}, l \leq n, l \neq p} l^{\omega_l(n)} \right) = 1.$$

These integers may be computed with the Euclidean algorithm. The polynomial $C_n(X) = \sum_{p \in \mathbf{P}, p \leq n} a_n^p C_n^p(X)$ is suitable since its leading coefficient is $\frac{1}{d_n}$ and it belongs to $\mathrm{Int}(\mathbf{P}, \mathbf{Z})$ (because each $C_n^p(X)$ belongs to $\mathrm{Int}(\mathbf{P}, \mathbf{Z})$).

Example 4.3 For $n = 5$,

$$C_5(X) = \frac{(X-1)(X-2)(X-3)(X-5)}{2^7 \cdot 3^2 \cdot 5}[3^2 \cdot 5 \cdot a(X-7) + 2^7 \cdot b(X-4)].$$

A direct application of the Euclidean algorithm leads to $a = 37$ and $b = -13$, so that

$$C_5(X) = \frac{(X-1)(X-2)(X-3)(X-5)(X-4999)}{2^7 \cdot 3^2 \cdot 5}.$$

A basis obtained with this algorithm can vary by changing the factors of the polynomials $C_n^p(X)$ according to Lemma 3.7. The calculations in this algorithm can become tedious and it should remain basically a theoretical tool. We describe a faster method to obtain a basis of $\mathrm{Int}(\mathbf{P}, \mathbf{Z})$ from the polynomials $C_n^p(X)$.

Note that $4999 \equiv 7 \pmod{2^3}$, $4999 \equiv 4 \pmod 3$, and $4999 \equiv 4 \pmod 5$ according to Lemma 3.7. Since we may use any integer m such that $m \equiv 7 \pmod{2^3}$, $m \equiv 4 \pmod 3$, and $m \equiv 4 \pmod 5$, we may, in fact, use this lemma to get smaller coefficients for $C_5(X)$. The smallest feasible m is -41, and the corresponding $C_5(X)$ is

$$\frac{(X-1)(X-2)(X-3)(X-5)(X+41)}{2^7 \cdot 3^2 \cdot 5}.$$

If we only consider positive integers, the smallest such m is 79, and then $C_5(X)$ is

$$\frac{(X-1)(X-2)(X-3)(X-5)(X-79)}{2^7 \cdot 3^2 \cdot 5}.$$

To compute $C_5(X)$, we had only to solve for one integer. It is possible that we may have to solve for two or more integers, as the next example illustrates.

Example 4.4 We construct the polynomial $C_6(X)$. The needed values of $u_{p,k}$ are listed in the following table.

p	$u_{p,1}$	$u_{p,2}$	$u_{p,3}$	$u_{p,4}$	$u_{p,5}$
2	1	3	5	7	9
3	1	2	4	5	7
5	1	2	3	4	6

The integers $1, 2, 3$, and 5 appear in each row, and according to Lemma 3.7, we must find two integers m and m' such that for each of the following three systems of congruences, m satisfies one of the two given congruences, and m' satisfies the remaining congruences.

1. $X \equiv 7 \pmod 8$ and $X \equiv 1 \pmod 2$,

2. $X \equiv 4 \pmod 9$ and $X \equiv 7 \pmod 1$,

3. $X \equiv 4 \pmod 5$ and $X \equiv 6 \pmod 1$.

These conditions may be mixed. For example, if we choose $m = 7$, then m' has to satisfy $m' \equiv 1 \pmod 2, m' \equiv 4 \pmod 9$, and $m' \equiv 4 \pmod 5$. The smallest such integer is $m' = -41$. Hence we can choose

$$C_6(X) = \frac{(X-1)(X-2)(X-3)(X-5)(X-7)(X+41)}{2^8 \cdot 3^2 \cdot 5}.$$

We could also choose $m' = 49$, which is the smallest positive integer which satisfies the above congruences when $m = 7$.

The Second Algorithmic Construction 4.5 For any n, the construction of $C_n(X)$ is analogous to that of $C_6(X)$:

$$C_n(X) = \prod_{p \in P, p \le n} \frac{1}{p^{\omega_p(n)}} \prod_{k=0}^{n-1} (X - m_k)$$

where the m_k are integers that we must compute. First, each integer m appearing in each row of the table $\{u_{p,k}\}_{p \le n, 0 \le k < n}$ is a m_k (each prime number $p \le n$ occurs in this table). The other integers m_k must satisfy a system of congruences according to Lemma 3.7.

Of course, there is a choice in determining a solution to this system. For instance, in the case of $C_6(X)$, if we consider the tree of the different choices, we get the following pairs for (m, m'):

$(-1+40k, 1+6h), (-1+10k, 7+24h), (-1+8k, 19+30h)$, and $(-1+2k, 79+120h)$.

We will study the question of whether or not there is a best choice in the following section.

5   THE SEARCH FOR A CANONICAL BASIS

If $f(X)$ is a polynomial of degree n in Int(Z), then $f(X) = \sum_{i=0}^{n} a_n \begin{pmatrix} X \\ n \end{pmatrix}$

for unique integers $a_0, \ldots, a_n$. Using the facts that

1. $\begin{pmatrix} k \\ n \end{pmatrix} = 0$ for each $0 \leq k \leq n - 1$, and

2. $\begin{pmatrix} n \\ n \end{pmatrix} = 1$,

it is easy to see that the coefficients a_i can be computed with a difference table argument. Can a basis with similar algebraic properties be constructed for Int($\mathbf{P}, \mathbf{Z}$)? At each step in the process outlined in Section 4 there are several choices. What are the optimal choices in these steps?

In order to search for an analogous basis for Int($\mathbf{P}, \mathbf{Z}$), let us see whether a polynomial which takes integral values on prime numbers necessarily takes integral values at other integers.

Proposition 5.1 *Let $f(X)$ be a polynomial with rational coefficients. If $f(X)$ is integral-valued on prime numbers, then $f(1)$ and $f(-1)$ are integers.*

Proof: Let $a = \pm 1$ and let $f(X)$ be in Int($\mathbf{P}, \mathbf{Z}$). For each prime number p, a belongs to X_p and $f(X)$ belongs to Int($X_p, \mathbf{Z}_{(p)}$). Hence, $f(a)$ belongs to $\bigcap_{p \in \mathbf{P}} \mathbf{Z}_{(p)} = \mathbf{Z}$.$\square$

Proposition 5.2 *If a is neither a prime number nor ± 1, then there exists $f(X)$ in Int($\mathbf{P}, \mathbf{Z}$) such that $f(a)$ is not an integer.*

Proof: Let a be an integer such that $a \notin \mathbf{P}$ and $a \neq \pm 1$. Let p be a prime number such that $v_p(a) \geq 1$. Let $r = v_p(a - p) + 1$ and let

$$f(X) = (X - p)\frac{((X - 1)\cdots(X - p + 1))^r}{p^r}.$$

Then $f(p) = 0$ and, for each $q \in \mathbf{P}$ with $q \neq p$, we have that $f(q) \in \mathbf{Z}$. Hence $f(X) \in$ Int($\mathbf{P}, \mathbf{Z}$) with $v_p(f(a)) = v_p(a - p) - r < 0$.$\square$

We are now able to prove that there does not exist a basis for Int($\mathbf{P}, \mathbf{Z}$) analogous to the basis $\left\{ \begin{pmatrix} X \\ n \end{pmatrix} \right\}_{n=0}^{\infty}$ of Int(Z). Suppose there exist a basis of Int($\mathbf{P}, \mathbf{Z}$) of the form $\{C_n(X)\}_{n \geq 0}$ such that each

$$C_n(X) = \frac{1}{d_n} \prod_{0 \leq k < n} (X - p_k)$$

with $C_n(p_n) = \pm 1$. Then $f(X) \in \mathbf{Q}[X]$ of degree n is in $\mathrm{Int}(\mathbf{P}, \mathbf{Z})$ if and only if $f(p_0), f(p_1), \ldots, f(p_n)$ are integers. Propositions 5.1 and 5.2 imply that $\{p_k | k \in \mathbf{N}\} \subset \mathbf{P} \cup \{\pm 1\}$. Then the condition $\prod_{0 \leq k < n}(p_n - p_k) = \pm d_n$ obviously implies $\{p_0, p_1\} = \{1, 2\}$ or $\{2, 3\}$ and $\{p_0, p_1, p_2\} = \{1, 2, 3\}$. Hence, there is really no choice for $C_3(X)$ other than $C_3(X) = \frac{(X-1)(X-2)(X-3)}{24}$. Then $C_3(5) = 1$ and $C_3(-1) = -1$, and thus $p_3 = 5$ or -1. With either choice for p_3, there is no prime p_4 such that $C_4(p_4) = \pm 1$. Consequently, there is no such basis.

The question of determining good bases is still open. With respect to this question, notice that for a fixed positive integer n, each prime integer p with $p \leq n$ is in the sequence $\{u_{p,k}\}_{0 \leq k \leq n}$. If we do not select p in the construction of $C_n(X)$, we would have to replace it by an element m such that $v_p(m - p) \geq \omega_p(n)$ (i.e., $m = p + kp^{\omega_p(n)}$ where k is some integer). Such elements rapidly increase with n. For example, the smallest positive integers we can use to replace $p = 2$ for $n = 2, 3, 4, 5, 6,$ and 7 are respectively $4, 10, 18, 130, 258,$ and 1026.

Since 1 is also common to each sequence $\{u_{p,k}\}_{0 \leq k \leq n}$ for any prime p, it is reasonable to assume that it should always be chosen in the construction of $C_n(X)$. Using the last two conditions, the first four elements of a basis of $\mathrm{Int}(\mathbf{P}, \mathbf{Z})$ are completely determined:

$$C_0(X) = 1, \ C_1(X) = X - 1, \ C_2(X) = \frac{(X-1)(X-2)}{2},$$

$$\text{and } C_3(X) = \frac{(X-1)(X-2)(X-3)}{24}.$$

There is then a choice for $C_4(X) = \frac{(X-1)(X-2)(X-3)(X-m)}{48}$ where $m \equiv 1$ (mod 2). We chose 5 in our earlier list, but there are other possibilities, $m = \pm 1$ for example.

Thus we may choose $C_4(X) =$

$$\frac{(X-1)^2(X-2)(X-3)}{48} \text{ or } \frac{(X^2-1)(X-2)(X-3)}{48}.$$

Our earlier choice of 5 seems justified by the fact that

1. $C_3(5) = 1$, and

2. 5 will be necessarily chosen in the construction of C_5.

But, as previously noted, $C_3(-1) = -1$ is invertible in $\mathbf{Z}$; hence $m = -1$ could be a good choice because for any polynomial of degree 3, $f(X) \in \mathrm{Int}(\mathbf{P}, \mathbf{Z})$ if and only if $f(1), f(2), f(3),$ and $f(-1)$ are integers.

As seen in Example 4.2, in constructing $C_5(X)$ we only need choose an element m such that $m \equiv -41$ (mod 120). We chose -41 because this was the smallest such m, but Proposition 5.2 seems to indicate that the best choice for the integer m comes from the set $P \cup \{\pm 1\}$. Hence, for $n = 5$ we choose $m = -41 + 120 = 79$ (which is a prime number) and thus,

$$C_5(X) = \frac{(X - 1)(X - 2)(X - 3)(X - 5)(X - 79)}{5760}.$$

Combining our previous results, we suggest the following 5 criteria for choosing the integers p_k which appear in the basis element

$$C_n(X) = \frac{1}{d_n} \prod_{k=0}^{n-1} (X - p_k),$$

where n is a given positive integer:

1. each p_k belongs to $P \cup \{\pm 1\}$,

2. $p_0 = 1$,

3. $\{p \in P | p \leq n\} \subset \{p_k | 1 \leq k < n\}$,

4. all the p_k are distinct,

5. the product $\prod_{1 \leq k < n} p_k$ is as small as possible.

The previous conditions lead to

$$C_6(X) = \frac{(X - 1)(X - 2)(X - 3)(X - 5)(X + 1)(X - 7)}{11,520},$$

$$C_7(X) = \frac{(X - 1)(X - 2)(X - 3)(X - 5)(X - 7)(X + 1)(X - 193)}{2,903,040},$$

and

$$C_8(X) = \frac{(X - 1)(X - 2)(X - 3)(X - 5)(X - 7)(X - 11)(X - 13)(X + 1)}{2^{11} \cdot 3^4 \cdot 5 \cdot 7}.$$

And so on ...

We see that the p_k which are obtained for n are not necessarily obtained for $n + 1$. Of course, for a fixed degree n, we could find n integers $p_0, p_1, \ldots, p_{n-1}$ such that

$$C_r(X) = \frac{1}{d_r} \prod_{k=0}^{r-1} (X - p_k)$$

for $r = 0, 1, \ldots n$. But these p_k would be very large and the entire computation would have to be repeated for $n + 1$.

References

[1] P.-J. Cahen, *Integer-valued polynomials on a subset*, Proc. Amer. Math. Soc. **117**: 919–929, (1993).

[2] P.-J. Cahen and J.-L. Chabert, *Integer-valued polynomials*, Monograph to appear.

[3] J.-L. Chabert, *Une caracterisation des polynômes prenant des valeurs entières sur tous les nombres premiers*, Canad. Math. Bull., to appear.

[4] R. Gilmer, *Sets that determine integer-valued polynomials*, J. Number Theory **33**: 95-100, (1989).

[5] J. Jones, D. Sato, H. Wada, and D. Wiens, *Diophantine representation of the set of prime numbers*, Amer. Math. Monthly **83**: 449-474, (1976).

[6] G. Lejeune-Dirichlet, *Beweis des Satzes dass jede unbegrenzte arithmetische Progression, deren erstes Glied und Differenz ganze Zahlen ohne gemeinschaftlichen Factor sind, unendlich viele Primzahlen enthält*, Abh. Königl. Preuss. Akad. Wiss. 45-81, (1837); *Werke*, vol. I, 313-342, Berlin, (1889).

[7] D. McQuillan, *On a Theorem of R. Gilmer*, J. Number Theory **39**: 245–250, (1991).

[8] G. Pólya, *Über ganzwertige Polynome in algbraischen Zahlkörpern*, J. Reine Angew. Math. **149**: 97-116, (1919).

Factorization of Bonds and Other Cash Flows

DOUGLAS L. COSTA Department of Mathematics, University of Virginia, Charlottesville, VA 22903-3199

Abstract This paper attempts to analyze bond portfolios through polynomial algebra. Viewing such a portfolio as a series of future payments, we argue that it is completely determined by its price-yield function. In a discrete-time model with smallest time unit δ, the price-yield function is a polynomial with real coefficients. It is a well-known algebraic result that such polynomials factor as a finite product of irreducible ones. Each factor represents a short duration portfolio, and these must be somehow related to the original portfolio. With this in mind, we define an instrument which has the effect of multiplying one bond portfolio by another. These "bond products" can be created using forward contracts. In effect, they amount to investing first in one portfolio and at maturity reinvesting the proceeds in the second. This concept gives meaning to polynomial factorizations of bond portfolios. Any such factorization will have the same price-yield function as the portfolio it factors. A consequence of this is that the duration and convexity of a portfolio can be computed in terms of the duration and convexity of the factors. In particular, the duration of a bond product is the sum of the durations of its factors. This raises the possibility of using factorizations to estimate sensitivity to changes in the shape of the yield curve, but limited numerical experimentation has so far proved inconclusive.

1 CASH FLOWS VIEWED AS POLYNOMIALS

A bond portfolio is a package of short and long positions in a collection of current and future payments. As such it is a finite set of ordered pairs of real numbers $\{(c_i, t_i) \mid c_i, t_i \in \mathbf{R},\ 1 \leq i \leq n,\ 0 \leq t_1 < \cdots < t_n\}$, where c_i is the amount to be received at time t_i. If B is the present value of the package, its *yield* is the value y implicitly defined by the relation $B = \sum_{i=1}^{n} c_i e^{-yt_i}$. The function $f(y) = \sum_{i=1}^{n} c_i e^{-yt_i}$ in which y is taken to be a variable, is the *price-yield function* of the portfolio. We are ready for our first observation.

THEOREM 1.1 A bond portfolio is completely determined by its price-yield function.

Proof: We are claiming that if two portfolios have identical price-yield functions, then the portfolios are identical. To see this we may take their difference (go long one and short the other). This will be a portfolio $\{(c_i, t_i)\}$ having $f(y)$ identically zero, and we must prove that $c_i = 0$ for $1 \leq i \leq n$. But this is merely the claim that the functions $e^{-yt_1}, \ldots, e^{-yt_n}$ are linearly independent over $\mathbf{R}$.

To verify this, note that if $\sum_{i=1}^{n} c_i e^{-yt_i} \equiv 0$, then setting $y = j \in \{0, 1, \ldots, n-1\}$, we get n linear equations in $c_1, \ldots, c_n$,

$$\sum_{i=1}^{n} e^{-jt_i} c_i = 0, \qquad 0 \leq j \leq n-1,$$

with coefficient matrix $\left((e^{-t_i})^j\right)$, the Vandermonde matrix in $e^{-t_1}, \ldots, e^{-t_n}$. Since $t_1, \ldots, t_n$ are distinct, the matrix is invertible and hence $c_1 = \cdots = c_n = 0$.

So the set of all bond portfolios is equivalent to the set of all price-yield functions $\{f(y) = \sum_{i=1}^{n} c_i e^{-yt_i} \mid n \geq 1, 0 \leq t_1 < \cdots < t_n\}$. Setting $Y = e^{-y}$, this is the set of all formal expressions $\sum_{i=1}^{n} c_i Y^{t_i}$ with $c_i \in \mathbf{R}$ and $t_i \in \mathbf{R}_+$.

Since the sum and product of any two such expressions is also of the same form, this is a ring. (It is the semigroup-ring of the additive semigroup $\mathbf{R}_+$ over $\mathbf{R}$. See [Gilmer, 1984].) The point is that multiplying two portfolios, and hence also factoring a portfolio, is possible. The aim of this article is to interpret the meaning of such products and factorizations. At the same time we will attempt to indicate their potential usefulness.

Before proceeding, we make some simplying assumptions. Instead of allowing the times of payment t_i to have any value in $\mathbf{R}_+$, we will discretize the situation by assuming that there is a smallest time unit δ (e.g., $\delta = 1$ day), and that every t_i is a non-negative integral multiple of δ. We refer to δ as one *period*. In this case, by setting as many $c_i = 0$ as necessary, we may write every price-yield function in the form $f(y) = \sum_{i=0}^{n} c_i e^{-y(i\delta)} = \sum_{i=0}^{n} c_i x^i$, where $x = e^{-y\delta}$ is the one-period discount factor. We shall call $g(x) = \sum_{i=0}^{n} c_i x^i$ the *price-discount* function. By Theorem (1.1) we may identify a bond B with the corresponding function $f(y)$ or $g(x)$. Thus, the set of all discrete-time bond portfolios is the polynomial ring $\mathbf{R}[x]$.

If we allow the possibility of perpetual cash flows, we obtain price-discount functions of the form $g(x) = \sum_{i=0}^{\infty} c_i x^i$, i.e., formal power series. These, too, form a ring $\mathbf{R}[[x]]$ of which $\mathbf{R}[x]$ is a subring. Here we are really only interested in those power series $g(x)$ which are convergent for $|x| < 1$.

To illustrate these correspondences, note that the monomial cx^n represents an n-period zero-coupon bond with face value c. The polynomial $1 - x^n$ represents an n-period interest-free loan of one monetary unit; and the power series $1 + x + \cdots + x^n + \cdots$ represents a perpetual annuity paying one unit at the beginning of each period.

The ring $\mathbf{R}[x]$ of discrete-time bond portfolios is a unique factorization domain [Zariski, 1958]. Every polynomial $g(x) = \mathbf{R}[x]$ can be factored into a finite product of irreducible polynomials and the factorization is unique to within scalar factors. The irreducible polynomials are of two types:

(1) Linear polynomials $a + bx$, $b \neq 0$. Note that these are all one-period cash flows and that they include the discount bond x.

(2) Irreducible quadratic polynomials $a + bx + cx^2$, $b^2 - 4ac < 0$. These are all two-period cash flows. So every bond (portfolio) can be expressed as a product of one- and two-period cash flows.

2 WHAT IT MEANS TO MULTIPLY CASH FLOWS

To develop this theme we need to have a term-structure for interest rates. For this, let $r_i = r_{oi}$ be the risk-free continuous rate of return for the time period from 0 to $i\delta$. Let r_{ij} be the forward rate applicable to the time period from $i\delta$ to $j\delta$, $i < j$. Then the following basic relationships hold:

(1) $ir_i + (j - i)r_{ij} = jr_j$;

(2) $(j - i)r_{ij} = \displaystyle\sum_{k=i}^{j-1} r_{k,k+1}.$

Consider a cash flow $g(x) = \sum_{i=0}^{n} c_i x^i$. The present value is $B = \sum_{i=0}^{n} c_i e^{-r_i(i\delta)}$. We assume that at time 0 one enters into forward contracts to reinvest c_i at the forward rate r_{in} from time $i\delta$ to time $n\delta$ (or borrow, if $c_i < 0$). Entering these forward contracts adds nothing to the cost of $g(x)$, theoretically. An easy calculation using the fundamental relations (1), (2) shows that the value of $g(x)$ will grow to $Be^{r_n(n\delta)}$ at time $n\delta$. Notice also, that if y_0 is the yield corresponding to the present value B and $x_0 = e^{-y_0\delta}$, then $B = g(x_0)$.

Next consider two cash flows $g(x) = \sum_{i=0}^{n} c_i x^i$ and $h(x) = \sum_{j=0}^{m} d_j x^j$. We define their *bond product*, denoted $g(x) * h(x)$, to be an instrument constructed as follows:

(1) Compute the present value $B_g = \sum_{i=0}^{n} c_i e^{-r_i(i\delta)}$ of $g(x)$ and the forward price of h as a cash flow starting at time $n\delta$ and running to time $(n + m)\delta$. This forward price is $F_h = \sum_{j=n}^{n+m} d_{j-n} e^{-r_{nj}(j-n)\delta}$.

(2) Purchase F_h units of cash flow $g(x)$ at time 0, along with all the associated reinvestment forward contracts, as in the previous paragraph.

(3) Enter into $B_g e^{r_n(n\delta)}$ long forward contracts at price F_h on cash flow $h(x)$ to begin at time $n\delta$, along with all associated reinvestment forward contracts.

Since the forward contracts are no-cost, the present value of this investment is the cost $B_g F_h$ of the purchase in step 2. Note that when the first cash flow matures at time $n\delta$ the accumulated value will be $B_g e^{r_n(n\delta)} F_h$, which will be reinvested through the contract in step 3 in $B_g e^{r_n(n\delta)}$ of $h(x)$ and which will mature to value $B_g e^{r_n(n\delta)} F_h e^{r_{n,n+m} m\delta} = (B_g F_h) e^{r_{m+n}(m+n)\delta}$ at time $(m + n)\delta$.

Thus, a bond product is defined by investing in cash flows serially with all proceeds at maturity of one reinvested in the next. This definition extends easily to a finite bond product $g_1(x) * g_2(x) * \cdots * g_s(x)$, whose present value (using the obvious extension of the notation above) will be $B_{g_1} F_{g_2} \cdots F_{g_n}$.

If one were to make the assumption of a flat term structure, then there would be a single risk-free rate r with $r_{ij} = r$ for all i, j. In this case, the present value of

$g(x) * h(x)$ would be

$$B_g F_h = \left(\sum_{i=0}^{n} c_i e^{-ri\delta} \right) \left(\sum_{j=0}^{m} d_j e^{-rj\delta} \right) = g(x)h(x) \text{ with } x = e^{-r\delta}.$$

Thus, we may define the *product yield* of $g(x) * h(x)$ implicitly as the value of y for which $B_g F_h = g(e^{-y\delta})h(e^{-y\delta})$. In other words, we have established that the price-discount function of $g(x) * h(x)$ is the polynomial $g(x)h(x)$. It is in this sense that we can think of the bond product $g(x) * h(x)$ as a factorization of the cash flow represented by $g(x)h(x)$.

Some examples are now in order.

EXAMPLES (1) The n-period discount bond x^n factors as $x * x * \cdots * x$. This says, in effect, that we can realize a future payment of one unit at time $n\delta$ as a current payment of x^n units invested and reinvested in a series of one-period forward contracts at rates $r_1, r_{12}, r_{23}, \ldots, r_{n-1,n}$, respectively.

(2) The n-period interest-free loan of one unit, $1 - x^n$, factors as $(1 - x) * (1 + x + \cdots + x^{n-1})$. This says that the n-period loan can be realized as a one-period interest-free loan of $(1 + x + \cdots + x^{n-1})$ units. It will mature at time 1δ to have value $\frac{1}{x}(1 + x + \cdots + x^{n-1}) - (1 + x + \cdots + x^{n-1}) = \left(\frac{1}{x} - 1\right)(1 + x + \cdots + x^{n-1})$, which is then used to purchase an annuity paying $\left(\frac{1}{x} - 1\right)$ units at time $1\delta, 2\delta, \ldots, n\delta$. Since $\left(\frac{1}{x} - 1\right)$ is the one-period interest on one unit, this exactly tracks the cash flow produced by the n-period interest-free loan.

(3) Carrying example (2) to the limit as $n \to \infty$, we have $1 = (1 - x) * (1 + x + \cdots + x^n + \cdots)$. This expresses the idea that the future stream of payments of $\left(\frac{1}{x} - 1\right)$ units every period which can be generated by a current unit can also be financed by a one-period interest-free loan of $(1 + x + \cdots + x^n + \cdots)$ units with the proceeds used to purchase a perpetual annuity.

[This example illustrates the fact that all cash flows $g(x) = c_0 + c_n x^n$ having $c_0 \neq 0$ are invertible in $\mathbf{R}[[x]]$. This means that in $\mathbf{R}[[x]]$, division of cash flows is possible except that, in general, one can't divide by zeroes.]

(4) An n-period coupon bond with coupon payments c and face value $1 - c$ has the form $c + cx + \cdots + cx^{n-1} + x^n$ cum coupon. For example, if $c = .05(1 - c)$ the bondholder is receiving 5% every period and $c = \frac{1}{21} = .0476190476$. Using this value of c and $n = 5$, and factoring using MAPLE, we get

$$c + cx + cx^2 + cx^3 + cx^4 + x^5$$
$$= (.504511 + x) * (.270783 + .398781x + x^2) * (.348569 - .855673x + x^2).$$

Note the large negative payment due at time 4δ.

3 RELATION TO DURATION AND CONVEXITY

Recall that the (modified) duration D of a bond B is defined by

$$D = -\frac{\partial B}{\partial y} / B, \text{ where } y \text{ is the yield.}$$

Applying this to a cash flow $g(x)$, the price-yield function is $f(y) = g(e^{-y\delta})$, so the duration D_g of $g(x)$ is given by

$$D_g = -f'(y)/f(y) = -\frac{d}{dy}(\ell n \, f(y)).$$

But $f(y) = g(x)$ with $x = e^{-y\delta}$, so that we also have

$$D_g = \delta x g'(x)/g(x) = \delta x \cdot \frac{d}{dx}(\ell n \, g(x)).$$

We have seen that the price-discount function of a bond product is the product of the price-discount functions of its factors. Using this and either of the preceding formulas, it is easy to verify the following result.

THEOREM 3.1 Let $g_1(x), \ldots, g_s(x)$ be discrete-time cash flows. Then

$$D_{g_1 * \cdots * g_s} = \sum_{i=1}^{s} D_{g_i}.$$

This makes sense in that investing in $g_1, \ldots, g_s$ serially suggests intuitively that the total duration should be the sum of their separate durations.

We note here that a sum $g_1(x) + \cdots + g_s(x)$ of cash flows has duration $\delta x(g_1'(x) + \cdots + g_s'(x))/(g_1(x) + \cdots + g_s(x))$. This can be rewritten algebraically to produce the usual formula:

$$D_{g_1 + \cdots + g_s} = \sum_{i=1}^{s} w_i D_{g_i}, \text{ where } w_i = g_i(x)/(g_1(x) + \cdots + g_s(x)).$$

The convexity C of a bond B is defined by $C = \frac{\partial^2 B}{\partial y^2}/B$. For a cash flow $g(x)$ with price-yield function $f(y) = g(e^{-y\delta})$, this becomes

$$C = f''(y)/f(y) = \delta D_g + \delta^2 x^2 g''(x)/g(x).$$

Using the Leibniz rule and some algebra, one gets the following result.

THEOREM 3.2 Let $g_1(x), \ldots, g_s(x)$ be discrete-time cash flows. Then

(i) $C_{g_1 + \cdots + g_s} = \sum_{i=1}^{s} w_i C_{g_i}, \quad w_i = g_i(x)/(g_1(x) + \cdots + g_s(x)).$

(ii) $C_{g_1 * \cdots * g_s} = (D_{g_1 * \cdots * g_s})^2 + \sum_{i=1}^{s}(C_{g_i} - D_{g_i}^2).$

[Note: If we think of $C_g - D_g^2$ as $\text{Var}_g(t) = E(t^2) - E(t)^2$, where t is the time of payment of a random unit of g's value, then (ii) says that $\text{Var}_{g_1 * \cdots * g_s}(t) = \sum_{i=1}^{s} \text{Var}_{g_i}(t)$.]

The formula in Theorem (3.1) suggests the possibility of using bond factorizations to analyze bond price sensitivity to changes in the shape of the yield curve. Let $g(x)$ be a cash flow such that $g(x) = g_1(x)g_2(x)\cdots g_s(x)$, and consider the bond product $g_1(x) * g_2(x) * \cdots * g_s(x)$. Using B_g to denote the present value of instrument g, the present value of the bond product is

$$B_{g_1 * \cdots * g_s} = B_{g_1} F_{g_2} \cdots F_{g_s}.$$

Now suppose that y_i is the current forward yield in the time period covered by g_i and that this forward yield changes by Δy_i. Let $D_{g_i}(y_i)$ be the duration of g_i computed at yield y_i. Then each F_{g_i} changes by approximately $-F_{g_i}D_{g_i}(y_i)\Delta y_i$. Incorporating these changes into $B_{g_1 * \cdots * g_s}$ and dropping second order terms, we get, approximately,

$$\Delta B_{g_1 * \cdots * g_s} = -B_{g_1 * \cdots * g_s} \left(\sum_{i=1}^{s} D_{g_i}(y_i)\Delta y_i \right).$$

This "bond product price sensitivity" formula works well in simple numerical experiments to estimate price changes for bond products. It does not do so well as an estimate of ΔB_g. On the other hand, if the yield curve changes so that Δr_i is the change in the spot rate r_i for the time period ending when g_i terminates, then $-B_g \left(\sum_{i=1}^{s} D_{g_i}(y_i)\Delta r_i \right)$ seems to estimate ΔB_g to the right order of magnitude.

Finally, we note that some care must be exercised in working with bond products. For instance, in most examples of cash flows $g(x)$, $g_1(x)$, $g_2(x)$ such that as polynomials $g(x) = g_1(x)g_2(x)$, it is easy to construct a term structure such that B_g, $B_{g_1 * g_2}$, and $B_{g_2 * g_1}$ are all different.

REFERENCES

1. R. Gilmer, 1984, *Commutative Semigroup Rings*, U. of Chicago Press.

2. J. Hull, 1993, *Options, Futures, and other Derivative Securities*, 2nd ed., Prentice Hall.

3. O. Zariski and P. Samuel, 1958, *Commutative Algebra*, vol. I, Van Nostrand.

A Characterization of Polynomial Rings with the Half-Factorial Property

JIM COYKENDALL   Department of Mathematics, Lehigh University, Bethlehem, Pennsylvania 18015

1. INTRODUCTION

In this paper, R will be an integral domain with quotient field K and x an indeterminate. Following Zaks [4], we define R to be a half-factorial domain (HFD) if R is atomic, and given any two irreducible factorizations of an element a$\in$ R

$$a = \pi_1 \pi_2 ... \pi_k = \xi_1 \xi_2 ... \xi_m,$$

then $k = m$.

Unlike unique factorization domains (UFD's), HFD's do not behave well with respect to polynomial extensions in general (see [1], [2, Example 5.4]). In [4, Theorem 2.4] it was established that if R is a Krull domain with divisor class group Cl(R), then R[x] is an HFD if and only if |Cl(R)| $\leq$2. However, there exist Krull domains with the HFD property that have |Cl(R)| >2, so in general, the HFD property is lost in the polynomial extension.

We shall presently show that for R[x] to be an HFD, it is necessary for R to be integrally closed, and from this we will deduce a characterization for "polynomially stable" Noetherian HFD's.

2. THE MAIN THEOREM

Let R and K be as above. We first record a useful lemma.

LEMMA 2.1. Let $p(x)$ be irreducible in $R[x]$, and let $0 \neq r \in R$. If $rp(x) = r_1 r_2 ... r_t f_1 f_2 ... f_k$ with $r_i \in R$ for $1 \leq i \leq t$ and $f_i \in R[x]$ with $0 < \deg(f_i) < \deg(p)$ for $1 \leq i \leq k$, then no f_i is monic.

Proof: Suppose that $rp(x) = r(q_{n+m} x^{n+m} + q_{n+m-1} x^{n+m-1} + ... + q_1 x + q_0) = (r_n x^n + ... + r_0)(x^m + s_{m-1} x^{m-1} + ... + s_0) = g_1(x) g_2(x)$ with $n \geq 1$.

From this we obtain the following system of equations:

$$r_n = r q_{n+m}$$

$$r_{n-1} + r_n s_{m-1} = r q_{n+m-1}$$

$$\cdot$$
$$\cdot$$
$$\cdot$$

$$r_0 + r_1 s_{m-1} + ... + r_m s_0 = r q_m$$

Inductively from these equations, we get that $r | r_i$ for every i. Therefore, $g_1(x) = r g(x)$ with $g(x) \in R[x]$. This shows that $p(x) = g(x) g_2(x)$, which is a contradiction.

With this lemma in hand, we can now prove the main theorem of this paper.

THEOREM 2.2. Let R be an integral domain. If $R[x]$ is an HFD, then R is integrally closed.

Proof: Assume that R is not integrally closed. We shall show that $R[x]$ is not an HFD. We note that we can also assume that R is an HFD, for if not, then $R[x]$ is certainly not an HFD (see [1]).

Let K be the quotient field of R, and let $\omega \in K\backslash R$ such that ω satisfies the monic irreducible polynomial $p(x) = x^n + p_{n-1}x^{n-1} + \ldots + p_1 x + p_0 \in R[x]$. Also assume that $\omega = r/s$ with $r, s \in R$ such that r and s have no factor in common (which is valid since R is an HFD). Consider the following element of $R[x]$:

$$s^n p(x) = s^n x^n + p_{n-1}s^n x^{n-1} + \ldots + p_1 s^n x + p_0 s^n$$
$$= (sx - r)q(x) \text{ with } q(x) \in R[x].$$

By assumption we have the following facts.

1. The number of factors of one irreducible factorization of the left hand side is $mn + 1$, where m is the number of irreducible factors of s.

2. The polynomial $(sx - r)$ is irreducible.

So we will investigate the number of factors of $q(x)$.

Notice that the leading coefficient of $q(x)$ is s^{n-1}. Assume that $q(x) = f_1(x)\ldots f_k(x)r_1\ldots r_t$ where each $f_i \in R[x]$ is irreducible of positive degree and each r_i is irreducible in R. As $p(x)$ is irreducible in $R[x]$, Lemma 2.1 shows that none of the f_i's is monic, and so we obtain the equation

$$s^{n-1} = L_1\ldots L_k r_1\ldots r_t,$$

where L_i is the leading coefficient of $f_i(x)$ and is a nonunit.

As R is an HFD, we have that $k + t \leq m(n-1)$. We conclude that the number of irreducible factors of $s^n p(x)$ (from this point of view) is $k + t + 1 \leq m(n-1) + 1 \leq mn + 1$. For $R[x]$ to be an HFD, the last inequality must be an equality, and hence $m=0$. This contradicts the fact that $\omega \in K\backslash R$.

We now give a corollary to this theorem which completely classifies all Noetherian HFD's that have "polynomial stability".

COROLLARY 2.3. Let R be a Noetherian ring. Then the following conditions are equivalent:

1. R is a Krull domain with $|Cl(R)| \leq 2$.

2. $R[x]$ is an HFD.

3. $R[x_1, ..., x_n]$ is an HFD for all $n \geq 1$.

4. $R[x_1, ..., x_n]$ is an HFD for some $n \geq 1$.

Proof: We first observe that the implications (3) implies (4) and (4) implies (2) are obvious. We will show that (1) implies (3) and (2) implies (1).

For the implication (1) implies (3), since R is a Krull domain with $|Cl(R)| \leq 2$, then $R[x_1, ..., x_n]$ is also a Krull domain with $|Cl(R[x_1, ..., x_n])| = |Cl(R)|$ (cf [3, Chapter 7, Proposition 13]). Since if R is a Krull domain, then $R[x]$ is an HFD if and only if $|Cl(R)| \leq 2$ [4, Theorem 2.4], we obtain the result inductively.

For (2) implies (1), we assume that $R[x]$ is an HFD. The previous theorem shows that R is integrally closed and hence a Krull domain (as R is Noetherian). Once again applying a result of Zaks [4, Theorem 2.4], we obtain that R must have $|Cl(R)| \leq 2$. This concludes the proof.

REFERENCES

1. D. D. Anderson, D. F. Anderson, and M. Zafrullah, *Factorization in integral domains*, J. Pure Appl. Algebra **69** (1990), 1-19.
2. D. D. Anderson, D. F. Anderson, and M. Zafrullah, *Rings between D[X] an K[X]*, Houston J. Math **17** (1991), 109-129.
3. N. Bourbaki, "Commutative Algebra," Addison-Wesley, Reading, 1972.
4. A. Zaks, *Half factorial domains*, Israel J. Math. **37** (1980), 281-302.

On Characterizations of Prüfer Domains Using Polynomials with Unit Content

DAVID E. DOBBS Department of Mathematics, University of Tennessee, Knoxville, Tennessee 37996-1300.

1 INTRODUCTION

In [10, Theorem 5], Gilmer and Hoffmann established that if R is a (commutative integral) domain with integral closure R' and quotient field K, then R' is a Prüfer domain if and only if each element of K is a root of a polynomial in $R[X]$ having a unit coefficient. A rather different proof of this fact was given a few years later by the author [7, Corollary 5], as a consequence of a result concerning the INC property for extensions of commutative rings [7, Theorem, p. 38]. During the Special Session on Commutative Ring Theory in Iowa City in March 1996, Dong Je Kwak gave a talk on his recent paper [1] with D.D. Anderson and noted that the equivalence of (1) and (2) in [1, Theorem 9] gave another proof of [10, Theorem 5]. Moreover, the talk revealed (cf. [1, Theorem 6]) that if R is integrally closed, then the equivalent conditions in [10, Theorem 5] are also equivalent to the condition that each element of K is a root of a quadratic polynomial in $R[X]$ having a unit coefficient of X.

My interest was initially piqued because the authors of [1] seemed unaware of the existence of [7]. More importantly, I realized during Professor Kwak's talk that some aspects of my proof of [7, Theorem] could be modified to produce a proof of a generalization of the fragment of [1, Theorem 6] which was quoted above; see Theorems 2.3 and 2.5 below. Following Professor Kwak's talk, I offered to write a short note summarizing this insight, and Professor Anderson graciously agreed to consider it for publication in these Proceedings, which he is editing.

This is that note, but it is not as short as originally foreseen. The additional material has to do with new applications of Theorem 2.5. These relate to an example at the close of [1] which shows that the naïve analogue of [10, Theorem 5] is false; specifically, "R' is a Prüfer domain" is not equivalent to "Each element of K is a root of a quadratic polynomial in $R[X]$ having a unit coefficient of X." The counterexample in [1] concerns a one-dimensional Noetherian local analytically irreducible residually rational domain R. Our final contribution in this note is a positive result, Theorem 2.8, showing how the concepts in Lemma 2.1 - Theorem 2.5 give a class of one-dimensional Noetherian local analytically irreducible (but not residually rational) domains R for which "R' is a Prüfer domain" is equivalent to a bounded-degree variant of the Anderson-Kwak condition involving quadratic polynomials.

2 RESULTS

During Professor Kwak's talk, I recalled two items from the proof of [7, Theorem]. The first of these was the identification $R[u]_{R\backslash M} = R_M[u_M]$, where R denotes any (commutative) ring, u is an element of a (unital commutative) ring extension T of R, M is a maximal ideal of R, and u_M is the canonical image of u in $T_{R\backslash M}$. The second item recalled from [7, Theorem] was the use of globalization in analyzing content. Together, these memories led me at once to the proof of Lemma 2.1 and, thus, to Theorem 2.3. Before presenting these generalizations of [1, Theorem 6], it is convenient to make the following definition.

Consider a ring extension $A \subseteq B$, an element $v \in B$, and integers $0 \leq j \leq n$. We say that property $P(A, B, v, n, j)$ [or, if no confusion is possible, property $P(v, n, j)$] is satisfied if v is a root of a polynomial $f \in A[X]$ such that $\deg(f) \leq n$ and 1 is the coefficient of X^j in f.

LEMMA 2.1   Let $A \subseteq B$ be a ring extension, let $v \in B$, and let $0 \leq j \leq n$ be integers. Then property $P(A, B, v, n, j)$ is satisfied if and only if property $P(A_M, B_{A\backslash M}, v_M, n, j)$ is satisfied for each maximal ideal M of A.
Proof: The "only if" assertion follows by considering the commutative diagram and identification

where the horizontal algebra maps are evaluation at v and v_M and the vertical homomorphisms are the canonical structure maps given by localization.

For the "if" assertion, let $E := \sum\{Av^i : 0 \leq i \leq n, i \neq j\}$. By taking common denominators, we have the identification (inside $B_{A\backslash M}$)

$$E_{A\backslash M} = \sum_{0 \leq i \leq n,\ i \neq j} A_M v_M{}^i$$

for each maximal ideal M of A. By hypothesis, $P(v_M, n, j)$ is satisfied for each M; that is, $v^j{}_M = v_M{}^j \in E_{A\backslash M}$ for each M. Hence, by globalization (as in [4, Corollary 1, p. 88]), $v^j \in E$; that is, $P(v, n, j)$ is satisfied. $\Box$

LEMMA 2.2   Let R be a valuation domain, with quotient field K. Then $P(v, n, j)$ is satisfied for each $v \in K$ and for all positive integers $j < n$.

Proof: If $v \in R$, then v is a root of $X^{j-1}(X - v)$ and so $P(R, K, v, n, j)$ is satisfied. If $v \notin R$, then $v^{-1} \in R$ since R is a valuation domain, in which case the conclusion follows by observing that v is a root of $X^j(-v^{-1}X + 1)$. $\square$

We pause to observe why the cases $j = 0, n$ were omitted from the statement of Lemma 2.2. If either of these cases were asserted, then the conclusion would ensure that K is integral over R, a contradiction if R is not a field.

We next generalize the following formulation of [1, Theorem 6, (1) $\Leftrightarrow$ (2)]: if R is an integrally closed domain with quotient field K, then R is a Prüfer domain if and only if $P(R, K, v, 2, 1)$ is satisfied for each $v \in K$.

THEOREM 2.3   For an integrally closed domain R with quotient field K, the following conditions are equivalent:

(1) R is a Prüfer domain;

(2) For each $v \in K$, there exist integers $0 \le j \le n$ such that $P(R, K, v, n, j)$ is satisfied;

(3) There exist positive integers $j < n$ such that, for each $v \in K$, the property $P(R, K, v, n, j)$ is satisfied;

(4) For all positive integers $j < n$ and for all $v \in K$, the property $P(R, K, v, n, j)$ is satisfied.

Proof: The implications (4) $\Rightarrow$ (3) $\Rightarrow$ (2) are trivial. As for (2) $\Rightarrow$ (1), one may cite any of [10, Theorem 2 or Theorem 5], [7, Corollary 5], or [1, Theorem 6 or Theorem 9]. It remains only to prove that (1) $\Rightarrow$ (4). Assume (1), let $j < n$ be positive integers, and let $v \in K$; our task is to show that $P(R, K, v, n, j)$ is satisfied. By Lemma 2.1, we may suppose that R is quasilocal; that is, that R is a valuation domain. An application of Lemma 2.2 completes the proof. $\square$

One may summarize the above recapture of [1, Theorem 6, (1) $\Rightarrow$ (2)], namely the proof that (1) implies the case $j = 1$, $n = 2$ of (4) in Theorem 2.3, as follows. If $v \in K$, then $(R + Rv^2)_{R \backslash M} = R_M + R_M v^2$ for each maximal ideal M of R; thus, by globalization, R may be assumed valuation; and if R is a valuation domain, the easy case analysis in the proof of Lemma 2.2 establishes the result. This summary is part of what occurred to the author during Professor Kwak's talk. Our consideration of additional generality (replacing 1, 2 with j, n) will be rewarded in new positive results below.

A casual reading of [1] may lead one to question whether the property $P(v, 2, 1)$ captures the essential content of condition (2) in [1, Theorem 6]. After all, that condition in [1] refers to "quadratic" polynomials and if $v \in R$, the proof of Lemma 2.2 does not produce a second-degree polynomial having a unit coefficient of X and v as a root. In response to this critique, several comments are in order. First, despite the statement of condition (2) in [1, Theorem 6], the "quadratic" polynomials produced there and in several proofs in [1] may, in fact, have degree less than 2. In the following paragraph, we review the underlying construction from [1], identify a problem with the presentation in [1], and then show how to fine-tune that presentation so as to produce suitable genuinely quadratic (that is, second-degree) polynomials. Second, with the "quadratic" formulation of [1] then fully proved, we will revisit the above definitions and results, so as to generalize the "genuinely quadratic" version of [1, Theorem 6], once again replacing $(1,2)$ with arbitrary (j,n). Third, if all one wanted in a variant of Lemma 2.2 were a proof of the "genuinely quadratic" case, all that need be added to our earlier proof (in case $v \in R$) is consideration of the polynomial $X^2 + X + (-v^2 - v)$.

The relevant construction from [1], given explicitly in the proof of [1, Lemma 5 (2)], pertains to a Prüfer domain R with quotient field K and an element $v \in K$, expressed as $v = ab^{-1}$, with $a, b \in R$ and $b \neq 0$. Since R is a Prüfer domain, the ideal $I := (a,b)$ is invertible, and so $\alpha a + \beta b = 1$ for some $\alpha, \beta \in I^{-1}$. Then v is a root of the polynomial $f := (bX - a)(\alpha X - \beta) = b\alpha X^2 - X + a\beta \in R[X]$. (Hence, $g := -f \in R[X]$, $g(v) = 0$, and 1 is the coefficient of X in g.) This reasoning, drawn directly from [1], is valid. But, is f (or g) necessarily "quadratic"? The answer is in the negative, for $b\alpha$ may be 0. This unfortunate case occurs precisely when α has been chosen as 0. Then $\beta b = 1$, whence $\beta = b^{-1}$ and $v = ab^{-1} = a\beta \in II^{-1} = R$, a case that can be handled as in the preceding paragraph, by considering $X^2 + X + (-v^2 - v)$. In this way, the proof of [1, Lemma 5 (2)] (and with it, the proof of the "quadratic" assertion in [1, Theorem 6, (1) $\Rightarrow$ (2)]) may be rectified. (Incidentally, whenever $v \in R$, some such "rectification" as above is needed to supplement the reasoning in [1], for the choices $a = v$, $b = 1 = \beta$ and $\alpha = 0$ meet the conditions specified in [1] but produce only a degree-one f. A different "rectification", which may be more in the spirit of [1], can also be achieved in case $v \in R$, by choosing $b = 1 = \alpha$ and $\beta = 1 - a$, for these choices also produce second-degree f.)

Now that we have shown how to prove what was stated in [1, Theorem 6, (1) $\Leftrightarrow$ (2)], we proceed to generalize this "genuinely quadratic" result in the spirit of Lemma 2.1 - Theorem 2.3. The first step is to make the following definition. Consider a ring extension $A \subseteq B$, an element $v \in B$, and integers $0 \leq j \leq n$. We say that property $Q(A, B, v, n, j)$ [or, if no confusion is possible, property $Q(v, n, j)$] is satisfied if v is a root of a polynomial $f \in A[X]$ such that $\deg(f) = n$ and 1 is the coefficient of X^j in f.

Observe that property $Q(A, B, v, n, j)$ is stronger than property $P(A, B, v, n, j)$. Our interest in this new property arises because the "rectification" achieved for [1, Theorem 6, (1) $\Leftrightarrow$ (2)] two paragraphs ago may be formulated as follows: if R is an integrally closed domain with quotient field K, then R is a Prüfer domain if and only if $Q(R, K, v, 2, 1)$ is satisfied for each $v \in K$. Theorem 2.5 presents the promised generalizations of this formulation in terms of arbitrary (j, n).

LEMMA 2.4   If $Q(A, B, v, 2, 1)$ is satisfied, then $Q(A, B, v, n, j)$ is satisfied for all positive integers $j < n$.

Proof: We claim that it suffices to establish the case $j = 1$. Indeed, assuming this case, suppose $2 \leq j < n$. By the assumed case, $f(v) = 0$ for some $f \in A[X]$ such that $\deg(f) = n - j + 1$ (≥ 2) and 1 is the coefficient of X in f. Then $g := X^{j-1}f \in A[X]$, $g(v) = 0$, $\deg(g) = n$, and 1 is the coefficient of X^j in g. This proves the above claim.

It remains to show that $Q(A, B, v, n, 1)$ is satisfied for all integers $n \geq 2$. The case $n = 2$ holding by hypothesis, we proceed by induction on n, assuming that $Q(A, B, v, n - 1, 1)$ is satisfied for some $n \geq 3$. Thus

$$av^{n-1} + \cdots + v + b = 0$$

for some coefficients $a, \ldots, b \in A$, with $a \neq 0$. Multiplying the above-displayed equation by v produces the polynomial relation

$$av^n + \cdots + v^2 + bv = 0.$$

Moreover, $cv^2 + v + d = 0$ for some $c, d \in A$ with $c \neq 0$, since $Q(A, B, v, 2, 1)$ is satisfied. Multiplying this equation by $b - 1$ and subtracting the result from the last-displayed equation, we obtain

$$av^n + \cdots + (1 - (b-1)c)v^2 + v - (b-1)d = 0$$

whence $Q(A, B, v, n, 1)$ is satisfied, completing the proof. $\square$

We next obtain the promised strengthening of Theorem 2.3.

THEOREM 2.5   For an integrally closed domain R with quotient field K, the following conditions are equivalent:

(1) R is a Prüfer domain;
(2) For each $v \in K$, there exist integers $0 \leq j \leq n$ such that $Q(R, K, v, n, j)$ is satisfied;
(3) There exist positive integers $j < n$ such that, for each $v \in K$, the property $Q(R, K, v, n, j)$ is satisfied;
(4) For all positive integers $j < n$ and for all $v \in K$, the property $Q(R, K, v, n, j)$ is satisfied.

Proof: The implications (4) $\Rightarrow$ (3) $\Rightarrow$ (2) are trivial. As for (2) $\Rightarrow$ (1), one may cite any of [10, Theorem 2 or Theorem 5], [7, Corollary 5], or [1, Theorem 6 or Theorem 9]. It remains only to prove that (1) $\Rightarrow$ (4). Assume (1), let $j < n$ be positive integers, and let $v \in K$; our task is to show that $Q(R, K, v, n, j)$ is satisfied. By Lemma 2.4, it suffices to show that $Q(R, K, v, 2, 1)$ is satisfied. This, in turn, follows from the "rectification" of [1, Theorem 6] which was carried out in the third paragraph following the proof of Theorem 2.3. The proof is complete. $\square$

REMARK 2.6   It is of some interest to give a direct proof that (1) $\Rightarrow$ (4) in Theorem 2.5 in case R is a valuation domain. This may be done in the spirit of the proof of Lemma 2.2. Indeed, if $v \in R$, consider the polynomial $X^n + X^j + (-v^n - v^j)$; while if $v \notin R$ (so that $v^{-1} \in R$), consider $-(v^{-1})^{n-j}X^n + X^j$. $\square$

Let R be a domain with integral closure R' and quotient field K. Consider the following two conditions:

(i) There exists a positive integer n such that, for each $v \in K$, there exists an integer $j \leq n$ such that $P(R, K, v, n, j)$ is satisfied;

(ii) There exists a positive integer n such that, for each $v \in K$, there exist integers $j \leq m \leq n$ such that $Q(R, K, v, m, j)$ is satisfied.

It is evident from the definitions that (i) $\Leftrightarrow$ (ii). Moreover, if R is integrally closed, it follows from Theorems 2.3 and 2.5 that (i) and (ii) are equivalent to R being a Prüfer domain. What if R is not integrally closed? It follows from [10, Theorem 5] (or [7, Corollary 5] or [1, Theorem 9]) that (i) and (ii) each imply that R' is a Prüfer domain. It is natural to ask whether the converse is valid. Evidence for a negative answer was provided by Anderson-Kwak who noted at the close of [1] that $R := \mathbb{F}_3[[Y^3, Y^4, Y^5]]$ has Prüferian integral closure (indeed, $R' = \mathbb{F}_3[[Y]]$ is a DVR) although $Q(R, K, Y, 2, j)$ is not satisfied for any $j \leq 2$. In fact, the answer is in the negative, as the next example shows that R' being a Prüfer domain does not imply (i) or (ii).

EXAMPLE 2.7   Let $R = \mathbb{Q} + Y\overline{\mathbb{Q}}[[Y]]$, where $\overline{\mathbb{Q}}$ is an algebraic closure of $\mathbb{Q}$. Then $R' = \overline{\mathbb{Q}}[[Y]]$ is a Prüfer domain, but for each positive integer n, there exists $v \in R'$ such that $P(v, n, j)$ is not satisfied for any $j \leq n$. (In other words, R does not satisfy the above condition (ii).) To see this, let v be an algebraic number such that $[\mathbb{Q}(v) : \mathbb{Q}] > n$; for instance, take $v = 2^{\frac{1}{n+1}}$. If v is a root of a nontrivial degree-m polynomial with coefficients in R and some coefficient equal to 1, then by focussing on the constant terms in the resulting power series relation, we see that v is a root of some polynomial $g \in \mathbb{Q}[X]$ such that $\deg(g) \leq m$ and 1 is a coefficient of g. Hence, $n < [\mathbb{Q}(v) : \mathbb{Q}] \leq \deg(g) \leq m$, to complete the proof. $\square$

To make further headway, we next compare some of the ring-theoretic properties of the ring in Example 2.7 and the Anderson-Kwak ring, $\mathbb{F}_3[[Y^3, Y^4, Y^5]]$, that was discussed above. The latter domain is Noetherian (quasi) local, of (Krull) dimension 1, and analytically irreducible (i.e., its integral closure is a DVR which is a finitely generated module). Moreover, it is a quasilocal i-domain, in the sense of [14]; i.e., its integral closure is a valuation domain. In addition, the Anderson-Kwak ring is residually rational, in the sense of [3]; i.e., its residue class field ($\mathbb{F}_3$) is canonically isomorphic to the residue class field of its integral closure.

On the other hand, the ring R in Example 2.7 is quasilocal and one-dimensional but not Noetherian (indeed, it is not even coherent). This R is an i-domain, but it is not analytically irreducible (notice that R' is not finitely generated over R). Nor is R residually rational (this term is often reserved for certain Noetherian local one-dimensional analytically irreducible domains, anyway), since the residue class fields of R and R' (namely, $\mathbb{Q}$ and $\overline{\mathbb{Q}}$) are not canonically isomorphic.

In contrast to Example 2.7, the final goal of this note is to develop a positive result which applies, in particular, to a class of analytically irreducible Noetherian rings, such as $\mathbb{Q} + Y\mathbb{Q}(\sqrt{2})[[Y]]$, sharing some of the properties common to the ring in Example 2.7 and the Anderson-Kwak ring. For this purpose, we next recall some background on pseudovaluation domains (PVDs).

Recall from [12] that a domain R is said to be a PVD if R has a (uniquely determined) valuation overring V such that $\mathrm{Spec}(R) = \mathrm{Spec}(V)$ as sets. Equivalently [2, Proposition 2.6], R is a PVD if and only if $R \cong V \times_F k$ where (V, M) is a valuation domain, $F = V/M$, and k is a subfield of F; in this (essentially uniquely determined) pullback description, V is the valuation overring such that

$\text{Spec}(R) = \text{Spec}(V)$ and $k = R/M$. Now, let R be a PVD, with V, M, F and k as above. Then $\dim(R) = \dim(V)$; R is an i-domain $\Leftrightarrow R' = V$ (cf. [5, Remark 4.8(a)]) $\Leftrightarrow F$ is algebraic over k (cf. [8, Corollary 1.5(5)]). If R is an i-domain, then R is residually rational $\Leftrightarrow R = V \Leftrightarrow R$ is a valuation domain. Moreover, R is Noetherian $\Leftrightarrow [F : k] < \infty$ and V is a DVR (cf. [8, Proposition 1.8]). Finally, R is coherent $\Leftrightarrow$ either $[F : k] < \infty$ and M is a principal ideal of V or R is a valuation domain (cf. [6, Proposition 3.5], [2, Corollary A.9]).

The focus of Theorem 2.8 is on PVD i-domains. Care is needed because the ring in Example 2.7 is also a PVD i-domain; notice that it induces an infinite-dimensional extension of residue class fields. At the meeting, Michael Gilbert gave a talk explaining how some of [9] generalizes a result of Gilmer-Huckaba [11, Proposition 9] on PVD i-domains.

THEOREM 2.8   Let R be a PVD with associated valuation overring (V, M) and quotient field K. Set $k = R/M$ and $F = V/M$. Put $N := [F : k]$. Then:

   (a) Suppose R is a valuation domain (i.e., $N = 1$). Then for each $v \in K$ and for all positive integers $j < n$, the properties $P(R, K, v, n, j)$ and $Q(R, K, v, n, j)$ are satisfied.

   (b) Suppose that R is not a valuation domain (i.e., $N > 1$). Let n be a positive integer. Consider the following three conditions:

   (1) For each $v \in K$, there exists an integer $j \le n$ so that $P(R, K, v, n, j)$ is satisfied;

   (2) For each $v \in K$, there exist integers $j \le m \le n$ so that $Q(R, K, v, m, j)$ is satisfied;

   (3) $N \le n$.

Then $(3) \Rightarrow (1) \Leftrightarrow (2)$. Moreover, if F is a separable extension of k, then (1), (2), and (3) are equivalent.

Proof: (a) Apply Theorems 2.3 and 2.5.

   (b) It was observed that $(1) \Leftrightarrow (2)$ in the comments following Remark 2.6. Next, assume (3). In particular, $[F : k] < \infty$, whence F is algebraic over k and, by the above remarks, R is an i-domain and $V = R'$, the integral closure of R. Let $v \in K$; we shall show that $P(R, K, v, n, j)$ is satisfied for some integer $j \le n$.

Without loss of generality, $v \ne 0$. We may suppose that $v \in R'$. (Otherwise, $v^{-1} \in R'$. If $P(v^{-1}, n, j)$ is satisfied for some $j \le n$, consider $f \in R[X]$ such that $f(v^{-1}) = 0$, $\deg(f) = m \le n$, and 1 is the coefficient of X^j in f. Then the polynomial relation $v^m f(v^{-1}) = 0$ produces $g \in R[X]$ such that $g(v) = 0$, $\deg(g) \le m \le n$, and 1 is the coefficient of X^{m-j} in g. Hence, $P(v, n, m - j)$ is satisfied.)

Let $\pi : V \to F$ be the canonical surjection. Put $\alpha := \pi(v)$. Observe that there is a canonical isomorphism $R[X]/M R[X] \to k[X]$. Consider the commutative diagram

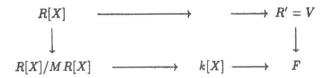

where the vertical homomorphisms are the canonical surjections, the top horizontal algebra map is evaluation at v, and the bottom horizontal maps are the above-mentioned canonical isomorphism and evaluation at α. Since

$$[k(\alpha):k] \leq [F:k] = N \leq n$$

there exists a monic polynomial $\overline{h} \in k[X]$ such that $\overline{h}(\alpha) = 0$ and $m := \deg(\overline{h}) \leq n$. Choose a monic degree-$m$ polynomial $h \in R[X]$ such that the composite of surjections $R[X] \to R[X]/MR[X] \to k[X]$ sends h to $\overline{h}$. Observe that $\deg(h) = \deg(\overline{h}) = m \leq n$ and, by chasing h through the above commutative diagram, we have that $\pi(h(v)) = \overline{h}(\alpha) = 0$, whence $h(v) \in M \subseteq R$. Then $g := h - h(v) \in R[X]$, $g(v) = 0$, $\deg(g) = \deg(h) = m \leq n$, and g is monic. In particular, $Q(v, m, m)$ is satisfied; thus, so is $P(v, n, m)$, and (1) follows.

Finally, suppose that (1) and (2) are satisfied. By the oft-cited motivation for this note ([10, Theorem 5], [7, Corollary 5], [1, Theorem 9]), R' is a Prüfer domain; it follows (cf. [12, Theorem 1.7] and [5, Remark 4.8 (a)]) that $R' = V$ and F is algebraic over k. Assume next that F is separable over k. Then, in order to establish (3), it suffices, by [13, Lemma 1.7, p. 263], to show that $[k(\alpha):k] \leq n$ for each $\alpha \in F$. Let $\pi : V \to F$ be the canonical surjection, and choose $v \in V$ such that $\pi(v) = \alpha$. By hypothesis, $Q(v, m, j)$ is satisfied, for some integers $j \leq m \leq n$. Thus, there exists $h \in R[X]$ such that $h(v) = 0$, $\deg(h) = m$, and 1 is the coefficient of X^j in h. Let $\overline{h} \in k[X]$ be the image of h under the composite $R[X] \to R[X]/MR[X] \to k[X]$ considered above. Observe that $\deg(\overline{h}) \leq \deg(h) = m \leq n$ and 1 is the coefficient of X^j in $\overline{h}$. By chasing h through the above commutative diagram, we have that $\overline{h}(\alpha) = \pi(h(v)) = \pi(0) = 0$. Therefore, $[k(\alpha):k] \leq \deg(\overline{h}) \leq n$, to complete the proof. $\square$

REMARK 2.9 (a) We begin with a few observations about the conditions in the statement of Theorem 2.8 (b). If (the PVD) R is coherent (for instance, Noetherian), then (3) is satisfied for some positive integer n, since $[F:k] < \infty$. Moreover, if (1) or (2) is satisfied, then (the PVD) R is an i-domain. Finally, Theorem 2.8 (b) pinpoints the reason that made Example 2.7 possible, namely the fact that $[\overline{\mathbb{Q}}:\mathbb{Q}] = \infty$.

(b) Next, we show how to construct some Noetherian local one-dimensional analytically irreducible (but not residually rational) domains R which do not share the behavior of the ring in Example 2.7. Let R be a Noetherian PVD which is not a field; i.e., $R \cong V \times_F k$ where (V, M) is a DVR (but not a field), $F = V/M$ and k is a subfield of F such that $[F:k] < \infty$. By the results recalled prior to the statement of Theorem 2.8, it is easy to verify that R has the asserted properties. Moreover, if F is separable over k, then each element in the quotient field of R is a root of some polynomial in $R[X]$ which has 1 as a coefficient and whose degree is at most $[F:k]$. Thus, in contrast to the Anderson-Kwak example of $\mathbb{F}_3[[Y^3, Y^4, Y^5]]$, we see that if $R := \mathbb{F}_3 + Y\mathbb{F}_9[[Y]]$, then each element in the quotient field of R is a root of some polynomial in $R[X]$ having 1 as a coefficient and degree at most 2. For a similar example in characteristic 0, consider $R := \mathbb{Q} + Y\mathbb{Q}(\sqrt{2})[[Y]]$.

(c) We close by stressing the degree-sensitive nature of Theorem 2.8 (b), in contrast to the emphasis on quadratic polynomials in [1]. Let R be a PVD i-domain, with V, M, K, k and F as in Theorem 2.8, such that F is separable over k. For each $v \in K$, let

$$m(v) := \min \{\deg(h) : h \in R[X], h(v) = 0, 1 \text{ is a coefficient of } h\}.$$

Example 2.7 showed that there need not be a positive integer n such that $m(v) \leq n$ for all $v \in K$. However, Theorem 2.8 (b) showed that such an integer n exists if and only if $[F : k] < \infty$; and, moreover, in this case, $[F : k]$ is the minimal such n. $\square$

REFERENCES

1. D.D. Anderson and D.J. Kwak, The u, u^{-1} lemma revisited, *Comm. Algebra* *24*: 2447–2454 (1996).
2. D.F. Anderson and D.E. Dobbs, Pairs of rings with the same prime ideals, *Can. J. Math. 32*: 362-384 (1980).
3. V. Barucci and R. Fröberg, Maximality properties for one-dimensional analytically irreducible local Gorenstein and Kunz rings, *Math. Scand.*, to appear.
4. N. Bourbaki, *Commutative Algebra*, Addison-Wesley, Reading (1972).
5. D.E. Dobbs, Coherence, ascent of going-down and pseudo-valuation domains, *Houston J. Math. 4*: 551-567 (1978).
6. D.E. Dobbs, On the weak global dimension of pseudo-valuation domains, *Canad. Math. Bull. 21*: 159-164 (1978).
7. D.E. Dobbs, On INC-extensions and polynomials with unit content, *Canad. Math. Bull. 23*: 37-42 (1980).
8. M. Fontana, Topologically defined classes of commutative rings, *Ann. Mat. Pura Appl. 123*: 331-355 (1980).
9. M.S. Gilbert, Extensions of commutative rings with linearly ordered intermediate rings, in preparation.
10. R. Gilmer and J.F. Hoffmann, A characterization of Prüfer domains in terms of polynomials, *Pac. J. Math. 60*: 81-85 (1975).
11. R. Gilmer and J.A. Huckaba, Δ-rings, *J. Algebra 28*: 414-432 (1974).
12. J. Hedstrom and E.G. Houston, Pseudovaluation domains, *Pac. J. Math. 75*: 137-147 (1978).
13. S. Lang, *Algebra*, third edition, Addison-Wesley, Reading (1993).
14. I.J. Papick, Topologically defined classes of going-down domains, *Trans. Amer. Math. Soc. 219*: 1-37 (1976).

On Flat Divided Prime Ideals

DAVID E. DOBBS Department of Mathematics, University of Tennessee, Knoxville, Tennessee 37996-1300.

To Ira Papick, on his 50th birthday

1 INTRODUCTION

Throughout this note, all rings considered are commutative, with identity. One motivation for the present work arises from the classical result that Tor may be used to find the global dimensions of a Noetherian ring (cf. [28, Corollaire 2, p. IV-35]). An immediate consequence of the cited result is that if R is a Noetherian ring, then gl. $\dim(R) = $ w. gl. $\dim(R) = \sup wd_R(R/M)$, where M ranges over Max(R), the set of all the maximal ideal of R. By way of contrast, if R is an arbitrary ring, then w. gl. $\dim(R) = \sup wd_R(R/I)$, where I ranges over the set of all the ideals of R. It is not evident for which rings R, besides Noetherian rings, the above supremum may be taken over Max(R), though [20, Corollaries 2.5.5 and 2.5.10] essentially examines the extent to which the techniques in [28] generalize to coherent rings. Our first result, Theorem 2.1, asserts that if R is a certain type of LPVD (i.e., a locally pseudo-valuation domain, in the sense of [14]), then w. gl. $\dim(R) = \sup\{wd_R(R/M) : M \in \text{Max}(R)\}$. (For background, recall from [23] that a domain R is called a PVD, or pseudo-valuation domain, if R has a valuation overring V such that Spec(R) = Spec(V) as sets. Any valuation domain is evidently a PVD; and, in fact, PVDs may be characterized as certain pullbacks involving valuation domains and field extensions [1, Proposition 2.6]. A domain R is called an LPVD if R_P is a PVD for each $P \in$ Spec(R) (equivalently, for each $P \in$ Max(R)). Any PVD is an LPVD, as is any Prüfer domain.)

The key to proving Theorem 2.1 lies in applying the main result of [10]. According to this result [10, Theorem 2.3], if (R, M) is a PVD but not a valuation domain, then w. gl. $\dim(R) = 2 \Leftrightarrow M$ is R-flat $\Leftrightarrow$ each prime ideal of R is R-flat; and, under the same hypotheses, w. gl. $\dim(R) = \infty \Leftrightarrow M$ is not R-flat $\Leftrightarrow wd_R(M) = \infty$. Accordingly, it is natural to ask about the weak dimensions of the nonmaximal prime ideals of a PVD. In Corollary 2.6 and Remark 2.7(a), we recover [10, Corollary 3.3], namely that each of these weak dimensions is 0; i.e., each nonmaximal prime ideal of a PVD is flat. More is true. In Corollary 2.5, the preceding assertion is generalized to the context of Q-primary ideals for nonmaximal primes Q in an LPVD. Our main technical result, Theorem 2.3, is couched still more generally, assuming only a primary ideal contained in a divided prime of a domain. (For background, recall from [8] that a prime ideal P of a domain R is called divided in R if $PR_P = P$; i.e., if P is comparable to each ideal of R with respect to inclusion. A domain R is called a divided domain if each of its prime ideals is divided in R; and "locally divided domains" are defined in the natural way. Each PVD is a divided domain, and so each LPVD is a locally divided domain.)

Divided prime ideals have arisen in, i.a., the Greenberg-Vasconcelos theory of umbrella rings and the quasilocal rings of global dimension 2 (cf. [29], [21], [31]). For this reason, it has often been convenient to assume that the relevant divided prime ideals in certain pullbacks are flat (see the riding hypotheses in [21] and [22], for instance). Theorem 2.3 is designed to clarify some of this work and, in particular, it is used below to sharpen some criteria for coherence (as in [20], as modified from [22]). The pullbacks of chief interest below are CPI-extensions, in the sense of [3]. (For some other recent work on CPI-extensions, see [13].) While flatness of the relevant divided primes P is posited in the coherence criteria for certain pullbacks R in [22] and [20], one may ask to compare such results with [19, Proposition 4.6 and Theorem 4.7], in which one posits instead that R_P is a valuation domain. ([19] is a recent paper tracing back, in part, to the characterization of coherence for the classical $D + M$ construction [17, Theorem 3] and the subsequent extension of those methods to the general $D + M$ construction [4, Theorem 3].) The relationship between the coherence results in [22] and [19] is examined in Remark 2.10, and thus constitutes the principal motivation for Theorem 2.3. The paper concludes by merging several themes in Theorem 2.11 and Corollary 2.12, which characterize when CPI-extensions are LPVDs, coherent LPVDs, PVDs, or coherent PVDs.

2 RESULTS

We begin by showing how the results of [10] may be used to show that, with regard to the calculation of weak global dimension, certain LPVDs behave like Noetherian rings. For background on weak global dimension, see [5].

THEOREM 2.1 If R is an LPVD (for instance, a PVD), then w. gl. $\dim(R) = \sup(\delta_M + wd_R(R/M)) = \delta + \sup wd_R(R/M)$, where M ranges over $\mathrm{Max}(R)$,

$$\delta_M := \begin{cases} 1, & \text{if } MR_M = (MR_M)^2 \text{ and } R_M \text{ is not a valuation domain (or a field)} \\ 0, & \text{otherwise} \end{cases}$$

and

$$\delta := \sup \delta_M = \begin{cases} 1, & \text{if there exists } N \in \text{Max}(R) \text{ such that} \\ & NR_N = (NR_N)^2 \text{ and } R_N \text{ is not a valuation} \\ & \text{domain (or a field)} \\ 0, & \text{otherwise.} \end{cases}$$

Proof: Suppose first that (R, M) is a PVD. If R is a field, then w. gl. $\dim(R) = 0 = wd_R(R/M)$ since each module (vector space) over a field is flat. If R is a valuation domain but not a field, then $wd_R(R/I) \leq 1$ for each ideal I of R since I is R-flat, whence w. gl. $\dim(R) = \sup wd_R(R/I) \leq 1$; the assertion now follows in this case since R/M is not R-flat (indeed, not even torsion-free over R, as a consequence of $M \neq 0$). We may assume henceforth that (the PVD) R is not a valuation domain. If $M = M^2$, then M is R-flat and w. gl. $\dim(R) = 2$ [10, Theorem 2.3(a)]; the assertion follows in this case since $wd_R(R/M) = 1$ and $\delta_M = 1$. Finally, suppose that (R is a PVD, but not a valuation domain, and) $M \neq M^2$. Then w. gl. $\dim(R) = \infty = wd_R(M)$, by [10, Theorem 2.3(b)]. As R/M is not R-flat, $wd_R(R/M) = 1 + wd_R(M) = 1 + \infty = \infty$, and so $\delta_M + wd_R(R/M) = 0 + \infty = \infty =$ w. gl. $\dim(R)$, thus completing the proof in case R is a PVD.

For the general case (R an LPVD), consider $M \in \text{Max}(R)$. Observe that $R/M \otimes_R R_M \cong R_M/MR_M \cong R/M$. Since R_M is R-flat, we thus see, by applying $\cdot \otimes_R R_M$ to an R-flat resolution of R/M, that $wd_{R_M}(R/M) \leq wd_R(R/M)$. The reverse inequality holds by the transitivity of flatness, and so $wd_{R_M}(R/M) = wd_R(R/M)$. By the PVD case considered above, we thus have w. gl. $\dim(R_M) = \delta_M + wd_R(R/M)$. Taking suprema as M ranges over $\text{Max}(R)$, we conclude that

$$\begin{aligned} \sup wd_R(R/M) = \sup wd_{R_M}(R/M) &\leq \sup \text{ w. gl. } \dim(R_M) \\ &= \text{w. gl. } \dim(R) = \sup(\delta_M + wd_R(R/M)) \\ &\leq \sup \delta_M + \sup wd_R(R/M) = \delta + \sup wd_R(R/M). \end{aligned}$$

It remains only to prove that w. gl. $\dim(R)$ is not less than $\delta + \sup wd_R(R/M)$. Deny. Then, by the above analysis, R is not quasilocal, $wd_R(R/M) = 1$ for all $M \in \text{Max}(R)$, and there exists $N \in \text{Max}(R)$ such that $NR_N = (NR_N)^2$ and R_N is not a valuation domain (or a field). As the work cited above from [10] ensures that w. gl. $\dim(R_N) = 2$, we have

$$2 = \text{w. gl. } \dim(R_N) \leq \sup \text{w. gl. } \dim(R_M) < \delta + \sup wd_R(R/M) = 1 + 1 = 2,$$

the desired contradiction. $\square$

In view of the definitions of δ_M and δ in Theorem 2.1, it is natural to observe that if P is a prime ideal of a domain R, then $P^n R_P = (PR_P)^n$ for each positive integer n. In particular, $P^n R_P = P^n$ if P is divided in R. This fact motivates Proposition 2.2(b). To motivate Proposition 2.2(a), recall (cf. [2, Example 3, p. 51]) that P^n need not be P-primary if P is an arbitrary prime ideal of a domain.

PROPOSITION 2.2   (a) If P is a divided prime of a domain R, then P^n is a P-primary ideal of R, for each positive integer n.

(b) Let R be a domain, $Q \subseteq P$ prime ideals of R, and A a Q-primary ideal of R. If P is divided in R, then $AR_P = A$.

Proof: (a) We show that if $a, b \in R$ satisfy $ab \in P^n$ and $a \notin \mathrm{rad}_R(P^n) = P$, then $b \in P^n$. As $ab \in P$ and P is prime, $b \in P$. Thus $b = (ab)a^{-1} \in P^n R_P = P^n$ by the above remarks. (Alternate proof of (a): Since $\mathrm{rad}_{R_P}(P^n R_P) = PR_P \in \mathrm{Max}(R_P)$, it follows that $P^n R_P$ is a $PR_P = P$-primary ideal of R_P [2, Proposition 4.2]. Hence, $P^n = P^n R_P \cap R$ is a P-primary ideal of R [2, Proposition 4.8].)

(b) Evidently, $A \subseteq AR_P$. It thus suffices to show that if $u := az^{-1}$ with $a \in A$ and $z \in R \backslash P$, then $u \in A$. Observe that $u \in AR_P \subseteq QR_P \subseteq PR_P = P \subseteq R$. Moreover, $zu = a \in A$. As $z \in R \backslash P \subseteq R \backslash Q = R \backslash \mathrm{rad}_R(A)$ and A is a primary ideal, it follows that $u \in A$. $\square$

The referee has observed the following sharpening of Proposition 2.2(b) (and the reader is invited to reformulate variants of (2.3)-(2.5) using it). If P is a divided prime of a domain R and A is an ideal of R such that $A \subseteq P$ and $AR_P \cap R = A$, then $AR_P = A$. For a proof, observe that $A = AR_P \cap R = AR_P \cap PR_P \cap R = AR_P \cap PR_P = (A \cap P)R_P = AR_P$.

We next present our main technical result.

THEOREM 2.3   Let R be a domain, $Q \subseteq P$ prime ideals of R such that P is divided in R, and A a Q-primary ideal of R. Then:

(a) $AR_P = A$.

(b) A is R-flat if (and only if) A is R_P-flat.

Proof: Of course, (a) is just the second assertion in Proposition 2.2, repeated here for convenience. As for (b), the "only if" assertion follows from (a) and the fact that flatness is preserved by base change, since $A = AR_P \cong A \otimes_R R_P$. Conversely, we show that if A is R_P-flat, then the multiplication map $m : I \otimes_R A \to R$ is a monomorphism for each ideal I of R. Now, since $I \otimes_R R_P \cong IR_P$ is an ideal of R_P and $A \otimes_R R_P \cong A$ is a flat ideal of R_P, the multiplication map $\mu : (I \otimes_R R_P) \otimes_{R_P} (A \otimes_R R_P) \to R_P$ is a monomorphism. Consider the commutative diagram

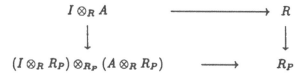

where the horizontal maps are m and μ, the left vertical map is the canonical isomorphism, and the right vertical map is the inclusion map. By a diagram chase, m is a monomorphism. $\square$

We next generalize a result of Greenberg [22, p. 19] which states that if P is a divided prime ideal of a domain R such that R_P is a valuation domain, then P is R-flat.

COROLLARY 2.4 Let R be a domain, $Q \subseteq P$ prime ideals of R such that P is divided in R, and A a Q-primary ideal of R. If R_P is a valuation domain, then A is R-flat.

Proof: By Theorem 2.3(a) or Proposition 2.2(b), A is an ideal of R_P. As *any* ideal of a valuation domain is flat, an application of Theorem 2.3(b) completes the proof. $\square$

COROLLARY 2.5 Let R be an LPVD (for instance, a PVD), Q a nonmaximal prime ideal of R, and A a Q-primary ideal of R. Then A is R-flat.

Proof: Suppose first that R is a PVD. Then R is a divided domain [9, p. 560]. Corollary 2.4 may now be applied, with $P := Q$, as the nonmaximality of P ensures that R_P is a valuation domain [23, Proposition 2.6]. The assertion is thus established if R is quasilocal.

For the general case, we need only show that AR_M is R_M-flat for each $M \in$ $\mathrm{Max}(R)$. Without loss of generality, $A \subseteq M$ (for, otherwise, $AR_M = R_M$ is trivially R_M-flat). As $Q = \mathrm{rad}_R(A) \subseteq \mathrm{rad}_R(M) = M$, it follows easily that AR_M is a QR_M-primary ideal of R_M (cf. [2, Proposition 4.8(ii), p. 53]). Since QR_M is a nonmaximal ideal of the pseudo-valuation domain R_M, the case established in the preceding paragraph yields that AR_M is R_M-flat, as desired. $\square$

The quasilocal case of Corollary 2.6 was obtained in [10, Corollary 3.3] by other means.

COROLLARY 2.6 If R is an LPVD (for instance, a PVD), then each nonmaximal prime ideal of R is R-flat.

Proof: If $Q \in \mathrm{Spec}(R) \backslash \mathrm{Max}(R)$, apply Corollary 2.5, with $A := Q$. $\square$

REMARK 2.7 (a) In addition to Corollary 2.6 and [10, Corollary 3.3], there is another natural way to show that each nonmaximal prime ideal of a PVD is flat. To wit, one need only rework the proof of [17, Theorem 7]. As written, that proof in [17] is given for the classical $D + M$ construction, with $R := k + M \subseteq V := K + M$, where (V, M) is a nontrivial valuation domain and $k \subseteq K$ is a field extension. The proof adapts *verbatim* to the more general case of a pseudo-valuation domain (R, M), with canonically associated valuation domain V, $K = V/M$, and $k = R/M$. To carry out the proof in this more general case, one also needs to recall the standard fact that a nonzero nonmaximal prime ideal of a valuation domain V cannot be a principal ideal of V. (For a generalization of this standard fact, in which V is replaced by an arbitrary going-down domain, see [9, Corollary 2.4]. For background on going-down domains, see [7], [16].)

(b) Another (in fact, the main) result in [17] is [17, Theorem 3], the characterization of coherence for the classical $D + M$ construction which was mentioned in the Introduction (and motivates most of the remainder of the present paper). A key step in the proof of [17, Theorem 3] referred to a result of Ferrand [18, Lemme] on the descent of flatness. (More recent studies of coherence, such as [19], have developed ideal-theoretic techniques which avoid such homological methods.) It is interesting to note what results if one applies [18, Lemme] to the context of a divided prime ideal P of a domain R (which is part of the riding assumptions for the work on coherence in [22, Proposition 4.1] and its reformulation in [20]). For such P and R, [18, Lemme] yields that P is R-flat if (and only if) P is R_P-flat and P/P^2 is R/P-flat. We now see that the condition "P/P^2 is R/P-flat" is redundant. Indeed, for P and R as above, it follows from Theorem 2.3(b), with $A := Q := P$, that P is R-flat if (and only if) P is R_P-flat.

Prior to the statement of Corollary 2.4, we had cause to discuss Greenberg's studies [22, p.19] involving a divided prime P of a domain R and the interplay with R_P. The same context arose in the final part of Remark 2.7(b). This context is more general than may appear at first blush, as it was noted in [15, Lemma 2.5(iii),(iv),(v)] that if (T, P) is a quasilocal domain and $R = D \times_{T/P} T$, where D is a domain having quotient field T/P, then $D \cong R/P$, $T = R_P$, and P is a divided prime ideal of R. It can be shown by standard gluing methods that under the above conditions, $P \in \operatorname{Max}(R)$ if and only if D is a field.

It seems natural to ask what happens if one modifies the above construction, now assuming only that P is a prime ideal of a domain R. The resulting pullback, $R/P \times_{R_P/PR_P} R_P$, is just $R + PR_P$; in [3], this ring was introduced and called the CPI-extension of R with respect to P. Evidently, $R + PR_P = R$ if and only if P is divided in R. In general, setting $D := R + PR_P$ and $Q := PR_P$ for convenience, we have that Q is a divided prime of D [3, Proposition 2.5, Theorem 2.4] (cf. also the proof of [8, Lemma 2.4(b),(c)]). Moreover, $D/Q \cong R/P$ and $D_Q = R_P$ by [11, Lemma 2.2(a)].

It is convenient to record the following facts next.

PROPOSITION 2.8   Let P be a prime ideal of a domain R. Then:

(a) $R + PR_P = R_P$ if and only if $P \in \operatorname{Max}(R)$.

(b) PR_P is $(R + PR_P)$-flat if and only if PR_P is R_P-flat. In particular, these conditions hold if R_P is a valuation domain.

Proof: (a) Suppose $P \in \operatorname{Max}(R)$. Then the inclusion map $R/P \to R_P/PR_P$ is an isomorphism. But the pullback of an isomorphism is an isomorphism, and so the inclusion map $R/P \times_{R_P/PR_P} R_P \to R_P$ is an isomorphism; that is, $R + PR_P = R_P$.

We pause to give a second, more computational, proof of the "if" assertion. It suffices to prove, for P maximal, that $z^{-1} \in R + PR_P$ for each $z \in R \backslash P$. As $P + Rz = R$, we have $p + rz = 1$ for some $p \in P$, $r \in R$. Then $z^{-1} = r + pz^{-1} \in R + PR_P$, completing the second proof.

For the "only if" assertion, let $D := R + PR_P$ and $Q := PR_P$ as above, and assume that $D = R_P$. As $R/P \cong D/Q = R_P/PR_P$ is a field, $P \in \operatorname{Max}(R)$. The interested reader may verify that an alternate proof of the "only if" assertion is available by applying [3, Theorem 2.6] to obtain an order-isomorphism between the posets $\operatorname{Spec}(D)$ and $\{J \in \operatorname{Spec}(R) : \text{either } J \subseteq P \text{ or } P \subseteq J\}$.

(b) As above, let $D := R + PR_P$ and $Q := PR_P$. As Q is a divided prime of D, the final observation in Remark 2.7(b) ensures that Q is D-flat if (and only if) Q is D_Q-flat. Since $D_Q = R_P$, the proof of the equivalence is complete. One proof of the final assertion follows since each ideal of a valuation domain is flat. For more baroque proofs of the final assertion, apply Corollary 2.4 or the result of Greenberg mentioned prior to Corollary 2.4.   □

We next reformulate some results on coherence in terms of CPI-extensions. The core results, drawn from [22] and [19], are then compared in Remark 2.10(a), with the aid of Theorem 2.3(b).

PROPOSITION 2.9   Let P be a prime ideal of a domain R. Then:

(a) Suppose that PR_P is $(R+PR_P)$-flat. Then $R+PR_P$ is coherent if and only if R/P and R_P are both coherent.

(b) Suppose that P is not a maximal ideal of R. Then $R + PR_P$ is coherent if and only if R/P is coherent and R_P is a valuation domain.

Proof: (a) As usual, let $D := R + PR_P$ and $Q := PR_P$. Since $D/Q \cong R/P$ and $D_Q = R_P$, Proposition 2.8(b) allows us to replace R with D and P with Q. Thus, without loss of generality, P is a flat divided prime ideal of R. The assertion in (a) is now seen as a restatement of [22, Proposition 4.1] (or [20, Theorem 5.1.5]).

(b) The assertion is evident if $P = 0$, and so we assume henceforth that $P \neq 0$. By using D and Q as in (a), as well as invoking the order-isomorphism [3, Theorem 2.6] for prime spectra of CPI-extensions, we may also assume that P is divided (and nonmaximal) in R. Since any valuation domain is coherent, the assertion in (b) is now seen as the result of applying [19, Theorem 4.7] to the pullback defining the CPI-extension of R with respect to P.   □

Proposition 2.8(a) shows that one cannot delete the hypothesis in Proposition 2.9(b) that P is nonmaximal in R, since a quasilocal coherent domain need not be a valuation domain.

REMARK 2.10   (a) It is natural to ask to what extent the results in parts (a) and (b) of Proposition 2.9 may be viewed as the "same" result. (Their proofs are certainly not the same!) As a partial answer, we offer the following result. Let P be a divided prime ideal of a coherent domain R. Then P is R-flat if and only if R_P is a valuation domain.

For a proof, note first that R_P inherits coherence from R (cf. [20, Theorem 2.4.2]). If P is R-flat, then R_P is a quasilocal coherent domain with flat maximal ideal and so, by a result of Vasconcelos [30, Lemma 3.9], R_P is a valuation domain. Conversely, if R_P is a valuation domain, then $P(= PR_P)$ is R_P-flat, and so by applying Theorem 2.3(b) [with $A := Q := P$], we have that P is R-flat, completing the proof.

Observe that the coherence of R was not used in the proof of the "if" assertion in the preceding paragraph. It would be interesting to find other connections between the results in Proposition 2.9 which, in this spirit, also avoid assuming the coherence of R.

(b) In the result established in (a), the hypothesis that R is coherent may be weakened to "R is a finite-conductor domain". To see this, one need only apply the above argument, with its appeal to [30, Lemma 3.9] now replaced by reasoning as in the proof of [12, Theorem 3.8].

We next digress by recounting some history concerning the "finite-conductor" concept. Recall that a domain R is called a finite-conductor domain if $Ra \cap Rb$ is a finitely generated ideal of R for all $a, b \in R$. The natural examples of finite-conductor domains are arbitrary coherent domains and arbitrary GCD-domains. Beginning in [17] and continuing through papers such as [19], it has often been convenient to examine coherence for certain pullbacks in terms of the weaker property of "finite-conductor". This property apparently first appeared (with the abbreviation "FC") in [6]. The abbreviation "FC" was ill-advised because of its prior meaning of "finitely many conjugates" in group theory (cf. [27, p. 441]), and so "FC" was soon abandoned in reference to the "finite-conductor" concept. Whatever credit may be appropriate for introducing the "finite-conductor" terminology should, we believe, be assigned to Steve McAdam, for the author first saw this terminology (and "FC") in a preprint of [25] which was consulted while writing [6]. It should be noted that neither the terminology nor the abbreviation appear in the printed version of [25], though that article seems to be the one making the first serious use of the "finite-conductor" concept.

If (R, M) is a PVD with canonically associated valuation overring V, then R is the pullback $R/M \times_{V/M} V$ (cf. [1, Proposition 2.6]). In view of the work reported in Proposition 2.9, it is perhaps not surprising that coherence was studied for PVDs some time ago. (Such studies appeared independently in [24] and [9].) It now seems natural to ask which CPI-extensions are coherent PVDs. Our final two results address related issues.

THEOREM 2.11   Let P be a nonmaximal prime ideal of a domain R. Then the following conditions are equivalent:

(1) $R + PR_P$ is a PVD (resp., a coherent PVD);

(2) R/P is a PVD (resp., a coherent PVD) and R_P is a valuation domain.

Proof: The parenthetical assertion follows by combining the other assertion with Proposition 2.9(b). As for the "other" assertion, we may replace R and P with $R + PR_P$ and PR_P, as in the proof of Proposition 2.9(b), and so we may assume, without loss of generality, that P is divided (and nonmaximal) in R.

(1) $\Rightarrow$ (2): Suppose that R is a PVD. Then, by [9, Lemma 4.5(i)], so is R/P. Moreover, since P is nonmaximal, [23, Proposition 2.6] yields that R_P is a valuation domain.

(2) $\Rightarrow$ (1): Assume (2). Let W be the canonically associated valuation overring of the pseudo-valuation domain R/P; view $W \subseteq R_P/PR_P = R_P/P$ in the usual way. Put $\mathcal{W} := \pi^{-1}(W)$, where $\pi : R_P \to R_P/P$ denotes the canonical surjection. Observe that R_P and W are valuation domains and that R_P/P is the quotient field of W. Since $\mathcal{W} = W \times_{R_P/P} R_P$, it follows by "Nagata composition" [26, (11.4), p. 35] that $\mathcal{W}$ is a valuation domain. Let M denote the maximal ideal of $\mathcal{W}$. As $W = \pi(\mathcal{W}) = \mathcal{W}/P$, the maximal ideal of W is M/P. Since $\text{Spec}(R/P) = \text{Spec}(W)$, M/P is also the maximal ideal of R/P. Thus, $M \in \text{Max}(R)$. We next indicate three ways to complete the proof from this point.

First, since R/M is a field and $R = R/M \times_{\mathcal{W}/M} \mathcal{W}$, it follows from the pullback characterization of PVDs [1, Proposition 2.6] that R is a PVD (with canonically associated valuation overring $\mathcal{W}$), as required.

For a second proof, note that R/P (being a PVD) is quasilocal. As P is divided in R, P is contained in each maximal ideal of R, and so R is quasilocal, necessarily with maximal ideal M. By citing [23, Theorem 2.7], we may thus conclude a second proof that R is a PVD.

The third proof is the most "diagrammatic" (and the most baroque). Like the first proof, it aims to establish that $R = R/M \times_{W/M} W$ (and then invoke [1, Proposition 2.6]). To do so, we obtain the pullback diagram (Cartesian square)

$$
\begin{array}{ccc}
R & \longrightarrow & \mathcal{W} \\
\downarrow & & \downarrow \\
R/M & \longrightarrow & \mathcal{W}/M
\end{array}
$$

by juxtaposing the following two pullback diagrams:

$$
\begin{array}{ccc}
R/P & \longrightarrow & \mathcal{W} \\
\downarrow & & \downarrow \\
R/M \cong (R/P)/(M/P) & \longrightarrow & \mathcal{W}/(M/P) = (\mathcal{W}/P)/(M/P) \cong \mathcal{W}/M
\end{array}
$$

$$
\begin{array}{ccc}
R & \longrightarrow & \mathcal{W} \\
\downarrow & & \downarrow \\
R/P & \longrightarrow & \mathcal{W}/P = \mathcal{W}.
\end{array}
$$

$\square$

If the "nonmaximal" hypothesis is deleted from Theorem 2.11, then (2) $\Rightarrow$ (1) [by combining Proposition 2.8(a) with the fact that each valuation domain is a coherent PVD], but (1) $\not\Rightarrow$ (2) [by combining Proposition 2.8(a) with the fact that not every coherent PVD is a valuation domain]. Similarly, one cannot delete the "nonmaximal" hypothesis in Corollary 2.12.

COROLLARY 2.12   Let P be a nonmaximal prime ideal of a domain R. Then the following conditions are equivalent:
(1) $R + PR_P$ is an LPVD (resp., a coherent LPVD);
(2) R/P is an LPVD (resp., a coherent LPVD) and R_P is a valuation domain.
Proof: As in the proof of Theorem 2.11, an appeal to Proposition 2.9(b) reduces our task to proving the nonparenthetical assertion. Also as in the proof of Theorem 2.11, the basic facts about CPI-extensions allow us to assume, without loss of generality, that P is divided (and nonmaximal) in R. Now, consider $M \in \mathrm{Max}(R)$. Since P is divided in R, P and M are comparable, and so P is properly contained in M. In fact, PR_M is a nonmaximal divided prime of R_M, for

$$
PR_M \subseteq PR_M(R_M)_{PR_M} = PR_M R_P = PR_P = P \subseteq PR_M.
$$

Hence, letting M vary over $\text{Max}(R)$ and applying Theorem 2.11 to the pseudo-valuation domains R_M, we have that $R(= R + PR_P)$ is an LPVD $\Leftrightarrow R_M$ is a PVD for each $M \in \text{Max}(R) \Leftrightarrow R_M/PR_M \ (\cong (R/P)_{M/P})$ is a PVD and $(R_M)_{PR_M}$ $(= R_P)$ is a valuation domain for each $M \in \text{Max}(R) \Leftrightarrow R/P$ is an LPVD and R_P is a valuation domain. Thus, $(1) \Leftrightarrow (2)$. $\square$

REFERENCES

1. D.F. Anderson and D.E. Dobbs, Pairs of rings with the same prime ideals, *Can. J. Math. 32*: 362–384 (1980).
2. M.F. Atiyah and I.G. Macdonald, *Introduction to Commutative Algebra*, Addison-Wesley, Reading (1969).
3. M.B. Boisen, Jr. and P.B. Sheldon, CPI-extensions: overrings of integral domains with special prime spectrums, *Can. J. Math. 29*: 722–737 (1977).
4. J.W. Brewer and E.A. Rutter, $D + M$ constructions with general overrings, *Michigan J. Math 23*: 33–41 (1976).
5. H. Cartan and S. Eilenberg, *Homological Algebra*, Princeton University Press, Princeton (1956).
6. D.E. Dobbs, On going-down for simple overrings, *Proc. Amer. Math. Soc. 39*: 515–519 (1973).
7. D.E. Dobbs, On going-down for simple overrings, II, *Comm. Algebra 1*: 439–458 (1974).
8. D.E. Dobbs, Divided rings and going-down, *Pacific J. Math. 67*: 353–363 (1976).
9. D.E. Dobbs, Coherence, ascent of going-down and pseudo-valuation domains, *Houston J. Math. 4*: 551–567 (1978).
10. D.E. Dobbs, On the weak global dimension of pseudo-valuation domains, *Can. Math. Bull. 21*: 159–164 (1978).
11. D.E. Dobbs, On locally divided integral domains and CPI-overrings, *Inter. J. Math. and Math. Sci. 4*: 119–135 (1981).
12. D.E. Dobbs, On the criteria of D.D. Anderson for invertible and flat ideals, *Can. Math. Bull. 29*: 25–32 (1986).
13. D.E. Dobbs, On Henselian pullbacks, these Proceedings.
14. D.E. Dobbs and M. Fontana, Locally pseudo-valuation domains, *Ann. Mat. Pura. Appl. 134*: 147–168 (1983).
15. D.E. Dobbs, M. Fontana, J.A. Huckaba and I.J. Papick, Strong ring extensions and pseudo-valuation domains, *Houston J. Math. 8*: 167–184 (1982).
16. D.E. Dobbs and I.J. Papick, On going-down for simple overrings, III, *Proc. Amer. Math. Soc. 54*: 35–38 (1976).
17. D.E. Dobbs and I.J. Papick, When is $D + M$ coherent?, *Proc. Amer. Math. Soc. 56*: 51–54 (1976).
18. D. Ferrand, Descente de la platitude par un homomorphisme fini, *C.R. Acad. Sci. Paris Sér. A-B 269*: A946–A949 (1969).
19. S. Gabelli and E. Houston, Coherent-like conditions in pullbacks, *Michigan J. Math.*, to appear.
20. S. Glaz, *Commutative Coherent Rings*, Lecture Notes in Math., vol. 1371, Springer-Verlag, Berlin (1989).

21. B. Greenberg, Global dimension of Cartesian squares, *J. Algebra* *32*: 31–43 (1974).

22. B. Greenberg, Coherence in Cartesian squares, *J. Algebra* *50*: 12–25 (1978).

23. J.R. Hedstrom and E.G. Houston, Pseudo-valuation domains, *Pacific J. Math.* *75*: 137–147 (1978).

24. J.R. Hedstrom and E.G. Houston, Pseudo-valuation domains (II), *Houston J. Math.* *4*: 199–207 (1978).

25. S. McAdam, Two conductor theorems, *J. Algebra* *23*: 239–240 (1972).

26. M. Nagata, *Local Rings*, Interscience, New York (1962).

27. W.R. Scott, *Group Theory*, Prentice-Hall, Englewood Cliffs (1964).

28. J.-P. Serre, *Algèbre Locale-Multiplicités*, Lecture Notes in Math., vol. 11, Springer-Verlag, Berlin (1965).

29. W.V. Vasconcelos, The local rings of global dimension two, *Proc. Amer. Math. Soc.* *35*: 381–386 (1972).

30. W.V. Vasconcelos, *Divisor Theory in Module Categories*, Math. Studies, vol. 14, North-Holland, Amsterdam (1974).

31. W.V. Vasconcelos, *The Rings of Dimension Two*, Marcel Dekker, New York (1976).

On Henselian Pullbacks

DAVID E. DOBBS Department of Mathematics, University of Tennessee, Knoxville, Tennessee 37996-1300.

1 INTRODUCTION

Our starting point is the following exercise of Nagata [19, Exercise 5.11.1, p. 206]: if a valuation domain R has a prime ideal P such that both R/P and R_P are Henselian, then R is Henselian. In this note, we develop a number of analogous results. Theorem 2.5(a) has three distinct generalizations of the motivating exercise. In these generalizations, the hypothesis that R is a valuation domain is weakened to the condition that the (commutative integral) domain R is quasilocal and integrally closed; or that R is a quasilocal going-down domain, in the sense of [5]; or that P is divided in R, in the sense of [6]. Another type of generalization, seeking to characterize Henselian valuation domains, appears in Corollaries 2.10 and 2.11, and is based on the characterization [4, Corollary 4] of valuation domains as the quasilocal integrally closed coherent going-down domains. As the possible transfer from R/P and R_P to R of the "quasilocal," "integrally closed," and "coherent" properties is rather well understood (cf. [12], [13]), we first develop a similar transfer result, Theorem 2.2, for going-down domains.

Our work is set in the context of CPI-extensions, in the sense of [3]. Recall that if P is a prime ideal of a domain R, then *the CPI-extension of R with respect to P is* $R+PR_P$, which may, of course, be viewed as the pullback $R_P \times_{R_P/PR_P} R/P$. Notice that R is this pullback if and only if $PR_P = P$; that is, if and only if P is divided in R (for instance, if R is a valuation domain). Our main result on the transfer of the "going-down domain" property is Corollary 2.3: if P is a prime ideal of a domain R, then R/P and R_P are going-down domains if and only if $R+PR_P$ is a going-down domain. (There seem to have been no earlier studies determining when a pullback is a going-down domain. To be sure, some types of going-down domains may be characterized via pullbacks, such as pseudo-valuation domains [1, Proposition 2.6] and GPVDs [10, Theorem 3.1, p. 155].) CPI-extensions arise naturally in our study, as the conditions in Corollary 2.3 are not equivalent to "R is a going-down domain." Indeed, Remark 2.4 reviews certain two-dimensional (possibly treed) domains which

317

are not going-down domains; the key insight, in view of Corollary 2.3, is that no such domain has a height 1 divided prime ideal.

A companion for Corollary 2.3 is Theorem 2.5(b), our main result on the transfer of the "Henselian" property: if P is a prime ideal of a quasilocal domain R such that R/P and R_P are Henselian, then $R + PR_P$ is Henselian. (There have been few studies of Henselian pullbacks. [9, Proposition 2.10] is noteworthy in this regard, although its natural application is to $k + M$ constructions, not to CPI-extensions.) Beginning with Corollary 2.6, a number of results are derived by suitably combining our main results (Corollary 2.3 and Theorem 2.5(b)) with known material. These corollaries prepare for Corollaries 2.10 and 2.11 (whose purpose was explained in the first paragraph). Finally, we note that Proposition 2.12 gives a characterization, in the spirit of Corollary 2.3, determining when a CPI-extension is either a divided or a locally divided domain. In this regard, recall that divided domains and locally divided domains are types of going-down domains which were introduced in [6]. Some additional background on these and other topics is recalled as needed below. Any unexplained terminology is standard, as in [15].

2   RESULTS

Recall from [16, p. 674] that if $D \subseteq E$ are domains and $P \in \mathrm{Spec}(D)$, then the extension $D \subseteq E$ is said to satisfy *going-down to P* in case the following condition holds: if $P \subseteq P_1$ are prime ideals of D and $Q_1 \in \mathrm{Spec}(E)$ satisfies $Q_1 \cap D = P_1$, then there exists $Q \in \mathrm{Spec}(E)$ such that $Q \subseteq Q_1$ and $Q \cap D = P$. Thus, an extension of domains $D \subseteq E$ satisfies GD (the going-down property, as in [15, p. 28]) if and only if $D \subseteq E$ satisfies going-down to P for each $P \in \mathrm{Spec}(R)$. A sharper criterion is available for overrings of Noetherian domains, for McAdam [16, Theorem 2] has shown that if E is an overring of Noetherian domain D, then $D \subseteq E$ satisfies GD if (and only if) $D \subseteq E$ satisfies GD to P for each height 1 prime ideal P of D. Lemma 2.1 provides a criterion which is somewhat in the same spirit but avoids the "Noetherian" hypothesis. First, recall from [6] that if D is a domain and $P \in \mathrm{Spec}(D)$, then P is said to be *divided in D* if $PD_P = P$; that is, if P is comparable with each principal ideal (equivalently, with each ideal) of D with respect to inclusion.

LEMMA 2.1   Let R be a domain and let $P \in \mathrm{Spec}(R)$ be divided in R. Then $R \subseteq T$ satisfies going-down to P for each extension domain T of R.

Proof: Deny. Then there exist a prime ideal $P_1 \supseteq P$ of D and $Q_1 \in \mathrm{Spec}(T)$ such that $Q_1 \cap R = P_1$ and no $Q \in \mathrm{Spec}(T)$ satisfies both $Q_1 \supseteq Q$ and $Q \cap R = P$. Choose a valuation overring (V, N_1) of T such that $N_1 \cap T = Q_1$ (cf. [15, Theorem 56]). Then $R \subseteq V$ does not satisfy going-down to P; for if, on the contrary, $N \in \mathrm{Spec}(V)$ satisfies $N \cap R = P$, then $Q := N \cap T$ would provide a contradiction. Accordingly, by [15, Exercise 37, (iii) $\Rightarrow$ (i), p. 44], there exists $J \in \mathrm{Spec}(V)$ such that J is a minimal prime over PV and

$$p_1 v_1 + \cdots + p_n v_n = rz$$

for some $p_i \in P$, $v_i \in V$, $r \in R\backslash P$, and $z \in V\backslash J$. Since V is a valuation domain, we may relabel so that $Vp_i \subseteq Vp_1$ for all i. Thus $p_i = a_i p_1$, with $a_i \in V$. It follows that $p_1 v = rz$, with

$$v = v_1 + a_2 v_2 + \cdots + a_n v_n \in V.$$

Now, $p := p_1 r^{-1} \in PR_P = P$, whence $z = pv \in PV \cap (V\backslash J) = \emptyset$, a contradiction. $\square$

Recall from [5] and [11] that a domain D is called a *going-down domain* in case $D \subseteq E$ satisfies GD for each extension domain E of D; equivalently, by [11, Theorem 1], for each valuation overring E of D. Examples of going-down domains include arbitrary Prüfer domains and domains of (Krull) dimension at most 1. The class of going-down domains is stable under the formation of rings of fractions (cf. [5, Lemma 2.1]) and factor domains [6, Remarks 2.11 and 3.2 (a), (b)]. Theorem 2.2 essentially presents a converse, since any domain is its own CPI-extension with respect to the divided prime 0.

THEOREM 2.2   For a domain R, the following conditions are equivalent:
   (1)   There exists $P \in \text{Spec}(R)$ such that P is divided in R and both R/P and R_P are going-down domains;
   (2) R is a going-down domain.
Proof: (2) $\Rightarrow$ (1): Use $P = 0$.
   (1) $\Rightarrow$ (2): Assume (1). By the result cited above from [11], it suffices to show that $R \subseteq V$ satisfies GD for each valuation overring V of R. Let $P_2 \subseteq P_1$ be prime ideals of R and let $Q_1 \in \text{Spec}(V)$ such that $Q_1 \cap R = P_1$. We shall produce $Q_2 \in \text{Spec}(V)$ such that $Q_2 \subseteq Q_1$ and $Q_2 \cap R = P_2$. Observe that P is comparable to P_1 and P_2, since P is divided in R. We next consider three cases.
   Case 1: $P \subseteq P_2$. By Lemma 2.1, $R \subseteq V$ satisfies going-down to P, and so there exists $J \in \text{Spec}(V)$ such that $J \subseteq Q_1$ and $J \cap R = P$. Observe that the extension of domains $R/P \subseteq V/J$ satisfies GD since R/P is a going-down domain. As $Q_1/J \cap R/P = P_1/P$, there exists $Q_2 \in \text{Spec}(V)$ such that $J \subseteq Q_2 \subseteq Q_1$ and $Q_2/J \cap R/P = P_2/P$. Then $Q_2 \cap R = P_2$, as desired.
   Case 2: $P_2 \subseteq P \subseteq P_1$. By Lemma 2.1 (or Case 1), there exists $I \in \text{Spec}(V)$ such that $I \subseteq Q_1$ and $I \cap R = P$. Observe that the extension of domains $R_P \subseteq V_I$ satisfies GD since R_P is a going-down domain. As $IV_I \cap R_P = PR_P$, there exists $Q_2 \in \text{Spec}(V)$ such that $Q_2 \subseteq I$ and $Q_2 V_I \cap R_P = P_2 R_P$. Then $Q_2 \cap R = P_2$, as desired.
   Case 3: $P_1 \subseteq P$. Argue as in the proof for Case 2, using the fact that the extension $R_P \subseteq V_{R\backslash P}$ satisfies GD, to produce a suitable Q_2. $\square$

COROLLARY 2.3   For a domain R and $P \in \text{Spec}(R)$, the following are equivalent:
   (1) R/P and R_P are going-down domains;
   (2) $R + PR_P$, the CPI-extension of R with respect to P, is a going-down domain.
Proof: Let $D = R + PR_P$ and $Q = PR_P$. Then $Q \in \text{Spec}(D)$ and Q is divided in D by [3, Proposition 2.5, Theorem 2.4] (cf. also the proof of [6, Lemma 2.4 (b),

(c)]). Moreover, $D/Q \cong R/P$ and $D_Q = R_P$ by [7, Lemma 2.2 (a)]. Thus, (1) $\Rightarrow$ (2) by applying Theorem 2.2 to D; and (2) $\Rightarrow$ (1) by the results on stability which were recalled prior to Theorem 2.2. $\square$

REMARK 2.4   One cannot delete the hypothesis "P is divided in R" from condition (1) in Theorem 2.2. Indeed, consider a two-dimensional domain R, and choose a height 1 nonmaximal prime ideal P of R. Then R/P and R_P are each one-dimensional, and hence going-down, domains, but R need not be a going-down domain. For instance, no two-dimensional Noetherian domain is a going-down domain. Indeed, [5, Theorem 2.2] established that each going-down domain D must be *treed*, in the sense that Spec(D), as a poset under inclusion, is a tree; and, as a consequence of the Principal Ideal Theorem, any two-dimensional Noetherian domain has infinitely many height 1 primes (cf. [15, Theorem 144]) and, thus, is not treed.

It should be pointed out that a two-dimensional quasilocal treed domain need not be a going-down domain. The first example of this phenomenon, due to W.J. Lewis, was reprised in [8, Example 2.1 (a)]. A more exotic example, produced in [8, Example 2.3], consists of a two-dimensional quasilocal domain R which is not a going-down domain although R and all its overrings are treed.

Theorem 2.2 permits a conclusion which applies to the diverse domains discussed in the preceding two paragraphs. To wit: if a two-dimensional domain R is not a going-down domain and if P is any height 1 prime ideal of R, then P is not divided in R. Thus, even if R is quasilocal and treed (as in the two examples cited from [8] above), such P fails to compare with some principal ideal of R, even though P is comparable with each prime ideal of R.

We turn now to the titular focus on Henselian domains. For our purposes, the most useful characterization of such (necessarily quasilocal) domains is given by the following result of Nagata (cf. [18, Theorem 43.12]): a domain D is Henselian if and only if E is quasilocal for each domain E which contains and is integral over D. Any complete local (Noetherian) domain is Henselian (cf. [18, Theorem 30.3]), and the class of Henselian domains is stable under the formation of factor domains (cf. [18, (43.4)]). Also, as noted in [9, Lemma 2.3], a result of Nagata [17, Theorem 10] may be reformulated, with the aid of [6, Theorem 2.5], to yield the following result: if D is an integrally closed (quasilocal) Henselian going-down domain, then D_Q is Henselian for each $Q \in$ Spec(D).

We next give our second main result. It is in the spirit of Theorem 2.2 and Corollary 2.3, and it generalizes the motivating exercise of Nagata [19] which was mentioned in the Introduction.

THEOREM 2.5   Let R be a domain and let $P \in$ Spec(R), such that both R/P and R_P are Henselian. Suppose further that R is quasilocal. (Quasilocality of R is automatic if P is divided in R, given that R/P is Henselian.) Then:

(a)   If either (i) R is integrally closed or (ii) R is a going-down domain or (iii) P is divided in R, then R is Henselian.

(b) $R + PR_P$, the CPI-extension of R with respect to P, is Henselian.

Proof: We establish the parenthetical assertion first. As R/P is Henselian (hence quasilocal), P is contained in a unique maximal ideal of R. If P is divided in R, then R must be quasilocal, for P is then comparable with, and hence contained in, each maximal ideal of R.

Next, we explain how the assertion regarding $D := R + PR_P$ follows from the other assertions. Let $Q = PR_P$. As in the proof of Corollary 2.3, the hypotheses ensure that $D/Q \cong R/P$ and $D_Q = R_P$ are Henselian. Moreover, since Q is divided in D, the previous paragraph yields that D is quasilocal. Thus, the assertion regarding D in (b) would follow by applying case (iii) of (a) to D and Q in place of R and P.

It remains only to show that if (i) R is integrally closed or (ii) R is a going-down domain or (iii) P is divided in R, then R is Henselian. Deny. Then, by the above comments, there exists a domain T which contains and is integral over R such that T is not quasilocal. Choose distinct maximal ideals, N_1 and N_2, of T. By the Going-up Theorem [18, Corollary 10.9], $N_1 \cap R = N_2 \cap R = M$, the unique maximal ideal of R. Now, the hypotheses ensure that $R \subseteq T$ satisfies going-down to P (by using (i) the Going-Down Theorem [18, Theorem 10.13], (ii) the definition of a going-down domain, or (iii) Lemma 2.1). As $P \subseteq M$, it follows that there exists $Q_i \in \mathrm{Spec}(T)$ such that $Q_i \subseteq N_i$ and $Q_i \cap R = P$, for $i = 1, 2$. However, since $R_P \subseteq T_{R \setminus P}$ is an integral extension and R_P is Henselian, $T_{R \setminus P}$ must be quasilocal. As the $Q_i T_{R \setminus P}$ are maximal ideals of $T_{R \setminus P}$ (by what is termed the Lying-over Theorem in [18, Corollary 10.8] and INC in [15, Theorem 44]), we have $Q_1 T_{R \setminus P} = Q_2 T_{R \setminus P}$, and so $Q_1 = Q_2 (=, \text{say}, Q)$. Next, consider T/Q, viewed as an integral extension of R/P. As R/P is Henselian, T/Q is quasilocal. However, N_1/Q and N_2/Q are distinct maximal ideals of T/Q, the desired contradiction. $\square$

It follows from the above proof that the hypothesis in Theorem 2.5 (a) could be generalized to "$R \subseteq T$ satisfies going-down to P for each domain T which contains and is integral over R." The above formulation was chosen to facilitate references in the following applications.

COROLLARY 2.6   Let R be a quasilocal domain, and let $P \in \mathrm{Spec}(R)$ such that R_P is Henselian. Then the following conditions are equivalent:
   (1) R/P is a Henselian going-down domain and R_P is a going-down domain;
   (2) $R + PR_P$ is a Henselian going-down domain.
Proof: Since any factor domain of a Henselian domain is Henselian, one need only combine Theorem 2.5 (b) and Corollary 2.3. $\square$

It is convenient to record a characterization of integrally closed CPI-extensions. Let R be a domain and $P \in \mathrm{Spec}(R)$. Then $R + PR_P$ is integrally closed if and only if both R/P and R_P are integrally closed. A proof of this fact was given in the second paragraph of the proof of [2, Proposition 2.3 (a)], with the aid of the standard description of the integral closure of a pullback (cf. [12, Corollary 1.5 (5)]). The characterization was first noted by Greenberg [14, Proposition 4.7] in case P is divided in R and P is R-flat.

COROLLARY 2.7   Let R be a quasilocal domain and let $P \in \mathrm{Spec}(R)$. Then the following conditions are equivalent:
   (1) R/P and R_P are integrally closed Henselian going-down domains;
   (2) $R + PR_P$ is an integrally closed Henselian going-down domain.

Proof: Observe via [3, Corollary 2.8] (cf. also [12, Theorem 1.4]) that $R + PR_P$ inherits the quasilocal property from R. Thus, one need only combine the above remark, Corollary 2.6, and the reformulation of [17, Theorem 10] which was recalled prior to the statement of Theorem 2.5. $\square$

We pause to note some variants of Corollaries 2.6 and 2.7. (A similar comment applies to Corollary 2.10 (a) as well.) In those results, if P is assumed to be divided in R, then R need not be assumed quasilocal. For a proof, simply apply the parenthetical assertion in Theorem 2.5.

Let D be a domain, with integral closure D'. Recall from [9] that D is said to be *1-split* in case the canonical map $\text{Spec}(E) \to \text{Spec}(D)$ is an injection (resp., bijection) for each domain E which contains and is integral over D. According to [9, Theorem 2.4 (a)], D is 1-split if and only if D_Q is Henselian for each $Q \in \text{Spec}(D)$. Moreover, by [9, Theorem 2.4 (c)], if D is a going-down domain, then D is 1-split if and only if D_N is Henselian for each maximal ideal N of D and $D \subseteq D'$ is unibranched; and if D is a pseudo-valuation domain with canonically associated valuation overring V then, by [9, Corollary 2.6 (a)], D is 1-split if and only if V is Henselian. This material may be used to reformulate the above results. We next rework Corollary 2.7 in this way, leaving the similar rephrasing of Corollary 2.6 for the interested reader.

COROLLARY 2.8   Let R be a quasilocal domain and let $P \in \text{Spec}(R)$. Then the following conditions are equivalent:
  (1) R/P and R_P are integrally closed 1-split going-down domains;
  (2) $R + PR_P$ is an integrally closed 1-split going-down domain.
Proof: By the above remarks, a quasilocal integrally closed going-down domain is 1-split if and only if it is Henselian. Accordingly, the assertion is now seen as a reformulation of Corollary 2.7. $\square$

We pause to illustrate Corollaries 2.7 and 2.8 in the case of a pseudo-valuation domain which is not a valuation domain.

EXAMPLE 2.9   Let $k \subseteq K$ be a proper field extension such that k is algebraically closed in K, let X be an analytic indeterminate over K, let $P = XK[[X]]$, and let $R = k + P$. Then R is a pseudo-valuation (hence, quasilocal going-down) domain, but not a valuation domain, with unique maximal ideal P, such that conditions (1) and (2) of Corollary 2.7 are satisfied.

Indeed, $V := K[[X]]$ is a DVR and $R = V \times_K k$ is therefore a pseudo-valuation domain with canonically associated valuation overring V, by [1, Proposition 2.6]. Then P is the unique maximal ideal of R, whence P is divided in R and $R + PR_P = R$ (cf. [12, Theorem 1.4]); moreover, R is integrally closed but is not a valuation domain (cf. [12, Corollary 1.5 (5), Example 2.6]). Finally, since V is a complete local (Noetherian) domain, V is Henselian, and so R is (1-split and) Henselian, by the above remarks concerning [9, Corollary 2.6 (a)]. $\square$

Theorem 2.5 (a) gave three different generalizations of the motivating exercise of Nagata. We return now to the context of that exercise, namely valuation domains. Our prevailing philosophy for such studies was given more than twenty years ago in a characterization of Prüfer domains [4, Corollary 4], whose "quasilocal" case is the following assertion: a domain D is a valuation domain if and only if D is a quasilocal integrally closed coherent going-down domain. By now, one understands well the transfer through pullbacks, such as CPI-extensions, of properties like "quasilocal" and "integrally closed" (cf. [12]) and "coherent" (in a series of papers, including [13]). By combining this work with our transfer results on "going-down domain" (Corollary 2.3) and "Henselian" (Theorem 2.5), we obtain the following two additional generalizations of [19, Exercise 5.11.1, p. 206].

COROLLARY 2.10 Let R be a domain and let $P \in \mathrm{Spec}(R)$. Then:
  (a) Suppose that R is quasilocal. Then R/P and R_P are Henselian valuation domains if and only if $R + PR_P$ is a Henselian valuation domain.
  (b) P is divided in R and both R/P and R_P are Henselian valuation domains if and only if R is a Henselian valuation domain.
Proof: (a) R/P, R_P and $R+PR_P$ are quasilocal, the last by virtue of [3, Corollary 2.8] (and the hypothesis that R is quasilocal). The "integrally closed Henselian going-down domain" property also transfers, by Corollary 2.7. It remains only to give the slightly more subtle analysis regarding the transfer of "coherent." Without loss of generality, $P \neq 0$ and P is not the unique maximal ideal of R (in both cases, the point being that fields are coherent). By applying [13, Proposition 4.6] to the pullback description $R+PR_P = R_P \times_{R_P/PR_P} R/P$, we obtain the following: if R_P is a valuation domain, then $R + PR_P$ is coherent if and only if R/P is coherent. This is a sufficiently strong "transfer of coherence" result for our present purposes, since localizations of valuation domains are valuation domains.
  (b): Apply (a), bearing in mind the parenthetical assertion in Theorem 2.5 and the fact that any prime of a valuation domain D is divided in D. $\square$

COROLLARY 2.11 For a valuation domain R, the following are equivalent:
  (1) There exists $P \in \mathrm{Spec}(R)$ such that R/P and R_P are Henselian;
  (2) For each $P \in \mathrm{Spec}(R)$, R/P and R_P are Henselian;
  (3) R is Henselian.
Proof: Since R is a valuation domain, so are R/P and R_P, for each $P \in \mathrm{Spec}(R)$. As each prime ideal of (the valuation domain) R is divided in R, the assertions now follow from Corollary 2.10 (b). $\square$

We turn next to analogues of Theorem 2.2 and Corollary 2.3 for two classes of going-down domains introduced in [6]. Recall that a *divided domain* is a domain D such that each $Q \in \mathrm{Spec}(D)$ is divided in D; and that a *locally divided domain* is a domain D such that D_Q is a divided domain for each prime (resp., maximal) ideal Q of D. Arbitrary Prüfer domains and one-dimensional domains are examples of locally divided domains; and any locally divided domain is a going-down domain [6, Remark 2.7 (b)]. A domain D is locally divided if and only if each of its CPI-extensions is D-flat [7, Theorem 2.4]. Also, the class of divided domains (resp., locally divided domains) is stable under the formation of rings of fractions and factor domains: see [6, Lemma 2.2 (a), (c)] (resp., [6, Remark 2.7 (b)]).

PROPOSITION 2.12 Let R be a domain and let $P \in \mathrm{Spec}(R)$. Then both R/P and R_P are divided (resp., locally divided) domains if and only if $R + PR_P$ is a divided (resp., locally divided) domain.

Proof: The "if" assertions follow from the above remarks on stability. For the "only if" assertions, we may show, as in the proof of Corollary 2.3, that R and P may be replaced by $R + PR_P$ and PR_P; with this *abus de langage*, we may assume henceforth that P is divided in R. We show first how the "locally divided" assertion follows from the "divided" assertion. To this end, suppose that R/P and R_P are locally divided domains (and that $PR_P = P$). We show that R_M is divided for each maximal ideal M of R. Observe that $P \subseteq M$ since P is divided in R; thus, $PR_M \in \mathrm{Spec}(R_M)$. Moreover, $R_M/PR_M \cong (R/P)_{M/P}$ is a localization of the locally divided domain R/P, and so by the above stability remarks, is a (quasilocal locally) divided domain. As $(R_M)_{PR_M} = R_P$ is also a (quasilocal locally) divided domain, it follows from the "divided" assertion that R_M is divided, as desired.

It remains only to prove that if R/P and R_P are divided domains and $PR_P = P$, then R is divided; that is, that $QR_Q = Q$ for each $Q \in \mathrm{Spec}(R)$. Since P is divided in R, there are only two cases.

Case 1: $Q \subseteq P$. It suffices to prove that $q \in Rz$ for each $q \in Q$ and $z \in R \backslash Q$. Now, since R_P is a divided domain, QR_P is divided in R_P; hence

$$qz^{-1} \in QR_Q = QR_P R_Q = QR_P(R_P)_{QR_P} = QR_P.$$

Thus, without loss of generality, $z \in R \backslash P$. Then, since P is divided in R and $z \notin P$, we have $P \subseteq Rz$, whence $q \in Q \subseteq P \subseteq Rz$, as desired.

Case 2: $P \subseteq Q$. Once again, we show that $q \in Rz$ for all $q \in Q$, $z \in R \backslash Q$. Now, since R/P is a divided domain, Q/P is divided in R/P; hence

$$\frac{q+P}{z+P} \in (Q/P)(R/P)_{Q/P} = Q/P$$

in the quotient field of R/P. Thus, there exists $q^* \in Q$ such that

$$\frac{q+P}{z+P} = \frac{q^*+P}{1+P}.$$

It follows that $q - zq^* \in P$. Moreover, since P is divided in R and $z \notin P$, we have $P \subseteq Rz$. Thus, since P is prime, $P \subseteq Pz \subseteq Qz$, and so $q - zq^* \in Qz$, whence $q \in Qz \subseteq Rz$, to complete the proof. $\square$

The interested reader may combine Proposition 2.12 and Theorem 2.5 to formulate "divided" and "locally divided" analogues of Corollaries 2.6-2.8. In closing, we give an application of our central result on Henselian pullbacks, Theorem 2.5, which is far removed from the context of going-down domains.

COROLLARY 2.13 Let R be a nonzero commutative ring with unity such that R is neither a regular local (Noetherian) ring nor a valuation domain. Let $P \in \mathrm{Spec}(R)$ such that R_P is Henselian. Then the following conditions are equivalent:

(1) R is a domain, P is divided in R, R/P is a two-dimensional Henselian regular local (Noetherian) ring, R_P is a (Henselian) valuation domain in which each ideal is countably generated, and R_P/P is countably generated as an R/P-module;
(2) gl. dim$(R) = 2$ and R is Henselian.

Proof: Note that (1) and (2) each imply that R is quasilocal (by the parenthetical assertion in Theorem 2.5 and by the definition of "Henselian"). If one deletes "Henselian" from (1) and (2) and from the hypothesis on R_P but inserts "R is quasilocal" into the remnant of (1), the resulting remnants of (1) and (2) are equivalent: this summarizes the major content of the Vasconcelos-Greenberg structure theory of umbrella rings. (Cf. [20, Theorem 2.2, Corollary 4.19, and (4.15)]. To eliminate any ostensibly homological conditions from the statement of (1), we have paraphrased the assertions in [20] as follows. According to a theorem of Osofsky, a valuation domain has global dimension at most 2 if and only if each of its ideals is countably generated; and according to a theorem of Kaplansky, if a nonmaximal prime ideal P of a domain R is divided in R, then the projective dimension of R_P/P as an R/P-module is 1 if and only if R_P/P is countably generated as an R/P-module.) Hence, by Theorem 2.5 (a) or (b), when the occurrences of "Henselian" are reinserted into the remnants of (1) and (2) and into the hypothesis on R_P, we obtain the equivalence of (1) and (2). $\square$

REFERENCES

1. D.F. Anderson and D.E. Dobbs, Pairs of rings with the same prime ideals, *Can. J. Math. 32*: 362-384 (1980).

2. V. Barucci, D.E. Dobbs and S.B. Mulay, Integrally closed factor domains, *Bull. Austral. Math. Soc. 37*: 353-366 (1988).

3. M.B. Boisen, Jr. and P.B. Sheldon, CPI-extensions: overrings of integral domains with special prime spectrums, *Can. J. Math. 29*: 722-737 (1977).

4. D.E. Dobbs, On going-down for simple overrings, *Proc. Amer. Math. Soc. 39*: 515-519 (1973).

5. D.E. Dobbs, On going-down for simple overrings, II, *Comm. Algebra 1*: 439-458 (1974).

6. D.E. Dobbs, Divided rings and going-down, *Pacific J. Math. 67*: 353-363 (1976).

7. D.E. Dobbs, On locally divided integral domains and CPI-overrings, *Internat. J. Math. and Math. Sciences 4*: 119-135 (1981).

8. D.E. Dobbs, On treed overrings and going-down domains, *Rend. Mat. 7*: 317-322 (1987).

9. D.E. Dobbs, Locally Henselian going-down domains, *Comm. Algebra 24*: 1621-1635 (1996).

10. D.E. Dobbs and M. Fontana, Locally pseudo-valuation domains, *Ann. Mat. Pura Appl. 134*: 147-168 (1983).

11. D.E. Dobbs and I.J. Papick, On going-down for simple overrings, III, *Proc. Amer. Math. Soc. 54*: 35-38 (1976).

12. M. Fontana, Topologically defined classes of commutative rings, *Ann. Mat. Pura Appl. 123*: 331-355 (1980).

13. S. Gabelli and E. Houston, Coherent-like conditions in pullbacks, *Michigan*

J. Math., to appear.

14. B. Greenberg, Coherence in Cartesian squares, *J. Algebra 50*: 12-25 (1978).

15. I. Kaplansky, *Commutative Rings*, rev. edition, Univ. Chicago Press, Chicago (1974).

16. S. McAdam, 1-going down, *J. London Math. Soc. 8*: 674-680 (1974).

17. M. Nagata, On the theory of Henselian rings, *Nagoya Math. J. 5*: 45-57 (1953).

18. M. Nagata, *Local Rings*, Interscience, New York (1962).

19. M. Nagata, *Field Theory*, Marcel Dekker, New York (1977).

20. W.V. Vasconcelos, *The Rings of Dimension Two*, Marcel Dekker, New York (1976).

An Intersection Condition for Prime Ideals

ROBERT GILMER, Department of Mathematics 3027, Florida State University, Tallahassee, Florida 32306–3027

1. INTRODUCTION

All rings considered in this paper are assumed to be commutative and to contain a unity element. A widely used result in commutative ring theory is the Prime Avoidance Theorem, which states that an ideal contained in a finite union of prime ideals is contained in one of the prime ideals. (Cf. [M_1, p. 186], [M_2], or [K, Theorem 81].) Reis and Viswanathan [RV] and Smith [S] considered rings satisfying the conclusion of the PAT even if the set of prime ideals of the union is infinite. Following Smith's terminology/notation, a ring R is said to *satisfy* (∗) if for any ideal I of R and any nonempty subset $\{P_a\}_{a \in A}$ of Spec(R), the inclusion $I \subseteq \cup_{a \in A} P_a$ holds only if I is contained in some P_a. Generalizing Theorem 1.1 of [RV] from the case where R is Noetherian, Smith showed that R satisfies (∗) if and only if each prime ideal of R is the radical of a principal ideal. G. Fu has asked about a characterization of rings satisfying the following condition (#), which is a natural dual of (∗).

1991 *Mathematics Subject Classification.* Primary 13A17; Secondary 13A15.

Key words and phrases: Prime ideal, intersection of ideals, prime avoidance theorem, compactly packed ring.

($\#$) *If $P \in \text{Spec}(R)$ and if $\{I_a\}_{a \in A}$ is any nonempty family of ideals of R, then P contains $\cap_{a \in A} I_a$ only if P contains some I_a.*

We show in Theorem 2 that R satisfies ($\#$) if and only if R is zero-dimensional and semiquasilocal. We then consider both local and global variants of condition ($*$) that were partially motivated by Smith's observation that a ring S satisfies ($*$) if an inclusion $P \subseteq \cup_{a \in A} P_a$ of prime ideals P, P_a in S holds only if P is contained in some P_a. In particular, Proposition 5 gives equivalent conditions on a fixed prime ideal Q in order that Q contains the intersection of a family of prime ideals only if Q contains one of the prime ideals of the family.

2. MAIN RESULTS

In characterizing rings satisfying ($\#$), we begin with a basic local result concerning this condition.

LEMMA 1. *Assume that $R = S \oplus T$, where S is quasilocal with maximal ideal M. The prime ideal $P = M \oplus T$ of R has the property that P contains the intersection of a family of ideals of R only if P contains one of the ideals of the family.*

Proof. Let $\{I_a\}_{a \in A}$ be a family of ideals of R such that $\cap_{a \in A} I_a \subseteq P$. Each I_a is of the form $B_a \oplus C_a$, where B_a is an ideal of S and C_a is an ideal of T. We have $\cap I_a = (\cap B_a) \oplus (\cap C_a) \subseteq P = M \oplus T$. Hence $\cap B_a \subseteq M$, and since (S, M) is quasilocal, $B_a \subseteq M$ for some $a \in A$. Consequently, $I_a = B_a \oplus C_a \subseteq M \oplus T = P$.   □

THEOREM 2. *The ring R satisfies condition ($\#$) if and only if R is zero-dimensional and semiquasilocal.*

Proof. ($\Leftarrow$): Assume that R is zero-dimensional with maximal ideals $M_1, \ldots, M_n$. If Q_i is the isolated M_i-primary component of (0), then $(0) = \cap_{i=1}^n Q_i$, where the ideals Q_j and Q_k are comaximal if $j \neq k$. Therefore if $1 \leq k \leq n$ and if $I_k = \cap_{j \neq k} Q_j$, then $R = Q_k \oplus I_k$. The ring $I_k \simeq R/Q_k$ is quasilocal, and $M_k = Q_k \oplus P_k$, where P_k is the maximal ideal of I_k. It follows from Lemma 1 that M_k satisfies the condition required of P in the statement of ($\#$), and hence R satisfies ($\#$).

($\Rightarrow$): We assume that R satisfies ($\#$). Suppose that R has positive dimension, and choose $P_1, P_2 \in \text{Spec}(R)$ such that $P_1 < P_2$. It is well-known that there exist primes $P_1^*, P_2^* \in \text{Spec}(R)$ such that $P_1 \subseteq P_1^* < P_2^* \subseteq P_2$ and no prime ideal of R lies properly between P_1^* and P_2^* (see, for example, [K, Theorem 11]). Moreover, there exists a family $\{Q_a\}_{a \in A}$ of P_2^*-primary ideals of R such that $P_1^* = \cap_{a \in A} Q_a$ [G, (17.4), p. 192]. Since $P_1^* \not\supseteq Q_a$ for each $a \in A$, this contradicts the assumption that R satisfies ($\#$). Therefore R is zero-dimensional. Let M be a maximal ideal of R and let $\{M_b\}_{b \in B}$ be the set of maximal ideals of R distinct from M. Condition ($\#$) implies that $M \not\supseteq \cap_{b \in B} M_b$. Choose an element $y_M \in \cap M_b$, $y_M \notin M$. Then $I = (\{y_M : M \in \text{Spec}(R)\}) = R$ since I is contained in no maximal ideal of R. Therefore $(y_{M_1}, \ldots, y_{M_n}) = R$ for some $M_1, \ldots, M_n \in \text{Spec}(R)$, and by choice of the elements y_{M_i}, it follows that $\text{Spec}(R) = \{M_i\}_{i=1}^n$. This completes the proof of Theorem 2. $\square$

Smith in [S] observed that condition ($*$) in a ring R is equivalent to the following apparently weaker condition ($**$):

($**$) *If a prime ideal P of R is contained in $\cup_{a \in A} P_a$, where $P_a \in$* $\text{Spec}(R)$ *for each $a \in A$, then $P \subseteq P_a$ for some $a \in A$.*

The natural analogue of ($**$) in terms of ($\#$) is the following condition ($\#\#$):

($\#\#$): *An inclusion $P \supseteq \cap_{a \in A} P_a$, with $P, P_a \in \text{Spec}(R)$, holds in* R *only if $P \supseteq P_a$ for some $a \in A$.*

However, ($\#\#$) is strictly weaker than ($\#$); for example, any ring with finite spectrum satisfies ($\#\#$), but such a ring need not be zero-dimensional. In fact, a ring satisfying ($\#\#$) need not be finite-dimensional.

Condition ($\#$) can also be considered in the case of a fixed prime ideal P; thus we say that $P \in \text{Spec}(R)$ *satisfies* ($\#$) if P has the property that an inclusion $P \supseteq \cap_{a \in A} I_a$ holds for ideals I_a of R only if P contains some I_a. In this connection we can show that the converse of Lemma 1 is valid.

PROPOSITION 3. *If $P \in \text{Spec}(R)$ satisfies ($\#$), then there exists a decomposition $R = S \oplus T$ of R such that S is quasilocal and $P = M \oplus T$, where M is the maximal ideal of S.*

Proof. Let $\{I_a\}_{a\in A}$ be the set of ideals of R that are not contained in P and choose $x \in \cap I_a$, $x \notin P$. Since $(x^2) \not\subseteq P$, we have $x \in (x^2)$, and hence (x) is idempotent, generated by an idempotent element e. Let $S = Re$, $T = R(1-e)$. Then $R = S \oplus T$ and $P = M \oplus T$ for some prime ideal M of S. By choice of x, each ideal of S not contained in M contains e — that is, S is the only ideal of S not contained in M. Consequently, S is quasilocal and M is the maximal ideal of S. $\square$

Condition $(*)$ can be considered for a fixed ideal I in an analogous way. Thus, we say that I *satisfies* $(*)$ if an inclusion $I \subseteq \cup_{a\in A} P_a$ holds, for $P_a \in \text{Spec}(R)$, only if $I \subseteq P_a$ for some $a \in A$. A slight modification of Smith's methods yields the following result, whose proof we omit.

PROPOSITION 4. *An ideal I of a ring R satisfies $(*)$ if and only if $\text{rad}(I)$ is the radical of a principal ideal of R.*

Similarly, condition $(\#\#)$ is meaningful for a fixed prime ideal of a ring. Standard techniques can be used to establish the following result.

PROPOSITION 5. *Let P be a fixed proper prime ideal of a ring R and let $\{P_a\}_{a\in A}$ be the set of primes of R that are not contained in P. The following conditions are equivalent.*
(1) If P contains the intersection of a family of primes of R, then P contains one of the primes of the family.
(2) $P \not\supseteq \cap_{a\in A} P_a$.
(3) There exists $r \in R - P$ such that P is the unique prime ideal of R maximal with respect to failure to contain r.
(4) There exists $r \in R - P$ such that if $N = \{r^i\}_{i=1}^{\infty}$, then R_N is quasilocal with maximal ideal P_N.

It is natural to ask for a characterization of rings satisfying condition $(\#\#)$ globally — that is, rings R such that each $P \in \text{Spec}(R)$ satisfies the conditions of Proposition 5. It follows as in the proof of Theorem 2 that such an R must be semiquasilocal, and clearly R satisfies d.c.c. for prime ideals. Thus, if $\mathcal{S}$ is any nonempty subset of $\text{Spec}(R)$, then the set $M(\mathcal{S})$ of minimal elements of $\mathcal{S}$ is nonempty and each element of $\mathcal{S}$ contains an element of $M(\mathcal{S})$. Consequently, $\cap \mathcal{S} = \cap M(\mathcal{S})$. In the direction of a converse, it is apparent that the following condition (C) is sufficient for R to satisfy $(\#\#)$:

(C) R satisfies d.c.c on prime ideals, and for each nonempty subset S of $\mathrm{Spec}(R)$, the set $M(S)$ of minimal elements of S is finite.

The following example shows, however, that (C) is strictly stronger than (##).

EXAMPLE 6.   Let K be a field, let $X = \{X_i\}_{i=1}^{\infty}$ be an infinite set of indeterminates over K, and let D be the power series ring $K[[X]]_1$ over K (see [G, p. 6]). Thus D is the union of the ascending sequence $\{K[[X_1,\ldots,X_n]]\}_{n=1}^{\infty}$ of power series rings in finitely many indeterminates over K; D is a quasilocal domain with maximal ideal M generated by the set X. Let $I = (\{X_iX_j : i \neq j\})$. For any positive integer k, the ideal $P_k = (X - \{X_k\})$ is prime in D and P_k contains I. Moreover, $D/P_k \simeq K[[X_k]]$ so M is the unique proper prime ideal of D that properly contains P_k. Conversely, if $P \neq M$ is a prime ideal of D containing I, then $X_j \notin P$ for some j, and consequently, $P_j \subseteq P$. Because D/P_j is one-dimensional it then follows that $P = P_j$. Therefore $\{M\} \cup \{P_j\}_{j=1}^{\infty}$ is the set of proper prime ideals of D that contain I. Thus $R = D/I$ is a one-dimensional quasilocal ring that satisfies (##) but not (C). That R fails to satisfy (C) is clear, as is the fact that M/I satisfies (##); the prime P_j/I satisfies (##) because $X_j + I$ is not in P_j/I, but is in each prime of R distinct from P_j/I.□

REFERENCES

[G] R. Gilmer, Multiplicative Ideal Theory, Queen's Series in Pure and Appl. Math. Vol 90, Kingston, Ontario, 1992.

[K] I. Kaplansky, Commutative Rings, Allyn & Bacon, Boston, 1970.

[M₁] N.H. McCoy, Rings and Ideals, MAA Carus Mathematical Monographs, No. 8, 1948.

[M₂] _____, A note on finite unions of ideals and subgroups, Proc. Amer. Math. Soc. 8 (1957), 633–637.

[RV] C.M. Reis and T.M. Viswanathan, A compactness property for prime ideals in Noetherian rings, Proc. Amer. Math. Soc. 25 (1970), 353–356.

[S] W.W. Smith, A covering condition for prime ideals, Proc. Amer. Math. Soc. 30 (1971), 451–2.

Genus Class Groups and Separable Base Change

ROBERT GURALNICK Department of Mathematics, University of Southern California, Los Angeles, CA 90089-1113

ROGER WIEGAND Department of Mathematics, University of Nebraska, Lincoln, NE 68588-0323

Let k be a field and R the coordinate ring of an affine curve over k. Thus R is a finitely generated reduced k-algebra of Krull dimension 1. Given a finitely generated torsion-free R-module, one can define an abelian group structure on the set Genus(M) of isomorphism classes of modules N such that $M_{\mathfrak{m}} \cong N_{\mathfrak{m}}$ for every maximal ideal $\mathfrak{m}$. The goal of this paper is to study the structure of the *genus class group* Genus(M) and its behavior under separable base change.

Base change can be a powerful tool in situations where one has some control on the amount of collapsing that can occur. The idea, roughly speaking, is to tensor with a suitable splitting field, thereby replacing field extensions (which can be rather intractable) by splitting (direct products), which can be analyzed combinatorially. Typically, one would like to know that one cannot have infinitely many non-isomorphic modules that become isomor-

---

Both authors were partially supported by grants from the National Science Foundation.

phic after the base change. This technique was used, for example, in [GR] and [CWW] to classify the one-dimensional Noetherian local rings of finite Cohen-Macaulay type, and in [HW] to study the multiplicative structure of fields.

The genus is the appropriate gadget to study in such questions, thanks to Grothendieck's faithfully flat descent theorem [EGA, (2.5.8)], a special case of which says that two finitely generated modules over a local ring that become isomorphic after faithully flat base change must already be isomorphic. Thus, in the general set-up, any two modules that become isomorphic after extension of scalars must already be in the same genus.

If L/k is a separable field extension, there is a homomorphism $\Phi_M :$ Genus$(M) \to$ Genus$(L \otimes_k M)$ taking N to $L \otimes_k N$. (We view $L \otimes_k N$ as a module over $L \otimes_k R$.) In the special case $M = R$, the genus class group is just the Picard group Pic(R) of isomorphism classes of invertible ideals of R. The group Genus(M) is not in general as well behaved as the Picard group. It is known, for example, that, for any integer n prime to char(k), Pic(R) has only finitely many elements of order n. Also, if L/k is separable the map $\Phi_R :$ Pic$(R) \to$ Pic$(L \otimes_k R)$ has finite kernel. (In fact, [GJRW], both of these results hold in higher dimensions as well.) We will see that the analogous results can fail for Genus(M).

On the positive side, we show that if M is isomorphic to an ideal of R then Genus(M) has both of the finiteness properties mentioned above, and we conclude that the finiteness results are valid for *all* lattices, provided the only singularities of Spec(R) are double points.

It is reasonable to restrict our attention to separable extensions. For one thing, inseparable base change can introduce nilpotents. That, however, is not the extent of the problem. In [GJRW] we showed that one can have a purely inseparable extension L/k of degree 3 and a Dedekind domain D, finitely generated as a k-algebra, such that $L \otimes_k D \cong L[X]$ but D has infinite class group.

1 THE GENUS CLASS GROUP

Here we review the basic tools for working with lattices and construct the genus class group. We do not need the assumption that R is a finitely generated algebra over a field.

Assumptions, Notation and Terminology We assume throughout this section that R is a *ring-order*, that is, a one-dimensional, reduced commutative Noetherian ring such that the integral closure $\tilde{R}$ of R in its total quotient ring K is finitely generated as an R-module (equivalently, as an R-algebra). ("Reduced" means there are no non-zero nilpotents. The term "ring-order" is due to Haefner and Levy [HL].) An R-module M is said to be *torsion-free* provided the non-zero-divisors of R act as non-zero-divisors on M. By an R-*lattice* we mean a finitely generated torsion-free R-module. The group of units of a ring A will always be denoted by $A^\bullet$, and $[M]$ denotes the isomorphism class of the module M.

Our basic tool for studying ring-orders and their lattices will be the pullback diagram

$$
\begin{array}{ccc}
R & \longrightarrow & \tilde{R} \\
\downarrow & & \downarrow \pi \\
R/c & \longrightarrow & \tilde{R}/c
\end{array}
\tag{1.1}
$$

which we call the *standard pullback* for R. Here c is the *conductor* of R in $\tilde{R}$, that is, the largest ideal of $\tilde{R}$ that is contained in R. Since $\tilde{R}$ is module-finite over R, the conductor contains a non-zero-divisor, and hence both R/c and $\tilde{R}/c$ are Artinian.

Suppose now that M is a lattice over a ring-order R. Let $\tilde{R}M$ denote the $\tilde{R}$-submodule of $K \otimes_R M$ generated by the image of M. Equivalently, $\tilde{R}M = (\tilde{R} \otimes_R M)/\text{torsion}$. We have the following "standard pullback" representation of the lattice M:

$$
\begin{array}{ccc}
M & \longrightarrow & \tilde{R}M \\
\downarrow & & \downarrow \\
M/cM & \xrightarrow{j} & \tilde{R}M/cM
\end{array}
\tag{1.2}
$$

Following [W], we will define an action of $(\tilde{R}/c)^\bullet$, the group of units of $\tilde{R}/c$, on the set of isomorphism classes of R-lattices. Let M be an R-lattice, with standard pullback (1.2). We assume throughout that M is faithful, that is, $(0 : M) = (0)$. (This is not a serious restriction, since if N is any non-zero R-lattice, then $R/(0 : N)$ is also a ring-order (or, if $\dim(N) = 0$, a direct product of fields, a case of no interest to us) and N is a faithful $R/(0 : N)$-lattice.) Given any $x \in \tilde{R}^\bullet$ we can choose an $\tilde{R}/c$-automorphism ϕ of $\tilde{R}M/cM$ with $\det(\phi) = x$. (The fact that M is faithful is essential here.) We define a new R-lattice M^x by the following pullback diagram:

$$
\begin{array}{ccc}
M^x & \longrightarrow & \tilde{R}M \\
\downarrow & & \downarrow \\
M/cM \xrightarrow{\ j\ } \tilde{R}M/cM & \xrightarrow{\ \phi\ } & \tilde{R}M/cM
\end{array}
\qquad (1.3)
$$

where j is the same map as in (1.2) (and the vertical arrow $\tilde{R}M \to \tilde{R}M/cM$ is the canonical map).

It turns out [W] that M^x is an R-lattice whose isomorphism class depends only on the unit x and not on the the the choice of ϕ. Moreover, this construction yields an action of the group $(\tilde{R}/c)^\bullet$ on the set of isomorphism classes of R-lattices. In (1.5) we will identify the orbits of this action and its stabilizers, but first we need some notation.

1.4 NOTATION   Let M be an R-lattice. Let Δ_M be the subgroup of $(\tilde{R}/c)^\bullet$ consisting of determinants of $\tilde{R}/c$-automorphism of $\tilde{R}M/cM$ that carry M/cM into M/cM. (Think of Δ_M as the group of determinants of automorphisms of the "bottom line" of the standard pullback (1.2).) The *restricted genus* of M is the set RestGenus(M) of isomorphism classes of lattices N such that N and M are locally isomorphic and $\tilde{R}N \cong \tilde{R}M$. (Once we define the abelian group structure on Genus(M) we will see that RestGenus(M) is a subgroup of Genus(M).)

1.5 THEOREM [W, (2.2), (2.3)], [WW, (1.6), (1.7)]   Let M and N be faithful R-lattices, and let Q be an arbitrary R-lattice.

(1) $Q \cong M^x$ for some $x \in (\tilde{R}/c)^\bullet$ if and only if $[Q] \in \text{RestGenus}(M)$.

(2) $M^x \cong M$ if and only if $x \in \Delta_M \pi(\tilde{R}^\bullet)$ (where $\pi : \tilde{R} \to \tilde{R}/c$ is the natural map).

(3) $\Delta_{M \oplus N} = \Delta_M \Delta_N$.

(4) $(M \oplus Q)^x \cong M^x \oplus Q$.

Using these results it is easy to cook up examples showing failure of direct-sum cancellation for lattices. In fact, [GW2, (1.9)], ring-orders for which cancellation holds for all lattices tend to be very special. Here, however, we will show that cancellation *does* hold for lattices within the same genus.

1.6 COROLLARY (cancellation within the genus)   Let M and N be R-lattices in the same genus. If $M \oplus N \cong M' \oplus N$ for some R-lattice M', then $M \cong M'$.

Proof: We may assume M is faithful. Since cancellation holds over local rings [E] and Dedekind domains, M and M' are in the same restricted genus. Therefore by (1.5)(1) there is a unit $x \in (\widetilde{R}/c)^{\bullet}$ such that $M^x \cong M'$. Now $M \oplus N \cong M^x \oplus N \cong (M \oplus N)^x$, whence $x \in \Delta_{M \oplus N} \pi(\widetilde{R}^{\bullet}) = \Delta_M \Delta_N \pi(\widetilde{R}^{\bullet})$ by (4), (2) and (3) of (1.5). Since M and N are locally isomorphic, the bottom lines of their pullbacks are isomorphic, whence $\Delta_M = \Delta_N$. Therefore $x \in \Delta_M \pi(\widetilde{R}^{\bullet})$, and $M^x \cong M$ by (1.5)(2).

Given a projective $\widetilde{R}$-module P, let r_i be the rank of $e_i P$, where the e_i are the primitive idempotents of $\widetilde{R}$. We define $\det(P)$ to be the direct sum of the modules $\bigwedge^{r_i}(e_i P)$. (If P is not faithful, we apply the convention that the 0th exterior power of the 0-module is the whole ring; so if $r_i = 0$ we take $\bigwedge^{r_i}(e_i P) = Re_i$. Thus $\det(P)$ is always an invertible ideal of $\widetilde{R}$.) We note that the isomorphism class of P is determined by the r_i and the isomorphism class of the ideal $\det(P)$.

1.7 LEMMA [W, (2.9)]   Let M be a faithful R-lattice, and let δ be an invertible ideal of $\widetilde{R}$. Then there is a lattice N in $\text{Genus}(M)$ such that $\det(\widetilde{R}N) \cong \delta$.

1.8 LEMMA   Let N_1, N_2 and N_3 be R-lattices in the same genus. Then there is an R-lattice N, unique up to isomorphism (and necessarily in the same genus), such that $N_1 \oplus N_2 \cong N_3 \oplus N$.

Proof: Since the N_i have the same annihilator we may assume that they are faithful. By (1.7) there is an R-lattice M in the same genus as the N_i such that $\det(M) = (\det(N_1))(\det(N_2))(\det(N_3))^{-1}$. By (1.5)(1) there is a unit $x \in (\widetilde{R}/c)^{\bullet}$ such that $N_1 \oplus N_2 \cong (N_3 \oplus M)^x$. Putting $N = M^x$ and using (1.5)(4), we get the desired isomorphism. Uniqueness of N follows from (1.6).

1.9 The Group Structure on the Genus

Given $[N_1], [N_2] \in \text{Genus}(M)$, we let $[N_1] \circ [N_2] = [N_3]$, where N_3 is the lattice defined (up to isomorphism) by the isomorphism $N_1 \oplus N_2 \cong M \oplus N_3$.

It is easy to verify that this operation makes Genus(M) into a commutative monoid with identity $[M]$. An application of (1.8) then shows that it is a group.

Note that the group structure is not really canonical, as it depends on the choice of $[M]$ as the identity element. One could of course use any other lattice in the genus, thereby changing the multiplication. We will always assume that M is the identity when we write "Genus(M)", despite the notational imprecision.

1.10 THEOREM   Let M be a faithful R-lattice. The following sequence, in which $\alpha : x \mapsto [M^x]$ and $\beta : [N] \mapsto [\det(\widetilde{R}N)]$, is exact:

$$1 \to \Delta_M \pi(\widetilde{R}^\bullet) \overset{\subseteq}{\longrightarrow} (\widetilde{R}/c)^\bullet \overset{\alpha}{\longrightarrow} \text{Genus}(M) \overset{\beta}{\longrightarrow} \text{Pic}(\widetilde{R}) \to 1.$$

Proof: Since $M^x \oplus M^y \cong M \oplus M^{xy}$ by (1.5)(4), α is a homomorphism. Exactness at $(\widetilde{R}/c)^\bullet$ and Genus(M) come follow from (1.5)(2) and (1.5)(1), respectively; and β is surjective by (1.7) since M is faithful.

2 FINITENESS THEOREMS

In this section we assume, in addition to the assumptions set out at the beginning of §1, that R is a finitely generated algebra over a field k. We recall the following finiteness results, valid for k-algebras of any dimension:

2.1 THEOREM [GJRW]   Let k be a field and A a finitely generated commutative k-algebra.

  (1) If n is prime to char(k), then Pic(A) has only finitely many elements of order n.

  (2) If L/k is a separable algebraic field extension (not necessarily finite-dimensional), then the kernel of the natural map $\text{Pic}(A) \to \text{Pic}(L \otimes_k A)$ is finite.

We will see that the analogous results can fail for the genus class group of an R-lattice. It is the arithmetic of field extensions, rather than geometry, that causes failure of (1). Indeed, we have the following slight generalization of the result [G2, (5.3)].

2.2 THEOREM   Let k be a separably closed field and R a ring-order finitely generated as a k-algebra. Let M be an R-lattice. For any integer n prime to char(k), Genus(M) has only finitely many elements of order n.

Proof: First assume that k is algebraically closed. By (1.10) and (2.1)(1) it suffices to prove that RestGenus(M) has this property. Then, as noted in [G2], RestGenus(M) $\cong G/F$ where G is a commutative algebraic group and F is finitely generated, whence the result.

Now suppose that k is separably closed of positive characteristic p, and let L/k be the algebraic closure. It will suffice to show that the kernel of $\Phi_M : \text{Genus}(M) \to \text{Genus}(LM)$ consists of elements of p-power order. If $[N] \in \text{Ker}(\Phi_M)$ there is a finite extension K/k, say of degree $q = p^e$ such that $KM \cong KN$. Then $\oplus^q M \cong \oplus^q N$, which implies that $[N]^q = [M]$ in Genus(M).

Pappacena [P] has shown that for arbitrary k there exists a positive integer m (depending on the lattice M) such that if n is prime to m, then the n-torsion subgroup of Genus(M) is finite.

Over arbitary fields, the genus class group of an ideal behaves very much like the Picard group:

2.3 THEOREM   Let I be an ideal of the ring-order R, which is finitely generated as a k-algebra.

  (1) If n is prime to char(k), then Genus(I) has only finitely many elements of order n.

  (2) If L/k is a separable algebraic field extension (not necessarily finite-dimensional), then the kernel of the natural map from Genus(I) to Genus($L \otimes_k I$) is finite.

Note that (1) follows from (2) and (2.2). (Take L to be the separable closure of k in an algebraic closure.) Before embarking on the proof of (2), we set up the machinery necessary for studying the behavior of genus class groups under separable base change. If A is a k-vector space, we will often write LA instead of $L \otimes_k A$.

2.4 LEMMA   Let L/k be a separable algebraic field extension, and let R be a ring-order finitely generated as a k-algebra. Then LR is a ring-order finitely generated as an L-algebra, and its standard pullback is obtained from the standard pullback for R by tensoring with L:

$$
\begin{array}{ccc}
LR & \longrightarrow & L\tilde{R} \\
\downarrow & & \downarrow \pi \\
LR/Lc & \longrightarrow & L\tilde{R}/Lc
\end{array}
\qquad (2.4.1)
$$

Proof: The main point is that $L\tilde{R}$ is a product of finitely many Dedekind domains, and we will give one of many ways to see this. Let F be the total quotient ring of $\tilde{R}$. Then LF is the total quotient ring of $L\tilde{R}$. Since L/k is separable LF is reduced. Also, $L\tilde{R}$ is Noetherian, being a finitely generated L-algebra. Of course $L\tilde{R}$ has dimension one, being integral over $\tilde{R}$. To show that $L\tilde{R}$ is a product of finitely many Dedekind domains, it will suffice to show that $L\tilde{R}$ is integrally closed in LF. By a direct limit argument we may assume that L/k is finite, and then the usual trace argument does the trick. To complete the proof, we point out that $\dim(LR) = 1$ since LR is integral over R, and that by faithful flatness Lc is exactly the conductor of LR in $L\tilde{R}$.

Now we prove (2) of (2.3). We may assume that I is faithful. By (2.1) the homomorphism $\mathrm{Pic}(\tilde{R}) \to \mathrm{Pic}(L\tilde{R})$ has finite kernel. Therefore, to prove (2) it is enough to show that the kernel of the map $\mathrm{RestGenus}(I) \to \mathrm{RestGenus}(LI)$ is finite. By (1.10) this amounts to showing that the map

$$
\frac{\tilde{R}^{\bullet}}{\Delta_I \pi(\tilde{R}^{\bullet})} \to \frac{(L\tilde{R})^{\bullet}}{\Delta_{LI} \pi((L\tilde{R})^{\bullet})}
$$

has finite kernel.

Let E be the endomorphism ring of I (viewed as multiplications). Then $R \subseteq E \subseteq \tilde{R}$, and one checks easily that $\Delta_I = (E/c)^{\bullet}$. Also, LE is the endomorphism ring of LI, and hence $\Delta_{LI} = (LI/Lc)^{\bullet}$. Our problem now reduces to showing that the map

$$
\frac{\tilde{R}^{\bullet}}{(E/c)^{\bullet}\pi(\tilde{R}^{\bullet})} \to \frac{(L\tilde{R})^{\bullet}}{(LI/Lc)^{\bullet}\pi((L\tilde{R})^{\bullet})}
$$

has finite kernel. But by [B2, Chap. IX, (5.3)] this map is the same as the map from $\mathrm{Ker}(\mathrm{Pic}(E) \to \mathrm{Pic}(\tilde{R}))$ to $\mathrm{Ker}(\mathrm{Pic}(LE) \to \mathrm{Pic}(L\tilde{R}))$. Another application of (2.1) completes the proof of (2).

We have already pointed out that (1) follows from (2) and (2.2). One can also deduce (1) directly from (2.1)(1) as follows: By (1.10) and (2.1)(1) it suffices to show that RestGenus(I) has finite n-torsion. The argument above shows that RestGenus(I) is isomorphic to the kernel of the map Pic(E) → Pic($\widetilde{R}$). Since Pic(E) has finite n-torsion by (2.1)(1), the same holds for RestGenus(I)

Note that we appealed to the usual Mayer-Vietoris for Pic rather than the exact sequence (1.10). The reason is that c might not be the whole conductor of E in its integral closure $\widetilde{R}$, so (1.10) as stated here might not apply directly to our situation. However, everything in this paper would go through if we were to replace the conductor by any ideal c of $\widetilde{R}$ contained in R such that R/c is Artinian.

There is another approach to (2.3) using ideas going back to Dress's 1969 paper [D]: Assuming I is a faithful ideal and again letting $E = \text{End}(I)$, one can show that the map $J \mapsto \text{Hom}_R(I, J)$ gives an isomorphism between Genus(I) and Pic(E). See, for example, [G1, §3].

Recall that a *Bass ring* is a ring-order R in which every ideal can be generated by two elements. Equivalently, each of the local rings of R has multiplicity at most 2. Another characterization is that $\widetilde{R}$ can be generated by 2 elements as an R-module. We refer the reader to [LW] for basic properties of Bass rings and their lattices. In particular, one can assign a faithful ideal cl(M) to each R-lattice in such a way that for R-lattices M and N one has

(1) cl($M \oplus N$) $\cong$ (cl(M))(cl(N)), and

(2) $M \cong N \iff$ Genus(M) = Genus(N) and cl(M) $\cong$ cl(N).

(If M has constant rank r then cl(M) is just $(\bigwedge^r M)$/torsion.)

Given a faithful ideal I in a Bass ring R, let $\rho(I) = \{r \in \widetilde{R} | rI \subseteq I\}$, the endomorphsim ring of I. Then I is an invertible ideal of $\rho(I)$ by [B1, §7].

2.5 LEMMA   Let R be a Bass ring, and let M be a faithful R-lattice. Put $S = \rho(\text{cl}(M))$. There is a natural isomorphism Genus(M) $\cong$ Pic(S) taking $[M]$ to $[S]$.

Proof: By [LW, (5.2)] the map $[N] \mapsto [\rho(\text{cl}(N))]$ is a bijection between Genus(M) and Pic(S). Therefore, so is the map $\Psi : [N] \mapsto [\rho(\text{cl}(N))][\rho(\text{cl}(M))]^{-1}$. Using (1.9) and item (1) above, one checks easily that Ψ is a homomorphism.

2.6 PROPOSITION   Let k be a field, let R be a Bass ring finitely generated as a k-algebra, and let M be an R-lattice.

(1) If n is prime to $\text{char}(k)$, then $\text{Genus}(M)$ has only finitely many elements of order n.

(2) For any separable algebraic extension L/k, the kernel of the natural map $\text{Genus}(M) \to \text{Genus}(LM)$ is finite.

Proof: As usual, we may assume that M is faithful. Now apply (2.1) and (2.5).

3 A FAMILY OF EXAMPLES

In this section we construct examples showing that, unlike the case of $\text{Pic}(R)$, the kernel of the map on genus class groups can be infinite, even for a finite separable extension.

Let k be an infinite field and A/k be a finite-dimensional commutative separable algebra over k. Assume that G is a finite group of automorphisms of A with fixed ring k. Let $C_0, \ldots, C_r$ be subgroups of G. Let A_i denote the fixed algebra of C_i. Choose a primitive element α for A/k, and let $\pi : A_0[X] \to A$ be the A_0-algebra homomorphism taking X to α. Define R by the pullback diagram

$$
\begin{array}{ccc}
R & \longrightarrow & A_0[X] \\
\downarrow & & \downarrow{\scriptstyle \pi} \\
k & \overset{\subseteq}{\longrightarrow} & A
\end{array}
\quad . \tag{3.1}
$$

Then R is a ring-order and is finitely generated as a k-algebra. Moreover, (3.1) is naturally isomorphic to the standard pullback (1.1) for R. (See [WW, (3.1)].)

We define R-lattices $M_i, 1 \le i \le r$ (which are fractional ideals of R) by the following pullback diagram:

$$
\begin{array}{ccc}
M_i & \longrightarrow & A_0[X] \\
\downarrow & & \downarrow{\scriptstyle \pi} \\
A_i & \overset{\subseteq}{\longrightarrow} & A
\end{array}
\quad . \tag{3.2}
$$

By [W, (2.1)] this diagram is naturally isomorphic to the standard pullback (1.2) for M_i.

To compute Δ_M, suppose ϕ is an A-automorphism of A carrying A_i into itself. Clearly, ϕ must be multiplication by a unit of A_i. So we see that

$$\Delta_{M_i} = A_i^\bullet. \tag{3.3}$$

Now $\pi(\widetilde{R}^\bullet) = A_0{}^\bullet$, and since $\text{Pic}(A_0[X])$ is trivial we see from the exact sequence (1.10) that

$$\text{Genus}(M_i) = \text{RestGenus}(M_i) \cong A^\bullet/A_0^\bullet A_i^\bullet.$$

Now let $M = \oplus_{i=1}^r M_i$. By (1.5)(3), we see that

$$\text{Genus}(M) = \text{RestGenus}(M) \cong A^\bullet/A_0^\bullet A_1^\bullet \cdots A_r^\bullet.$$

If U is any G-module, let us define $\lambda(U)$ to be the submodule generated by the fixed points of $C_i, 0 \le i \le r$. So we see that

$$\text{Genus}(M) = \text{RestGenus}(M) \cong A^\bullet/A_0^\bullet A_1^\bullet \cdots A_r^\bullet \cong A^\bullet/\lambda(A^\bullet). \tag{3.4}$$

Now let L/k be a separable field extension and let $B = LA$ and $B_i = LA_i$. Then all the diagrams above apply as well with A replaced by B and A_i by B_i. Note that G acts on B (by letting G act trivially on L) and B_i is precisely the set of C_i fixed points on B. Set $N = LM$. So we see that $\text{Genus}(N) \cong B^\bullet/\lambda(B^\bullet)$.

In particular, if $f : A \to LA$ is the natural injection, we see that the kernel of $\Phi_M : \text{Genus}(M) \to \text{Genus}(LM)$ is isomorphic to the kernel of the induced map $f^\bullet : A^\bullet/\lambda(A^\bullet) \to B^\bullet/\lambda(B^\bullet)$.

Assume, in particular, that A/k is a Galois field extension with Galois group G. Let L be a separable extension containing A. Then, we can identify $B^\bullet$ with $L^\bullet \otimes_{\mathbf{Z}} \mathbf{Z}G$ as a G-module where G acts trivially on $L^\bullet$ (note that B is a direct sum of $|G|$ copies of L (one for each primitive idempotent) and that G permutes these copies of L). With this identification, $f(a) = \sum_{g \in G} g^{-1}(a) \otimes g$.

We investigate two cases in particular with $r = 2$. Thus, M is isomorphic to the direct sum of 2 ideals of R.

3.5 EXAMPLE   Let A/k be a degree-4 non-cyclic Galois extension of number fields, and let $A_i, 0 \le i \le 2$, be the proper intermediate fields. Let $M := M_1 \oplus M_2$ as above.

(1) Genus(M) is an infinite elementary abelian 2-group.
(2) If L/k is any finite algebraic extension, then Ker(Φ_M : Genus(M) → Genus(LM)) is finite.
(3) If L/k is the algebraic closure, then Genus(LM) is trivial (whence Ker(Φ_M : Genus(M) → Genus(LM)) is infinite.

Proof: By [CGW, (6.3)], $A^\bullet/\lambda(A^\bullet)$ and $B^\bullet/\lambda(B^\bullet)$ are elementary abelian 2-groups. Moreover, since k is a number field, $A^\bullet/\lambda(A^\bullet)$ is infinite. (See [GW, (1.2), (1.5), (1.8)].)

If L is the algebraic closure of k, then $B^\bullet$ is divisible (being a direct sum of 4 copies of $L^\bullet$), and hence so is $B^\bullet/\lambda(B^\bullet)$. Thus Φ_M is the trivial map with infinite kernel.

On the other hand, if L is a finite extension containing A, it follows either by [CGW] or directly that the kernel of Φ_M is isomorphic to $S/\lambda(A^\bullet)$ where S consists of the elements in $A^\bullet$ whose norms are squares in L. If $L = A$, then the kernel is finite as computed in [CGW, (6.3)]. Since the elements in $L^\bullet/A^\bullet$ of order 1 or 2 form a finite subgroup by [CGW, (4.1)], the kernel is finite whenever L/k is finite.

This shows the contrast with the case of an ideal I. In that case, the kernel is finite for any separable extension (infinite or not) and even in the inseparable case, there is some finite extension beyond which there is no further kernel (see [GJRW, (5.2)]).

3.6 EXAMPLE   Let A/k be a Galois extension of number fields with Galois group G elementary abelian of order 8. Let $A_i, 0 \le i \le 2$ be degree-4 extensions of k with intersection k, and let $M := M_1 \oplus M_2$ as above. If $L = A$, the kernel of Φ_M : Genus(M) → Genus(LM) is infinite and is equal to the kernel of the map Φ_M : Genus(M) → Genus(k^aM), where k^a is an algebraic closure of k.

Proof: We use the results of [CGW, §2]. In the notation of that section, S and T are finitely generated $\mathbb{Z}G$-modules defined in terms of G and the C_i. In our situation, T is an infinite cyclic group, and G acts on T by letting the generator for C_i act by multiplication by -1 on T (see [CGW, (2.9)]). It follows easily from the proof of [CWG, (2.2)] that we can identify the kernel of f^* with the kernel of the natural map $H^1(G, S \otimes_{\mathbf{Z}} A^\bullet) \to H^1(G, S \otimes_{\mathbf{Z}} B^\bullet)$ (the map on cohomology is induced by f). Since $B^\bullet$ is an induced module, it follows by Frobenius reciprocity that the righthand term is 0 (and in particular, there is no further kernel obtained if L is a proper extension of A). So in this case, the kernel to the algebraic closure is exactly the same as the kernel obtained by extending scalars to A. This will be infinite if and only if $H^1(G, S \otimes_{\mathbf{Z}} A^\bullet)$ is infinite. By Cebatorev density, $A^\bullet$ has a G-summand isomorphic to the direct sum of infinitely many copies of the permutation lattice $\mathbf{Z}[G/C]$ where C is the subgroup of G generated by the product of the generators of the C_i. It follows ([CGW, (2.9)]) that $|H^1(C, S)| = 2$ and so $H^1(G, S \otimes_{\mathbf{Z}} A^\bullet)$ is infinite.

So in this example the kernel is infinite even though L/k is finite, and, in contrast to (3.5), there is no increase in the kernel upon passage to the algebraic closure.

Both of these examples demonstrate the strange behavior of the genus with respect to direct sums: Both M_1 and M_2 are ideals of R, yet the genus of $M_1 \oplus M_2$ shares none of the nice properties established for the genus of an ideal in §2. Still, one can say something about the genus of a direct sum:

3.7 PROPOSITION Let R be a ring-order and M a faithful R-lattice. Let N be any R-lattice. Then the map $\text{Genus}(M) \to \text{Genus}(M \oplus N)$ taking $[X] \to [X \oplus N]$ is a surjective homomorphism. If $[N] \in \text{Genus}(M)$ the map is an isomorphism.

Proof: It is easy to check that the map is a homomorphism. To see that it is surjective, let $[Z] \in \text{Genus}(M \oplus N)$. Choose, using (1.7), a lattice $Y \in \text{Genus}(M)$ such that $(\det(\widetilde{R}Y))(\det(\widetilde{R}N)) \cong \det(\widetilde{R}Z)$. Then $Y \oplus N$ and Z are in the same restriced genus, so by (1.5)(1) there is an element $x \in (\widetilde{R}/c)^\bullet$ such that $(Y \oplus N)^x \cong Z$. Putting $X = Y^x$, we have $X \oplus N \cong Z$ by (1.5)(4). The last statement is clear from (1.6).

We close with an example showing that in Pappcena's result [P] mentioned after the proof of (2.2), one cannot find an integer m that works for all R-lattices.

3.8 EXAMPLE Let L/k be an extension of algebraic number fields with an intermediate field K of degree 2 over k. Assume that $[L : K] \geq 4$. Choose a k-algebra surjection $\pi : k[X] \to L$, and form the pullback

$$\begin{array}{ccc} R & \longrightarrow & k[X] \\ \downarrow & & \downarrow \pi \\ K & \stackrel{\subseteq}{\longrightarrow} & L \end{array}.$$

Fix a prime p. Then there is an R-lattice M (depending on p) such that Genus(M) has infinitely many elements of order p.

Proof: Since $[L : K] \geq 4$ a construction going back to Drozd and Roĭter [DR] (cf. proof of case (3) of [CWW, (2.6)]) there is an R-lattice M such that $\Delta_M = K^{\bullet p}$, the group of pth powers of elements of elements of $K^{\bullet}$. Then Genus(M) $\cong L^{\bullet}/k^{\bullet}K^{\bullet p}$, which contains $\Gamma := K^{\bullet}/k^{\bullet}K^{\bullet p}$. We show that Γ is infinite. If $|\Gamma| = n < \infty$, choose r with $p^r > n$. Let $G = \{1, \tau\}$ be the Galois group of K/k. By the proof of [GW1, (1.5)] there are G-module homomorphisms $\phi_i : K^{\bullet} \to \mathbf{Z}G$ such that the induced map ϕ from $K^{\bullet}$ to the direct sum $(\mathbf{Z}G)^{(r)}$ is surjective. (In fact $K^{\bullet}$ maps onto a free $\mathbf{Z}G$-module of infinite rank, by the proof of [CGW. (7.2)].) For each i, the image of k under ϕ_i is the set of fixed points of G on $\mathbf{Z}G$, that is, the ideal generated by $1 + \tau$. The image of $K^{\bullet p}$ is $p\mathbf{Z}G$. It follows that Γ maps onto the direct sum of r copies of $\frac{(\mathbf{Z}/p\mathbf{Z})G}{(1+\tau)}$, and since $\frac{(\mathbf{Z}/p\mathbf{Z})G}{(1+\tau)}$ has order p this is a contradiction.

REFERENCES

[B1] H. Bass, On the ubiquity of Gorenstein rings, *Math. Z., 82*: 8–28 (1963).

[B2] H. Bass, *Algebraic K-Theory*, W. A. Benjamin, New York (1968).

[CGW] J.-L. Colliot-Thélène, R. Guralnick and R. Wiegand, The multiplicative group of a field modulo the product of subfields, *J. Pure Appl. Algebra, 106*: 233–262 (1996).

[CWW] N. Cimen, R. Wiegand and S. Wiegand, One-dimensional rings of finite representation type, *Abelian Groups and Modules* (A. Facchini and C. Menini, eds.), pp. 95–121, Kluwer, Dordrecht (1995).

[D] A. Dress, On the decomposition of modules, *Bull. Amer. Math. Soc., 75*: 984–986 (1969).

[DR] Ju. A. Drozd and A. V. Roĭter, Commutative rings with a finite number of indecomposable integral representations (Russian), *Izv. Akad.*

Nauk. SSSR, Ser. Mat., 31: 783–798 (1967).

[E] E. G. Evans, Jr., Krull-Schmidt and cancellation over local rings, *Pacific J. Math., 46*: 115–121 (1973).

[EGA] A. Grothendieck and J. A. Dieudonné *Éléments de géométrie algébrique, Chap. IV (Part 2)*, Inst. Hautes Études Sci. Publ. Math., 24 (1965).

[G1] R. Guralnick, The genus of a module, *J. Number Theory, 18*: 169–177 (1984).

[G2] R. Guralnick, Bimodules over PI rings, *Methods in Module Theory* (G. Abrams, J. Haefner and K. Rangaswamy, eds.), pp. 117–134, Marcel Dekker, New York (1992).

[GJRW] R. Guralnick, D. Jaffe, W. Raskind and R. Wiegand, On the Picard group: torsion and the kernel induced by a faithfully flat map, *J. Algebra, 183*: 420–455 (1996).

[GR] E. Green and I. Reiner, Integral representations and diagrams, *Michigan Math. J., 25*: 53–84 (1978).

[GW1] R. Guralnick and R. Wiegand, Galois groups and the multiplicative structure of field extensions, *Trans. Amer. Math. Soc., 331*: 563–584 (1992).

[GW2] R. Guralnick and R. Wiegand, Picard groups, cancellation and the multiplicative structure of fields, *Zero-Dimensional Commutative Rings* (D. Anderson and D. Dobbs, eds.), pp. 65–79, Marcel Dekker, New York (1995).

[HL] J. Haefner and L. S. Levy, Commutative orders whose lattices are direct sums of ideals, *J. Pure Appl. Algebra, 24*: 1–20 (1988).

[HW] D. Holley and R. Wiegand, Torsion in quotients of the multiplicative group of a number field, *Contemp. Math., 171*: 201–204 (1994).

[LW] L. S. Levy and R. Wiegand, Dedekind-like behavior of rings with 2-generated ideals, *J. Pure Appl. Algebra, 37*: 41–58 (1985).

[P] C. Pappacena, Torsion in the class group of a noncommutative curve, preprint.

[W] R. Wiegand, Cancellation over commutative rings of dimension one and two, *J. Algebra, 88*: 438–459 (1984).

[WW] R. Wiegand and S. Wiegand, Stable isomorphism of modules over one-dimensional rings, *J. Algebra, 107*: 425–435 (1987).

Generalized Integral Closures

Franz Halter-Koch
Institut für Mathematik, Karl-Franzens-Universität, A-8010 Graz, Austria.

1. Throughout this paper, let D be an integral domain and L a quotient field of D. We denote by $\mathfrak{F}(D)$ the set of all non-zero fractional ideals of D and by $\mathfrak{f}(D)$ the set of all non-zero finitely generated fractional ideals of D. For $A, B \subset L$, we set

$$(A:B) = \{x \in L \mid xB \subset A\}.$$

It is well known that an element $x \in L$ is integral over D if and only if $x \in (J : J)$ holds for some $J \in \mathfrak{F}(D)$. In [2], a modified concept, called pseudo-integrality, was defined in the following way: $x \in L$ is called pseudo-integral over D if and only if $x \in (J_v : J_v)$ holds for some $J \in \mathfrak{f}(D)$ (where J_v denotes the smallest divisional ideal containing J). In [8], this notion was generalized: Instead of J_v, the authors there considered J^* for an arbitrary star-operation $*$ on D.

In this note we investigate the notion of $*$-integrality in the sense of [8]. We consider its behaviour under localizations and its connections with valuation theory. The main result (Theorem 3) states that the $*$-integral closure of D is the intersection of all $*$-valuation overrings.

2. We start by fixing some terminology. We denote by $D^\times$ the group of invertible elements of D; in particular, $L^\times = L \setminus \{0\}$. For any set X, we denote by $\mathbb{P}_{\text{fin}}(X)$ the set of non-empty finite subsets of X. A subset $X \subset L$ is called *D-fractional* if $X \cap L^\times \neq \emptyset$ and $cX \subset D$ holds for some $c \in D \setminus \{0\}$. For *D*-fractional $X \subset L$, we denote by $\langle X \rangle = \langle X \rangle_D \in \mathfrak{F}(D)$ the fractional *D*-ideal generated by X.

Our main reference for star-operations is [6], §32. Throughout this paper, all star-operations are assumed to be of finite character, that is,

$$J^* = \bigcup_{\substack{B \in \mathfrak{f}(D) \\ B \subset J}} B^*$$

holds for all $J \in \mathfrak{F}(D)$. If

$$*: \begin{cases} \mathfrak{F}(D) & \to & \mathfrak{F}(D) \\ J & \mapsto & J^* \end{cases}$$

is a star-operation, we set $X^* = \langle X \rangle^*$ for every *D*-fractional subset X of L. The assignment $X \mapsto X^*$ is an ideal system on $L^\times$ with respect to $(L^\times)^+ = D \setminus \{0\}$ in the sense of [7], §3, which extends the finite ideal system $E \mapsto E^*$ (for $E \in \mathbb{P}_{\text{fin}}(L^\times)$).

We denote by $\circ$ the *trivial star-operation*, given by $J^\circ = J$ for all $J \in \mathfrak{F}(D)$ (and consequently $X^\circ = \langle X \rangle$ for all *D*-fractional $X \subset L$). The ideal system corresponding to $\circ$ is the *d*-system, and every finite ideal system on $L^\times$ which is finer that the *d*-system arises from a star-operation.

A readable account on star-operations is [1], §2. Historical background material and motivations (again in the language of ideal systems) may be found in [3].

The following two Propositions are basic for the construction of star-operations.

Proposition 1. A. *Let*

$$*: \begin{cases} \mathfrak{F}(D) & \to & \mathfrak{F}(D) \\ J & \mapsto & J^* \end{cases}$$

be a star-operation on D. *Then the following conditions hold for all D-fractional subsets* X, Y *of* L *and all* $c \in L^\times$:
   (I1) $\langle X \rangle \subset X^*$.
   (I2) $X \subset Y^*$ *implies* $X^* \subset Y^*$.

(I3) $\{c\}^* = \langle c \rangle = cD$.

(I4) $cX^* = (cX)^*$.

(I5) $X^* = \bigcup_{E \in \mathbb{P}_{\text{fin}}(X)} E^*$.

B. *Let*

$$*: \begin{cases} \mathbb{P}_{\text{fin}}(L^\times) & \to & \mathfrak{F}(D) \\ E & \mapsto & E^* \end{cases}$$

be a map such that **(I1)** *to* **(I4)** *hold for all* $X, Y \in \mathbb{P}_{\text{fin}}(L^\times)$ *and all* $c \in L^\times$. *Then* $*$ *has a unique extension to a star-operation on* D *(again denoted by* $*$*). It is given by* **(I5)** *for all* D-*fractional subsets* X *of* L.

Proof. Straightforward. □

Proposition 2. *Let* $*$ *be a star-operation on* D *and* $(J_\lambda)_{\lambda \in \Lambda}$ *a family of fractional* $*$-*ideals with the following properties:*

1) $D \subset J_\lambda$ *holds for all* $\lambda \in \Lambda$.

2) *For all* $\nu, \mu \in \Lambda$, *there exists some* $\lambda \in \Lambda$ *such that* $J_\nu J_\mu \subset J_\lambda$.

Then the family $(J_\lambda)_{\lambda \in \Lambda}$ *is upwards directed,*

$$J = \bigcup_{\lambda \in \Lambda} J_\lambda \subset L$$

is an overring of D, *and there is a unique star-operation* $\circledast$ *on* J *such that*

$$E^\circledast = \bigcup_{\lambda \in \Lambda} (EJ_\lambda)^*$$

for all $E \in \mathbb{P}_{\text{fin}}(L^\times)$. *If* $* = \circ$ *is the trivial star-operation on* D, *then so is* $\circledast$ *on* J.

Proof. By 1) and 2), the family $(J_\lambda)_{\lambda \in \Lambda}$ is upwards directed, and therefore J is an overring of D. If $E \in \mathbb{P}_{\text{fin}}(L^\times)$, then $((EJ_\lambda)^*)_{\lambda \in \Lambda}$ is an upwards directed family of fractional D-ideals containing E. Therefore $E^\circledast$ is a J-module containing E. If $c \in D \setminus \{0\}$ satisfies $cE \subset D$, then $cE^\circledast \subset J$ implies $E^\circledast \in \mathfrak{F}(J)$ and $\langle E \rangle_J \subset E^\circledast$. Now the assertion follows from Proposition 1: Condition **(I1)** has just been verified, and conditions **(I2)** to **(I4)** are easily checked.

If $* = \circ$ is the trivial star-operation, then every $z \in E^\circledast$ is a linear combination of elements of E with coefficients in some J_λ. Hence $E^\circledast$ is the fractional J-ideal generated by E. □

As an application of Proposition 2 we obtain a criterion for over-rings of D to be t-linked, a notion which is of importance in the theory of Prüfer-v-multiplication domains, see [5].

Corollary 1. *Let* $(J_\lambda)_{\lambda \in \Lambda}$ *be a family of fractional* t-*ideals of* D *satisfying conditions* 1) *and* 2) *of Proposition* 2. *Then*

$$J = \bigcup_{\lambda \in \Lambda} J_\lambda$$

is a t-*linked overring of* D.

Proof. By [5], Proposition 2.1 it is sufficient to prove that $(Q \cap D)_t = D$ implies $Q = J$ for every t-ideal Q of J. By Proposition 2, there exists a star-operation $\circledast$ on D such that

$$E^\circledast = \bigcup_{\lambda \in \Lambda} (EJ_\lambda)_t$$

holds for all $E \in \mathbb{P}_{\text{fin}}(L^\times)$. If Q is a t-ideal of J such that $(Q \cap D)_t = D$, then

$$1 \in (Q \cap D)_t = \bigcup_{E \in \mathbb{P}_{\text{fin}}(Q \cap D)} E_t$$

implies $1 \in E_t$ and hence $E_t = D$ for some $E \in \mathbb{P}_{\text{fin}}(Q \cap D)$. Thus $(EJ_\lambda)_t = J_\lambda$ holds for all $\lambda \in \Lambda$, and we obtain

$$J = E^\circledast \subset E_t \subset Q_t \subset J,$$

whence $Q_t = J$. $\square$

Corollary 2.
   i) *The complete integral closure and the pseudo-integral closure of* D *are* t-*linked over* D.
   ii) *Suppose that* $J_v \in \mathfrak{f}(D)$ *holds for all* $J \in \mathfrak{f}(D)$. *Then the integral closure of* D *is* t-*linked over* D.

Proof. i) The complete integral closure $\hat{D}$ and the pseudo-integral closure $\tilde{D}$ of D are given by

$$\hat{D} = \bigcup_{J \in \mathfrak{F}(D)} (J_v : J_v) \quad \text{and} \quad \tilde{D} = \bigcup_{J \in \mathfrak{f}(D)} (J_v : J_v).$$

Thus the assertion follows from Corollary 1.

ii) If $J_v \in \mathfrak{f}(D)$ holds for all $J \in \mathfrak{f}(D)$, then the pseudo-integral closure of D coincides with the integral closure and the assertion follows from **i)**. □

Remarks. Parts of Corollary 2 are already in [5] (Corollaries 2.3 and 2.14). The assumption made in **ii)** holds for quasi-coherent domains, see [4].

3. Let $*$ be a star-operation on D. An element $x \in L$ is called *$*$-integral* over D if $xJ^* \subset J^*$ holds for some $J \in \mathfrak{f}(D)$. The set $\mathrm{cl}^*(D)$ of all $*$-integral elements of L over D is called the *$*$-closure* of D, and D is called *$*$-closed* if $\mathrm{cl}^*(D) = D$. This notion coincides with the corresponding ones in [7] and [3].

It is proved in [8] that $\mathrm{cl}^*(D)$ is an integrally closed overring of D. In particular, if D is $*$-closed, then it is integrally closed. If $J = \langle E \rangle \in \mathfrak{f}(D)$, where $E \in \mathbb{P}_{\mathrm{fin}}(L^\times)$, then $(J^* : J^*) = (E^* : E)$ is a fractional $*$-ideal containing D. For $E, F \in \mathbb{P}_{\mathrm{fin}}(L^\times)$, we have

$$((EF)^* : (EF)) \supset (E^* : E)(F^* : F),$$

and therefore

$$\mathrm{cl}^*(D) = \bigcup_{J \in \mathfrak{f}(D)} (J^* : J^*) = \bigcup_{E \in \mathbb{P}_{\mathrm{fin}}(L^\times)} (E^* : E)$$

is the union of a family of fractional $*$-ideals of D possessing properties **1)** and **2)** of Proposition 2. We thus obtain a star-operation $\circledast$ on $\mathrm{cl}^*(D)$ satisfying

$$F^{\circledast} = \bigcup_{E \in \mathbb{P}_{\mathrm{fin}}(L^\times)} \left[(E^* : E)F\right]^*$$

for all $F \in \mathbb{P}_{\mathrm{fin}}(L^\times)$. We call $\circledast$ the *closure* of $*$ on $\mathrm{cl}^*(D)$. If $* = \circ$ is the trivial star-operation on D, then $\mathrm{cl}^*(D)$ is the integral closure of D and $\circledast$ is the trivial star-operation on $\mathrm{cl}^*(D)$.

Theorem 1. *Let* $*$ *be a star-operation on* D *and* $\circledast$ *the closure of* $*$ *on* $\mathrm{cl}^*(D)$. *Then* $\mathrm{cl}^*(D)$ *is* $\circledast$-*closed. In particular,* $\mathrm{cl}^*(D)$ *is integrally closed.*

Proof. Let $x \in L$ be $\circledast$-integral over $\mathrm{cl}^*(D)$. Then there exists some $F \in \mathbb{P}_{\mathrm{fin}}(L^\times)$ such that

$$xF \subset F^\circledast = \bigcup_{E \in \mathbb{P}_{\mathrm{fin}}(L^\times)} \left[(E^* : E)F \right]^*.$$

Since this is the union over an upwards directed family and xF is finite, we obtain $xF \subset \left[(E^* : E)F \right]^*$ for some $E \in \mathbb{P}_{\mathrm{fin}}(L^\times)$, and consequently

$$xEF \subset \left[(E^* : E)EF \right]^* \subset (E^*F)^* = (EF)^*,$$

which entails $x \in \big((EF)^* : (EF) \big) \subset \mathrm{cl}^*(D)$. $\square$

4. Next we consider the behaviour of the $*$-closure under localizations. Let $*$ be a star-operation on D, and let $\emptyset \neq T \subset D$ be a multiplicatively closed subset. By Proposition 1, there is a star-operation $(T^{-1}*)$ on $T^{-1}D$ such that

$$E^{(T^{-1}*)} = T^{-1}E^*$$

holds for all $E \in \mathbb{P}_{\mathrm{fin}}(L^\times)$. The star-operation $(T^{-1}*)$ is called the *quotient star-operation* of $*$ with respect to T.

Theorem 2. *Let* $*$ *be a star-operation on* D *and* $\emptyset \neq T \subset D$ *a multiplicatively closed subset. Then*

$$T^{-1}\mathrm{cl}^*(D) = \mathrm{cl}^{(T^{-1}*)}(T^{-1}D).$$

Proof. Suppose first that $a \in \mathrm{cl}^*(D)$ and $t \in T$. Then there exists some $E \in \mathbb{P}_{\mathrm{fin}}(L^\times)$ such that $aE \subset E^*$, and hence $t^{-1}aE \subset T^{-1}E^* = E^{(T^{-1}*)}$, which implies $t^{-1}a \in \mathrm{cl}^{(T^{-1}*)}(T^{-1}D)$.

If $z \in \mathrm{cl}^{(T^{-1}*)}(T^{-1}D)$, then $zE \subset E^{(T^{-1}*)} = T^{-1}E^*$ holds for some $E \in \mathbb{P}_{\mathrm{fin}}(L^\times)$, and since E is finite, there exists some $t \in T$ such that $ztE \subset E^*$. This implies $zt \in \mathrm{cl}^*(D)$ and hence $z \in T^{-1}\mathrm{cl}^*(D)$. $\square$

5. A *valuation overring* of D is a valuation ring V between D and L. Let $*$ be a star-operation on D and V a valuation overring of D. V is called a $*$-*valuation overring* of D if $E^* \subset EV$ holds for all $E \in \mathbb{P}_{\mathrm{fin}}(L^\times)$ (then $X^* \subset XV$ holds for all D-fractional subsets X of L). Note that EV is a fractional principal ideal of V. More precisely, if $v \colon L \to \Gamma \cup \{\infty\}$ is a Krull valuation with valuation ring V and $c \in E$ is such that $v(c) = \min v(E)$, then $EV = cV$. Observe that V is a $*$-valuation overring of D if and only if v is a $*$-valuation in the sense of [7], p. 46.

Proposition 3. *Let V be a valuation overring of D, $*$ a star-operation on D and $\circledast$ the closure of $*$ on $\mathrm{cl}^*(D)$.*
   i) *If V is a $*$-valuation overring of D, then $\mathrm{cl}^*(D) \subset V$, and V is a $\circledast$-valuation overring of $\mathrm{cl}^*(D)$.*
  ii) *Every $\circledast$-valuation overring of $\mathrm{cl}^*(D)$ is a $*$-valuation overring of D.*

Proof. i) Let V be a $*$-valuation overring of D and $z \in \mathrm{cl}^*(D)$. Then there exists some $E \in \mathbb{P}_{\mathrm{fin}}(L^\times)$ such that $zE \subset E^*$, and $EV = cV$ for some $c \in E$. Thus we obtain

$$zcV = zEV \subset E^*V = EV = cV \,,$$

and $z \in V$ follows.

In order to prove that V is a $\circledast$-valuation overring of $\mathrm{cl}^*(D)$, we must show that $F^\circledast \subset FV$ for all $F \in \mathbb{P}_{\mathrm{fin}}(L^\times)$. If $F \subset L^\times$ is finite, then

$$F^\circledast = \bigcup_{E \in \mathbb{P}_{\mathrm{fin}}(L^\times)} \left[(E^* : E)F\right]^* \,,$$

and for all $E \in \mathbb{P}_{\mathrm{fin}}(L^\times)$, we obtain

$$\left[(E^* : E)F\right]^* \subset (E^* : E)FV \subset \mathrm{cl}^*(D)FV = FV \,.$$

ii) Let V be a $\circledast$-valuation overring of $\mathrm{cl}^*(D)$. If $E \in \mathbb{P}_{\mathrm{fin}}(L^\times)$, then $E^* \subset E^\circledast \subset EV$ shows that V is a $*$-valuation overring of D. $\square$

6. A star-operation $*$ on D is called *eab* (*endlich arithmetisch brauchbar*) if $(EF_1)^* \subset (EF_2)^*$ implies $F_1^* \subset F_2^*$ for all $E, F_1, F_2 \in \mathbb{P}_{\mathrm{fin}}(L^\times)$. If $*$ is eab, then D is $*$-closed; see [7], Corollary on p. 25. The converse need not be true. However, if D is $*$-closed, then there is a star-operation $*_a$, closely connected with $*$, which is eab. The following Proposition 4 recalls the corresponding construction.

Proposition 4. *Let* $*$ *be a star-operation on* D, *and let* D *be* $*$-*closed. Then there exists a star-operation* $*_a$ *on* D *such that*

$$E^{*_a} = \bigcup_{B \in \mathbb{P}_{\text{fin}}(L^\times)} ((EB)^* : B)$$

holds for all $E \in \mathbb{P}_{\text{fin}}(L^\times)$. *It possesses the following properties:*

i) *For every* D-*fractional subset* X *of* L,

$$X^* \subset X^{*_a} = \bigcup_{B \in \mathbb{P}_{\text{fin}}(L^\times)} ((XB)^* : B).$$

ii) $*_a$ *is eab.*

iii) $*_a = *$ *holds if and only if* $*$ *is eab. In particular,* $(*_a)_a = *_a$.

iv) *A valuation overring* V *of* D *is a* $*$-*valuation overring if and only if it is a* $*_a$-*valuation overring.*

Proof. [7], p. 43, Théoreme 1 and Corollary 1, and [7], p. 48, Proposition 5. □

Associated with an eab star-operation on D, the Kronecker function ring D^* is defined as follows. Let $L(X)$ be a rational function field over L in an indeterminate X. For $f \in L[X] \setminus \{0\}$, let $A_f \in \mathfrak{F}(D)$ be the fractional ideal generated by the coefficients of f. Now define

$$D^* = \{0\} \cup \left\{ \frac{f}{g} \mid f, g \in L[X] \setminus \{0\}, A_f^* \subset A_g^* \right\}.$$

If $*$ is a star-operation on D for which D is merely $*$-closed, we set $D^* = D^{*_a}$.

Let D be integrally closed and $(V_\lambda)_{\lambda \in \Lambda}$ the family of all valuation overrings of D. The *basic eab star-operation* $\flat$ on D is defined by

$$E^\flat = \bigcap_{\lambda \in \Lambda} EV_\lambda$$

for all $E \in \mathbb{P}_{\text{fin}}(L^\times)$ (note that $\flat$ is equivalent to the b-operation considered in [6], §32; the latter is not of finite character). The Kronecker funtion ring $D^\flat$ is called the *basic Kronecker function ring* for D. It is contained in every D^* for an eab star-operation $*$ on D.

Proposition 5. *Let $*$ be a star-operation on D such that D is $*$-closed.*

   i) *$W \mapsto W \cap L$ is a bijection from the set of all valuation overrings of D^b onto the set of all valuation overrings of D.*

   ii) *Let W be a valuation overring of D^b. Then $D^* \subset W$ holds if and only if $W \cap L$ is a $*$-valuation overring of D.*

   iii) *For every $E \in \mathbb{P}_{\text{fin}}(L^\times)$, we have*

$$E^{*_\bullet} = \langle E \rangle_{D^\bullet} \cap L = \bigcap_{\lambda \in \Lambda} E V_\lambda,$$

where $(V_\lambda)_{\lambda \in \Lambda}$ is the family of all $*$-valuation overrings of D.

Proof. i) Let W be a valuation overring of D^b. By [6], Theorem 32.10, W is the trivial extension of $W \cap L$ to $L(X)$. Therefore $W \mapsto W \cap L$ is injective. If V is a valuation overring of D, then $V^b \supset D^b$ and $V^b \cap L = V$ by [6], Theorem 32.1. Morever, V^b is a valuation ring.

   ii) Since $D^* = D^{*_\bullet}$ and a $*$-valuation overring of D is the same as a $*_a$-valuation overring of D, we may assume that $* = *_a$ is eab. Let $w : L(X) \to \Gamma \cup \{\infty\}$ be a Krull valuation possessing the valuation ring W.

   Suppose first that $D^* \subset W$. We must prove that

$$\langle b_0, \ldots, b_m \rangle^* \subset \langle b_0, \ldots, b_m \rangle (W \cap L)$$

holds for all $b_0, \ldots, b_m \in L^\times$. Indeed, suppose that $c \in \langle b_0, \ldots, b_m \rangle^* = A_g^*$, where $g = b_0 + b_1 X + \cdots + b_m X^m$. Then $A_c^* = \langle c \rangle^* \subset A_g^*$ implies $g^{-1} c \in D^* \subset W$ and hence

$$0 \le w(g^{-1}c) = w(c) - w(g) = w(c) - \min\{w(b_0), \ldots, w(b_m)\}.$$

Therefore $c \in \langle b_0, \ldots, b_m \rangle (W \cap L)$ follows.

   Suppose now that $W \cap L$ is a $*$-valuation overring of D. A typical non-zero element of D^* is of the form $g^{-1} f$, where $f, g \in L[X] \setminus \{0\}$ and $A_f^* \subset A_g^*$, and we must prove that $w(f) \ge w(g)$, which means $g^{-1} f \in W$. Suppose that $f = a_0 + a_1 X + \cdots + a_n X^n$, $g = b_0 + b_1 X + \cdots + b_m X^m$ and

$$A_f^* = \langle a_0, \ldots, a_n \rangle^* \subset \langle b_0, \ldots, b_m \rangle^* = A_g^*.$$

If $i \in \{0,1,\ldots,n\}$ and $a_i \in \langle b_0,\ldots,b_m \rangle^* \subset \langle b_0,\ldots,b_m \rangle (W \cap L)$ implies $w(a_i) \geq \min\{w(b_0),\ldots,w(b_m)\} = w(g)$. Consequently, $w(f) = \min\{w(a_0),\ldots,w(a_m)\} \geq w(g)$ follows.

iii) If $E \in \mathbb{P}_{\mathrm{fin}}(L^\times)$, then $E^{*\bullet} = \langle E \rangle_{D^\bullet} \cap L$ holds by [6], Theorem 32.7. Let $(V_\lambda^*)_{\lambda \in \Lambda}$ be the family of all valuation overrings of D^*. By i) and ii), $(V_\lambda^* \cap L)_{\lambda \in \Lambda}$ is the family of all $*$-valuation overrings of D. By [6], Proposition 32.18, we have

$$\langle E \rangle_{D^\bullet} = \bigcap_{\lambda \in \Lambda} EV_\lambda,$$

and consequently

$$E^{*\bullet} = \langle E \rangle_{D^\bullet} \cap L = \bigcap_{\lambda \in \Lambda} EV_\lambda^* \cap L = \bigcap_{\lambda \in \Lambda} E(V_\lambda^* \cap L). \quad \square$$

Theorem 3. *If $*$ is any star-operation on D, then $\mathrm{cl}^*(D)$ is the intersection of all $*$-valuation overrings of D.*

Proof. Let $\circledast$ be the closure of $*$ on $\mathrm{cl}^*(D)$. By Theorem 1, $\mathrm{cl}^*(D)$ is $\circledast$-closed. By Proposition 5 iii) (applied with $E = \{1\}$), $\mathrm{cl}^*(D)$ is the intersection of all $\circledast$-valuation overrings of $\mathrm{cl}^*(D)$. By Proposition 3, the $\circledast$-valuation overrings of $\mathrm{cl}^*(D)$ are just the $*$-valuation overrings of D. $\quad \square$

REFERENCES

[1] D. D. Anderson, J. L. Mott and M. Zafrullah, *Finite Character Representations for Integral Domains*, Boll. UMI 6B: 613–630 (1992).

[2] D. F. Anderson, E. G. Houston and M. Zafrullah, *Pseudo-integrality*, Canad. Math. Bull 34: 15–22 (1991), 15–22.

[3] K. E. Aubert, *Divisors of finite character*, Annali di Mat. Pura et Appl. 33: 327–361 (1983).

[4] V. Barucci, D. F. Anderson and D. E. Dobbs, *Coherent Mori domains and the principal ideal theorem*, Comm. Algebra 15: 1119–1156 (1987).

[5] D. E. Dobbs, E. G. Houston, T. G. Lucas and M. Zafrullah, *t-linked overrings and Prüfer v-multiplication domains*, Comm. Algebra 17: 2835–2852 (1989).

[6] R. Gilmer, *Multiplicative ideal theory*, Marcel Dekker (1972).

[7] P. Jaffard, *Les systmes d' idéaux*, Dunod (1960).

[8] A. Okabe and R. Matsuda, *Star operations and generalized integral closures*, Bull. Fac. Sci. Ibaraki Univ. 24: 7–13 (1992).

Coefficient and Stable Ideals in Polynomial Rings

WILLIAM HEINZER   Department of Mathematics, Purdue University, West Lafayette, IN 47907

DAVID LANTZ   Department of Mathematics, Colgate University, Hamilton, NY 13346

Let $x_1, \ldots, x_d$ be indeterminates over an infinite field F, let R denote the polynomial ring $F[x_1, \ldots, x_d]$, and let M denote the maximal ideal $(x_1, \ldots, x_d)R$. If I is an M-primary ideal, the Hilbert polynomial

$$P_I(n) = e_0(I)\binom{n+d-1}{d} - e_1(I)\binom{n+d-2}{d-1} + \cdots + (-1)^d e_d(I)$$

gives the length of the R-module R/I^n for sufficiently large positive integers n. The integral closure I' of I is the unique largest ideal of R containing I and having the same coefficient e_0 (i.e., multiplicity) as I, and the *Ratliff-Rush ideal* $\tilde{I}$ of I is the unique largest ideal containing I and having the same Hilbert polynomial as I. Kishor Shah has shown in [S1] that there exists a unique chain of ideals[1]

$$I \subseteq \tilde{I} = I_{\{d\}} \subseteq \cdots \subseteq I_{\{k\}} \subseteq \cdots \subseteq I_{\{0\}} = I' \, .$$

where, for $0 \le k \le d$, the ideal $I_{\{k\}}$ is maximal with the property of having the same coefficients $e_0, \ldots, e_k$ of its Hilbert polynomial as those of I. The ideal $I_{\{k\}}$ is called the *k-th coefficient ideal* of I. If $I = I_{\{k\}}$, we say I is an e_k-*ideal*.

We are particularly interested in the case where R is of dimension two. In this setting, an M-primary ideal I has reduction number at most one (i.e., if J is a minimal reduction of I, then $JI = I^2$) if and only if the Rees algebra $R[It]$ is Cohen-Macaulay [HM, Prop. 2.6],[JV, Theorem 4.1], or [S2, Corollary 4(f)]. Moreover, the coefficients $e_1(I)$ and $e_2(I)$ are nonnegative, and it follows from [Hu, Theorem 2.1] that I has reduction number at most one if and only if $\lambda(R/I) = e_0(I) - e_1(I)$, and if this holds, then $e_2(I) = 0$. We say that an ideal with these properties is *stable*. Thus, I is stable if and only if $I = \tilde{I}$ and $e_2(I) = 0$. Stable ideals are e_1-ideals, but it is shown in [HJL, Example 5.4] that there exist e_1-ideals I for which $e_2(I) > 0$. We are interested in a better understanding of the features of e_1-ideals and the distinguishing aspects between e_1-ideals and the more restrictive subset of stable ideals.

Our purposes in this paper are:

(1) to present examples of first coefficient and stable ideals in dimension 2,

(2) to compare the description of the coefficient ideals given by Shah in [S1, Theorems 2 and 3] with that given in [HJLS, Theorem 3.17] involving the blowup of I,

(3) to present examples of coefficient ideals in higher dimensions,

(4) to present two results on the existence of stable ideals in dimension 2, and

---

[1] The existence of this unique chain of ideals is shown in [S1, Theorem 1] for an ideal primary for the maximal ideal of a quasi-unmixed local ring with infinite residue field. Since R_M is regular and so, in particular, quasi-unmixed, and since the length of R/I^n is equal to the length of $R_M/I^n R_M$, Shah's result also applies in our setting.

(5) to prove the e_1-closure of certain monomial ideals in dimension 2 are stable ideals.

In particular, in connection with (3) we present in more detail and with typographical corrections Example 3.22 of [HJLS], which establishes the existence of examples of ideals I in dimension d such that for all sufficiently large positive integers n one has

$$I^n = \widetilde{I^n} = (I^n)_{\{d\}} < \cdots < (I^n)_{\{k\}} < \cdots < (I^n)_{\{0\}} = (I^n)' ,$$

and thus gives examples where all the associated coefficient ideals are distinct. Also in Section 3 we use a suggestion made to us by Karen Smith to observe that if I is a monomial M-primary ideal, then all the associated coefficient ideals $I_{\{k\}}$ of I are monomial ideals.

1. EXAMPLES OF COEFFICIENT IDEALS IN DIMENSION 2.

Let x and y be indeterminates over the field F, let $R = F[x, y]$, and let $M = (x, y)R$. The following examples illustrate the associated coefficient ideals of various ideals I:

Consider the ideal $I = (x^6, x^2 y^4, y^6)R$. The form ring $G(I)$ of R with respect to I has depth one (as we have checked with Macaulay via the Rees algebra $R[It]$), so I and all its powers are Ratliff-Rush ideals [HLS, (1.2)]. We have the following data:

Ideal	Generators	e_0	e_1	e_2
$I = \widetilde{I}$	$x^6, x^2 y^4, y^6$	36	12	4
$I_{\{1\}}$	$I, x^4 y^2$	36	12	0
$I' = M^6$		36	15	0

Since $e_2(I_{\{1\}}) = 0$, this ideal is stable.

To obtain a strict inclusion between I and $\widetilde{I}$, as well as between the coefficient ideals, we modify as follows: Consider the ideals

Ideal	Generators	e_0	e_1	e_2
I	$x^{12}, x^8 y^4, x^6 y^6, x^2 y^{10}, y^{12}$	144	60	4
$\widetilde{I}$	$I, x^4 y^8$	144	60	4
$I_{\{1\}}$	$\widetilde{I}, x^{10} y^2$	144	60	0
$I' = M^{12}$		144	66	0

Note that we have $I^2 = \left(\widetilde{I}\right)^2 = (\widetilde{I^2})$. Also, we have again that the e_1-closure $I_{\{1\}}$ of I is a stable ideal.

Let us turn to an inspection of M-primary ideals of R of which the e_1-closure is not stable. Since the set of stable ideals integral over a given ideal is closed under intersection [HJL, Corollary 4.4], we can speak of the *stable closure* of the ideal I, which we denote $s(I)$. Again consider the ideals:

I	$x^7, x^5y^3 + x^3y^5, y^7$	49	10	1
$\widetilde{I} = I_{\{1\}}$	I, x^6y^4	49	10	1

Our verification that $I_{\{1\}} = (I, x^6y^4)$ is by using Macaulay; see [HJL, Example 5.4].[2] The element x^6y^4 is in $(I : M)$. It is the preimage in R of a socle element of R/I (as is x^4y^6). However, another element of $(I : M) - I$ is $x^6y^2 + x^4y^4 + x^2y^6$, and this element is not in $\widetilde{I}$. This leads us to ask: Is there a way to tell which elements of $(I : M)$ are in $\widetilde{I}$? In this example, $(I : M)$ is not even contained in $s(I)$, since:

$s(I)$	$\widetilde{I}$, all monomials of degree 9	49	11	0

It follows from Result 4.1 that this ideal is stable. To see that this is indeed the stable closure, we note that there is no intervening ideal with Hilbert coefficients 49,10,0, since $I_{\{1\}}$ is an e_1-ideal; so the Hilbert polynomial of this ideal is the next possible for a stable ideal. (It is also true that the set of e_1-ideals integral over a given ideal is closed under intersection [HJL, Prop. 4.5].)

Another ideal having the property that its Ratliff-Rush closure is an e_1-ideal that is not stable is

I	$x^8, y^8, x^6y^3 + x^3y^6$	64	14	1

If we adjoin to I the element x^7y^4, we get the same Hilbert polynomial; and we can verify that the resulting ideal and all its powers are Ratliff-Rush. Thus:

$\widetilde{I}$	I, x^7y^4	64	14	1

As in Example 5.4 of [HJL], we have shown using Macaulay that $\widetilde{I} = I_{\{1\}}$. We have

$s(I)$	I, x^3y^7, x^4y^6	64	15	0

because this ideal is stable and, as in the last case, in view of the Hilbert

---

[2] In Example 5.4 of [HJL], $I = J$.

coefficients, is smallest in I' containing I. In general, we would like to understand better the process of passing from an ideal to its stable closure.

With regard to Example (E5) on page 387 of [HJLS], we can now confirm that the linear and constant terms of Hilbert functions of ideals between $I = (x^3, y^3)$ and its integral closure $I' = (x, y)^3$ are as given there.

2. PASSING FROM I TO ITS COEFFICIENT IDEALS.

Let R be a d-dimensional Noetherian quasi-unmixed local ring, and let I be an ideal primary for the maximal ideal of R. Kishor Shah has shown that one way to attain all the coefficient ideals of I is as follows: For each integer k in $\{1, 2, \ldots, d\}$:

$$I_{\{k\}} = \bigcup (I^{n+1} : B) \, ,$$

where n varies over the positive integers and B varies over all the k-element subsets of sets of d generators of minimal reductions of I^n.

In particular, if I is such that I and all its powers are Ratliff-Rush, i.e., $G(I)$ has positive depth, then taking a minimal reduction $\mathbf{q}$ of I and considering $(I^{n+1} : \mathbf{q}^n)$ does not give us more than I, for the image of $\mathbf{q}^n$ in the n-th graded piece I^n/I^{n+1} of $G(I)$ contains a regular element of $G(I)$.

Let us see how Shah's description of the coefficient ideals gives the same results as the description given in [HJLS] involving the blowup of I. The description of the coefficient ideals given in [HJLS, Theorem 3.17] can be phrased as follows: The ideal $I_{\{k\}}$ is the intersection of R and the extensions of I to the following family of overrings: $R[I/a]_P$, as a varies over a fixed set A of d generators of a minimal reduction of I and P varies over the primes of height $\leq k$ in $R[I/a]$ that contain I (or equivalently a). Let P be such a prime in such a ring $R[I/a]$. Then since $R[I/a]$ is the degree-0 piece of the localization $R[It][1/(at)]$ of the Rees algebra $R[It]$ of I, and this localization is a Laurent polynomial ring in the indeterminate at over $R[I/a]$, it follows that $PR[It][1/(at)] \cap R[It]$ is a prime of height $\leq k$ in the Rees algebra. Since $a \in P$, we have $I \subseteq P$; so the image Q of P in the form ring $G(I) = R[It]/IR[It]$ is a prime of height $\leq k - 1$. Now for any set C of d generators of a minimal reduction of a power I^n of I, $G(I)$ is integral over $(R/I)[\bar{c} : c \in C]$, and the elements $\bar{c} = ct + IR[It]$ form a regular sequence

$G(I)$. Thus, for any k-element subset B of C, the prime Q cannot contain every $\bar{b}$ for $b \in B$ (for otherwise $\operatorname{ht}(Q) \geq k$). Taking preimage in the Rees algebra, there is some b in B for which $bt \notin PR[It]$, so that b/a is a unit in $R[I/a]_P$. Therefore, if $f \in (I^{n+1} : B)$, then $f = (b/a^n)^{-1}(bf/a^n) \in IR[I/a]_P$. In other words, the union of Shah's description is contained in the intersection of [HJLS].

For the reverse inclusion, given an element in the intersection of [HJLS], we must find a set B of the kind described by Shah so that $(I^{n+1} : B)$ contains that element. In the Rees algebra $R[It]$, take an irredundant primary decomposition of $IR[It]$, say $IR[It] = Q_1 \cap Q_2 \cap \ldots \cap Q_n$, where each Q_j is homogeneous. For each $k = 1, \ldots, d$, let J_k denote the intersection of those Q_j of which the radical has height at least $k+1$. Applying the "refined generalized prime avoidance lemma" [S1, Lemma 2(F)] to the images $J_k/IR[It]$ in $G(I)$, we see that we can find, for some positive integer N, elements $c_1, \ldots, c_d$ in I^N for which, for each $k = 1, \ldots, d$, we have $c_1 t^N, \ldots, c_k t^N \in J_k$ and $\dim(G(I)/(c_1 t^N + IR[It], \ldots, c_k t^N + IR[It])G(I)) = d - k$. Moreover, if $J_{k-1} < J_k$, then we must choose $c_n t^N$ from $J_k - J_{k-1}$. It follows that all minimal primes of $(I, c_1 t^N, \ldots, c_k t^N)R[It]$ have height at least $k+1$. Suppose f is in the intersection of [HJLS], i.e., f in R is also in $IR[I/a]_P$ for each a in A and each prime P in $R[I/a]$ containing I and of height $\leq k$. Then $f \in R[It] \cap \bigcap\{IR[It]_P : I \subseteq P, \operatorname{ht}(P) \leq k\} = \bigcap\{Q_j : \operatorname{ht}(\operatorname{rad}(Q_j)) \leq k\}$. Since $c_1 t^N, \ldots, c_k t^N$ are in the remaining Q_j (in fact, if $J_{k-1} < J_k$, then $c_k t^N$ has this property), we see that $f(c_1 t^N, \ldots, c_k t^N)$ is in the degree-N piece $I^{N+1} t$ of $IR[It]$, i.e., $f \in I^{N+1} : (c_1, \ldots, c_k)$. (In fact, if $J_{k-1} < J_k$, then $f \in I^{N+1} : c_k$. Thus, the distinct coefficient ideals of I are in fact colon ideals of I^{N+1} by a single element of the sequence $c_1, \ldots, c_d$. The above is a very slight reworking of Shah's proof, but he does not note this realization of coefficient ideals as colons by a single element.)

3. COEFFICIENT IDEALS IN HIGHER DIMENSIONS.

Let $R = F[x, y, z]$ where x, y, z are indeterminates over an infinite field F, and let $M = (x, y, z)R$. Consider the ideal

Ideal	Generators	e_0	e_1	e_2	e_3
I	$x^6, x^2 y^4, y^6, x^2 z^4, z^6$	216	144	88	32

Using Macaulay we see that the form ring $G(I)$ of R with respect to I has depth one, so I and all its powers are Ratliff-Rush ideals. We claim that

$I_{\{2\}}$	$I, x^4y^2z^2$	216	144	88	40
$I_{\{1\}}$	$(x^2, y^2, z^2)^3$	216	144	8	0
$I' = I_{\{0\}} = M^6$		216	180	20	0

so that

$$I = \tilde{I} < I_{\{2\}} < I_{\{1\}} < I' = M^6.$$

The Hilbert polynomial

$$P_I(n) = e_0(I)\binom{n+2}{3} - e_1(I)\binom{n+1}{2} + e_2(I)\binom{n}{1} - e_3(I)$$

gives the Hilbert function for all positive integers n, but not for $n = 0$, i.e., the postulation number of I is zero (as this term is used in [M]).

To find these coefficient ideals of I we examine the blowup of I. Dividing by x^6, we have the affine piece

$$R_x = F[x, y, z, (y/x)^4, (y/x)^6, (z/x)^4, (z/x)^6] \subset (R_x)' = F[x, y/x, z/x].$$

Then in the notation of [HJLS], $R_x^{(2,x)} = R_x[x^2(y/x)^2(z/x)^2]$ and $R_x^{(1,x)} = R_x[(y/x)^2, (z/x)^2]$. Both of the following assertions were checked by using Macaulay: The ring $R_x^{(2,x)}$ has depth 2 (i.e., the maximal ideal is not an associated prime of a principal ideal), and $R_x^{(1,x)}$ is a complete intersection.

Also, $R_y = F[x, y, z, (x/y)^2, (x/y)^2(z/y)^4, (z/y)^6]$. In this case $R_y = R_y^{(2,y)}$ (again, by Macaulay, it has depth 2), and $R_y^{(1)} = R_y[(z/y)^2]$ is a complete intersection. Similarly for R_z.

Now, $I_{\{2\}}$ is the intersection of the contractions of the extensions of I to $R_x^{(2,x)}$, $R_y^{(2,y)}$ and $R_z^{(2,z)}$. It is clear that this intersection contains $x^4y^2z^2$ and that the blowup of $(I, x^4y^2z^2)R$ has these rings as its affine pieces; checking that $(I, x^4y^2z^2)R$ is Ratliff-Rush, we conclude that it is $I_{\{2\}}$. Similarly for $I_{\{1\}}$.

INTERLUDE 3.1   We thought at first that $I_{\{1\}}$ was the ideal

$$(I, x^4y^2, x^4z^2, x^2y^2z^2)R.$$

— we missed some of the contraction from the affine pieces of the model

— so we wanted to show that this ideal is Ratliff-Rush. After 8.5 hours of

Macaulay run we obtained this verification by computing that the depth of the Rees algebra of this ideal is 3. Hence by [HM] the depth of the form ring with respect to this ideal is 2.

The verification that this ideal is Ratliff-Rush using Macaulay turned out to be very time- and memory-consuming; so we were led to seek a simplifying approach. We note the following: The generators of this ideal are all in the subring $A = F[x^2, y^2, z^2]$ of $B = F[x, y, z]$, and B is free over A. Let J denote the A-ideal generated by $x^6, y^6, z^6, x^2y^4, x^4y^2, x^2z^4, x^4z^2, x^2y^2z^2$. Then the ideal above is JB, and J is an ideal generated in degree 3 in the polynomial ring A. Since the form ring of B with respect to JB is free over the form ring of A with respect to J, if we show that the latter form ring has positive depth, then so has the former, so that all powers of JB are Ratliff-Rush. Thus, we set Macaulay to computing the projective dimension of the Rees algebra of $(x^3, y^3, z^3, xy^2, x^2y, xz^2, x^2z, xyz)B$. But even this turned out to challenge Macaulay.

INTERLUDE 3.2   Another method of simplifying the computations on the defining ideal of the Rees algebra, to obtain information on the depth of the form ring and the Cohen-Macaulay property of the blowup, was shown to us by Craig Huneke. Let I be a quasi-homogeneous ideal in the polynomial ring $R = F[x, y, \ldots]$; let $I = (f_1, \ldots, f_n)R$ where the f_i's are quasi-homogeneous. Then Macaulay can compute the kernel J of the R-algebra epimorphism from $T = R \otimes_F F[a_1, \ldots, a_n]$ (a_i's indeterminates) onto the Rees algebra $R[It]$ defined by $a_i \mapsto f_i t$. The program can then find a minimal projective resolution of J over T; the number of matrices in this resolution is the projective dimension of $R[It]$ over T, so that the Auslander-Buchsbaum formula yields the depth of $R[It]$. It follows from [HM, Theorem 2.1, page 262] that, if $G(I)$ is not Cohen-Macaulay, then the depth of $G(I)$ is one less than that of $R[It]$. Suppose the maximal minors of the last matrix in the resolution of J generate an ideal primary for $(x, y, \ldots, a_1, \ldots, a_n)T$. Then each ring in the blowup of I, i.e., $\text{Proj}(R[It])$, has projective dimension less than that of $R[It]$. Thus, in this case, if $\text{depth}(R[It]) = \dim(R)$, then $\text{Proj}(R[It])$ is Cohen-Macaulay.

But the computation of the projective resolution of J may be very long and difficult, even for Macaulay. Huneke suggests passing to the quotient ring T/xT of T obtained by setting $x = 0$. Since x is regular on both T and $R[It]$, [N, 27.2] the projective dimension of J over T is equal to the projective dimension of J/xJ over T/xT.

Let $R = F[x, y, z]$. The ideal $I = (x^4, y^4, z^4, x^2y^2z^2, x^3y^3z, x^3yz^3)R$ was pointed out to us by Les Reid to be an example of a monomial ideal that is 6-generated but has 7 "corner" elements, that is, there are 7 monomials in $I : (x, y, z)$ that are not in I. Is this property of the ideal I reflected in some way in the associated coefficient ideals of I?

EXAMPLE 3.22 OF [HJLS] REVISITED   Let F be an infinite field, and let $x, y_2, \ldots, y_d$ be indeterminates over F. Consider the affine domain $S = F[x, \{xy_i, y_i^4, y_i^6\}_{i=2}^d]$. Then

$$S^{(1,xS)} = S^{(1)} = F[x, \{xy_i, y_i^2\}_{i=2}^d] < F[x, \{y_i\}_{i=2}^d] = S' \ .$$

To see that, for k in $\{1, \ldots, d-1\}$, we have $S^{(k+1,xS)} < S^{(k,xS)}$, we note that the product of x^{2k-2} and y_i^2 for k distinct values of i is an element of $S^{(k,xS)}$, since any prime P in S of height at most k and containing x does not contain y_i^4, y_i^6 for at least one of the y_i's that appear in the product, so one of the factors y_i^2 is a unit in S_P, and the product of the remaining factors is an element of S. But this element is not in $S^{(k+1,xS)}$ because it is not in the localization of S at the prime ideal of height $k + 1$ generated by x and xy_i, y_i^4, y_i^6 for the y_i's that appear in the product.

The domain S is an affine piece of the blowup of the ideal I generated by $x^6, x^2(xy_i)^4, (xy_i)^6, i = 2, \ldots, d$ in the polynomial ring $R = F[x, \{xy_i\}_{i=2}^d]$. Forming the rings of fractions of R, S with respect to the complement in R of the maximal ideal $M = (x, \{xy_i\}_{i=2}^d)R$ yields a regular local ring R_M, and $(R - M)^{-1}S$ is an affine piece of the blowup of IR_M which retains the properties verified in the last paragraph.

Moreover, since the extensions of the affine piece of the blowup of I as described in [HJLS] are distinct, it follows that, for sufficiently high powers of I, the contractions of these powers from the various extensions of the blowup are distinct; i.e., for n sufficiently large,

$$\widetilde{(I^n)} = (I^n)_{\{d\}} < (I^n)_{\{d-1\}} < \ldots < (I^n)_{\{0\}} = (I^n)' \ .$$

In fact, we believe that these strict inclusions hold for $n = 1$.

OBSERVATION 3.3   Let $R = F[x_1, \ldots, x_d]$ be a polynomial ring in d variables $x_1, \ldots, x_d$ over an infinite field F, let $M = (x_1, \ldots, x_d)R$ and let I be an M-primary ideal. We say that I is a *monomial ideal* if I is generated by monomials in $x_1, \ldots, x_d$. Karen Smith suggested to us the following argument to show that if I is a monomial ideal, then all the associated coefficient ideals $I_{\{k\}}$ of I are also monomial ideals. For each d-tuple $a = (a_1, \ldots, a_d)$ in the algebraic d-torus $F^* \times \cdots \times F^*$, where $F^* = F - 0$ is the multiplicative group of units of F, define an F-automorphism $\phi_a : R \to R$ by setting $\phi_a(x_i) = a_i x_i$ for $1 \le i \le d$. To show $I_{\{k\}}$ is a monomial ideal it suffices to show $\phi_a(I_{\{k\}}) = I_{\{k\}}$ for each $a = (a_1, \ldots, a_d)$. By assumption $I = (f_1, \ldots, f_n)R$, where each f_i is a nonzero monomial. Let $R_i = R[f_1/f_i, \ldots, f_n/f_i]$. Then ϕ_a naturally extends to an F-automorphism of R_i which we continue to call ϕ_a, and $IR_i = f_i R_i$ is mapped to itself under ϕ_a. The invariance of $f_i R_i$ under ϕ_a implies that the union of the associated primes of $f_i R_i$ of height at most k in R_i is mapped onto itself under ϕ_a. Therefore ϕ_a extends to an automorphism of the localization T_{ik} of R_i at the complement of the union of the associated primes of $f_i R_i$ of height at most k. It follows that $f_i T_{ik} \cap R$ is mapped onto itself by ϕ_a. By [HJLS, Theorem 3.17], $I_{\{k\}} = \cap_{i=1}^n f_i T_{ik} \cap R$. Therefore the ideal $I_{\{k\}}$ is mapped onto itself by ϕ_a, so $I_{\{k\}}$ is a monomial ideal.

4. STABLE IDEALS IN DIMENSION 2.

Let us examine with $R = F[x, y]$ what ideals between $I = (x^n, y^n)R$ and its integral closure $I' = (x, y)^n R$ are stable. The following two results show that many of these ideals are stable:

RESULT 4.1   Let a, b be a minimal reduction of the height-2 ideal A in the ring R, and n be a positive integer; set $I = (a^n, b^n)$. Then I is a minimal reduction of A^n; suppose the reduction number is 1, i.e., $IA^n = A^{2n}$. Then for every nonnegative integer j we have $IA^{n+j} = A^{2n+j}$. Set $J = I + A^{n+i+1}$ for a positive integer i; then we have

$$J^2 = IJ + (A^{n+i+1})^2 = IJ + A^{2n+2i+2}$$
$$= IJ + IA^{n+2i+2} \subseteq I(J + A^{n+i+1}) = IJ \; ;$$

i.e., J is stable. Moreover, for each ideal $K \subseteq A^{n+i}$ the ideal $L = J + K$ is also stable. For, we have

$$L^2 = J^2 + JK + K^2 \subseteq IJ + (I + A^{n+i+1})K + A^{2n+2i}$$
$$\subseteq I(J + K) + A^{2n+2i+1} + A^{2n+2i}$$
$$= I(J + K) + A^{2n+2i} = I(J + K + A^{n+2i})$$
$$= I(J + K) = IL$$

where we have used $A^{n+2i} \subseteq J$.

RESULT 4.2   For r a real number, let $\lceil r \rceil$ denote the smallest integer greater than or equal to r. Suppose a, b form a regular sequence in a ring R and m and n are positive integers, and let $I = (a^m, b^n)R$. Then for any ideal J contained in $a^{\lceil m/2 \rceil}b^{\lceil n/2 \rceil}R$, the sum $I + J$ is stable. For, $(a^{\lceil m/2 \rceil}b^{\lceil n/2 \rceil})^2 \in I^2$, so $J^2 \subseteq I^2$, so $(I + J)^2 = I^2 + IJ + J^2 \subseteq I^2 + IJ = I(I + J)$.

Applying these paragraphs to x, y in $F[x, y]$, we see that: (1) for any integer $n \geq 2$, we can find stable ideals I between $(x^n, y^n)R$ and $(x, y)^n R$ such that $e_0(I) = n^2$ and $e_2(I) = 0$ (both necessarily) and $e_1(I)$ is any integer from 0 to $(n - 2)(n - 1)/2$; and (2) for any positive integers m, n, we can find stable ideals I between $(x^m, y^n)R$ and its integral closure with $e_0(I) = mn$, $e_2(I) = 0$ (again, both necessarily) and $e_1(I)$ is any integer from 0 to $\lceil \frac{m}{2} \rceil \lceil \frac{n}{2} \rceil$.

In (4.3) we prove that the e_1-closure of certain monomial ideals are stable. This shows that examples such as [HJL, Example 5.4] are of necessity not generated by monomials.

OBSERVATION 4.3   Let $R = F[x, y]$ where x and y are indeterminates over the infinite field F, let m and n be positive integers and suppose I is a monomial ideal of R integral over $(x^m, y^n)R$. We show that the e_1-closure J of I is stable. By [HJLS, Theorem 3.17] (see also [HJL, (3.2)]), $J = ID \cap R$, where D is the first coefficient domain of I. Then D is also the first coefficient domain of J and each of the rings $R[J/x^m]$ and $R[J/y^n]$ is contained in D. Therefore $JR[J^2/x^m y^n] \cap R = J$. It follows that $(J^3 : x^m y^n) = J$. To show that J is stable, it suffices to show that $J^2 \subseteq (x^m, y^n)J$. The Briançon-Skoda Theorem [LS, Theorem 1] implies $J^2 \subseteq (x^m, y^n)R$. By (3.3), the ideal J is a monomial ideal. Let $a \in J^2$ with a a monomial. Then $a \in (x^m, y^n)R$ implies either $a \in x^m R$ or $a \in y^n R$. Suppose $a = x^m b$ with $b \in R$. Then

$b \in (J^2 : x^m) \subseteq (J^3 : x^m y^n) = J$ which means $a \in (x^m, y^n)J$. A similar argument applies in case $a \in y^n R$. Therefore J is stable.

REFERENCES

[HJL] W. Heinzer, B. Johnston and D. Lantz, First coefficient domains and ideals of reduction number one, *Comm. Algebra : 21*, 3797-3827 (1993).

[HJLS] W. Heinzer, B. Johnston, D. Lantz and K. Shah, Coefficient ideals in and blowups of a commutative Noetherian domain, *J. Algebra : 162*, 355-391 (1993).

[HLS] W. Heinzer, D. Lantz and K. Shah, The Ratliff-Rush ideals in a Noetherian ring, *Comm. Algebra : 20*, 591-622 (1992).

[HM] S. Huckaba and T. Marley, Depth properties of Rees algebras and associated graded rings, *J. Algebra : 156*, 259-271 (1993).

[Hu] C. Huneke, Hilbert functions and symbolic powers, *Michigan Math. J. : 34*, 293-318 (1987).

[J] B. Johnston, Toward parametric Cohen-Macaulifications of 2-dimensional finite local cohomology domains, *Math. Z. : 207*, 569-581 (1991).

[JV] B. Johnston and J. Verma, On the length formula of Hoskin and Deligne and associated graded rings of two-dimensional regular local rings, *Math. Proc. Camb. Phil. Soc. : 111*, 423-432 (1992).

[LS] J. Lipman and A. Sathaye, Jacobian ideals and a theorem of Briançon-Skoda, *Michigan Math. J. : 28*, 199-222 (1981).

[M] T. Marley, The coefficients of the Hilbert polynomial and the reduction number of an ideal, *J. London Math. Soc. (2) : 156*, 1-8 (1989).

[N] M. Nagata, *Local Rings*, Interscience, New York (1962).

[S1] K. Shah, Coefficient ideals, *Trans. Amer. Math. Soc. : 327*, 373-384 (1991).

[S2] K. Shah, On the Cohen-Macaulayness of the fiber cone of an ideal, *J. Algebra : 143*, 156-172 (1991).

Almost Generalized GCD-Domains

REBECCA L. LEWIN Mathematics Department, University of Wisconsin -
La Crosse, La Crosse, Wisconsin

1 INTRODUCTION

GCD-domains are an important class of integral domains from classical ideal
theory. They can be characterized by the property that the intersection of any two
principal ideals is principal. This property can be generalized in various ways.
One such generalization was first introduced by Anderson [1]. He called an
integral domain a *generalized GCD-domain (GGCD-domain)* if the intersection
of any two integral invertible ideals is invertible. GGCD-domains were further
studied in [2].

A second type of generalization of a GCD-domain was given by Zafrullah [9].
He called an integral domain R an *almost GCD-domain (AGCD-domain)* if for a
and b $\in$ R-$\{0\}$, there is a natural number n = n(a,b) with $a^n R \cap b^n R$ principal
(or equivalently, $(a^n, b^n)_v$ principal). These domains were further studied in [3]
and several closely related classes of domains were introduced, including the
notions of almost Bezout domains and almost Prüfer domains. Here a domain R is
called an *almost Bezout domain or AB-domain* (respectively, *almost Prüfer* or
AP-domain) if for a, b $\in$ R-$\{0\}$, there is an n = n(a, b) with (a^n, b^n) principal
(respectively, invertible).

In the spirit of these definitions, Anderson and Zafrullah [3] introduced an additional generalization of a GCD-domain. A domain R is called an *almost generalized GCD-domain (AGGCD-domain)* if for a, b ∈ R-{0}, there is an n = n(a, b) with $a^n R \cap b^n R$ invertible (or equivalently, $(a^n, b^n)_v$ invertible). The purpose of this paper is to continue the study of AGGCD-domains.

The general reference for basic results from commutative ring theory and multiplicative ideal theory will be [7] and [5], respectively. Throughout this paper, n and m will represent natural numbers.

2 PRELIMINARIES

Let R be an integral domain with quotient field K. The set of all nonzero fractional ideals of R is denoted by F(R) and f(R) denotes the subset of F(R) consisting of finitely generated fractional ideals. The inverse of a fractional ideal I is $I^{-1} = \{y \in K \mid yI \subseteq R\}$. I^{-1} is also a fractional ideal of R and I is said to be invertible if $II^{-1} = R$. We denote $(I^{-1})^{-1}$ by I_v. The operation on F(R) which sends I to I_v is an example of a •-operation, called the v-operation. Recall that a •-operation is a function •:F(R) → F(R) satisfying (1) $(a)^* = (a)$, $(aA)^* = aA^*$; (2) $A \subseteq A^*$, if $A \subseteq B$ then $A^* \subseteq B^*$; and (3) $(A^*)^* = A$. A fractional ideal I is called a •-ideal if $I = I^*$, and a •-ideal I is said to be of finite type if $I = J^*$ for some $J \in f(R)$. For basic properties of •-operations, the reader is referred to [5, Sections 32, 34].

Another example of a •-operation which will be important in this paper is the t-operation. For $I \in F(R)$, I_t is defined by $I_t = \cup \{J_v \mid J \subseteq I \text{ and } J \in f(R)\}$. Thus for I finitely generated, $I_t = I_v$. A fractional ideal I is t-invertible if there is a fractional ideal J with $(IJ)_t = R$. In this case, we can take $J = I^{-1}$. Clearly, if I is invertible, then I is also t-invertible.

Let $f_t(R)$ be the set of finite type t-ideals of R. Then $f_t(R)$ is a semigroup under the "t-product" $I \cdot J = (IJ)_t$. Now $f_t(R)$ need not form a group. An integral domain R in which $f_t(R)$ does form a group is called a Prüfer v-multiplication domain (PVMD). The set of t-invertible t-ideals, denoted $I_t(R)$, is a subgroup of

$f_t(R)$. The set of principal ideals, $P(R)$, is a subgroup of $I_t(R)$ and the quotient group $Cl_t(R) = I_t(R)/P(R)$ is called the t-class group of R. It was shown in [9] that if R is a AGCD-domain, then $Cl_t(R)$ is torsion and if R is an integrally closed AGCD-domain, then R is a PVMD.

Let $R \subseteq S$ be an extension of commutative rings. We denote the integral closure of R in S by $\overline{R}$. If no S is indicated, $\overline{R}$ is the integral closure of R in its quotient field. If $R \subseteq S$ has the property that for each $s \in S$, there exists an $n = n(s)$ with $s^n \in R$, then the extension is called a root extension. It is obvious that a root extension is an integral extension. R is said to be root closed in S if for any $s \in S$, $s^n \in R$ implies $s \in R$. Zafrullah [9] showed that if R is an AGCD-domain, then $R \subseteq \overline{R}$ is a root extension and $\overline{R}$ is also an AGCD-domain. It will be shown in section 4 that the same result holds if we replace AGCD by AGGCD.

3 AGGCD-DOMAINS

In [1], Anderson introduced the notion of a generalized GCD-domain. An integral domain is called a *generalized GCD-domain (GGCD-domain)* if the intersection of two integral invertible ideals is invertible. Further study of these domains was done in [2]. There it was noted that examples of GGCD-domains include GCD-domains, Prüfer domains and π-domains. Several equivalent conditions for a domain to be a GGCD-domain were given. The following theorem summarizes a few of these equivalences and includes some extensions which follow immediately by induction. Note that if we replace "invertible" by "principal", then these conditions are all equivalent to R being a GCD-domain.

THEOREM 3.1 For an integral domain R, the following statements are equivalent.

1) R is a GGCD-domain.

2) Any finite intersection of integral invertible ideals of R is invertible.

3) For $a, b \in R-\{0\}$, $aR \cap bR$ is invertible.

4) For $a_1, a_2, ..., a_k \in R-\{0\}$, $a_1R \cap a_2R \cap ... \cap a_kR$ is invertible.

5) For $a, b \in R-\{0\}$, $aR:bR = \{r \in R \mid rb \in aR\}$ is invertible.

Another generalization of a GCD-domain was introduced by Zafrullah [9] and further studied by Anderson and Zafrullah [3]. A domain R is called an *almost GCD-domain (AGCD-domain)* if for a, b $\in$ R-{0}, there is an n = n(a, b) with $a^n R \cap b^n R$ principal. In the same spirit, the authors also defined an almost generalized GCD-domain. A domain R is called an *almost generalized GCD-domain (AGGCD-domain)* if for a, b $\in$ R-{0}, there is an n = n(a, b) with $a^n R \cap b^n R$ invertible. Examples of AGGCD-domains include the AGCD and AP domains studied in [3]. There are some interesting connections between these various generalizations. Many of the results concerning AGCD-domains and the related domains defined in [3] have analogues for AGGCD-domains.

We begin by giving several equivalent conditions for an integral domain to be an AGGCD-domain. Each of the statements in Theorem 3.1 for GGCD-domains has an analogous statement for AGGCD-domains. Two additional equivalences are also included. It is easily seen that similar statements could be added to Theorem 3.1. If we replace "invertible" by "principal", it is not hard to see that the following conditions are all equivalent to R being an AGCD-domain.

THEOREM 3.2 For an integral domain R, the following statements are equivalent.

1) R is an AGGCD-domain.

2) For $a_1, ..., a_k \in$ R-{0}, there is an n = n($a_1, ..., a_k$) with $a_1^n R \cap a_2^n R \cap ... \cap a_k^n R$ invertible.

3) For integral invertible ideals A and B of R, there is an n = n(A, B) with $A^n \cap B^n$ invertible.

4) For integral invertible ideals $A_1, ..., A_k$ of R, there is an n = n($A_1, ..., A_k$) with $A_1^n \cap ... \cap A_k^n$ invertible.

5) For a, b $\in$ R-{0}, there is an n = n(a, b) with $(a^n, b^n)_v$ invertible.

6) For $a_1, ..., a_k \in$ R-{0}, there is an n = n($a_1, ..., a_k$) with $(a_1^n, ..., a_k^n)_v$ invertible.

7) For a, b $\in$ R-{0}, there is an n = n(a, b) with $a^n R : b^n R$ invertible.

The proof of this result will use some properties of t-invertible ideals. A very useful result from [3] generalizes the well-known fact from commutative ring theory that for an invertible ideal $A = (\{a_\alpha\})$, we have $A^n = (\{a_\alpha^n\})$ for any n. Anderson and Zafrullah showed that for $\{a_\alpha\} \subseteq R\text{-}\{0\}$, R an integral domain, if $(\{a_\alpha\})$ is t-invertible, then $(\{a_\alpha^n\})_t = ((\{a_\alpha\})^n)_t$ for any n. Using the fact that for t-invertible ideals I and J, $(IJ)_t = (IJ_t)_t = (I_tJ_t)_t$, it is easily seen that for t-invertible ideals $A_1, ..., A_k$ $(A_1 ... A_k)^{-1} = \left(A_1^{-1} ... A_k^{-1}\right)_t$.

Proof:

(1) $\Rightarrow$ (3)   Let A and B be invertible ideals of R. Then A and B are finitely generated, say $A = (a_1, ..., a_k)$ and $B = (b_1, ..., b_r)$. Now for any m, $A^m = (a_1^m, ..., a_k^m)$ and $B^m = (b_1^m, ..., b_r^m)$ are also invertible. Since R is an AGGCD-domain, for each pair i,j, there is an n_{ij} with $a_i^{n_{ij}}R \cap b_j^{n_{ij}}R$ invertible. Let $n = \prod_{i,j} n_{ij}$. Then for each i,j, $a_i^n R \cap b_j^n R = a_i^{n_{ij}\hat{n}_{ij}}R \cap b_j^{n_{ij}\hat{n}_{ij}}R$ where $\hat{n}_{ij} = \frac{n}{n_{ij}}$. We will show that each $a_i^n R \cap b_j^n R = \left(a_i^{n_{ij}}R \cap b_j^{n_{ij}}R\right)^{\hat{n}_{ij}}$ and hence is invertible. Since for ideals I and J, I = J iff $I_M = J_M$ for all maximal ideals M of R, it is enough to show this in the local case. Hence $a_i^n R$, $b_j^n R$ and $a_i^n R \cap b_j^n R$ are principal for every i and j. For ease of notation, we will temporarily drop the subscripts. So consider $a, b \in R$, a quasi-local domain, with $aR \cap bR$ principal. It is easily seen that $(aR \cap bR) = \left(a^{-1}, b^{-1}\right)^{-1}$. Therefore for any n, we have $(aR \cap bR)^n = \left[\left(a^{-1}, b^{-1}\right)^{-1}\right]^n = \left(\left[\left(a^{-1}, b^{-1}\right)_v\right]^{-1}\right)^n$

$= \left[\left(\left[\left(a^{-1}, b^{-1}\right)_v\right]^{-1}\right)^n\right]_t$

$= \left(\left[\left(a^{-1}, b^{-1}\right)_v\right]^{-1} \cdots \left[\left(a^{-1}, b^{-1}\right)_v\right]^{-1}\right)_t$

$= \left(\left(a^{-1}, b^{-1}\right)_v \cdots \left(a^{-1}, b^{-1}\right)_v\right)^{-1} = \left(\left[\left(a^{-1}, b^{-1}\right)_v\right]^n\right)^{-1}$

$= \left(\left[\left(a^{-1}, b^{-1}\right)^n\right]_v\right)^{-1} = \left(a^{-n}, b^{-n}\right)^{-1}$

$= a^n R \cap b^n R$

Thus for each i,j, $a_i^n R \cap b_j^n R = \left(a_i^{n_{ij}}R \cap b_j^{n_{ij}}R\right)^{\hat{n}_{ij}}$ and hence is invertible. Now let M be a maximal ideal of R. Since A^n and B^n are both invertible, $(A^n)_M = a_{i_0}^n R_M$ and $(B^n)_M = b_{j_0}^n R_M$ for some i_0 and j_0.

Hence, we have $(A^n)_M \cap (B^n)_M = a_{i_0}^n R_M \cap b_{j_0}^n R_M = \left(a_{i_0}^n R \cap b_{j_0}^n R \right)_M$

$\subseteq \left(\sum_{i,j} \left(a_i^n R \cap b_j^n R \right) \right)_M \subseteq (A^n \cap B^n)_M$. Thus $A^n \cap B^n = $

$\sum_{i,j} \left(a_i^n R \cap b_j^n R \right)$ is a finite sum of invertible ideals and hence is

finitely generated. Since $A^n \cap B^n$ is locally principal, $A^n \cap B^n$ is

invertible.

$(3) \Rightarrow (4)$ Induction on k.

$(4) \Rightarrow (2) \Rightarrow (1)$ Clear.

$(2) \Rightarrow (6)$ Let $a_1, ..., a_k \in R - \{0\}$. Now $(a_1, ..., a_k)^{-1} = a_1^{-1} R \cap ... \cap a_k^{-1} R$ and

thus $a_1 a_2 \cdots a_k (a_1, ..., a_k)^{-1} = a_1 a_2 \cdots a_k \left(a_1^{-1} R \cap ... \cap a_k^{-1} R \right) = $

$\hat{a}_1 R \cap ... \cap \hat{a}_k R$ where $\hat{a}_i = \frac{a_1 a_2 \cdots a_k}{a_i} \in R$. By (2), there is an n with

$(\hat{a}_1)^n R \cap ... \cap (\hat{a}_k)^n R$ invertible. Thus $a_1^n a_2^n \cdots a_k^n (a_1^n, ..., a_k^n)^{-1}$

$= a_1^n a_2^n \cdots a_k^n \left((a_1^n)^{-1} R \cap ... \cap (a_k^n)^{-1} R \right) = (\hat{a}_1)^n R \cap ... \cap (\hat{a}_k)^n R$ is

invertible.   But this is true iff $(a_1^n, ..., a_k^n)^{-1}$ is invertible and hence

$(a_1^n, ..., a_k^n)_v$ is invertible.

$(6) \Rightarrow (2)$ Let $a_1, ..., a_k \in R - \{0\}$ and set $\hat{a}_i = \frac{a_1 a_2 \cdots a_k}{a_i}$ for each i. As in

$(2) \Rightarrow (6)$, $a_1 R \cap ... \cap a_k R = a_1 a_2 \cdots a_k (\hat{a}_1, ..., \hat{a}_k)^{-1}$. By (6), there is

an n with $((\hat{a}_1)^n, ..., (\hat{a}_k)^n)_v$ invertible and hence $a_1^n R \cap ... \cap a_k^n R = $

$a_1^n a_2^n \cdots a_k^n ((\hat{a}_1)^n, ..., (\hat{a}_k)^n)^{-1}$ is also invertible.

$(5) \Rightarrow (1)$ Let $a, b \in R - \{0\}$. For any n, $(a^n, b^n)^{-1} = a^{-n} R \cap b^{-n} R$ and hence

$a^n R \cap b^n R = a^n b^n (a^n, b^n)^{-1}$. So $a^n R \cap b^n R$ is invertible iff $(a^n, b^n)^{-1}$

is invertible iff $(a^n, b^n)_v$ is invertible.

$(1) \Rightarrow (7)$ This follows since for any n, $a^n R \cap b^n R = (a^n R : b^n R) b^n R$.

4 OVERRINGS OF AGGCD-DOMAINS

Zafrullah showed that if R is an AGCD-domain and $\overline{R}$ represents its integral

closure, then $R \subseteq \overline{R}$ is a root extension and $\overline{R}$ is also an AGCD-domain. The

same result holds if AGCD is replaced by AGGCD. The fact that for R an

AGGCD-domain, $R \subseteq \overline{R}$ is a root extension is an immediate consequence of a result of Anderson and Zafrullah [3, Prop 3.2].

In order to show that the integral closure of an AGGCD-domain is an AGGCD-domain, it will be helpful to know the relationship between $(I \cap J)\overline{R}$ and $I\overline{R} \cap J\overline{R}$ where I and J are invertible ideals in an AGGCD-domain R. Our first step in this direction is to generalize a result due to Zafrullah [9, Lemma 3.5].

LEMMA 4.1 Let R be an AGCD-domain. If a, b $\in$ R with $aR \cap bR$ principal, then $(aR \cap bR)\overline{R} = a\overline{R} \cap b\overline{R}$.

Proof: Suppose $aR \cap bR = cR$. Then $c = \text{lcm}(a, b)$ and $[(a,b)R]_v = dR$ where $d = \gcd(a, b)$. Now $\left[\left(\frac{a}{d}, \frac{b}{d}\right)R\right]_v = \left[\frac{1}{d}(a, b)R\right]_v = \frac{1}{d}[(a,b)R]_v = R$. Thus $\frac{a}{d}$ and $\frac{b}{d}$ have a least common multiple in R, and since $\left[\left(\frac{a}{d}, \frac{b}{d}\right)R\right]_v = R$, we have $\text{lcm}\left[\frac{a}{d}, \frac{b}{d}\right] = \frac{ab}{d^2}$. Therefore $\frac{a}{d}R \cap \frac{b}{d}R = \frac{ab}{d^2}R$. Thus $\frac{a}{d}\overline{R} \cap \frac{b}{d}\overline{R} = \frac{ab}{d^2}\overline{R}$ [9, Lemma 3.5]. Hence $a\overline{R} \cap b\overline{R} = d\left(\frac{a}{d}\overline{R} \cap \frac{b}{d}\overline{R}\right) = \frac{ab}{d}\overline{R}$. But since $\text{lcm}(a, b) = c$, we have that $d = \gcd(a, b) = \frac{ab}{c}$ and hence $c = \frac{ab}{d}$. Therefore $(aR \cap bR)\overline{R} = c\overline{R} = \frac{ab}{d}\overline{R} = a\overline{R} \cap b\overline{R}$ as desired.

In the AGGCD-domain case, this result can be extended to invertible rather than just principal ideals. In order to see this, we make two observations. First, recall that an overring S of a domain R is said to be LCM-stable over R if for all a, b $\in$ R, $(aR \cap bR)S = aS \cap bS$. This property was studied in [4] and [8]. Now if S is an LCM-stable overring of R, then for a, b $\in$ R, $aS \cap bS$ is principal (respectively, invertible), whenever $aR \cap bR$ is principal (respectively, invertible). This gives the following result. The proof is easy.

THEROEM 4.2 Let R be an AGCD-domain (AGGCD-domain) and S an LCM-stable overring of R. Then S is an AGCD-domain (AGGCD-domain).

In particular, since any localization of a domain R is an LCM-stable overring of R, we have the following immediate corollary.

COROLLARY 4.3 Let R be an AGCD-domain (AGGCD-domain). Then any localization of R is an AGCD-domain (AGGCD-domain).

For our second observation, recall that in a semi-quasilocal domain (that is, a domain with a finite number of maximal ideals), any invertible ideal is principal [7]. Thus the notions of AGCD and AGGCD coincide in the semi-quasilocal case. These observations are needed in the proof of the following result.

LEMMA 4.4 Let R be an AGGCD-domain. If I and J are invertible ideals of R with $I \cap J$ invertible, then $(I \cap J)\overline{R} = I\overline{R} \cap J\overline{R}$.

Proof: It suffices to show equality locally. Let $\mathcal{M}$ be a maximal ideal of $\overline{R}$ and $M = \mathcal{M} \cap R$. Then R_M is a quasilocal AGGCD-domain and hence an AGCD-domain. Since I, J, and $I \cap J$ are invertible ideals of R, I_M, J_M and $(I \cap J)_M$ are principal ideals of R_M. So by Lemma 4.1, $(I_M \cap J_M)\overline{R_M} = I_M\overline{R_M} \cap J_M\overline{R_M}$. Since $\overline{R_M} = \overline{R}_M$ we have $(I \cap J)\overline{R}_M = ((I \cap J)R_M)\overline{R}_M = (I_M \cap J_M)\overline{R}_M = I_M\overline{R}_M \cap J_M\overline{R}_M = I\overline{R}_M \cap J\overline{R}_M$. Since $\overline{R}_{\mathcal{M}}$ is a localization of $\overline{R}_M$, we get $(I \cap J)\overline{R}_{\mathcal{M}} = I\overline{R}_{\mathcal{M}} \cap J\overline{R}_{\mathcal{M}}$ as desired.

We are now ready to prove that the integral closure of an AGGCD-domain is also an AGGCD-domain.

THEOREM 4.5 If R is an AGGCD-domain, then $\overline{R}$ is an AGGCD-domain.

Proof: Let a, $b \in \overline{R}$. Since $R \subseteq \overline{R}$ is a root extension, we can find an n with a^n, $b^n \in R$. Then there exists an $m = m(a^n, b^n)$ with $(a^n)^m R \cap (b^n)^m R$ invertible. By the previous lemma, we have $(a^{nm}R \cap b^{nm}R)\overline{R} = a^{nm}\overline{R} \cap b^{nm}\overline{R}$. But since $I = a^{nm}R \cap b^{nm}R$ is invertible, $II^{-1} = R$. Hence $(I\overline{R})(I^{-1}\overline{R}) = (II^{-1}\overline{R})$ and so $I\overline{R}$ is also invertible. Thus $\overline{R}$ is an AGGCD-domain.

5 RELATED CLASSES OF DOMAINS

Since a principal ideal is invertible, it is obvious that every AGCD-domain is an AGGCD-domain. We have observed that in the semi-quasilocal case, they are equivalent. There are also other situations in which the two notions coincide. Recall that the t-class group of R, denoted $Cl_t(R)$, is the quotient group of the t-invertible t-ideals modulo the principal ideals. Similarly, the class group of R, denoted $C(R)$ is the quotient group of the invertible ideals modulo the principal ideals. It was shown in [3], that if $(a,b)_t$ is t-invertible, then $[(a,b)^n]_t = (a^n, b^n)_t$. Using this, we get the following equivalences.

THEOREM 5.1 For an integral domain R, the following statements are equivalent.
1) R is an AGCD-domain.
2) R is an AGGCD-domain with $Cl_t(R)$ torsion.
3) R is an AGGCD-domain with $C(R)$ torsion.

Proof:
1) $\Rightarrow$ 2)   It is shown in [3], that any AGCD-domain has torsion t-class group. Also, as remarked above, any AGCD-domain is an AGGCD-domain.

2) $\Rightarrow$ 3)   Follows since $C(R)$ is a subgroup of $Cl_t(R)$.

3) $\Rightarrow$ 1)   Let $a, b \in R-\{0\}$. Then there is an $n = n(a, b)$ with $(a^n, b^n)_v$ invertible. Since $C(R)$ is torsion, there is an m with $[(a^n, b^n)_v]^m$ principal. Therefore, $([(a^n, b^n)_v]^m)_v$ is also principal. Now
$([(a^n, b^n)_v]^m)_v = [(a^n, b^n)_v \cdots (a^n, b^n)_v]_v = [(a^n, b^n) \cdots (a^n, b^n)]_v$
$= [(a^n, b^n)^m]_v$. So $[(a^n, b^n)^m]_v = [(a^n, b^n)^m]_t$ is principal and hence t-invertible. Thus we have that $[(a^n, b^n)^m]_v = [(a^n, b^n)^m]_t$
$= (a^{nm}, b^{nm})_t = (a^{nm}, b^{nm})_v$. Therefore $(a^{nm}, b^{nm})_v$ is principal and R is an AGCD-domain.

Another class of examples of AGGCD-domains are the almost Prüfer domains introduced in [3]. Recall that an integral domain R is called an *almost Prüfer domain (AP-domain)* if for a, b ∈ R-{0}, there is an n = n(a, b) with (a^n, b^n) invertible. But if (a^n, b^n) is invertible, then so is $(a^n, b^n)_v$ and hence R is an AGGCD-domain. It was shown in [3] that an integral domain R is an AP-domain iff $\overline{R}$ is a Prüfer domain and R ⊆ $\overline{R}$ is a root extension. In particular, R is an integrally closed AP-domain iff R is Prüfer. Thus if D is a Dedekind domain whose class group is not torsion, D is an AP-domain but not an AGCD domain.

Most of the results in [3] concerning the connections between AGCD-domains and AB-domains carry over if AGCD is replaced by AGGCD. The proofs are nearly identical to the AGCD-domain case, so we will not state them but refer you to Lemma 5.1 - Theorem 5.6 in [3].

In [9], Zafrullah showed that an integrally closed AGCD-domain is a Prüfer v-multiplication domain (PVMD) with torsion t-class group. According to Theorem 5.1, an AGGCD-domain need not have a torsion t-class group. While any AP-domain is an AGGCD-domain, an AP-domain need not be an AGCD-domain. However, if R is an integrally closed AGGCD-domain, we can still conclude that R is a PVMD. It is shown in [3], that if a domain R is root closed and a, b ∈ R, then $(a^n, b^n)_t = [(a, b)^n]_t$. For our final result, we use this fact to characterize integrally closed AGGCD-domains.

THEOREM 5.2   Let R be an integral domain. Then the following statements are equivalent.

1) R is an integrally closed AGGCD-domain.
2) R is a root closed AGGCD-domain.
3) R is a PVMD with $Cl_t(R) = C(R)$, that is, every t-invertible t-ideal is invertible.

Proof:

1) ⇒ 2)   Clear.

2) $\Rightarrow$ 3)   To show R is a PVMD, it suffices to show that every 2-generated ideal is t-invertible [6]. Let a, b $\in$ R-{0}. There is an n = n(a, b) with $(a^n, n^n)_v = (a^n, b^n)_t$ invertible. Since R is root closed, $(a^n, b^n)_t = [(a, b)^n]_t$ is invertible. So there is an ideal I of R with $[(a^n, b^n)]_t I = R$. Therefore $[(a, b)^n I]_t = ([(a,b)^n]_t, I)_t = R_t = R$. Thus we have $\left[(a,b)\left((a,b)^{n-1}I\right)\right]_t = R$ and (a, b) is t-invertible. So R is a PVMD. Now suppose I is a t-invertible t-ideal. It is well-known that any t-invertible t-ideal has finite type, say $I = (a_1, ..., a_k)_t$. Since R is an AGGCD-domain, there is an n with $(a_1^n, ..., a_k^n)_t$ invertible. As above, $[(a_1, ..., a_k)^n]_t = (a_1^n, ..., a_k^n)_t$ is invertible. Since $I = (a_1, ..., a_k)_t$ is t-invertible, $(a_1, ..., a_k)$ is also t-invertible. Thus we have $I^n = [(a_1, ..., a_k)_t]^n = [(a_1, ..., a_k)_v]^n = \left(\left[(a_1, ..., a_k)^{-1}\right]^{-1}\right)^n = \left(\left[(a_1, ..., a_k)^{-1}\right]^n\right)^{-1} = \left(\left[(a_1, ..., a_k)^n\right]^{-1}\right)^{-1} = [(a_1, ..., a_k)^n]_v = [(a_1, ..., a_k)^n]_t$ is invertible. Therefore I is also invertible and thus every t-invertible t-ideal is invertible.

3) $\Rightarrow$ 1)   It is well-known that any PVMD is integrally closed. Let a,b $\in$ R-{0}. Then $(a, b)_t$ is a finite type t-ideal and hence is t-invertible. But since $Cl_t(R) = C(R)$, we have that $(a, b)_t$ is actually invertible. Thus using n(a, b) = 1, we see that R is an AGGCD-domain.

REFERENCES

1. D.D. Anderson, π-domains, divisorial ideals and overrings, Glasgow Math. J. 19(1978), 199-203.

2. D.D. Anderson and D.F. Anderson, Generalized GCD-domains, Comment. Math. Univ. St. Pauli 28(1979), 215-221.

3. D.D. Anderson and M. Zafrullah, Almost Bezout domains, J. Algebra 142(1991), 285 -309.

4. R. Gilmer, Finite element factorization in group rings, Lecture Notes in Pure and Applied Math., vol. 7, Marcel Dekker: New York, 1974.

5. R. Gilmer, *Multiplicative Ideal Theory*, Marcel Dekker: New York, 1972.

6. M. Griffin, Some results on v-multiplication domains, Canad. J. Math. 19(1967), 710-722.

7. I. Kaplansky, *Commutative Rings*, Revised Edition, University of Chicago Press: Chicago, 1974.

8. H. Uda, LCM-stableness in ring extensions, Hiroshima Math. J. 13(1983), 357-377.

9. M. Zafrullah, A general theory of almost factoriality, Manuscripta Math. 51(1985), 29-62.

Polynomial Behavior of Prime Ideals in Polynomial Rings and the Projective Line Over Z

AIHUA LI   Department of Mathematics, Loyola University, New Orleans, LA 70118

SYLVIA WIEGAND   Department of Mathematics, University of Nebraska, Lincoln, NE 68588-0323

Over the past forty years, considerable interest and research have been inspired by Kaplansky's question: "For which partially ordered sets U is there a Noetherian ring R so that $\mathrm{Spec}(R) \cong U$?" (For a Noetherian ring R, the topological structure of the set of prime ideals of R, called the spectrum of R or $\mathrm{Spec}(R)$, is determined by its structure as a partially ordered set ordered by inclusion.) An important related problem is to determine which partially ordered sets arise as spectra of two-dimensional domains related to polynomial rings. This problem appears to be more tractable. In particular, in [rW2] Roger Wiegand characterizes $\mathrm{Spec}(\mathbf{Z}[x])$, the set of prime ideals in the polynomial ring over the integers, as a partially ordered set satisfying certain axioms. Surprisingly, his axioms do not seem to depend upon the polynomial structure of $\mathrm{Spec}(\mathbf{Z}[x])$.

In this paper we analyze prime spectra of rings related to polynomial rings using the concept of a *coefficient* subset of a partially ordered set. Loosely speaking, a coefficient subset behaves like the set of prime ideals extended from a coefficient ring A in the ring of polynomials over A. We focus our attention particularly on coefficient subsets of $\mathrm{Spec}(\mathbf{Z}[x])$.

The partially ordered set of the projective line over $\mathbf{Z}$, $\mathrm{Proj}(\mathbf{Z}[h,k])$, is similar to $\mathrm{Spec}(\mathbf{Z}[x])$, but lacks an essential property, the existence of radical elements (defined in (1.4)). Thus $\mathrm{Proj}(\mathbf{Z}[h,k])$ is not order-isomorphic to $\mathrm{Spec}(\mathbf{Z}[x])$. Using "projective" coefficient subsets of $\mathrm{Proj}(\mathbf{Z}[h,k])$, which behave like prime ideals extended from $\mathbf{Z}$, we give a partial set of axioms for $\mathrm{Proj}(\mathbf{Z}[h,k])$.

For $R = \mathbf{Z}[x]$, the polynomial ring in one variable over the integers, Roger Wiegand in [rW1] specifies five properties, given below in (1.2), that characterize $\mathrm{Spec}(\mathbf{Z}[x])$. Every pair of countable partially ordered sets satisfying these properties are order-isomorphic. The class of rings A whose spectra satisfy these properties includes:

(1) Two-dimensional domains A finitely generated as a k-algebra, where k is an algebraic extension of a finite field, [rW2],

(2) Polynomial rings $A = D[x]$ over a domain D, where D is an order in an algebraic number field and D is finitely generated as a $\mathbf{Z}$-algebra, [rW2],

(3) Birational extensions A of $\mathbf{Z}[x]$ of the form $A = \mathbf{Z}[x][\frac{g_1}{f}, \ldots, \frac{g_t}{f}]$, where the $g_1, \ldots, g_t, f \in \mathbf{Z}[x]$, [AW1].

In a series of papers, William Heinzer, Dave Lantz, and Sylvia Wiegand study various two-dimensional partially ordered sets related to polynomials over a semilocal one-dimensional domain R: namely, the spectrum of the polynomial ring $R[x]$, the projective line over R and the spectrum of a birational extension of $R[x]$. In [HLW1], the projective line over R is shown to have a slightly different structure than $\mathrm{Spec}(R)$. Heitmann, McAdam and Shah [HM], [M], [MS] have studied the relationship between $\mathrm{Spec}(R)$ and $\mathrm{Spec}(R[x])$ for higher dimensions.

In this paper we analyze $\mathrm{Spec}(\mathbf{Z}[x])$ further in terms of behavior which reflects the polynomial structure of the prime ideals. We compare $\mathrm{Spec}(\mathbf{Z}[x])$ to the projective line over $\mathbf{Z}$, or $\mathrm{Proj}(\mathbf{Z}[h,k])$, which is described in (1.5). We show that the partially ordered set $\mathrm{Proj}(\mathbf{Z}[h,k])$ is not order-isomorphic to $\mathrm{Spec}(\mathbf{Z}[x])$; actually $\mathrm{Proj}(\mathbf{Z}[h,k])$ satisfies four of the five basic axioms of $\mathrm{Spec}(\mathbf{Z}[x])$, but the last one fails in $\mathrm{Proj}(\mathbf{Z}[h,k])$.

In section one we give some background and notation and we define the projective line over $\mathbf{Z}$. In section two we describe the polynomial behavior of $\mathrm{Spec}(\mathbf{Z}[x])$. In section three we give an analogous description of

$\mathrm{Proj}(\mathbf{Z}[h,k])$.

1 NOTATION, BACKGROUND, PROJECTIVE LINE OVER Z

The following notation will be used throughout this paper:

1.1. Notation for partially ordered sets

Let U be a partially ordered set satisfying the ascending and descending chain conditions. For $u \in U$, the *height* of u, $\mathrm{ht}(u)$, is the length of the longest chain to u from a minimal element in U; the *dimension* of U, or $\dim(U)$ is the $\max\{\mathrm{ht}(u) \mid u \in U\}$.

For all u and v in U, and for all subsets S and T of U, define

$$G_S(u) = \{y \in S \mid y > u\}, \qquad G_S(T) = \bigcap_{t \in T} G_S(t),$$
$$G_S(u,v) = G_S(u) \cap G_S(v), \quad \text{and} \quad H_i(U) = \{u \in U \mid \mathrm{ht}(u) = i\}.$$

We write $G(u), G(u,v)$, and $G(T)$ for $G_U(u), G_U(u,v)$, and $G_U(T)$, respectively.

Roger Wiegand has given the following list of axioms, which describe $\mathrm{Spec}(\mathbf{Z}[x])$:

1.2. The CZP axioms

Let U be a partially ordered set.

(P1) U is countable with a unique minimal element.

(P2) U has dimension two.

(P3) For each element u of height one, $G(u)$ is infinite.

(P4) For each pair u, v of distinct elements of height one, $G(u,v)$ is finite.

(P5) Given finite subsets $S \subset H_1(U)$ and $T \subset H_2(U)$ with $S \neq \emptyset$, there is a height-one element w in U such that (i) $w < t$, $\forall t \in T$, and (ii) whenever $t' \in U$ is greater than both w and s for some s in S, then $t' \in T$.

A partially ordered set which satisfies (P1)–(P5) is called **CZP** .

1.3 THEOREM [rW2, Theorem 1]   A partially ordered set U is **CZP** if and only if $U \cong \mathrm{Spec}(\mathbf{Z}[x])$.

The most significant of the axioms in (1.2), Axiom (P5) gives rise to a related concept, that of a radical element.

1.4. DEFINITION   Let U be a partially ordered set of dimension two. A U-*pair* is a pair (S, T) of finite subsets of U, where $T \subseteq H_2(U)$ and $\emptyset \neq S \subseteq H_1(U)$. If (S, T) is such a pair and $w \in H_1(U)$ satisfies $\bigcup_{s \in S}(G(w) \cap G(s)) \subseteq T \subset G(w)$, then w is called a *radical element* for (S, T).

There are two standard interpretations of the projective line over $\mathbf{Z}$ as a partially ordered set. The first is the interpretation from algebraic geometry; the second, more ring-theoretic, will be used in this paper as our working definition because it relates the projective line over $\mathbf{Z}$ to $\mathrm{Spec}(\mathbf{Z}[x])$.

1.5 NOTATION   The projective line over $\mathbf{Z}$, denoted $\mathrm{Proj}(\mathbf{Z}[h, k])$ is

(1) The partially ordered set $H-\mathrm{Spec}(\mathbf{Z}[h, k]) = \{$ prime ideals generated by homogeneous polynomials in variables h and k over $\mathbf{Z}\} - \{$ prime ideals of form (h, k, p), where p is a prime integer $\}$.

(2) The union $\mathrm{Spec}(\mathbf{Z}[x]) \cup \mathrm{Spec}(\mathbf{Z}[1/x])$, where $\mathrm{Spec}(\mathbf{Z}[x]) \cap \mathrm{Spec}(\mathbf{Z}[1/x])$ is identified with $\mathrm{Spec}(\mathbf{Z}[x, 1/x])$. In the intersection we identify $p\mathbf{Z}[x]$ with $p\mathbf{Z}[1/x]$ for every prime integer p, and we identify $f(x)\mathbf{Z}[x]$ with $x^{-\deg(f)}f(x)\mathbf{Z}[1/x]$ for an irreducible polynomial $f(x) \neq x$ with $\deg(f) > 0$. (An approximation is pictured below.)

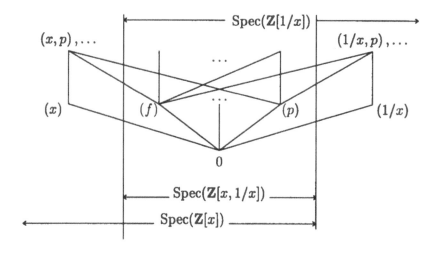

There is an order-isomorphism between $\operatorname{Spec}(\mathbf{Z}[x])$ and $\operatorname{Spec}(\mathbf{Z}[1/x])$, but it is not compatible with the identification; for example,

$(2x + 3) \subseteq (x, 3)$, but $(2 + 3(1/x)) \not\subseteq ((1/x), 3)$.

1.6. DEFINITION   Let A be a subset of a partially ordered set U, where U has a unique minimal element z_0. Following [RW1], we define the #-closure of A in U as

$$A^{\#} = A \cup \{G_U(x, y) \mid x, \ y \in A\} \cup \{z_0\}.$$

If $A^{\#} = A$, we say A is #-closed.

Note: $(A^{\#})^{\#} = A^{\#}$, so $A^{\#}$ is #-closed.

1.7 THEOREM [rW1, Theorem 1]   Suppose U and V are **CZP** partially ordered sets, and that A and B are nonempty finite #-closed subposets of U and V respectively. Then an order-preserving isomorphism from A to B can be extended to an order-preserving isomorphism from U to V.

2 THE POLYNOMIAL PROPERTY FOR $\operatorname{Spec}(\mathbf{Z}[x])$

As we mention in the introduction, the **CZP** axioms (1.2) for the partially ordered set $\operatorname{Spec}(\mathbf{Z}[x])$ are not related to the polynomial structure of $\mathbf{Z}[x]$ in an obvious way.

In this section we introduce a property which displays the structure of $\operatorname{Spec}(\mathbf{Z}[x])$ as a polynomial ring, that is, the property that some subset of primes should behave like the extensions of primes of the coefficient ring. Such subsets are called coefficient sets; this is formalized in Definition 2.1.

The height-one elements of $\operatorname{Spec}(\mathbf{Z}[x])$ are either *integer prime ideals*, generated by a prime integer p or *polynomial prime ideals* generated by an irreducible polynomial $f(x)$ of positive degree. The height-two primes are of form $(f(x), p)$ where p is a prime integer and $f(x)$ is a monic irreducible polynomial of $\mathbf{Z}[x]$ with an irreducible image in $(\mathbf{Z}/p)\mathbf{Z}[x]$. The special subset of all integer prime ideals is denoted C_0.

2.1 DEFINITION   Let U be a two-dimensional partially ordered set. A subset Δ of $H_1(U)$ is called a *coefficient subset* of U if
  (1) $\forall u \in \Delta, |G(u)| = \infty$;
  (2) $\forall u \neq v \in \Delta, \quad G(u, v) = \emptyset$;

(3) $\bigcup_{u \in \Delta} G(u) = H_2(U)$. A coefficient set Δ is *attached to* an element w in $H_1(U)$ if

$$|G(v) \cap G(w)| = 1, \quad \forall v \in \Delta.$$

For example, the set C_0 of integer prime ideals is a coefficient set of $\mathrm{Spec}(\mathbf{Z}[x])$ attached to (x), and C_0 is also a coefficient set of $\mathrm{Spec}(\mathbf{Z}[x])$ attached to $(x + q)$, for every prime integer q.

2.2 THEOREM   There are infinitely many coefficient sets attached to (x) in $\mathrm{Spec}(\mathbf{Z}[x])$. Thus there exist infinitely many coefficient sets in $\mathrm{Spec}(\mathbf{Z}[x])$. Proof: For a prime integer $p \neq 2$, consider two subsets A and B_p of $\mathrm{Spec}(\mathbf{Z}[x])$ with elements and relations shown in the left and right portions of Diagram 2.2.1:

Obviously, both A and B_p are #-closed in $\mathrm{Spec}(\mathbf{Z}[x])$ and $A \cong B_p$. Let ϕ_p be the isomorphism sending (2) to $(px + 2)$ and fixing the other elements. By Theorem 1.7, ϕ_p can be extended to an order-preserving isomorphism $\phi'_p: \mathrm{Spec}(\mathbf{Z}[x]) \longrightarrow \mathrm{Spec}(\mathbf{Z}[x])$.

Let $\Delta_p = \phi'_p(C_0)$. Since C_0 is a coefficient set of $\mathrm{Spec}(\mathbf{Z}[x])$, so is Δ_p. Furthermore for every $(q) \in C_0$,

$$|G(\phi'_p((q)), \phi'_p((x)))| = |G((q),\ (x))| = 1 \quad \text{and} \quad \phi'_p((x)) = \phi_p((x)) = (x).$$

Therefore Δ_p is a coefficient set attached to (x) of $\mathrm{Spec}(\mathbf{Z}[x])$.

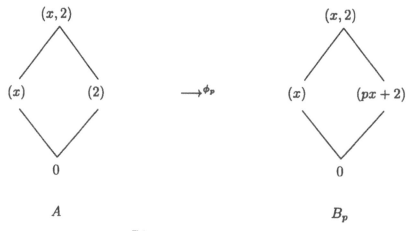

Diagram 2.2.1

Claim: If p, q are distinct prime integers, not equal to 2, then $\Delta_p \neq \Delta_q$.

Proof of Claim: We have that

$$\phi'_q((2)) = \phi_q((2)) = (qx + 2) \qquad \text{and} \qquad \phi'_p((2)) = \phi_p((2)) = (px + 2).$$

If $(qx+2) \in \Delta_p$ then $(qx+2) = \phi'_p((q'))$ for some prime number $q' \neq 2$. Thus $G((qx + 2),\ (px + 2)) = G(\phi'_p((q')), \phi'_p((2))) = G((q'), (2)) = \emptyset$. This yields a contradiction since $(x, 2)$ is above both $(qx + 2)$ and $(px + 2)$. Therefore $\Delta_p \neq \Delta_q$, proving the claim and the theorem.

2.3 COROLLARY   Every **CZP** set U has infinitely many coefficient sets. Furthermore, for every element u of height one there exist infinitely many coefficient sets attached to u. For every finite subset S_1 of C_0, there exists a coefficient set Δ attached to (x) with $S_1 \subseteq \Delta$, $\Delta \neq C_0$.

Proof: These statements follow from (1.3) and (1.7).

2.4 THEOREM   Let U be a **CZP** set and $\Psi : \mathrm{Spec}(\mathbf{Z}[x]) \longrightarrow U$ an order-isomorphism. Let $\Delta_0 = \Psi(C_0)$. Then there exist infinitely many coefficient sets of U attached to $\Psi((x))$ which are disjoint from Δ_0. For every finite subset F of C_0 there exists a coefficient set attached to (x) which contains F and is distinct from C_0.

Proof: We show that $\mathrm{Spec}(\mathbf{Z}[x])$ contains infinitely many coefficient sets consisting only of polynomial prime ideals.

Consider the subsets $A = B_0$ and B_n of $\mathrm{Spec}(\mathbf{Z}[x])$ in Diagram 2.2.2. Then $A^{\#} = A$ and $B_n^{\#} = B_n$ for all n. The order-preserving isomorphism $\varphi_n : A \longrightarrow B_n$ sends (2) to $(x^n + 2)$ and fixes the other elements. By Theorem 1.7, φ_n extends to an isomorphism $\varphi'_n : \mathrm{Spec}(\mathbf{Z}[x]) \longrightarrow \mathrm{Spec}(\mathbf{Z}[x])$. Then $\Delta_n = \varphi'_n(C_0)$ is a coefficient set. We claim that $\Delta_n \neq \Delta_m$ if $n \neq m$ and that each Δ_n contains only prime ideals generated by polynomials of degree greater than zero. For a fixed $n > 0$, $\varphi'_n((2)) = (x^n + 2) \notin C_0$. For $p \neq 2$, if $\varphi'_n((p)) = (q)$ for some prime number q, then

$$G((x^n + 2), (q)) = G(\varphi'_n(2), \varphi'_n((p))) = \varphi'_n(G((2), (p))) = \emptyset,$$

since φ'_n is an order-isomorphism. This is a contradiction because $x^n + 2$ has an irreducible factor $f(x)$ in $\mathbf{Z}/q\mathbf{Z}$, and so $(f(x), q) \in G((x^n + 2, q))$.

Therefore $\varphi_n'((p))$ is generated by a polynomial of positive degree. Thus

$$\Delta_n \cap C_0 = \emptyset, \quad \text{and} \quad \Psi(\Delta_n) \cap \Delta_0 = \Psi(\Delta_n) \cap \Psi(C_0) = \Psi(\Delta_n \cap C_0) = \emptyset.$$

Hence $\Psi(\Delta_n)$ is a coefficient set disjoint from Δ_0.

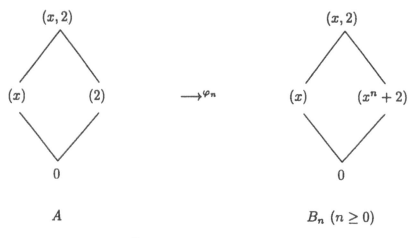

Diagram 2.2.2

Claim: If $n \neq m$, then $\Delta_n \neq \Delta_m$.

Proof of claim: We have

$$(x^n + 2) = \varphi_n'((2)) \in \Delta_n, \quad (x^m + 2) = \varphi_m'((2)) \in \Delta_m.$$

If $(x^n + 2) \in \Delta_m$, then $(x^n + 2) = \varphi_m'((p))$ for some prime number $p \neq 2$. But $G((2), (p)) = \emptyset$ implies $G((x^m + 2), (x^n + 2)) = G(\varphi_m'((2)), \varphi_m'((p))) = \emptyset$, a contradiction to $(x, 2) \in G((x^m + 2), (x^n + 2))$. Therefore $(x^n + 2) \in \Delta_n - \Delta_m$.

Completion of proof: Since C_0 is attached to (x), each set Δ_n is a coefficient set of U attached to $\Psi((x))$ and the first statement of the theorem follows. The second statement is proved similarly.

Question: Does every coefficient set other than C_0 contain only finitely many elements of C_0? The answer is "yes". This is a special case of the following result:

2.5. PROPOSITION   Let U, Ψ and Δ_0 be as in (2.4) and let Δ be a coefficient set in $\mathrm{Spec}(\mathbf{Z}[x])$ distinct from C_0. Then $\Psi(\Delta) \cap \Delta_0$ is a finite subset of U.

Proof: We first prove that Δ contains only finitely many integer prime ideals.

Note that Δ is not a proper subset of C_0, because if there were an element $(p) \in C_0 - \Delta$, and yet Δ contained no polynomial prime ideal, then height-two primes of form $(p, f(x))$, for $f(x)$ a polynomial of positive degree, contain no element of Δ, which contradicts (2.1.3). Thus if $\Delta \neq C_0$ then there exists a polynomial prime $(f(x)) \in \Delta - C_0$, where $f(x)$ is an irreducible polynomial of degree > 0. Let $f(x) = a_n x^n + \ldots + a_1 x + a_0$, $n \geq 1$, $a_i \in \mathbf{Z}$. Assume $(p) \in \Delta \cap C_0$ where p is a prime integer. Since Δ is a coefficient set, $G((p),(f)) = \emptyset$, and so $f(x)$ is a constant (mod p) and $p \mid (f(x) - a_0)$. But $f(x) - a_0$ has only finitely many factors, and so $\Delta \cap C_0$ is a finite set. Therefore $\Psi(\Delta) \cap \Delta_0 = \Psi(\Delta) \cap \Psi(C_0) = \Psi(\Delta \cap C_0)$ is a finite set because Ψ is an order-isomorphism.

2.6. PROPOSITION   If Δ is a coefficient set of $\mathrm{Spec}(\mathbf{Z}[x])$ with $\Delta \neq C_0$, then the set $N = \{\deg(f) \mid (f(x)) \in \Delta\}$ is unbounded.

Proof: By way of contradiction, assume N has an upper bound $M > 0$; that is, $\deg(f) \leq M$, $\forall (f) \in \Delta$. By 2.5, $|\Delta \cap C_0| < \infty$; therefore there exists an integer prime $(p) \in C_0 - \Delta$. There exists a polynomial $g(x) \in \mathbf{Z}[x]$ of degree $M + 1$ so that its image in $(\mathbf{Z}/p\mathbf{Z})[x]$ is irreducible. Thus $(g(x), p)$ is a height-two element of $\mathrm{Spec}(\mathbf{Z}[x])$. By the definition of coefficient set,

$$\bigcup_{s \in \Delta} G(s) = H_2(\mathrm{Spec}(\mathbf{Z}[x])), \quad \text{so} \quad (g(x), p) \in \bigcup_{s \in \Delta} G(s).$$

On the other hand, for every $(q) \in \Delta \cap C_0$, with q a prime integer, we have $q \neq p$ (since $(p) \notin \Delta$); thus $(g(x), p) \notin G((q))$. For all $(f(x)) \in \Delta - C_0$, $f(x) \notin (g(x), p)$, since $\deg(f) \leq M < M + 1 = \deg(g(x))$. Thus $(g(x), p) \notin G(s)$, for all $s \in \Delta$, a contradiction. Therefore N is unbounded.

3. PROPERTIES OF $\mathrm{Proj}(\mathbf{Z}[h, k])$

3.1 NOTATION   As in section two, C_0 denotes the special coefficient set of $\mathrm{Spec}(\mathbf{Z}[x])$ consisting of the integer prime ideals; in this section C_0 is considered to be a subset of $\text{Proj}(\mathbf{Z}[h,k])$ as in (1.5.2). If $f(x) = a_n x^n + \ldots + a_1 x + a_0 \in \mathbf{Z}[x]$ and $a_0 \neq 0$, then $(1/x^n)f(x) = a_0(1/x)^n + \ldots + a_{n-1}(1/x) + a_n$. We identify $f(x)$ with $(1/x^n)f(x)$, written $f(x) \sim \frac{1}{x^n}f(x)$. We denote the leading coefficient a_n of $f(x)$ by $l(f) = a_n$ and the constant term a_0 of $f(x)$ by $f(0)$.

3.2. REMARK   Let $f(x)$ be a polynomial in $\mathbf{Z}[x]$ with $\deg(f) = n > 0$ and $f \neq x$. Then

(1) In $\mathbf{Z}[x]$, $f(x) \in (x,p)\mathbf{Z}[x]$ if and only if $p \mid f(0)$;

(2) In $\mathbf{Z}[\frac{1}{x}]$, $f(x) \sim \frac{1}{x^n}f(x) \in (\frac{1}{x}, p)\mathbf{Z}[\frac{1}{x}]$ if and only if $p \mid l(f)$.

3.3. LEMMA   For all $P \in H_1(\text{Spec}(\mathbf{Z}[x]))$, the set $G_{\text{Proj}(\mathbf{Z}[h,k])}(P, (\frac{1}{x}))$ is finite; for every $Q \in H_1(\text{Spec}(\mathbf{Z}[1/x]))$, the set $G_{\text{Proj}(\mathbf{Z}[h,k])}(Q, (x))$ is finite.

Proof: If $P = (p)$, then $G_{\text{Proj}(\mathbf{Z}[h,k])}(P, (\frac{1}{x})) = \{(1/x, p)\}$. If $P = (f(x))$, where $f(x)$ is an irreducible polynomial in $\mathbf{Z}[x]$ of positive degree, then by (3.2), $P \subset (1/x, q)$ for some prime number q if and only if $q \mid l(f)$, i.e., q divides the leading coefficient of $f(x)$. Since $l(f)$ has finitely many prime numbers as factors, $G_{\text{Proj}(\mathbf{Z}[h,k])}(P, (\frac{1}{x}))$ is finite. Similarly $G_{\text{Proj}(\mathbf{Z}[h,k])}(Q, (x))$ is finite.

3.4. PROPOSITION $\text{Proj}(\mathbf{Z}[h,k])$ satisfies axioms (P1) – (P4) of (1.2).

Proof: Axioms (P1) – (P3) hold since $\text{Proj}(\mathbf{Z}[h,k]) = \text{Spec}(\mathbf{Z}[x]) \cup \text{Spec}(\mathbf{Z}[1/x])$ and $\text{Spec}(\mathbf{Z}[x])$, $\text{Spec}(\mathbf{Z}[1/x])$ both satisfy (P1) – (P5). Let $X = \text{Proj}(\mathbf{Z}[h,k])$, $X_1 = \text{Spec}(\mathbf{Z}[x])$, $X_2 = \text{Spec}(\mathbf{Z}[1/x])$, $X_0 = \text{Spec}(\mathbf{Z}[x, 1/x])$.   For (P4), we show: $\forall P, Q \in H_1(\text{Proj}(\mathbf{Z}[h,k]))$, $|G_X(P,Q)| < \infty$.

Case 1:   If $P, Q \in \text{Spec}(\mathbf{Z}[x])$, then $G_X(P,Q) = G_{X_1}(P,Q) \cup G_{X_2}(P,Q,(1/x))$. But $|G_{X_1}(P,Q)| < \infty$ since $\text{Spec}(\mathbf{Z}[x])$ satisfies (P4). and by Lemma 3.3, $|G_{X_2}(P,Q)| \leq |G_X(P,(1/x))| < \infty$.

Case 2: If $P, Q \in \text{Spec}(\mathbf{Z}[1/x])$, then, as in case 1, $|G_X(P,Q)| < \infty$.

Case 3: If $\{P,Q\} = \{(x), (1/x)\}$, then $G(P,Q) = \emptyset \implies |G(P,Q)| = 0 < \infty$.

3.5 THEOREM $\text{Proj}(\mathbf{Z}[h,k])$ does not satisfy (P5). Thus $\text{Proj}(\mathbf{Z}[h,k]) \not\cong \text{Spec}(\mathbf{Z}[x])$.

Proof: For $T_0 = \{(x,2), (1/x,2), (1/x,3)\}$,   and $S_0 = \{(5), (2), (1/x)\}$, we show that (S_0, T_0) has no radical elements in $\text{Proj}(\mathbf{Z}[h,k])$.

Suppose (S_0, T_0) has a height-one radical element P. Then $P \subset t$, $\forall t \in T_0$, and obviously P is not an integer prime ideal. Thus P is a polynomial prime ideal. However P cannot be (x) or $(1/x)$ since $(x) \not\subseteq (\frac{1}{x}, 2)$ and $(\frac{1}{x}) \not\subseteq (x, 2)$. Write $P = (f(x))$, where $f(x) = a_n x^n + \ldots + a_1 x + a_0$, $n \geq 1$, $a_0 \neq 0$, $a_i \in \mathbf{Z}$ and $f(x)$ is irreducible. Since P is a radical element, we have $\bigcup_{s \in S} G(P, s) \subset T$.

Case 1: Suppose there exists i, $1 \leq i \leq n$, such that 5 does not divide a_i. Then, modulo 5, f has positive degree. Thus the image $\bar{f}$ of f in $\mathbf{Z}/5\mathbf{Z}$ has at least one irreducible factor $\bar{f}_1$ over $\mathbf{Z}/5\mathbf{Z}$, where $f_1 \in \mathbf{Z}[x]$. Now $(5) \in (f_1, 5)$ and $P = (f) \in (f_1, 5)$. But $(f_1, 5) \notin T$, so $(f_1, 5) \in G(P, (5)) - T$. This contradicts $\bigcup_{s \in S} G(P, s) \subset T$. Thus Case 1 does not occur.

Case 2: $5 \mid a_i$, for every $i > 0$. Now 5 does not divide a_0 since $f(x)$ is irreducible in $\mathbf{Z}[x]$. Thus $P = (f) \subseteq (1/x, 5)$, since $5 \mid l(f)$ by (3.2). Also $(5) \subseteq (1/x, 5)$; thus $(1/x, 5) \in G(P, (5))$. But $(1/x, 5) \notin T$, again a contradiction. Therefore P is not a radical element. So (P5) does not hold in $\text{Proj}(\mathbf{Z}[h, k])$.

A natural question arises: For which pairs (S, T) in $\text{Proj}(\mathbf{Z}[h, k])$ do radical elements exist and for which pairs do they not exist?

3.6. LEMMA   The coefficient set C_0 of prime-generated ideals satisfies, $\forall P \in C_0$,

(1) $|G(P, (1/x))| = |G(P, (x))| = 1$ and

(2) $G(P, Q) \neq \emptyset$, $\forall Q \in H_1(\text{Proj}(\mathbf{Z}[h, k])) - C_0$.

Proof: Obviously $C_0 = \{(p) \mid p \text{ is a prime number}\}$ satisfies (1). For every $Q \in H_1(\text{Proj}(\mathbf{Z}[h, k])) - C_0$, Q is generated by a polynomial of degree greater than 0 in $\mathbf{Z}[x]$ or $\mathbf{Z}[\frac{1}{x}]$. Let $P = (p)$; then $G((x), P) = \{(x, p)\}$, $G((\frac{1}{x}), p) = \{(\frac{1}{x}, p)\}$. Let $Q = (f(x))$, for $f(x) \neq x$ irreducible in $\mathbf{Z}[x]$. Then

Case 1: $p \mid l(f) \implies (f) \subset (1/x, p)$, $p \subset (1/x, p)$ and so $(\frac{1}{x}, p) \in G(P, Q)$.

Case 2: p does not divide $l(f) \implies (f) \subset (f'(x), p)$, $p \subset (f'(x), p)$ where $f'(x)$ is an irreducible factor of $f(x)$ in $\mathbf{Z}/p\mathbf{Z}$. Thus $(f'(x), p) \in G(Q, P)$.

The counterexample in (3.5) can be generalized as follows:

3.7 PROPOSITION   Let (S, T) be a $\text{Proj}(\mathbf{Z}[h, k])$-pair and C_1 a coefficient set for $\text{Proj}(\mathbf{Z}[h, k])$ attached to (x) and to $(1/x)$ (that is, $\forall P \in C_1$, $|G(P, (1/x))| = |G(P, (x))| = 1$). Suppose also that $G(P, Q) \neq \emptyset$, for every $P \in C_1$, and for every $Q \in H_1(\text{Proj}(\mathbf{Z}[h, k])) - C_1$ and that there exist distinct elements P_0 and P_1 of C_1 such that $P_0 \in S$, $T \cap G(P_0) = \emptyset$, but $T \cap G(P_1) \neq \emptyset$. Then (S, T) has no radical element except P_1.

Proof: Suppose Q is a radical element for (S, T).

Case i: If $Q \in C_1$, then $Q \neq P_1 \implies G(Q, P_1) = \emptyset$, but then $\emptyset \neq T \cap G(P_1) \not\subseteq G(Q)$, and so $T \not\subseteq G(Q)$. This contradicts that Q is a radical element. Thus $Q = P_1$.

Case ii: Suppose $Q \notin C_1$. Then $G(P_0, Q) \neq \emptyset$; thus $G(P_0, Q) \not\subseteq G(P_0) \cap T = \emptyset$ and thus $G(P_0, Q) \not\subseteq T$, contradicting that Q is a radical element.

Thus S, T has no radical element except possibly P_1.

3.8. COROLLARY   Let (S, T) be a $\text{Proj}(\mathbf{Z}[h, k])$-pair such that $T \not\subseteq G(P)$ for every $P \in C_1$. Suppose that C_1 is a coefficient set for $\text{Proj}(\mathbf{Z}[h, k])$ attached to (x) and to $(1/x)$. Then S can be extended to $S' \supseteq S$ so that (S', T) has no radical elements in $\text{Proj}(\mathbf{Z}[h, k])$.

Proof: We have $T \subset H_2(\text{Proj}(\mathbf{Z}[h, k])) = \bigcup_{P \in C_1} G(P)$, but $T \not\subseteq G(P)$, $\forall P \in C_1$. So there exist P_1, $P_2 \in C_1$, $T \cap G(P_i) \neq \emptyset$, $i = 1, 2$. Since T is finite, there is a $P_0 \in C_1$ with $T \cap G(P_0) = \emptyset$. Let $S' = S \cup \{P_0\}$; then $P_0 \in S'$ and (S', T) satisfies the conditions in (3.7). Therefore (S', T) has no radical elements.

Note that $\text{Proj}(\mathbf{Z}[h, k]) = \text{Spec}(\mathbf{Z}[x]) \cup \text{Spec}(\mathbf{Z}[1/x])$, and both $\text{Spec}(\mathbf{Z}[x])$ and $\text{Spec}(\mathbf{Z}[1/x])$ are $\mathbf{CZP}$; thus, in $\text{Spec}(\mathbf{Z}[x])$ and $\text{Spec}(\mathbf{Z}[1/x])$, there are radical elements for every pair (S, T). In some cases, these radical elements are radical elements of the pair (S, T) as a $\text{Proj}(\mathbf{Z}[h, k])$-pair.

3.9 PROPOSITION   Let (S, T) be a $\text{Spec}(\mathbf{Z}[x])$-pair (respectively, $\text{Spec}(\mathbf{Z}[1/x])$-pair) such that $G(s) \subseteq \text{Spec}(\mathbf{Z}[x])$, (respectively, $G(s) \subseteq \text{Spec}(\mathbf{Z}[1/x])$), $\forall s \in S$. Then (S, T) has radical elements as a $\text{Proj}(\mathbf{Z}[h, k])$-pair.

Proof: Since $\text{Spec}(\mathbf{Z}[x])$ is $\mathbf{CZP}$, (S, T) has infinitely many radical elements as a $\text{Spec}(\mathbf{Z}[x])$-pair. We choose a radical element $w \neq (x)$ in $\text{Spec}(\mathbf{Z}[x])$. Let $X = \text{Proj}(\mathbf{Z}[h, k])$, $X_1 = \text{Spec}(\mathbf{Z}[x])$, $X_2 = \text{Spec}(\mathbf{Z}[1/x])$, $X_0 = \text{Spec}(\mathbf{Z}[x, 1/x])$. Then $w \subset t$, $\forall t \in T$, and $\bigcup_{s \in S} G_{X_1}(w, s) \subseteq T$. On the other hand, $\forall s \in S$, $G(w, s) \subseteq G(s) \subseteq \text{Spec}(\mathbf{Z}[x])$; thus $\bigcup_{s \in S} G_X(w, s) = \bigcup_{s \in S} G_{X_1}(w, s) \subseteq T$, so w is a radical element for (S, T) as a $\text{Proj}(\mathbf{Z}[h, k])$-pair.

There are many pairs (S, T) which satisfy the conditions in (3.9). For example, if S consists entirely of prime ideals generated by monic irreducible polynomials in $\mathbf{Z}[x]$ of degree ≥ 1, then $G(s) \subseteq \mathrm{Spec}(\mathbf{Z}[x])$, for all s. (A monic polynomial $f(x) \notin (1/x, p)$ because p doesn't divide $\mathrm{l}(f) = 1$.)

3.10 PROPOSITION   Let (S, T) be a $\mathrm{Proj}(\mathbf{Z}[h, k])$-pair where $T \neq \emptyset$, and let C_0 be the special coefficient set of $\mathrm{Proj}(\mathbf{Z}[h, k])$. Then (S, T) has a radical element if

(1) $S \cap C_0 = \bigcup_{m \in T}(L(m) \cap C_0)$, where $L(m) = \{p \in H_1(\mathrm{Proj}(\mathbf{Z}[h, k])) \mid p \subset m\}$.

(2) $\bigcup_{s \in S} G((s, (\frac{1}{x}))) \subseteq T$, and

(3) $(\frac{1}{x}) \notin S$.

Proof: Let $S \cap C_0 = \{(p_1), \ldots, (p_r)\}$, and $S - C_0 = \{(f_1), \ldots, (f_m)\}$ where $p_1, \ldots, p_r$ are prime integers in $\mathbf{Z}$ and $f_1, \ldots, f_m$ are irreducible polynomials in $\mathbf{Z}[x]$ with positive degree. By (3), $S \subset \mathrm{Spec}(\mathbf{Z}[x])$. Let $T' = T - G((\frac{1}{x}))$; then (S, T') is an $\mathrm{Spec}(\mathbf{Z}[x])$-pair. Since $\mathrm{Spec}(\mathbf{Z}[x])$ satisfies (P5), there are infinitely many radical elements of (S, T') in $\mathrm{Spec}(\mathbf{Z}[x])$. Take one radical element outside S, say P_0. Then $P_0 \in L(m), \forall m \in T'$. If $P_0 \in C_0$, then $P_0 \in \bigcup_{m \in T}(L(m) \cap C_0) \subseteq S$ by (1), a contradiction. So $P_0 \notin C_0$; say $P_0 = (f(x))$, where $f(x)$ is an irreducible polynomial of $\mathbf{Z}[x]$ other than $x, f_1, \ldots, f_m$. Let t be a positive integer greater than the degree of $f(x)$. Then $f(x)$ and the product $p_1 \cdots p_r \cdot x^t \cdot f_1 \cdots f_m$ are relatively prime in $\mathbf{Z}[x]$. By [K, page 102, exercise 3], $p_1 \cdots p_r \cdot x^t \cdot f_1 \cdots f_m + yf(x)$ is a prime ideal in $\mathbf{Z}[x, y]$. By Hilbert's Irreducibility Theorem [L], we can choose a prime number p such that

$$g(x) = p_1 \cdots p_r \cdot x^t \cdot f_1 \cdots f_m + pf(x)$$

is irreducible in $\mathbf{Z}[x]$; thus $w = (g(x))$ is a prime ideal of $\mathbf{Z}[x]$.

Claim: w is a radical element for (S, T).

Note that $(f(x)) \subset m$, for every $m \in T'$, and if $m' \in \mathrm{Spec}(\mathbf{Z}[x])$, $m' \supset (f(x))$, $m' \supset s$ for some s in S, then $m' \in T'$. This is because $(f(x))$ is a radical element of (S, T').

First, every m in T contains one of the $p_i's$ by condition (1). Also every m' in T' contains $f(x)$; thus the m' contain $g(x)$. So $w \subset m', \forall m' \in T'$. For every $m \in T - T'$, m has the form $m = (p_i, \frac{1}{x})$ where $(p_i) \in S \cap C_0$. But

p_i divides the leading coefficient of $g(x)$. That implies $g(x) \in (p_i, \frac{1}{x}) = m$; therefore $(g(x)) \subset m$.

Second, assume m is a height-two element in $\text{Proj}(\mathbf{Z}[h,k])$ such that $m \supset w$, $m \supset s$ for some $s \in S$. If $m \in G((\frac{1}{x}))$, then $m \in \bigcup_{s \in S} G((s,(\frac{1}{x})) \subseteq T$ by condition (2). If $m \notin G((\frac{1}{x}))$, then m is a prime ideal of $\mathbf{Z}[x]$ (i.e., $m \in \text{Spec}(\mathbf{Z}[x])$). But w is a radical element of (S,T') as a $\text{Spec}(\mathbf{Z}[x])$-pair, so $m \in T' \subseteq T$. We conclude that w is a radical element of (S,T).

3.11 NOTATION   Let U be a partially ordered set of dimension two.

A subset $C^{\mathbf{P}}$ of height-one elements of U is said to be *proj-coefficient* provided

(1) For every $p \neq q \in C^{\mathbf{P}}$, $G(p,q) = \emptyset$.

(2) The union $\bigcup_{p \in C^{\mathbf{P}}} G(p) = H_2(U)$.

(3) For every $p \in C^{\mathbf{P}}$, $u \in H_1(U) - C^{\mathbf{P}}$, $G(p,u) \neq \emptyset$.

For $S \subseteq H_1(U)$, a proj-coefficient subset $C^{\mathbf{P}}$ is said to be *attached to S* provided

(4) For every $p \in C^{\mathbf{P}}$, $u \in S, |G(p,u)| = 1$.

For example in $\text{Proj}(\mathbf{Z}[h,k])$, C_0 is a proj-coefficient subset attached to $\{(x),(\frac{1}{x})\}$.

3.12 PROPERTIES OF $\text{Proj}(\mathbf{Z}[h,k])$

In summary, $\text{Proj}(\mathbf{Z}[h,k]) = \text{Spec}(\mathbf{Z}[x]) \cup \text{Spec}(\mathbf{Z}[1/x])$ satisfies the following properties of a partially ordered set U:

(1) (P1) – (P4) of (1.2).

(2) There are two special height-one elements u_1, $u_2 \in U$ (in particular, (x) and $(1/x)$ of $\text{Proj}(\mathbf{Z}[h,k])$) such that $G(u_1,u_2) = \emptyset$. Let $U_1 = U - (\{u_2\} \cup G(u_2))$ and $U_2 = U - (\{u_1\} \cup G(u_1))$. Then $U_1 \cong U_2$; both are **CZP**.

(3) For u_1, u_2 as above , there exists a proj-coefficient set Δ_0 attached to u_1 and u_2.

(4) Although (P5) is not satisfied, many $\text{Proj}(\mathbf{Z}[h,k])$-pairs have radical elements (see (3.9) and (3.10)).

(5) There exists an order-isomorphism $\psi_0 : U_1 \longrightarrow U_2$ such that $\psi_0|_{\Delta_0} = 1|_{\Delta_0}$ and $\psi_0(\text{Spec}(\mathbf{Z}[x])) = \text{Spec}(\mathbf{Z}[1/x])$.

Remark: What we lack is a property like (P5) to tell us exactly when radical elements exist. We believe that such a property is needed to completely characterize $\text{Proj}(\mathbf{Z}[h, k])$.

Conjecture: $\text{Proj}(\mathbf{Z}[h, k])$ has the following property: If (S, T) is a $\text{Proj}(\mathbf{Z}[h, k])$-pair such that

(1) $S \cap C_0 = \bigcup_{m \in T}(L(m) \cap C_0)$, where $L(m) = \{p \in H_1(\text{Proj}(\mathbf{Z}[h, k])) \mid p \subset m\}$. and

(2) $(x) \in S$, and $(\frac{1}{x}) \in S$,

then there exist radical elements for (S, T)

If this conjecture can be proved, then we believe the properties in (3.12) together with this property will characterize $\text{Proj}(Z[h, k])$.

Next we give more conditions on a partially ordered set to guarantee that it is isomorphic to $\text{Proj}(\mathbf{Z}[h, k])$.

3.13 THEOREM   Let D be a partially ordered set satisfying the following properties:

(1) $\exists v_1,\ v_2 \in H_1(D)$ such that $G(v_1, v_2) = \emptyset$, and for $D_1 = D - (\{v_2\} \cup G(v_2))$ and $D_2 = D - (\{v_1\} \cup G(v_1))$ both D_1 and D_2 are **CZP** (that is, $D_1 \cong \text{Spec}(\mathbf{Z}[x]) \cong \text{Spec}(\mathbf{Z}[1/x]) \cong D_2$).

(2) There exist isomorphisms $\varphi :\ \text{Spec}(\mathbf{Z}[x]) \longrightarrow D_1$, and $\psi :\ D_1 \longrightarrow D_2$ such that $\psi(v_1) = v_2$. and $\psi\varphi\psi_0^{-1}|_{X_0} = \varphi|_{X_0}$, where $X_0 = \text{Spec}(\mathbf{Z}[x, 1/x])$.

Then

$$D \cong \text{Proj}(\mathbf{Z}[h, k]) = X .$$

Proof: We extend the isomorphism φ in condition (2) to an isomorphism $\varphi' :\ \text{Proj}(\mathbf{Z}[h, k]) \longrightarrow D$. Let $X = \text{Proj}(\mathbf{Z}[h, k])$, $X_1 = \text{Spec}(\mathbf{Z}[x])$, $X_2 = \text{Spec}(\mathbf{Z}[1/x])$. We define $\varphi'|_{X_1} = \varphi$; $\varphi'|_{X_2-X_1} = \psi\varphi\psi_0^{-1}|_{X_2-X_1}$. Then φ' is a one-to-one map. To check it is an order-preserving isomorphism, it is enough to check that, $\forall m \in G_X(\text{Spec}(\mathbf{Z}[1/x]))$, $\forall Q \in X_0$, $m \supset Q \iff \varphi'(m) \supset \varphi'(Q)$. Note that

$$\begin{aligned} m \supset Q\ (\text{in} \text{Spec}(\mathbf{Z}[1/x])) &\iff \psi_0^{-1}(m) \supset \psi_0^{-1}(Q)\ (\text{in} \text{Spec}(\mathbf{Z}[x])) \\ &\iff \varphi\psi_0^{-1}(m) \supset \varphi\psi_0^{-1}(Q)\ (\text{in} D_1) \\ &\iff \psi\varphi\psi_0^{-1}(m) \supset \psi\varphi\psi_0^{-1}(Q)\ (\text{in} D_2). \end{aligned}$$

Since $Q \in X_0$, by condition (2), $\psi\varphi\psi_0^{-1}(Q) = \varphi(Q) = \varphi'(Q)$. On the other hand, $m \in \text{Spec}(\mathbf{Z}[1/x]) - \text{Spec}(\mathbf{Z}[x])$, so the definition of φ' yields

$\psi\varphi\psi_0^{-1}(m) = \varphi'(m)$. Thus $m \supset Q \iff \varphi'(m) \supset \varphi'(Q)$. Finally, φ' is an isomorphism and $D \cong \text{Proj}(\mathbf{Z}[h, k])$

Remark: Here is a possible variation of the (P5) axiom which may hold and may yield (with (P1)-(P4) of (1.2)) a characterization of $U = \text{Proj}(\mathbf{Z}[h, k])$.

(P5a) There exist elements u_1, $u_2 \in U$ with $\text{G}(u_1, u_2) = \emptyset$.

(P5b) There exists a projective prime-coefficient set Γ attached to u_1 and u_2. (Thus $|\text{G}(p, u_i)| = 1, \forall p \in \Gamma$.)

(P5c) Suppose for every U-pair (S, T) such that $u_1, u_2, q_1, q_2 \in S$, where $q_1 \neq q_2 \in \Gamma$, that

(i) If $p_0 \in \Gamma - S$ is such that $\text{G}(p_0) \cap T \neq \emptyset$ and $\text{G}(p_0, s) \subseteq T \subseteq \text{G}(p_0)$, for every $s \in S$, then there exists a unique radical element, namely p_0. Otherwise there is no radical element.

(ii) If $\text{G}(p) \cap T \neq \emptyset, \forall p \in \Gamma \cap S$, then there exist infinitely many radical elements.

More Remarks: $\text{Proj}(\mathbf{Z}[h, k])$ has other interesting properties, such as that there exists an order-preserving map $\sigma : U \to U$ such that $\sigma^2 = $ the identity, and σ restricted to C_0 is the identity on C_0. This σ is defined by setting $\sigma((f(x))) = (f(1/x))$ and extending to all of $\text{Proj}(\mathbf{Z}[h, k])$.

However there is much less freedom for isomorphisms from $\text{Proj}(\mathbf{Z}[h, k])$ to itself than for $\text{Spec}(\mathbf{Z}[x])$.

REFERENCES

[AM1] M.F. Atiyah, I.G. Macdonald, *Introduction to Commutative Algebra*, Addison/Wesley Publishing Company (1969).

[HW] W. Heinzer and S. Wiegand, Prime ideals in two-dimensional polynomial rings, *Proc. Amer. Math. Soc., 107*, 572–586 (1989).

[HLW1] W. Heinzer, D. Lantz, S. Wiegand, Projective lines over one-dimensional semilocal domains and spectra of birational extensions, *Algebra Geometry and Applications* (C. Bajaj, ed.), Springer Verlag, New York 1994.

[HLW2] W. Heinzer, D. Lantz, S. Wiegand, Prime ideals in birational extensions of polynomial rings, *Contemporary Mathematics 159* (1994).

[HLW3] W. Heinzer, D. Lantz, S. Wiegand, Prime ideals in birational extensions of polynomial rings II, *Zero-Dimensional Commutative Rings* (D. Anderson and D. Dobbs, eds.), Marcel Dekker, New York (1995).

[HM] R. Heitmann and S. McAdam, Comaximizable primes, *Proc. Amer. Math. Soc 107*, 661-669 (1991).

[K] I. Kaplansky, *Commutative Rings*, Boston, Allyn and Bacon (1970).

[L] Serge Lang, *Diophantine Geometry*, Interscience, New York (1959).

[LW] A. Li and S. Wiegand, Prime ideals in birational extensions of polynomials over the integers, preprint.

[M] S. McAdam, Strongly comaximizable primes, *Journal of Algebra 170*, 206-228, (1994).

[MS] S. McAdam and C. Shah, Lifting chains of primes to integral extensions, preprint.

[rW1] R. Wiegand, Homomorphisms of affine surfaces over a finite field, *J. London Math. Soc. (2), 18*, 28-32, (1978).

[rW2] R. Wiegand, The prime spectrum of a two-dimensional affine domain *J. Pure. & Appl. Algebra 40*, 209-214, (1986).

Characterizing When $R(X)$ is Completely Integrally Closed

THOMAS G. LUCAS   University of North Carolina at Charlotte, Charlotte, North Carolina

INTRODUCTION

In what follows, all rings will be commutative with nonzero unit. As in [H], we let $T(R)$ denote the total quotient ring of the ring R and let $Q(R)$ denote the complete ring of quotients of R. For a polynomial $a(\mathrm{x}) \in Q(R)[\mathrm{x}]$, we denote the ($R$-)content of $a(\mathrm{x})$ by $C(a)$; i.e., $C(a)$ is the R-submodule of $Q(R)$ generated by the coefficients of $a(\mathrm{x})$. We let $\mathcal{U} = \{u(\mathrm{x}) \in R[\mathrm{x}] | C(u) = R\}$ denote the set of unit content polynomials in $R[\mathrm{x}]$.

Recall that for a pair of rings $R \subseteq S$, an element $s \in S$ is said to be <u>almost integral over</u> R if there is a finitely generated R-module $A \subseteq S$ which contains each positive power of s [G, page 132]. This definition does not rule out the possibility that if T is a ring containing S, then an element s of S might be almost integral over R if considered as an element of T but not if considered as an element of S. (For an example of such behavior see [G, Exercise 2, page 144].) For an element t in the total quotient ring of R, t is almost integral over R if and only if there is a regular element $r \in R$ such that $t^n r \in R$ for each positive integer n [G, Theorem 13.1]. Our main purpose is to characterize when $R(\mathrm{x}) = R[\mathrm{x}]_\mathcal{U}$ is completely integrally closed in $T(R[\mathrm{x}])$. We do this for reduced rings in Corollary 10, and for rings with nonzero nilpotents in Corollary 15.

For our purposes it will be convenient to extend the notion of "almost integral" to trios of rings. For a trio of rings $R_1 \subseteq R_2 \subseteq R_3$, we say that an element $s \in R_2$ is <u>almost integral over</u> R_1 <u>as an element of</u> R_3, if there is a finitely generated R_1-submodule A of R_3 which contains each positive power of s. As is the case for a pair of rings $R \subseteq S$, the set of all elements of R_2 which are almost integral over R_1 as elements of R_3 forms a subring of R_2. If no element of $R_2 \setminus R_1$ is almost integral over R_1 as an element of R_3, we say that R_1 is <u>completely</u> <u>integrally</u> <u>closed</u> <u>in</u> R_2 <u>as</u> <u>a</u> <u>subring</u> <u>of</u> R_3.

A number of the proofs which we give make use of the "content formula" [G, Theorem 28.1]. Specifically, we use the following: for a pair of polynomials $f(X), g(X) \in Q(R)$, there is an integer $k \geq 0$ such that $C(f)^k C(fg) = C(f)^{k+1} C(g)$.

We let N denote the nilradical of R, NT denote the nilradical of $T(R[X])$ and $\overline{R}$ denote the ring R/N.

1 WHEN R(X) IS COMPLETELY INTEGRALLY CLOSED

For a ring R with (nonzero) zero divisors, questions involving integrality properties often have answers which deal more with the relationship between the ring and its, so called, *ring of finite fractions* $Q_0(R)$ rather than with that between the ring and its total quotient ring. For example, if R is a reduced ring, then

(i) $R[X]$ is integrally closed if and only if R is integrally closed in $Q_0(R)$ [Lu1, Corollary 4].

(ii) $R[X]$ is completely integrally closed if and only if R is completely integrally closed in $Q_0(R)$ as a subring of $T(R[X])$ [Lu2, Theorem 1.6].

Moreover, it may be that $R[X]$ is completely integrally closed while $T(R)[X]$ is not [Lu2, Example 2.4]. With that said, we begin with the formal development of the ring of finite fractions. The first step in this development is the following lemma.

Lemma 1. (cf. [Lu1, Lemma 1])*Let* $J = (a_0, a_1, \ldots, a_n)$ *be a finitely generated ideal of* R *with no nonzero annihilators and let* $a(X) = a_n X^n + \cdots + a_1 X + a_0$. *If* f *is an* R-*module homomorphism from* J *into* R, *then* f *is induced by multiplication by the element* $(b_n X^n + \cdots + b_1 X + b_0)/a(X)$ *of* $T(R[X])$ *where* $b_i = f(a_i)$ *for each* i. *Conversely, if* $b(X) = b_n X^n + \cdots + b_1 X + b_0 \in R[X]$ *is such that* $a_i b_j = a_j b_i$ *for each* i *and* j, *then multiplication by* $b(X)/a(X)$ *defines an* R-*module homomorphism from* J *into* R.

Proof. The proof is quite elementary, for in either case we have $a_j[b(X)/a(X)] = b_j[a(X)/a(X)] = b_j$ for each j. □

Recall that an ideal I of R is said to be *semi-regular* if it contains a finitely generated ideal B with $Ann(B) = (0)$ (see, for example, [AAM]). We let $\mathcal{F}$ denote the set of semi-regular ideals of R. It is easy to see that $\mathcal{F}$ is closed with respect to finite products and (therefore) finite intersections. Moreover, for two semi-regular ideals J_1 and J_2, if $f_1 \in Hom(J_1, R)$ and $f_2 \in Hom(J_2, R)$, then both the sum $f_1 + f_2$ and the product $f_1 f_2$ can be considered as R-module homomorphisms from $J_1 J_2$ to R. Let J be a semi-regular ideal of R and let $A = (a_0, a_1, \ldots, a_n)$ be a finitely generated semi-regular ideal contained in J. For $f \in Hom(J, R)$, let $b_i = f(a_i)$ and let $a(X) = \sum a_i X^i$ and $b(X) = \sum b_i X^i$. Set $g = b(X)/a(X) \in T(R[X])$. Then for each $r \in J$, $rb_i = rf(a_i) = a_i f(r)$. Thus $gr = rb(X)/a(X) = f(r)a(X)/a(X) = f(r)$; i.e., as in Lemma 1, the homomorphism f is induced by multiplication by the element g of $T(R[X])$. [Note that from this point on we will write "fr" instead of "$f(r)$".]

As in the construction of the complete ring of quotients in [La], we define an equivalence relation on the set $\{f \in Hom(J, R)|J \in \mathcal{F}\}$ by setting two homomorphisms $f_1 \in Hom(J_1, R)$ and $f_2 \in Hom(J_2, R)$ equivalent if they agree on some semi-regular ideal of R. From [La, Lemma 1, p. 38] we see that f_1 and f_2 agree on some semi-regular ideal if and only if they agree on $J_1 \cap J_2$. Formally, the ring of finite fractions, $Q_0(R)$, is the set of equivalence classes of the set $\{f \in Hom(J, R)|J \in \mathcal{F}\}$. More practically we can simply view $Q_0(R)$ as the subring of $T(R[X])$ consisting of those elements $g = \sum b_i X^i / a_i X^i \in T(R[X])$ where $a_i b_j = a_j b_i$ for each i and j. [This can be done formally by first noting that R, $T(R)$ and $Q_0(R)$ all embed naturally into $Q(R)$ (see [La]). Next form the total quotient ring $T(Q(R)[X])$ and simply note that as subrings of $T(Q(R)[X])$, $Q_0(R)$ is contained in $T(R[X])$.]

Before going further we recall some useful lemmas from [Lu1] and [Lu2]. Note that in both Lemma 1.4 and Lemma 1.5 of [Lu2] the ring R is assumed to be reduced but neither proof uses this hypothesis.

Lemma 2. (cf. [Lu1, Lemma 2] and [Lu2, Lemma 1.4]) *Let R be a ring. Then*

(a) *If $s \in Q(R)$ is such that $sJ \subseteq Q_0(R)$ for some finitely generated semi-regular ideal J of R, then $s \in Q_0(R)$.*

(b) $Q(R) \cap T(R[X]) = Q_0(R)$ *and* $Q(R)[X] \cap T(R[X]) = Q_0(R)[X]$.

Lemma 3. (cf. [Lu2, Lemma 1.5]) *Let R be a ring and let $s(X) \in Q_0(R)[X]$ be almost integral over $R[X]$ as an element of $T(R[X])$. If $g(X) = g_m X^m + \cdots + g_1 X + g_0 \in R[X]$ is a polynomial of degree m such that $C(s)C(g)$ is contained in R and $s(X)^n g(X)$ is in $R[X]$ for each integer $n \geq 1$, then $s_0^n g_j g_k^j \in R$ for each $j = 0, 1, \ldots, m$ and $k = j, j+1, \ldots, m$.*

Theorem 4. *Let R be a ring and let $s(X) = s_n X^n + \cdots + s_1 X + s_0 \in Q_0(R)[X]$ be almost integral over $R[X]$ as an element of $T(R[X])$. If $g(X)$ is a regular element of $R[X]$ such that $s_i C(g) \subseteq R$ for each i, and $s(X)^k g(X) \in R[X]$ for each $k \geq 1$, then there is a semi-regular ideal J of R and a positive integer t such that $C(g^t) \subseteq J$ and $s_i J \subseteq J$ for each i.*

Proof. Let $g(X) = g_m X^m + \cdots + g_0$ satisfy the conditions above.

The proof will be by induction on the number of nonzero coefficients of $s(X)$.

For a monomial $s(X) = s_n X^n$, let $J = \bigcap J_k$ where $J_k = \{r \in R|s_n^k r \in R\}$. Then J contains $C(g)$ and is such that $s_n J \subseteq J$.

Assume the result holds for all polynomials with q or fewer nonzero coefficients and let $s(X)$ have $q + 1$ nonzero coefficients. If $s_0 = 0$, then $X^{-1}s(X)$ also satisfies the hypothesis. Hence we may assume that both s_0 and $r(X) = (s(X) - s_0)$ are nonzero.

Obviously, $s_0^k g_0 \in R$ for each k. Moreover, by Lemma 3, $s_0^k g_j^{j+1} \in R$ for each $j = 0, 1, \ldots, m$ and each $k \geq 1$. Hence, for a sufficiently large integer i, $s_0^k C(g)^i \subseteq R$ for each $k \geq 1$. Thus $s_0^k g(\mathsf{x})^i \in R[\mathsf{x}]$ for each $k \geq 1$. Let $H = \bigcap H_k$ where for each $k \geq 1$, $H_k = \{b \in R | s_0^k b \in R\}$. Then we have $C(g^i) \subseteq H$ and $s_0 H \subseteq H$.

For the polynomial $r(\mathsf{x})$, $r(\mathsf{x})^k g(\mathsf{x})^{i+1} \in R[\mathsf{x}]$ for each $k \geq 1$. Thus, by the induction hypothesis, there is a semi-regular ideal J and a positive integer p such that $C(g^p) \subseteq J$ and $s_i J \subseteq J$ for $i = 1, 2, \ldots, n$. It follows that the ideal HJ contains $C(g^t)$ for $t = i + p$ and $s_j HJ \subset HJ$ for each $j = 0, 1, \ldots, n$. $\square$

If R is a reduced ring, then the complete integral closure of R in $T(R[\mathsf{x}])$ is the ring $R^! = \bigcup\{Hom(J, J) | J \in \mathcal{F}\}$ and the complete integral closure of $R[\mathsf{x}]$ is $R^![\mathsf{x}]$ [Lu2, Theorem 1.6]. Using the same definition for $R^!$ no matter whether R is reduced or not, we next show that the ring $R^!$ is the complete integral closure of R in $Q_0(R)$ as a subring of $T(R[\mathsf{x}])$.

Theorem 5. *Let R be a ring. Then*

    (a) *$R^!$ is the complete integral closure of R in $Q_0(R)$ as a subring of $T(R[\mathsf{x}])$.*

    (b) *$R^![\mathsf{x}]$ is the complete integral closure of $R[\mathsf{x}]$ in $Q_0(R)[\mathsf{x}]$ as a subring of $T(R[\mathsf{x}])$.*

    (c) *$R^![\mathsf{x}]_{\mathcal{U}}$ is the complete integral closure of $R(\mathsf{x})$ in $Q_0(R)[\mathsf{x}]_{\mathcal{U}}$ as a subring of $T(R[\mathsf{x}])$.*

Proof. (a): We will first show that each element of $R^!$ is almost integral over R. Let $g \in R^!$. Then there is a semi-regular ideal J of R such that $gJ \subseteq J$. Let A be a finitely generated semi-regular ideal which is contained in J. Then for each integer $n \geq 1$, $g^n A \subset R$. Hence for each polynomial $a(\mathsf{x})$ with content equal to A and each $n \geq 1$, there is a polynomial $b_n(\mathsf{x})$ with degree less than or equal to the degree of $a(\mathsf{x})$ such that $g^n = b_n(\mathsf{x})/a(\mathsf{x})$. It follows that each positive power of g is contained in the R-submodule of $T(R[\mathsf{x}])$ generated by the set $\{1/a(\mathsf{x}), \mathsf{x}/a(\mathsf{x}), \ldots, \mathsf{x}^m/a(\mathsf{x})\}$ where m is the degree of $a(\mathsf{x})$. Hence g is almost integral over R as an element of $T(R[\mathsf{x}])$.

Now suppose that $f \in Q_0(R)$ is almost integral over R as an element of $T(R[\mathsf{x}])$. Then f is almost integral over $R[\mathsf{x}]$ as well. Hence there is a polynomial $a(\mathsf{x}) \in R[\mathsf{x}]$ such that $C(a)$ is semi-regular and $f^n a(\mathsf{x}) \in R[\mathsf{x}]$ for each $n \geq 1$. Thus each positive power of f is in $Hom(C(a), R)$. Let $J = \bigcap J_n$ where $J_n = \{r \in R | f^n r \in R\}$. Then J contains $C(a)$ and $fJ \subset J$.

(b): Let $s(\mathsf{x}) \in Q_0(R)[\mathsf{x}]$ be almost integral over $R[\mathsf{x}]$ as an element of $T(R[\mathsf{x}])$. Then there is a regular element $r(\mathsf{x}) \in R[\mathsf{x}]$ such that $s(\mathsf{x})^k r(\mathsf{x})$ is in $R[\mathsf{x}]$ for each integer $k \geq 1$. It may or may not be the case that each coefficient of $r(\mathsf{x})$ multiplies each coefficient of $s(\mathsf{x})$ into R, but the content formula guarantees that R contains $C(s)C(r)^m$ for some integer m. It follows that there is a semi-regular ideal J and an integer t such that $C(r^t) \subseteq J$ and $s_i J \subseteq J$ for each i. Hence $s(\mathsf{x})$ is in $R^![\mathsf{x}]$.

Thus $R^I[X]$ is the complete integral closure of $R[X]$ in $Q_0(R)[X]$ as a subring of $T(R[X])$.

(c): Let $f(X)/u(X) \in Q_0(R)[X]_{\mathcal{U}}$ be almost integral over $R(X)$ as an element of $T(R[X])$. As above, there is a regular element $r(X) \in R[X]$ such that $(f(X)/u(X))^n r(X)$ is in $R(X)$ for each $n \geq 1$. Thus there are elements $b_n(X) \in R[X]$ and $v_n(X) \in \mathcal{U}$ such that $f(X)^n r(X) v_n(X) = b_n(X)u(X)^n \in R[X]$. Since $C(v_n) = R$, the content formula implies that $C(f^n r) \subseteq R$ for each n. It follows that $f(X)^n r(X)$ is in $R[X]$. Thus by part (b) $R^I[X]_{\mathcal{U}}$ is the complete integral closure of $R(X)$ in $Q_0(R)[X]_{\mathcal{U}}$ as a subring of $T(R[X])$. $\square$

A number of corollaries can be derived from Theorem 5.

Corollary 6. *Let R be a ring. Then R is completely integrally closed in $Q_0(R)$ as a subring of $T(R[X])$ if and only if $R[X]$ is completely integrally closed in $Q_0(R)[X]$ as a subring of $T(R[X])$.*

Corollary 7. *Let D be a completely integrally closed domain. Then $D(X)$ is completely integrally closed.*

Proof. Let K be the quotient field of D. Then $K[X]_{\mathcal{U}}$ is completely integrally closed since $K[X]$ is a PID. It follows that $D(X)$ is completely integrally closed. $\square$

Lemma 8. *Let R be a von Neumann regular ring. Then for each regular multiplicative set S, $R[X]_S$ is completely integrally closed.*

Proof. Since R is locally a field, $R[X]_S$ is locally a PID. Hence, $R[X]_S$ is completely integrally closed. $\square$

If R is a reduced ring, then the complete integral closure of R in $T(R[X])$ is the ring $R^I = \bigcup\{Hom(A, A) | A \in \mathcal{F}\}$ and the complete integral closure of $R[X]$ is $R^I[X]$ [Lu2, Theorem 1.6]. Thus for reduced rings, by combining Lemma 8 and Lemma 2 with part (c) of Theorem 5 we can not only characterize when $R(X)$ is completely integrally closed but also completely describe the complete integral closure when it is not.

Corollary 9. *Let R be a reduced ring. Then $R^I[X]_{\mathcal{U}}$ is the complete integral closure of $R(X)$.*

Proof. Let $k(X)/u(X) \in T(R[X])$ be almost integral over $R(X)$. Since R is reduced, $Q(R)$ is von Neumann regular. Hence $Q(R)[X]_{\mathcal{U}}$ is completely integrally closed. This, together with the fact that $Q(R)[X]_{\mathcal{U}} \cap T(R[X]) = Q_0(R)[X]_{\mathcal{U}}$, implies $k(X)/u(X)$ is in $Q_0(R)[X]_{\mathcal{U}}$. It follows that $R^I[X]_{\mathcal{U}}$ is the complete integral closure of $R(X)$. $\square$

Corollary 10. *Let R be a reduced ring. Then the following are equivalent.*

(1) *R is completely integrally closed in $T(R[X])$.*
(2) *R is completely integrally closed in $Q_0(R)$ as a subring of $T(R[X])$.*

(3) $R[X]$ *is completely integrally closed.*

(4) $R(X)$ *is completely integrally closed.*

(5) $R = R^l$.

Things are more complicated for rings with nonzero nilpotents. The most obvious complication is that to be completely integrally closed, $R(X)$ must contain NT, the nilradical of $T(R[X])$. Necessary and sufficient conditions for $R(X)$ to contain NT have been established in [Lu4, Theorem 8 and Corollary 9]. Theorem 11 below essentially combines Theorem 8 and Corollary 9 of [Lu4]. For completeness we will provide a fairly detailed sketch of a proof. Two definitions are in order before we present this theorem. First, a semi-regular ideal I of R is said to be Q_0-invertible if $I Hom(I, R) = R$. This condition is equivalent to $IR(X)$ being an invertible ideal of $R(X)$ and to I being a finitely generated semi-regular ideal which is locally principal (see [A] and [Lu3, Lemma 6]). Second, the Q_0-integral closure of a ring R is simply the integral closure of R in $Q_0(R)$.

Theorem 11. (cf. [Lu4, Theorem 8 and Corollary 9]) *Let R be a ring. If R is integrally closed in $Q_0(R)$, then the following are equivalent:*

(1) $NT \subset R(X)$.

(2) *For each finitely generated semi-regular ideal A and each nonzero nilpotent $n \in N$, there is a finitely generated ideal B of R and an integer $k \geq 0$ such that $n \in AB$ and $nA^k = BA^{k+1}$.*

(3) *For each finitely generated semi-regular ideal A and each nonzero nilpotent $n \in N$, there are finitely generated ideals B and C of R such that $n \in AB$, C is semi-regular and $nC = BCA$.*

(4) *For each finitely generated semi-regular ideal A and each nonzero nilpotent $n \in N$, $N \subset A$ and $A/Ann(n)$ generates a Q_0-invertible ideal of the Q_0-integral closure of $R/Ann(n)$.*

Proof. Obviously (2) implies (3). That (1) implies (2) is a consequence of the content formula.

For the remainder of the proof we let $A = (a_0, a_1, \ldots, a_m)$ be a semi-regular ideal of R and let n be a nilpotent element of R. We also use $\widehat{R}$ to denote the ring $R/Ann(n)$ and $\widehat{R}'$ to denote the integral closure of $\widehat{R}$ in $Q_0(\widehat{R})$.

(3)$\Rightarrow$(4): Let B and $C = (c_1, c_2, \ldots, c_r)$ be finitely generated ideals of R with C semi-regular, $n \in AB$ and $nC = BCA$. Let $\{d_0, d_1, \ldots, d_p\}$ generate AC. Then for each $b \in B$ and each d_i there is an $f_i \in C$ such that $nf_i = bd_i$. Hence for each i and j, $nf_id_j = bd_id_j = nf_jd_i$. Let $f(X) = \sum f_iX^i$ and $d(X) = \sum d_iX^i$. Since both A and C are semi-regular, $d(X)$ is regular. Hence $nf(X)/d(X) = b$. While it need not be the case that $f_id_j = f_jd_i$, the difference is an annihilator of n. Hence modulo $Ann(n)$, multiplication by $f(X)/d(X)$ does define a homomorphism on $AC\widehat{R}$. Denote this homomorphism by g.

Since $nC = BCA$ and $nT(R[X])$ is a faithful $\widehat{R}$-module, a Cramer's Rule argument leads to the conclusion that each element of $gA\widehat{R}$ is integral over $\widehat{R}$.

Write $n = b_0 a_0 + \cdots + b_m a_m$ where $b_0, b_1, \ldots, b_m \in B$, and, for each i, let $g_i \in Q_0(\widehat{R})$ correspond to the element $e_i(X)/d(X) \in T(R[X])$ where $ne_i(X)/d(X) = b_i$. Thus $n = n\sum a_i e_i(X)/d(X)$. Modulo the annihilator of n, this equation yields $1 = g_0\widehat{a}_0 + g_1\widehat{a}_1 + \cdots + g_m\widehat{a}_m$. That $A\widehat{R}'$ is Q_0-invertible follows from the fact that each $g_i\widehat{a}_j$ is integral over $\widehat{R}$.

$(4)\Rightarrow(1)$: Let $g_0, g_1, \ldots, g_m \in Q_0(\widehat{R})$ be such that $\sum g_j\widehat{a}_{m-j} = 1$ and $g_j\widehat{a}_i \in \widehat{R}'$ for each i and j. Let $g(X) = \sum g_i X^i$. Then $g(X)\widehat{a}(X) \in U(\widehat{R}')$. Since $U(\widehat{R}')$ is the saturation of $U(\widehat{R})$ in $\widehat{R}'$ [GH, Theorem 3], there are polynomials $u(X) \in U(R)$ and $v(X) \in U(\widehat{R}')$ such that $g(X)\widehat{a}(X)v(X) = \widehat{u}(X)$. Thus $1/\widehat{a}(X) = g(X)v(X)/\widehat{u}(X)$ in $T(\widehat{R}[X])$. Let $h(X) = g(X)v(X)$ and let H denote the content of $h(X) = \sum h_i X^i$ as an $\widehat{R}$-submodule of $Q_0(\widehat{R})$. By the content formula there is an integer $k \geq 0$ such that $\widehat{A}^k = \widehat{A}^{k+1}H$. Thus $H \subseteq Hom(\widehat{A}^{k+1}, \widehat{A}^k)$.

Let $\{d_0, d_1, \ldots, d_p\}$ be a generating set for A^{k+1}. Then for each i and j, there is an element $e_{ij} \in A^k$ such that $h_i\widehat{d}_j = \widehat{e}_{ij}$. Let $d(X) = \sum d_j X^j$ and $e_i(X) = \sum e_{ij} X^j$. Since $\widehat{d}_j\widehat{e}_{ir} = \widehat{d}_r\widehat{e}_{ij}$, $h_i = \widehat{e}_i(X)/\widehat{d}(X)$ and $nd_j e_{ir} = nd_r e_{ij}$. In $\widehat{R}$, we have $1/\widehat{a}(X) = \sum\widehat{e}_i(X)X^i/\widehat{d}(X)\widehat{u}(X)$. Since n is in A and each e_{ij} is in A^k, $s_i = ne_i(X)/d(X)$ is an R-module homomorphism from A^{k+1} into A^{k+1}. As R is integrally closed in $Q_0(R)$, s_i is in R. Now setting $s(X) = \sum s_i X^i$ gives a polynomial in $R[X]$ with the property that $n/a(X) = s(X)/u(X) \in R(X)$. $\square$

A larger ring of finite fractions comes into play when R has nonzero nilpotents; namely, the ring $Q_0(R + NT)$—the ring of finite fractions over $R + NT$. Note that if $J = (a_1 + b_1, a_2 + b_2, \ldots, a_m + b_m)$ is a semi-regular ideal of $R + NT$ with each a_i in R and each b_i in NT, then the ideal $A = (a_1, a_2, \ldots, a_m)$ is a semi-regular ideal of R and there is an integer $k \geq 0$ such that $a_i^k \in J$ for each i. It follows that each $f \in Hom(J, R + NT)$ can be identified with multiplication by a quotient $\sum s_i X^i / \sum a_i^k X^i \in T(R[X])$ where $s_i = fa_i^k \in R + NT$. Thus, as with $Q_0(R)$, $Q_0(R + NT)$ embeds naturally in $T(R[X])$.

From Corollary 10 it is easy to see that if R is a reduced ring, then $Q_0(R)[X]_\mathcal{U}$ is always completely integrally closed. To characterize when $R(X)$ is completely integrally closed for a nonreduced ring R, it will help to know when $Q_0(R)[X]_\mathcal{U}$ is completely integrally closed. In the next two theorems we assume $R = Q_0(R)$. The first of these two results is from [Lu4] and concerns when $R(X)$ is integrally closed.

Theorem 12. (cf. [Lu4, Theorem 13]) *Let R be a ring. If $R = Q_0(R)$, then the following are equivalent:*

(1) $R(X)$ *is integrally closed.*

(2) $NT \subset R(X)$.

(3) $R + NT = Q_0(R + NT)$ *and* $NT \subset R(X)$.

Recall from above that $\overline{R}$ denotes the (reduced) ring R/N.

Theorem 13. *Let $R = Q_0(R)$ be its own ring of finite fractions and let S be a subset of $\mathcal{U}$. Then $R[\mathrm{x}]_S$ is completely integrally closed if and only if $NT \subset R[\mathrm{x}]_S$.*

Proof. Obviously, if $R[\mathrm{x}]_S$ is completely integrally closed, then $NT \subset R[\mathrm{x}]_S$. Conversely, if $NT \subset R[\mathrm{x}]_S$, then $NT \subset R(\mathrm{x})$ since $S \subseteq \mathcal{U}$. Hence $R + NT = Q_0(R + NT)$ [Lu4, Theorem 13].

Assume $NT \subset R[\mathrm{x}]_S$ and let $f(\mathrm{x}) \in T(R[\mathrm{x}])$ be almost integral over $R[\mathrm{x}]_S$. Then there is a polynomial $a(\mathrm{x}) \in R[\mathrm{x}]$ with semi-regular content such that $f(\mathrm{x})^n a(\mathrm{x}) \in R[\mathrm{x}]_S$ for each integer $n \geq 0$. We may assume $f(\mathrm{x})a(\mathrm{x}) = b(\mathrm{x}) \in R[\mathrm{x}]$. We also have that $\overline{f}(\mathrm{x})$ is almost integral over $\overline{R}[\mathrm{x}]_S$. Since $\overline{R}$ is reduced, $Q_0(\overline{R})[\mathrm{x}]_S$ is completely integrally closed. It follows that there are polynomials $k(\mathrm{x}) \in Q_0(\overline{R})[\mathrm{x}]$ and $u(\mathrm{x}) \in S$ such that $\overline{f}(\mathrm{x}) = k(\mathrm{x})/\overline{u}(\mathrm{x})$ and the coefficients of $k(\mathrm{x})$ are almost integral over $\overline{R}$. As in the proof of Theorem 5, $k(\mathrm{x})^n a(\mathrm{x})$ is in $\overline{R}[\mathrm{x}]$ for each integer n and there is an integer t such that $C(\overline{a}^t)C(k) \subseteq \overline{R}$. Thus by Theorem 4 there is a semi-regular ideal $\overline{J}$ of $\overline{R}$ which contains $C(\overline{a}^s)$ for some s and is such that $k_i \overline{J} \subseteq \overline{J}$ for each i. It does no harm to assume $s = 1$ and we shall do so.

Write $k(\mathrm{x}) = k_m \mathrm{x}^m + \cdots + k_1 \mathrm{x} + k_0$. For each i, let $g_i(\mathrm{x}) \in R[\mathrm{x}]$ be such that $k_i = \overline{g}_i(\mathrm{x})/\overline{a}(\mathrm{x})$. Then $g_{ir}a_s \equiv g_{is}a_r \bmod N$ for each r and s. Let $d_{irs} = a_r g_{is} - a_s g_{ir} (\in N)$ and let $d_{ir}(\mathrm{x}) = d_{irt}\mathrm{x}^t + \cdots + d_{ir1}\mathrm{x} + d_{ir0} \in NR[\mathrm{x}]$. Then, for each i and r, $a_r g_i(\mathrm{x})/a(\mathrm{x}) = g_{ir} + d_{ir}(\mathrm{x})/a(\mathrm{x}) \in R + NT$. Thus $g_i(\mathrm{x})/a(\mathrm{x}) \in Q_0(R + NT) = R + NT$. Set $g(\mathrm{x}) = \sum g_i(\mathrm{x})\mathrm{x}^i$. Then $g(\mathrm{x})/a(\mathrm{x})$ is in $R[\mathrm{x}]_S$ since $NT \subset R[\mathrm{x}]_S$. Also, $k(\mathrm{x})\overline{a}(\mathrm{x}) = \overline{g}(\mathrm{x})$ and, therefore, $\overline{f}(\mathrm{x})\overline{u}(\mathrm{x}) = \overline{g}(\mathrm{x})/\overline{a}(\mathrm{x})$. It follows that $f(\mathrm{x})u(\mathrm{x}) - g(\mathrm{x})/a(\mathrm{x}) \in NT \subset R[\mathrm{x}]_S$. Hence, $f(\mathrm{x}) \in R[\mathrm{x}]_S$ and $R[\mathrm{x}]_S$ is completely integrally closed. $\square$

Theorem 14. *Let R be a ring and let S be a subset of $\mathcal{U}$. Then $R[\mathrm{x}]_S$ is completely integrally closed if and only if $NT \subset R[\mathrm{x}]_S$ and R is completely integrally closed in $Q_0(R)$ as a subring of $T(R[\mathrm{x}])$.*

Proof. Obviously, if $R[\mathrm{x}]_S$ is completely integrally closed, then NT is contained in $R[\mathrm{x}]_S$. That R is completely integrally closed in $Q_0(R)$ as a subring of $T(R[\mathrm{x}])$ follows from the fact that $R(\mathrm{x}) \cap Q_0(R) = R$.

For the converse, assume $R[\mathrm{x}]_S$ contains NT and that R is completely integrally closed in $Q_0(R)$ as a subring of $T(R[\mathrm{x}])$. Thus $R[\mathrm{x}]$ is completely integrally closed in $Q_0(R)[\mathrm{x}]$ as a subring of $T(R[\mathrm{x}])$ by Corollary 10. Since S is a subset of $\mathcal{U}$, $Q_0(R)[\mathrm{x}]_S$ is completely integrally closed. Hence, if $f(\mathrm{x}) \in T(R[\mathrm{x}])$ is almost integral over $R[\mathrm{x}]_S$, then there are polynomials $k(\mathrm{x}) \in Q_0(R)[\mathrm{x}]$ and $u(\mathrm{x}) \in S$ such that $f(\mathrm{x}) = k(\mathrm{x})/u(\mathrm{x})$. From the proof of Theorem 5(c), we have that $k(\mathrm{x})$ is almost integral over $R[\mathrm{x}]$. Thus $f(\mathrm{x})$ is in $R[\mathrm{x}]_S$. $\square$

Corollary 15. *Let R be a ring. Then the following are equivalent.*

(1) $R(\mathrm{x})$ is completely integrally closed.

(2) $NT \subset R(X)$ and R is completely integrally closed in $Q_0(R)$ as a subring of $T(R[X])$.

(3) R is completely integrally closed in $Q_0(R)$ as a subring of $T(R[X])$ and for each finitely generated semi-regular ideal A of R and each nilpotent $n \in N$, $N \subset A$ and there is a finitely generated ideal B of R and an integer $k \geq 0$ such that $nA^k = BA^{k+1}$.

2 THE COMPLETE INTEGRAL CLOSURE OF R(X)

Let S be a finitely generated R-submodule of $T(R[X])$. Then it is easy to see that there is a polynomial $g(X) \in R[X]$ with semi-regular content and an integer $m \geq 0$ such that S is contained in the R-module $(1/g(X))R + (X/g(X))R + \cdots + (X^m/g(X))R$. It follows that an element $f \in T(R[X])$ is almost integral over R (as an element of $T(R[X])$) if and only if there is a polynomial $a(X) \in R[X]$ with semi-regular content and an integer $m \geq 0$ such that for each integer $n \geq 1$, $f^n a(X)$ is a polynomial over R of degree less than or equal to m. If we let $\widetilde{R}$ denote the complete integral closure of R in $T(R[X])$, then it is easy to show that each element of $\widetilde{R}[X]_{\mathcal{U}}$ is almost integral over $R(X)$. In Example 20, we show that $\widetilde{R}$ can be equal to $Q_0(R + NT)$ even if $Q_0(R + NT)$ properly contains $Q_0(R) + NT$. By way of Theorem 17, we may conclude that in this case $\widetilde{R}[X]_{\mathcal{U}}$ is the complete integral closure of $R(X)$. We leave as an open question whether it is always the case that $\widetilde{R}[X]_{\mathcal{U}}$ is the complete integral closure of $R(X)$.

The following lemma from [Lu4] will be useful. In the lemma, $\overline{Q}_0$ denotes the ring $Q_0(R + NT)/NT$.

Lemma 16. (cf. [Lu4, Lemma 11]) *Let R be a ring. Then:*

(a) *For each $f \in Q_0(R + NT)$ there is a finitely generated semi-regular ideal A of R such that $fA \subseteq R + NT$.*

(b) $\overline{Q}_0 = Q_0(\overline{R}) \cap T(R[X])/NT$.

(c) $\overline{Q}_0[X] = Q_0(\overline{R})[X] \cap T(R[X])/NT$.

(d) $Q_0(R + NT) \cap R(X) \subseteq R + NT$.

Our next result concerns the complete integral closures of $R + NT$, $(R + NT)[X]$ and $(R + NT)[X]_{\mathcal{U}}$ in $T(R[X])$. Specifically, we show that these closures are, respectively, the rings $R^\diamond = \bigcup\{Hom(A + NT, A + NT) | A \in \mathcal{F}\}$, $R^\diamond[X]$ and $R^\diamond[X]_{\mathcal{U}}$. It follows that $\widetilde{R}$ is a subring of $Q_0(R + NT)$.

Theorem 17. *Let R be a ring. Then*

(a) $R^\diamond$ *is the complete integral closure of $R + NT$ in $T(R[X])$.*

(b) $R^\diamond[X]$ *is the complete integral closure of $(R + NT)[X]$.*

(c) $R^\diamond[X]_{\mathcal{U}}$ *is the complete integral closure of $(R + NT)[X]_{\mathcal{U}}$.*

Proof. (a): First, let $f \in Hom(A + NT, A + NT)$ for some $A \in \mathcal{F}$ and let $a(X)$ be a polynomial in $R[X]$ with semi-regular content contained in A. Then for each coefficient a_j and each positive integer n, $f^n a_j$ is in $A + NT$. It follows that each positive power of f is in $(1/a(\mathbf{x}))(R+NT)+(\mathbf{x}/a(\mathbf{x}))(R+NT)+\ldots(\mathbf{x}^m)/a(\mathbf{x}))(R+NT)$ where m is the degree of $a(\mathbf{x})$. Thus each element of $R^\diamond$ is almost integral over $R + NT$.

Now let $g \in T(R[\mathbf{x}])$ be almost integral over $R + NT$ and let S be a finitely generated $R + NT$-submodule of $T(R[\mathbf{x}])$ which contains each positive power of g. Let $\{b_1(\mathbf{x})/a(\mathbf{x}), b_2(\mathbf{x})/a(\mathbf{x}), \ldots, b_k(\mathbf{x})/a(\mathbf{x})\}$ be a generating set for S with $a(\mathbf{x}), b_1(\mathbf{x}), b_2(\mathbf{x}), \ldots, b_k(\mathbf{x}) \in R[\mathbf{x}]$ (and $a(\mathbf{x})$ regular) and let m be the maximum degree of the b_is. Then S, and hence each positive power of g, is contained in the $R+NT$-module $(1/a(\mathbf{x}))(R+NT)+(\mathbf{x}/a(\mathbf{x}))(R+NT)+\cdots+(\mathbf{x}^m/a(\mathbf{x}))(R+NT)$. Moding out by NT, we see that $\overline{g}$ is almost integral over $\overline{R}$. It follows that $\overline{g}$ is in $Q_0(\overline{R}) \cap T(R[\mathbf{x}])/NT$ and $\overline{g}^n\overline{a}(\mathbf{x}) \in \overline{R}[\mathbf{x}]$ for each $n \geq 1$. Hence g is in $Q_0(R+NT)$ and $\overline{g}^n\overline{C}(a) \subseteq \overline{R}$ for each $n \geq 1$. It follows that $g^nC(a) \subset R + NT$ for each n. As in the proof of Theorem 4, there is a semi-regular ideal A of R which contains $C(a)$ and is such that $g \in Hom(A + NT, A + NT)$.

(b): Let $g \in T(R(\mathbf{x}))$ be almost integral over $(R + NT)[\mathbf{x}]$. Modulo NT, g can be written as a polynomial over $\overline{Q}_0$ with each coefficient almost integral over $\overline{R}$ (as an element of $T(\overline{R}[\mathbf{x}])$). Let $\overline{g} = \overline{g}_n\mathbf{x}^n + \cdots + \overline{g}_0$. Since $\overline{Q}_0 = Q_0(R+NT)/NT$ and NT is common ideal of $R + NT$ and $T(R[\mathbf{x}])$, for each i there is an element g_i of $Q_0(R+NT)$ which is almost integral over $R + NT$ and is such that $\overline{g}_i = g_i + NT$. It follows that $R^\diamond[\mathbf{x}]$ is the complete integral closure of $(R + NT)[\mathbf{x}]$.

(c): From Lemma 16 we have that $\overline{Q}_0[\mathbf{x}]_{\mathcal{U}} = Q_0(\overline{R})[\mathbf{x}]_{\mathcal{U}} \cap T(R[\mathbf{x}])/NT$. That $R^\diamond[\mathbf{x}]_{\mathcal{U}}$ is the complete integral closure of $(R + NT)[\mathbf{x}]_{\mathcal{U}}$ follows from this result and part (b). $\square$

Theorem 18. *If $Q_0(R) + NT = Q_0(R + NT)$, then $R^!(\mathbf{x})]_{\mathcal{U}} + NT$ is the complete integral closure of $R(\mathbf{x})$. Moreover, if, in addition, $N = NT \cap Q_0(R)$, then $R^\diamond = R^! + NT$ and $R^\diamond[\mathbf{x}]_{\mathcal{U}}$ is the complete integral closure of $R(\mathbf{x})$.*

Proof. Assume $Q_0(R) + NT = Q_0(R + NT)$ and let $f(\mathbf{x}) \in T(R[\mathbf{x}])$ be almost integral over $R(\mathbf{x})$. By Theorem 17, $Q_0(R)[\mathbf{x}]_{\mathcal{U}} + NT = Q_0(R + NT)[\mathbf{x}]_{\mathcal{U}}$ is the complete integral closure of $Q_0(R)[\mathbf{x}]_{\mathcal{U}}$. Thus there are polynomials $g(\mathbf{x}) \in Q_0(R)$ and $u(\mathbf{x}) \in \mathcal{U}$ and an element $m(\mathbf{x}) \in NT$ such that $f(\mathbf{x}) = (g(\mathbf{x})/u(\mathbf{x})) + m(\mathbf{x})$. Since each element of NT is integral over $R(\mathbf{x})$, $g(\mathbf{x})/u(\mathbf{x})$ must be almost integral over $R(\mathbf{x})$. By Theorem 5, we have $g(\mathbf{x})/u(\mathbf{x}) \in R^![\mathbf{x}]_{\mathcal{U}}$. Hence $R^![\mathbf{x}]_{\mathcal{U}} + NT$ is the complete integral closure of $R(\mathbf{x})$.

Obviously, $R^\diamond$ contains $R^! + NT$ even if $Q_0(R) + NT$ is not equal to $Q_0(R+NT)$. Let $f \in R^\diamond$ and assume $Q_0(R) + NT = Q_0(R + NT)$ and $N = NT \cap Q_0(R)$. Then there is a semi-regular ideal A of R such that $fA \subset A + NT$. It does no harm to assume that A contains N, and we shall do so. Under this assumption we have that $(A + NT) \cap Q_0(R) = A$. Since $Q_0(R) + NT = Q_0(R + NT)$, $f = g + m$ for some $g \in Q_0(R)$ and some $m \in NT$. To complete the proof, it suffices to show that

g is in $R^!$. But since N is the nilradical of both R and $Q_0(R)$, $gN \subseteq N$. Hence $gA \subseteq (A + NT) \cap Q_0(R) = A$ and, therefore, g is in $R^!$. $\square$

As we will see in Example 20, it is not always the case that $Q_0(R) + NT = Q_0(R + NT)$. To construct the ring in Example 20 we use a variation on the technique used to construct rings of the form $A + B$ (see [H, section 26]). The basic idea is to start with a ring D and then to let $\mathcal{P}$ be a nonempty subset of $Spec\, D$. Next let $\mathcal{A} = \{i = (\alpha, n)|\, P_\alpha \in \mathcal{P}$ and $n \geq 1\}$, and then, for each $i = (\alpha, n) \in \mathcal{A}$, let K_i denote the quotient field of D/P_α. Now let I be an ideal of D which is contained in the intersection of the P_αs. Then $B = \sum K_i$ is a D/I-module and we can form a ring $R = (D/I) + B$ from the direct sum of D/I and B by defining multiplication as $(r, a)(s, b) = (rs, rb + sa + ab)$. Given a ring D, a nonempty subset $\mathcal{P}$ of $Spec\, D$ and an ideal I which is contained in each $P_\alpha \in \mathcal{P}$, we will say that $R = (D/I) + B$ is the $A + B$ ring corresponding to D, $\mathcal{P}$ and I. In the event that $I = \bigcap P_\alpha$, then the $A + B$ ring R is isomorphic to the ring $A' + B$ where A' denotes the canonical image of D in the direct product $\prod K_i$ (and $A' + B$ is simply the sum of A' and B as subrings of $\prod K_i$). In [H], it is the ring $A' + B$ that is constructed.

To simplify the proof that the ring presented in Example 20 has the various properties attributed to it, we give the following theorem which establishes some of the basic properties of $A + B$ rings.

Theorem 19. *Let D be a ring, $\mathcal{P} = \{P_\alpha\}$ be a nonempty subset of $Spec\, D$ and I be an ideal of D which is contained in $\bigcap P_\alpha$. Let $R = (D/I) + B$ be the $A + B$ ring corresponding to D, $\mathcal{P}$ and I. Then*

(1) *A finitely generated ideal $J = ((r_1, a_1), (r_2, a_2), \ldots, (r_m, a_m))$ is semi-regular if and only if no P_α/I contains the ideal $J' = (r_1, r_2, \ldots, r_m)D/I$, $J = J'R$ and J' is a semi-regular ideal of D/I.*

(2) *$T(R)$ can be identified with the ring $(D/I)_S + B$ where $S = D/I \setminus (Z(D/I) \cup (\bigcup P_\alpha/I))$.*

(3) *$Q_0(R)$ can be identified with the ring $E + B$ where $E = \bigcup \{Hom(J', D/I)|\, J'$ is an ideal of D/I for which $J'R$ is semi-regular $\}$.*

Proof. For an element (r, a) of R, let $(r)_i$ denote the image of r in K_i and let $(a)_i$ denote the ith component of a. For each $i \in \mathcal{A}$ we let e_i denote the element of B for which $(e_i)_i = 1$ and all other components are 0.

For each $i = (\alpha, n)$, K_i is naturally isomorphic to the quotient field $\overline{K}_i$ of $(D/I)/(P_\alpha/I)$. Thus B is naturally isomorphic to $\overline{B} = \sum \overline{K}_i$. It follows that $(D/I) + B$ is naturally isomorphic to $(D/I) + \overline{B}$. Hence in establishing this result we may assume $I = (0)$.

(1): Let $J = ((r_1, a_1), (r_2, a_2), \ldots, (r_m, a_m))$ be a finitely generated ideal of R and let $J' = (r_1, r_2, \ldots, r_m)D$. If some P_α contains J', then there are infinitely many $i = (\alpha, n)$ such that $(a_j)_i = 0$ for each j. For each such i, $(0, e_i)$ is a nonzero annihilator of J. Hence we may assume no P_α contains J' and, therefore, the annihilator of J' is contained in each P_α. It follows that if s is a nonzero annihilator of J' in D, then $(s,0)$ is a nonzero annihilator of J in R. Thus we may further assume that J' is a semi-regular ideal of D.

Since no P_α contains J', for each $i = (\alpha, n)$ there is a j such that r_j is not contained in P_α. It follows that $r_j K_i = K_i$. Hence $J'B = B$ and $J'R = J' + B$ is a finitely generated semi-regular ideal of R.

To complete the proof, all we need show is that J semi-regular implies $J = J'R$. Since J is semi-regular, for each i there is a j such that $(0, e_i)(r_j, a_j) \neq (0,0)$; i.e., $(r_j)_i \neq -(a_j)_i$. It follows that $B \subset J$. Thus $J = J' + B = J'R$.

(2): By (1), an element (r,a) is a regular element of R if and only if r is a regular element of D which is contained in no P_α and for each i, $(r)_i + (a)_i \neq 0$. It follows that $T(R)$ can be identified with the ring $D_S + B$ where $S = D \setminus (Z(D) \cup (\bigcup P_\alpha))$.

(3): Let $J = J'R$ be a finitely generated semi-regular ideal of R and let $f \in Hom(J', D)$ and $b \in B$. For each j let $fr_j = s_j$ and fix an $i = (\alpha, n)$. Since J' is not contained in P_α, we may assume that some r_j, say r_1, survives in D/P_α. Since P_α is prime and $r_1 s_k = r_k s_1$ for each k, $s_1 \in P_\alpha$ implies each $s_k \in P_\alpha$. Moreover, if $r_j \in P_\alpha$, then $s_j \in P_\alpha$. Hence we can set $fe_i = (s_1/r_1)e_i$ and then extend f to a map on all of B. It follows that $(f,b)(r,a) = (fr, fa + rb + ab)$ is in R for each (r,a) in J. Thus (f,b) is an R-module homomorphism from J into R.

Now let g be an R-module homomorphism from J to R. Since $B \subset J$, $g(0, e_i) = (0, e_i)(g(0, e_i))$ for each i. It follows that $gB \subset B$. For each j, let $g(r_j, 0) = (s_j, c_j)$. Then the function f defined by $fr_j = s_j$ is a D-module homomorphism from J' to D. Hence $h = g - (f,0)$ is an R-module homomorphism which maps J into B. Let $(0, b_j) = h(r_j, 0)$. Since J' is finitely generated, there are only finitely many i for which $(b_j)_i \neq 0$ for some j. Moreover, $(b_j)_i \neq 0$ implies $(r_j)_i \neq 0$. Since $r_j b_k = r_k b_j$ for each j and k, there is an element $b \in B$ such that $(b)_i = 0$ whenever $(b_j)_i = 0$ for each j and $(b)_i = (b_j/r_j)_i$ for some nonzero $(b_j)_i$. Then h is equivalent to multiplication by $(0, b)$ and g can be identified with multiplication by (f, b). $\square$

Example 20. Let $D = K[u, v, \{uy^n, vz^n | n \geq 1\}]$, $M = (u, v, y, z)K[u, v, y, z] \cap D$ and $I = uv(y - z)^2 K[u, v, y, z]$ where K is a field with characteristic 0. Let $\mathcal{P} = \{M_\alpha \in Max\, D | I \subset M_\alpha \neq M\}$ and let $R = (D/I) + B$ be the $A + B$ ring corresponding to D, $\mathcal{P}$ and I. Then:

(1) $R = T(R)$ and the only semi-regular ideals of R are those which contain a power of $uR + vR$.

(2) For each $i \geq 1$, $f^i = (uy^i X + vz^i)/(uX + v)$ and is an element of $Q_0(R + NT)$.

(3) $Q_0(R + NT) = R[f] + NT$ and is the complete integral closure of R in $T(R[X])$. Thus $Q_0(R + NT)[X]_{\mathcal{U}}$ is the complete integral closure of $R(X)$.

(4) $R = Q_0(R)$

(5) $R + NT = Q_0(R) + NT$ is not equal to $Q_0(R + NT)$.

Proof. In both D and $K[u, v, y, z]$, the radical of I is the ideal $uv(y-z)K[u, v, y, z]$. As an ideal of D, $\sqrt{I}$ is a prime ideal; as an ideal of R, $\sqrt{I}R$ is the nilradical. Thus $\overline{D} = D/\sqrt{I}$ is an integral domain and $\overline{R} = \overline{D} + B$.

(1): For an element (r, a) of R, if r is not a unit of D/I, then some M_α contains r. Thus $R = T(R)$.

The only ideals of D which contain I and are not contained in some M_α are those which contain a power of $(u, v)D$. Hence by Theorem 19, the ideal $J = (r_1, b_1), (r_2, b_2), \ldots, (r_n, b_n))$ is a finitely generated semi-regular ideal of R if and only if $J' = (r_1, r_2, \ldots, r_n)D/I$ is an (semi-regular) ideal of D/I which contains a power of $(u, v)D/I$.

(2): Simple calculations show that $fu = uy + (uv(z-y)/(uX+v)) \in R + NT$ and $fv = vz + (uv(y-z)X/(uX+v)) \in R + NT$. Thus f is an element of $Q_0(R + NT)$.

That $f^i = (uy^iX+vz^i)/(uX+v)$ follows from induction. Assume true for i and compute f^{i+1}. By the induction hypothesis, $f^{i+1} = f(uy^iX+vz^i)/(uX+v) = (u^2y^{i+1}X^2 + (uvyz^i + uvy^iz)X+v^2z^{i+1})/(uX+v)^2$. Computing the difference between this expression for f^{i+1} and $(uy^{i+1}X+vz^{i+1})/(uX+v)$, we get $uv(yz^i + y^iz - y^{i+1} - z^{i+1})/(uX+v)^2$. Note that the numerator factors as $uv(y-z)(z^i - y^i)$ which is an element of I. Thus $f^i = (uy^iX+vz^i)/(uX+v)$ for each $i \geq 1$. Moreover, for each $n > 1$, f^i can also be written in the form $(u^ny^iX+v^nz^i)/(u^nX+v^n)$.

(3): From the above expression for f^i, we have that each positive power of f is contained in the two generated R-module $(1/(uX+v))R + (X/(uX+v))R$. Hence f is almost integral over R. By Theorem 17, all we need show is that $Q_0(R+NT) = R[f] + NT$. For this equality it suffices to show that $Q_0(\overline{R}) = \overline{R}[f]$.

Since $fu = uy + (uv(z-y)/(uX+v))$ and $fv = vz + (uv(y-z)X/(uX+v))$, multiplication by f defines an $\overline{R}$-module homomorphism from $u\overline{R}+v\overline{R}$ into $\overline{R}$ but f itself is not in $\overline{R}$. Viewed as an element of the quotient field of $\overline{D}$, $f = uy/u = vz/v$. Set $w = uy/u = vz/v$. Then $\overline{D}$ can be viewed as the domain $K[u, v, \{uw^n\}, \{vw^n\}]$. As in R, the semi-regular ideals of $\overline{R}$ must contain a power of $u\overline{R}+v\overline{R}$. Hence $Q_0(\overline{R})$ can be identified with the ring $\overline{E} + B$ where $\overline{E}$ is the ideal transform of $(u, v)\overline{D}$ (in the quotient field of $\overline{D}$). Since $K[u, v, w]$ is a Krull domain and $w^n(u, v) \subset \overline{D}$, $\overline{E} = K[u, v, w]$ and $Q_0(\overline{R}) = \overline{R}[f]$.

(4): We first consider positive powers of f. For an integer $m \geq 1$, $f^mu = uy^m + uv(z^m - y^m)/(uX+v)$ and $f^mv = vz^m + uv(y^m - z^m)X/(uX+v)$. Since K has characteristic 0, $uv(y^m - z^m) \in \sqrt{I} \setminus I$. Thus (f^m, b) is not in $Q_0(R)$.

Note that the nilradical of D/I is also the complete set of zero divisors of D/I. Hence $Q_0(D/I) = T(D/I)$ and both u and v are regular elements of D/I. Thus by Theorem 19, each element of $Q_0(R)$ can be written in the form (g, b) where $b \in B$ and $g \in T(D/I)$ is such that $g(u, v)^kD/I \subseteq D/I$ for some integer $k \geq 1$; i.e., $g \in (u, v)^kD/I)^{-1}$. To complete the proof it suffices to prove that $((u, v)D/I)^{-1} = D/I$.

Let g be in $((u,v)D/I)^{-1}$. Since $Q_0(R+NT) = R[f] + NT$, there are elements $r_0, r_1, \ldots, r_m \in D$ and $t_0, t_0' \in K[u,v,y,z]$ such that g can be written in the form $g = n + r_0 + r_1 f + r_2 f^2 + \cdots + r_m f^m$ where $n = uv(y-z)(t_0\mathrm{x}+t_0')/(u\mathrm{x}+v)$. Since the goal is show $g \in D/I$, we may assume $r_0 = 0$. Writing g in the form of a finite fraction, we have

$$g = \frac{uv(y-z)(t_0\mathrm{x}+t_0') + \sum r_i(uy^i\mathrm{x}+vz^i)}{u\mathrm{x}+v}.$$

Using this form and calculating gu and gv we get

$$gu = uvt_0(y-z) + \sum r_i uy^i + \frac{uv[u(y-z)t_0' - v(y-z)t_0 + \sum r_i(z^i - y^i)]}{u\mathrm{x}+v}$$

and

$$gv = uvt_0'(y-z) + \sum r_i vz^i + \frac{uv[v(y-z)t_0 - u(y-z)t_0' + \sum r_i(y^i - z^i)]\mathrm{x}}{u\mathrm{x}+v}.$$

Since both u and v are regular elements of D/I, we must have $uv[v(y-z)t_0 - u(y-z)t_0' + \sum r_i(y^i - z^i)]$ in I.

Factoring out $uv(y-z)$, we see that $h = vt_0 - ut_0' + \sum r_i(y^{i-1} + y^{i-2}z + \cdots + z^{i-1})$ must be in $(y-z)K[u,v,y,z]$. As elements of $K[u,v,y,z]$, $t_0 = a_0(y,z) + ub_0(u,y,z) + vc_0(v,y,z) + uvd_0(u,v,y,z)$, $t_0' = a_0'(y,z) + ub_0'(u,y,z) + vc_0'(v,y,z) + uvd_0'(u,v,y,z)$ and each r_i is of the form $r_i = a_i + ub_i(u,y) + vc_i(v,z) + uvd_i(u,v,y,z)$ where a_0, a_0', b_i, c_i and d_i are polynomials over K in the indicated indeterminates and a_i is in K for each $i \geq 1$.

Now specify both $u = 0$ and $v = 0$ in the expression for h. Since h is in $(y-z)K[u,v,y,z]$, we also have $a_1 + a_2(y+z) + \cdots + a_m(y^{m-1} + y^{m-2}z + \cdots + z^{m-1}) \in (y-z)K[u,v,y,z]$. Since each a_i is in K, each a_i must be zero.

Next specify only $v = 0$. By doing so we must have $-ua_0' - u^2b_0' + \sum ub_i(y^{i-1} + \cdots + z^{i-1}) \in (y-z)K[u,y,z]$. Similarly, if we specify only $u = 0$, we must have $va_0 + v^2c_0 + \sum vc_i(y^{i-1} + \cdots + z^{i-1}) \in (y-z)K[v,y,z]$.

Set $n_b = (uva_0' + u^2vb_0')(y-z)/(u\mathrm{x}+v)$ and $g_b = n_b + \sum ub_i f^i$. Dividing the numerator of g_b by $u\mathrm{x}+v$, we get $ub_1y + ub_2y^2 + \cdots + ub_my^m$ with remainder $uv(y-z)[a_0' + ub_0' - \sum b_i(y^{i-1} + \cdots + z^{i-1})]$ which is in I. Thus g_b is in D/I. Similarly, set $n_c = (uva_0 + uv^2c_0)(y-z)\mathrm{x}/(u\mathrm{x}+v)$ and $g_c = n_c + \sum vc_i f^i$. As with g_b, g_c is in D/I. Set $g' = g - g_b - g_c$. Since g_b and g_c are in D/I, $g' \in ((u,v)D/I)^{-1}$. The numerator of g' has the form $uv(y-z)[(ub_0 + uvd_0)\mathrm{x} + (vc_0' + uvd_0')] + uv\sum d_i(uy^i\mathrm{x}+vz^i)$. Note that the coefficient on x is a multiple of u^2v and the constant term is a multiple of uv^2. It follows that if both $g'u$ and $g'v$ are in D/I, then so is g'. Hence $g \in D/I$.

(5): Since f is not in $\overline{R}$, $\overline{R} \neq \overline{Q}_0$. Hence $R + NT = Q_0(R) + NT \neq Q_0(R + NT)$. $\square$

REFERENCES

[A] D.D. Anderson, Some remarks on the ring $R(X)$, *Comment. Math. Univ. St. Pauli* **26** (1977) 137–140.

[AAM] D.D. Anderson, D.F. Anderson and R. Markanda, The rings $R(X)$ and $R\langle X\rangle$, *J. Algebra* **95** (1985), 96–115.

[G] R. Gilmer, *Multiplicative Theory of Ideals*, Marcel Dekker, New York, 1972.

[GH] R. Gilmer and J. Hoffman, A characterization of Prüfer domains in terms of polynomials, *Pac. J. Math.* **60** (1975), 81–85.

[H] J. Huckaba, *Commutative Rings with Zero Divisors*, Marcel Dekker, New York, 1988.

[K] I. Kaplansky, *Commutative Rings, revised edition*, The University of Chicago Press, Chicago, 1974.

[La] J. Lambek, *Lectures on Rings and Modules*, Chelsea, New York, 1986.

[Lu1] T. Lucas, Characterizing when $R[X]$ is integrally closed, II, *J. Pure Appl. Algebra* **61** (1989), 49–52.

[Lu2] T. Lucas, The complete integral closure of $R[X]$, *Trans. Amer. Math. Soc.* **330** (1992), 757–768.

[Lu3] T. Lucas, Strong Prüfer rings and the ring of finite fractions, *J. Pure Appl. Algebra* **84** (1993), 59–71.

[Lu4] T. Lucas, The integral closure of $R(X)$ and $R\langle X\rangle$, *Comm. Alg.*, to appear.

On Root Closure in Noetherian Domains

MOSHE ROITMAN Department of Mathematics and Computer Science, University of Haifa, Mount Carmel, Haifa 31905, Israel
E-mail: mroitman@mathcs2.haifa.ac.il

ABSTRACT. We realize any multiplicative monoid of positive integers generated by primes as the root closure semigroup of a noetherian domain. We show that generally, the root closure of a noetherian domain is not obtainable in finitely many steps. We also present a simple proof of Brewer-Costa-McCrimmon's theorem on root closure of polynomial extensions.

0 INTRODUCTION

We consider here just commutative rings with identity and all ring homomorphisms preserve the identity. If $A \subseteq B$ are rings (with the same identity), and $n \geq 1$ is an integer we say that A is n-root closed in B if $b^n \in A$ with $b \in B$ implies $b \in A$; more generally, if S is a set of positive integers, A is S-root closed in B if A is n-root closed in B for all n in S.

The ring A is root closed in B if A is n-root closed in B for all $n \geq 1$. Throughout this paper, we will formulate definitions for pairs of rings $A \subseteq B$ as above. If A is an integral domain, and no ring B is mentioned, we intend B to be the quotient field of A.

As in [3], $\mathcal{C}(A, B)$ (the root closure semigroup of A in B) is the set of all positive integers n for which A is n-root closed in B. Thus $\mathcal{C}(A, B)$ is a submonoid of the multiplicative monoid of positive integers and it is generated by primes. In this paper all monoids will be submonoids of the multiplicative monoid of positive integers. If A is a domain with quotient field K, by our convention, we define $\mathcal{C}(A)$ to be $\mathcal{C}(A, K)$.

I thank Professors David F. Anderson and David E. Dobbs (University of Tennessee at Knoxville) for their useful suggestions and remarks.

We now recall some definitions from [6]. Let $A \subseteq B$ be rings, S a set of positive integers. The smallest subring of B which contains A and is S-root closed in B is called the *(total) S-root closure* of A in B and is denoted by $\mathcal{R}^S(A, B)$ or by $\mathcal{R}^S_\infty(A, B)$. The *root closure* of A in B is $\mathcal{R}^S(A, B)$, where S is the set of all positive integers.

For $0 \le j < \infty$ define $\mathcal{R}^S_j(A, B)$ inductively as follows: $\mathcal{R}^S_0(A, B) = A$; and $\mathcal{R}^S_j(A, B)$ is the subring of B generated by the elements $b \in B$ such that $b^m \in \mathcal{R}^S_{j-1}(A, B)$ for some $m \in S$. Thus for $0 < j < \infty$ we have

$$\mathcal{R}^S_j(A, B) = \mathcal{R}^S_1(\mathcal{R}^S_{j-1}(A, B), B).$$

By [6, Proposition 1.1] we have

$$\mathcal{R}^S_\infty(A, B) = \bigcup_{j=0}^{\infty} \mathcal{R}^S_j(A, B),$$

an ascending union of subrings. These facts allow us to reduce proofs for $0 \le j \le \infty$ to the case $j = 1$, using induction on finite j. We say that the S-root closure of A in B is *obtainable in n steps* if $\mathcal{R}^S_\infty(A, B) = \mathcal{R}^S_n(A, B)$. For $0 \le j \le \infty$ we let $\mathcal{R}_j(A, B) = \mathcal{R}^S_j(A, B)$, where S is the set of all positive integers.

As in [7], the ring A is said to be *seminormal* in B if $b \in B$ and $b^2, b^3 \in A$ implies $b \in A$. If A is n-root closed in B for some $n \ge 2$, then A is seminormal in B, but the converse is false.

By Brewer-Costa-McCrimmon's Theorem [8], in case of polynomial extensions, we always have

$$\mathcal{C}(A[\mathbf{X}], B[\mathbf{X}]) = \mathcal{C}(A, B),$$

where $\mathbf{X}$ is a set of independent indeterminates over B. We present a simple proof of this theorem in Section 1 in the more general setting of graded rings (cf. [1, 2, 3, 5]). In Section 2, using the results of Section 1, we continue the work in [6]. We show that generally, the root closure of a noetherian domain is not obtainable in finitely many steps, even if the domain is one dimensional, thus answering a question in [6]. To prove this result we construct a noetherian domain A for fields $k \subseteq K$ satisfying certain assumptions. This construction is useful since some properties of the pair $k \subseteq K$ are mimicked by the pair $A \subseteq Q$, where Q is the quotient field of A. We also use this construction to show that any (multiplicative) monoid of positive integers generated by primes can be realized as $\mathcal{C}(A)$ for some noetherian domain A, thus answering a question of David F. Anderson. This result was proved by Anderson under the additional assumption that 2 is not in the generating set of primes; in this case A can be chosen to be a seminormal affine or a one-dimensional seminormal local domain [4, Theorem 6]. The analogous result with a general domain A (no restrictions on the set of primes) was also proved by D.F. Anderson in [3, Theorem 2.7]. Moreover, he has also shown that A may be chosen to be quasilocal.

1 ROOT CLOSURE IN GRADED RINGS

By a *graded* ring we mean an N-graded ring. If $A = \bigoplus_{i=0}^{\infty} A_i$ and $B = \bigoplus_{i=0}^{\infty} B_i$ are graded rings, the notation $A \subseteq B$ means that A is a graded subring of B, thus $A_i \subseteq B_i$ for all i. The proofs in this section fit also the more general case of Γ-graded rings, where Γ is a totally ordered monoid (cf. [1, 2, 3]).

LEMMA 1.1. *Let $A \subseteq B$ be graded rings. Let*

$$f = b_0 + \cdots + b_d$$

be the homogeneous decomposition of a nonzero element $f \in B$ satisfying $f^m \in A$ for some $m \geq 1$. Then there exists an integer $N \geq 0$ such that $(mb_d)^N b_i \in A[b_d]$ for all $0 \leq i \leq d$.

Proof. If the lemma is false, there exists a minimal integer r so that $(mb_d)^k b_r \notin A[b_d]$ for all $k \geq 0$, thus $r < d$. We have

$$f^m = (b_0 + \cdots + b_r + \cdots + b_d)^m = (b_0 + \cdots + b_{r-1})^m + mb_r b_0^{m-1} + h,$$

where h is an element in B of degree $> r$. There is a $k \geq 2$ such that $(mb_0)^k(b_0 + \cdots + b_{r-1})^m \in A[b_d]$. Since $A[b_d]$ is a graded subring of B we obtain $(mb_0)^k(mb_r b_0^{m-1}) \in A[b_d]$, so $(mb_0)^{k+m-1} b_r \in A[b_d]$, a contradiction. $\square$

Let $A \subseteq B$ be rings and S be a nonempty set of positive integers. Then A is called *S-closed* in B if any element $b \in B$ satisfying $b^n \in A$ for all $n \in S$ belongs to A [3, §3]. Assume that S contains an integer > 1. Let d be the greatest common divisor of S. Clearly A is S-closed in B iff any element $b \in B$ satisfying $b^{di} \in A$ for $i >> 0$ belongs to A (cf. [3]) . This means that if $d = 1$, then A is S-closed in B iff A is seminormal in B; if $d > 1$ then A is S-closed in B iff A is d-root closed in B iff A is p-root closed in B for any prime factor p of d [3, Theorem 3.2]. This allows us to reduce the property of S-closure to the case that S consists of one prime (p-root closure for p prime) or two primes (seminormality). Note also that if A is S-closed in B and S contains an integer > 1, then A is seminormal in B.

If f is a nonzero element of a graded ring B with homogeneous decomposition $f = b_r + \cdots + b_d$, where b_r, b_d are nonzero, then the *length* $\ell(f)$ is defined as $d - r + 1$; $\ell(0) = 0$.

LEMMA 1.2 (cf. [5, Theorem 2] and [3, Theorem 2.4]). *Let $A \subseteq B$ be graded rings. Let S be a nonempty set of positive integers. Assume that if b is a homogeneous element of B such that $b^n \in A$ for all $n \in S$ then $b \in A$. Then A is S-closed in B.*

Proof. Using the preceding remarks and also specializing them to homogeneous elements we may assume that S consists of one or two primes.

First we assume that $S = \{2, 3\}$ and prove that A is seminormal in B.

For $f \in B$ we show by induction on $\ell(f)$ that if $f^2, f^3 \in A$, then $f \in A$. Let $\ell(f) > 0$, that is, $f \neq 0$. Let Let

$$f = b_0 + \cdots + b_d$$

be the homogeneous decomposition of f with $b_d \neq 0$. Assume that $f^2, f^3 \in A$, thus $b_d \in A$. Set $g = f - b_d = b_0 + \cdots + b_{d-1}$. Since $b_d \in A$, Lemma 1.1 implies that there exists an integer $N \geq 0$ such that $b_d^N g \in A$. Let N be minimal, and assume that $N > 0$. Since $f^2, f^3, b_d \in A$, we obtain that $g^2 + 2b_d g$ and $g^3 + 3b_d g^2 + 3b_d^2 g$ belong to A. Multiply successively each of these two expressions by b_d^{N-1} to infer that $b_d^{N-1} g^2, b_d^{N-1} g^3 \in A$. Thus $(b_d^{N-1} g)^2, (b_d^{N-1} g)^3 \in A$. By the inductive assumption also $b_d^{N-1} g \in A$, a contradiction. It follows that $N = 0$, and that $g \in A$. We conclude that $f \in A$ and that A is seminormal in B.

In the general case let f be an element of B satisfying $f^p \in A$ for all $p \in S$. We prove that $f \in A$ by induction on $\deg f$ starting with $\deg f = 0$. Let $f \neq 0$ and let $f = b_0 + \cdots + b_d$ be the homogeneous decomposition of f with $b_d \neq 0$. Since $b_d^p \in A$ for all $p \in S$, we obtain $b_d \in A$. Let $g = f - b_d$.

By Lemma 1.1 there exists an integer $N \geq 0$ such that $(pb_d)^N f^i \in A$ for all $p \in S$ and $0 \leq i \leq p$. Thus $(pb_d f^i)^j \in A$ for all $p \in S$, $0 \leq i \leq p$ and $j \geq N$. Since A is seminormal in B we see that $pb_d f^i \in A$ for all $p \in S$ and $0 \leq i \leq p$. For any prime p in S we obtain by the binomial expansion for $g^p = (f - b_d)^p$ that $g^p \in A$. By the inductive assumption, $g \in A$. We conclude that $f \in A$ and that A is S-root closed in B. $\square$

To apply results like Lemma 1.2 to domains one uses the fact that if R is a graded domain and S is the set of nonzero homogeneous elements of R, then the domain R_S is integrally closed [1, Proposition 2.1]; e.g., R is seminormal iff R is seminormal in R_S (cf. [3, Theorem 2.4] and its proof).

COROLLARY 1.3. [5, Theorem 2] *Let $A \subseteq B$ be graded rings. Then A is seminormal in B iff any homogeneous element $b \in B$ satisfying $b^2 \in A, b^3 \in A$ belongs to A.*

Corollary 1.3 implies that if $A \subseteq B$ are any rings such that A is seminormal in B then $A[\mathbf{X}]$ is seminormal in $B[\mathbf{X}]$ ([8]): the general case follows from the case that $\mathbf{X}$ is finite, and by induction, the finite case is reduced to the case of one indeterminate, which is a direct consequence of Corollary 1.3. For an alternative proof, note that Corollary 1.3 holds also for Γ-graded rings, where Γ is a totally ordered monoid (cf. [2, Theorem 6.1]).

As a consequence if A is a domain, then A is seminormal iff $A[\mathbf{X}]$ is seminormal. See also [2, Corollary 6.2].

If $A \subseteq B$ is a pair of rings, then the seminormalization of A in B is the smallest subring of B containing A which is seminormal in B.

COROLLARY 1.4. *If $A \subseteq B$ are graded rings, then the seminormalization of A in B is the smallest subring R of B containing A and satisfying the* *property that if b is homogeneous element in B such that* $b^2, b^3 \in R$ *then* $b \in R$.

COROLLARY 1.5. [3, Theorem 2.4] *Let* $A \subseteq B$ *be graded rings and let* S *be a nonempty set of positive integers. Then* $\mathcal{R}^S(A, B)$ *is a graded subring of* B; *it is the smallest subring* C *of* B *containing all the m-th roots in* B *of its homogeneous elements for* $m \in S$.

Let S be a nonempty set of positive integers. Lemma 1.2 and Corollary 1.5 imply that if $A \subseteq B$ are any rings, then $\mathcal{R}^S(A[\mathbf{X}], B[\mathbf{X}]) = \mathcal{R}^S(A, B)[\mathbf{X}]$ and that $\mathcal{C}(A[\mathbf{X}], B[\mathbf{X}]) = \mathcal{C}(A, B)$. In particular, if A is S-root closed in B, then $A[\mathbf{X}]$ is S-root closed in $B[\mathbf{X}]$. If A is a domain, then $\mathcal{R}^S(A[\mathbf{X}]) = R^S(A)[\mathbf{X}]$, and $\mathcal{C}(A[\mathbf{X}]) = \mathcal{C}(A)$. These results are in [8]. For a generalization to domains see [3, Corollary 2.5].

PROPOSITION 1.6. *Let* $k \subseteq K$ *be fields of zero characteristic. Then for any nonempty set* S *of positive integers and for all* $0 \leq j \leq \infty$ *we have*

1. $\mathcal{R}_j^S(k[\mathbf{X}], K[\mathbf{X}]) = \mathcal{R}_j^S(k, K)[\mathbf{X}]$.
2. $\mathcal{R}_j^S(k(\mathbf{X}), K(\mathbf{X})) = \mathcal{R}_j^S(k, K)(\mathbf{X})$.

Proof. It is enough to consider the case $j = 1$. First assume that $\mathbf{X}$ contains just one indeterminate X. Let $0 \neq f \in K[X]$ such that $f^m \in k[X]$ for some integer $m \in S$. Let

$$f = a_0 + \cdots + a_n X^n$$

with $a_i \in K$ and $a_n \neq 0$. By Lemma 1.1 we have $(ma_n X^n)^N f \in k(a_n)[X]$ for some $N \geq 0$. Since $m \neq 0$ in K, we obtain that $f \in k(a_n)[X] \subseteq \mathcal{R}_1^S(k, K)[X]$. It follows that $\mathcal{R}_1^S(k[X], K[X]) \subseteq \mathcal{R}_1^S(k, K)[X]$; so, we have equality.

Now let $0 \neq g \in K(X)$ such that $g^m \in k(X)$ for some integer $m \in S$. Thus $g = c\frac{g_1}{g_2}$, where $c \in K$ and g_1, g_2 are coprime monic polynomials in $K[X]$. Also $g^m = a\frac{h_1}{h_2}$ where $a \in k$ and h_1, h_2 are coprime monic polynomials in $k[X]$. Since h_1 and h_1 are coprime in $K[X]$ we obtain that $c^m = a$ and that $g_1^m = h_1, g_2^m = h_2$. Thus $g \in R_1^S(k, K)(X)$. It follows that

$$\mathcal{R}_1^S(k(X), K(X)) = \mathcal{R}_1^S(k, K)(X).$$

(2) holds by induction for any finite $\mathbf{X}$, and hence for any $\mathbf{X}$.

For any $\mathbf{X}$ we have $\mathcal{R}_1^S(k[\mathbf{X}], K[\mathbf{X}]) \subseteq \mathcal{R}_1^S(k(\mathbf{X}), K(\mathbf{X})) = \mathcal{R}_1^S(k, K)(\mathbf{X})$. Thus $\mathcal{R}_1^S(k[\mathbf{X}], K[\mathbf{X}]) \subseteq \mathcal{R}_1^S(k, K)(\mathbf{X}) \cap K[\mathbf{X}] = \mathcal{R}_1^S(k, K)[\mathbf{X}]$. Hence (1) holds. $\square$

2 EXAMPLES

For the following construction cf. [11, Example 1]. Let $k \subseteq K$ be fields, and let $\mathcal{L}$ be a family of finite field extensions of k which are contained in K, and which generate K as a subfield. Let $\mathbf{X} = \{X_L \,|\, L \in \mathcal{L}\}$ be a set of independent indeterminates over K. Let D be the subring of $K[\mathbf{X}]$ generated by $\bigcup\{k + X_L L[X_L] \,|\, L \in \mathcal{L}\}$. Let T be the multiplicative set of all nonzero polynomials in $k[\mathbf{X}]$ which are not divisible in this ring by any of the indeterminates in $\mathbf{X}$. Set $A = D_T$; we write $A = \mathcal{A}(k, K, \mathcal{L})$. If $L_1, \ldots, L_n$ are subfields of K, then the subfield of K generated by $L_1 \cup \cdots \cup L_n$ equals $\prod_{i=1}^n L_i$, that is, the additive subgroup of K generated by the products $\prod_{i=1}^k c_i$, where $c_i \in L_i$ for all $1 \le i \le n$. Thus we have

REMARK 2.1. *D is the k-subspace of $K[\mathbf{X}]$ generated by 1 and by the monomials of the form $c \prod_{i=1}^n X_{L_i}$ for $L_1, \ldots, L_n$ in $\mathcal{L}$ and $c \in \prod_{i=1}^n L_i$.*

LEMMA 2.2. *Any nonzero ideal of D contains a nonzero polynomial in $k[\mathbf{X}]$.*

Proof. Let f be a nonzero element of D. Let $X_{L_1}, \ldots, X_{L_n}$ be all the indeterminates occurring in f. Let $L = \prod_{i=1}^n L_i$. Consider the norm $g = \mathrm{N}_{L(X_{L_1}, \ldots, X_{L_n})/k(X_{L_1}, \ldots, X_{L_n})}(f)$. Thus $0 \ne g \in k[X_{L_1}, \ldots, X_{L_n}]$, and $g = fh$, where $h \in L[X_{L_1}, \ldots, X_{L_n}]$. Set $g_0 = g \prod_{i=1}^n X_{L_i}$, and $h_0 = h \prod_{i=1}^n X_{L_i}$. By Remark 2.1, $h_0 \in D$. Since $g_0 = fh_0 \in k[\mathbf{X}]$, the Lemma follows. $\qquad\square$

COROLLARY 2.3. *Any nonzero ideal in A contains a product of indeterminates in $\mathbf{X}$. Thus any element in D which is not divisible in $K[\mathbf{X}]$ by any indeterminate is invertible in A.*

LEMMA 2.4. *A is a one-dimensional noetherian domain.*

Proof. For $L \in \mathcal{L}$ let

$$P_L = X_L K[\mathbf{X}] \cap D,$$

thus P_L is a prime ideal of D. The ideal P_L of D is generated by the set LX_L. Since L is a finite extension of k, we see that the ideal P_L is finitely generated.

To complete the proof it is enough to show that the ideals $P_L A$ ($L \in \mathcal{L}$) are all the nonzero prime ideals of A. Indeed, let P be a nonzero prime ideal of A. By Corollary 2.3 the ideal P contains an indeterminate X_L. For any $c \in L$ we have $(cX_L)^2 = (c^2 X_L)X_L \in X_L A$. Thus $(P_L A)^2 \subseteq P$. Hence, $P_L A \subseteq P$. Assume that $P \ne P_L A$. Hence there is an element $f \in (P \cap D) \setminus P_L A$. By Corollary 2.3 the element $X_L + f$ is invertible in A, a contradiction. $\qquad\square$

If K_1, K_2 are subfields of K, we denote by $\langle K_1, K_2 \rangle$ the subfield generated by $K_1 \cup K_2$.

We extend the definition of $\mathcal{A}(k, K, \mathcal{L})$ for a *family* $\mathcal{L} = \{L_i\}_{i \in I}$ of subfields of K using indeterminates X_i ($i \in I$); the subfields L_i are not necessarily distinct. Remark 2.1, which we use below, still holds with this definition. Thus we have

LEMMA 2.5. *Let $k \subseteq \Delta \subseteq K$ be a field. For any $L \in \mathcal{L}$ let $\tilde{L} = \langle \Delta, L \rangle$. Let $\tilde{\mathcal{L}}$ be the family of all subfields $\tilde{L}$. Then the subring $A[\Delta]$ of $K(\mathbf{X})$ can be canonically identified with $\mathcal{A}(\Delta, K, \tilde{\mathcal{L}})$.*

Proof. Let $\theta : K(X_L \mid L \in \mathcal{L}) \to K(X_L \mid L \in \tilde{\mathcal{L}})$ be the isomorphism over K that sends X_L to $X_{\tilde{L}}$ for $L \in \mathcal{L}$. By Remark 2.1, $\theta(A[\Delta]) = \mathcal{A}(\Delta, K, \tilde{\mathcal{L}})$. $\square$

LEMMA 2.6. *Let S be a set of positive integers. Assume that $\operatorname{char} k = 0$, and that*

$$\mathcal{R}_j^S(\prod_{i=1}^{n} L_i, K) = \langle \mathcal{R}_j^S(k, K), \prod_{i=1}^{n} L_i \rangle$$

for all finite j and for any distinct subfields $L_1, \ldots, L_n$ in $\mathcal{L}$. For $0 \le i \le \infty$ let $\mathcal{L}_i$ be the family of subfields of K of the form $\langle \mathcal{R}_i^S(k, K), L \rangle$, where $L \in \mathcal{L}$.

Then, for any $0 \le i \le \infty$ the ring $\mathcal{R}_i^S(A)$ equals $A[\mathcal{R}_i^S(k, K)]$, and is canonically isomorphic to the ring $\mathcal{A}(\mathcal{R}_i^S(k, K), K, \mathcal{L}_i)$.

Proof. By Lemma 2.5, the equality $\mathcal{R}_i^S(A) = A[\mathcal{R}_i^S(k, K)]$ for a given i implies the existence of a canonical isomorphism as stated.

It is enough to prove the lemma for finite i. We proceed by induction. Let $i = 1$. Let f be an element of $K(\mathbf{X})$ such that $f^m \in A$ for some $m \in S$. We may assume that $f \in K[\mathbf{X}]$ and that $f^m \in D$. Let $X_{L_1}, \ldots, X_{L_n}$ be all the indeterminates occurring in f. We show that $f \in D[R_1^S(k, K)]$ by induction on n. If $n = 0$, that is, $f \in K$, then $f^m \in k$, so $f \in R_1^S(k, K)$. Let $n > 0$. For a given $1 \le r \le n$ put X_{L_r} to zero, and use the inductive assumption on $n - 1$ and Remark 2.1 to obtain that all monomials in f (with their coefficients) which do not contain the indeterminate X_{L_r} are in $D[R_1^S(k, K)]$. Let $L = \prod_{i=1}^{n} L_i$. Since $f^m \in L[X_{L_1}, \ldots, X_{L_n}]$, by Proposition 1.6 (1) we have $f \in \mathcal{R}_1^S(L, K)[\mathbf{X}] = \langle \mathcal{R}_1^S(k, K), L \rangle[\mathbf{X}]$. Thus the monomials in f containing all the indeterminates $X_{L_1}, \ldots, X L_n$ belong to $D[R_1^S(k, K)]$. Hence $f \in D[R_1^S(k, K)]$. Thus $\mathcal{R}_1^S(A) \subseteq A[R_1^S(k, K)]$, so we have equality.

Now let $i > 1$. We have

$$\mathcal{R}_i^S(A) = \mathcal{R}_{i-1}^S(A[\mathcal{R}_1^S(k, K)])$$

$$= A[\mathcal{R}_{i-1}^S(\mathcal{R}_1^S(k, K), K)] = A[\mathcal{R}_i^S(k, K)].$$

We have used the inductive assumption on $i-1$ with k replaced by $\mathcal{R}_1^S(k, K)$, and every field $L \in \tilde{L}$ replaced by $\langle \mathcal{R}_1^S(k, K), L \rangle$. Indeed, for $0 \le j < \infty$ and for $L_1, \ldots, L_n$ in $\mathcal{L}$ we have

$$\mathcal{R}_j^S(\langle \mathcal{R}_1^S(k, K), \prod_{i=1}^{k} L_i \rangle, K) = \mathcal{R}_j^S(\mathcal{R}_1^S(\prod_{i=1}^{k} L_i, K))$$

$$= \mathcal{R}_{j+1}^S(\prod_{i=1}^{k} L_i, K) = \langle \mathcal{R}_{j+1}^S(k, K), \prod_{i=1}^{k} L_i \rangle$$

$$= \langle \mathcal{R}_j^S(\mathcal{R}_1^S(k, K), K), \prod_{i=1}^k L_i \rangle.$$

□

Combining Lemmas 2.4 and 2.6 we obtain

COROLLARY 2.7. *Under the assumptions of Lemma 2.6, for any* $0 \leq j \leq \infty$, $\mathcal{R}_j^S(A)$ *is a one-dimensional noetherian domain.*

THEOREM 2.8. *Let* $k \subseteq K$ *be fields of zero characteristic, and let* $\mathcal{L}$ *be a set of finite field extensions of* k, *which generate the field* K. *Assume that the degrees* $[L_1 : k]$ *and* $[L_2 : k]$ *are coprime for any two distinct fields* L_1, L_2 *in* $\mathcal{L}$, *and that for any* n *distinct fields* $L_1, \ldots, L_n$ *in* $\mathcal{L}$ *and for all finite* j

$$\mathcal{R}_j^S(\prod_{i=1}^n L_i, K) = \langle \mathcal{R}_j^S(k, K), \prod_{i=1}^n L_i \rangle.$$

Then for any positive integer d *there exists a noetherian domain* A *of Krull dimension* d *with the following additional properties:*

(1) *A contains* k, *and the quotient field of* A *contains* K.
(2) *For any set* S *of positive integers, the domain* $\mathcal{R}^S(A)$ *is noetherian; if* $d = 1$, *then for any* $0 \leq j \leq \infty$, *the domain* $\mathcal{R}_j^S(A)$ *is noetherian one dimensional.*
(3) *For any set* S *of positive integers and for any* $0 \leq j \leq \infty$ *we have* $\mathcal{R}_j^S(A) \cap K = \mathcal{R}_j^S(k, K)$. *In particular,* $A \cap K = k$.
(4) *For any set* S *of positive integers*

$$\mathcal{C}(\mathcal{R}^S(A)) \subseteq \mathcal{C}(\mathcal{R}^S(k, K), K),$$

with equality if $\mathcal{C}(\mathcal{R}^S(k, K), K) = S$.

The domain A *can be defined as*

$$A = \mathcal{A}(k, K, \mathcal{L})[Z_1, \ldots, Z_{d-1}],$$

where $Z_1, \ldots, Z_{d-1}$ *are indeterminates over* A.

Proof. First assume that $d = 1$.

Let $A = \mathcal{A}(k, K, \mathcal{L}) = D_T$. Clearly (1) holds. Also (2) holds by Corollary 2.7 .

(3) We show that $A \cap K = k$. Indeed, let $c \in A \cap K$. Thus there is an element t in $T \subseteq k[\mathbf{X}]$ such that $ct \in D$. For any monomial $X_{L_1} \ldots X_{L_n}$ occurring in t we have $c \in \prod_{i=1}^n L_i$. Let $r = [\prod_{i=1}^n L_i : k] = \prod_{i=1}^n [L_i : k]$. Since no indeterminate divides t in $k[\mathbf{X}]$, it follows from our assumption on the coprimeness of degrees that the greatest common divisor of all such r's is 1. Hence $c \in k$, and $A \cap K = k$. (3) now follows from Lemma 2.6.

(4) Let $m \notin \mathcal{C}(\mathcal{R}^S(k, K), K)$ be a positive integer. Thus there exists an element $c \in K \setminus \mathcal{R}^S(k, K)$ such that $c^m \in \mathcal{R}^S(k, K)$. By (3) we have $c \notin \mathcal{R}^S(A)$. Hence $m \notin \mathcal{C}(\mathcal{R}^S(A))$. Thus $\mathcal{C}(\mathcal{R}^S(A)) \subseteq \mathcal{C}(\mathcal{R}^S(k, K), K)$.

If $C(\mathcal{R}^S(k,K),K) = S$, then $S \subseteq C(\mathcal{R}^S(A)) \subseteq C(\mathcal{R}^S(k,K),K) = S$, so $C(\mathcal{R}^S(A)) = S$.

Let $d > 1$. Let S be a set of positive integers. We have by [8] that

$$\mathcal{R}^S(A[Z_1,\ldots,Z_{d-1}]) = \mathcal{R}^S(A)[Z_1,\ldots,Z_{d-1}],$$

so the ring $\mathcal{R}^S(A[Z_1,\ldots,Z_{d-1}])$ is noetherian d-dimensional.

By sending all the indeterminates Z_i to 0 we obtain for any $0 \leq j \leq \infty$ that $\mathcal{R}_j^S(A[Z_1,\ldots,Z_{d-1}]) \cap K = \mathcal{R}_j^S(A) \cap K = \mathcal{R}_j^S(k,K)$, thus (3) holds for $A[Z_1,\ldots,Z_{d-1}]$. By [8], $C(\mathcal{R}^S(A[Z_1,\ldots,Z_{d-1}])) = C(\mathcal{R}^S(A)[Z_1,\ldots,Z_{d-1}]) = C(\mathcal{R}^S(A))$, and (4) follows. $\qquad\square$

PROPOSITION 2.9. *For any positive integer d there exists a noetherian domain A of Krull dimension d such that $\mathcal{R}_1(A)$ is a factorial domain (in particular, root closed), and such that for any monoid S of positive integers which is generated by primes, the ring $\mathcal{R}^S(A)$ is noetherian, and also*

$$C(\mathcal{R}^S(A)) = S.$$

Proof. Let t be a transcendental element over a given field F. Set $k = F(t)$. For any prime p let $\sqrt[p]{t}$ be a p-th root of t in an algebraic closure of k. Let $K = F(\{\sqrt[p]{t} \mid p \text{ prime}\})$. Let S be a monoid generated by primes. If $p \in S$, then $[k(\sqrt[p]{t}) : k] = p$ by [10, Theorem 9.1, Ch. VI]. Thus if Φ is a finite subset of S, then

$$[k\left(\{\sqrt[p]{t} \mid p \in \Phi\}\right) : k] = \prod_{p \in \Phi} p.$$

Now let q be a prime and let $x \in K$ such that $x^q \in k(\{\sqrt[p]{t} \mid p \in S\})$, but $x \notin k(\{\sqrt[p]{t} \mid p \in S\})$. Thus $x \in k(\{\sqrt[p]{t} \mid p \in \Phi\})$ for some finite subset Φ of S. Hence $q \mid \prod_{p \in \Phi} p$, and so $q \in \Phi \subseteq S$. We claim that $x \in k(\sqrt[q]{t})$; otherwise, the polynomial $X^q - x^q$ has no root in $k(\sqrt[q]{t})$ since any root of unity in K belong to k. Using [10] again, we obtain $[k(x, \sqrt[q]{t}) : k] = q^2$, contradicting the fact that the degree of any finite extension of k contained in K is a product of distinct primes. It follows that $k(\{\sqrt[p]{t} \mid p \in S\})$ is S-root closed, and that $\mathcal{R}^S(k,K) = k(\{\sqrt[p]{t} \mid p \in S\})$. Thus $\mathcal{R}_i^S(k,K) = \mathcal{R}^S(k,K)$ for all $0 \leq i \leq \infty$. Since $\sqrt[q]{t} \notin \mathcal{R}^S(k,K)$ for any prime $q \notin S$, we obtain that $C(\mathcal{R}^S(k,K),K) = S$.

Now let $\mathcal{L}$ be the set of all fields of the form $k(\sqrt[p]{t})$ for p prime. By Theorem 2.8 we obtain a d-dimensional noetherian domain

$$A = \mathcal{A}(k,K,\mathcal{L})[Z_1,\ldots,Z_{d-1}] = D_T[Z_1,\ldots,Z_{d-1}].$$

Since $\mathcal{R}_1(A) \supseteq K$, we obtain $\mathcal{R}_1(A) = K[Z_1,\ldots,Z_{d-1},\mathbf{X}]_T$, and so $\mathcal{R}_1(A)$ is factorial. $\qquad\square$

For the next lemma cf. [6, Example 2.1].

LEMMA 2.10. *Let p be a prime and let F be a field of characteristic different from p. Let t be an indeterminate over p. Define inductively:*

$$v_0 = t, \text{ and for } n > 0, v_n = \sqrt[p]{1 + v_{n-1}},$$

where the roots are taken in an algebraic closure of $k(t)$. Let $k = F(t)$, and $K = k(v_0, v_1, \dots)$. Then $\mathcal{R}_j(k, K) = k(v_j)$ for all $0 \le j < \infty$, and so,

$$\mathcal{R}_0(k, K) \subsetneqq \mathcal{R}_1(k, K) \subsetneqq \dots.$$

Proof. We have for all $n > 0$:

$$v_n = \sqrt[p]{1 + \sqrt[p]{\dots \sqrt[p]{1 + t}}} \quad (n \text{ radicals}).$$

Set $k = F(t)$ and $K = k(v_0, v_1, \dots)$. Clearly

$$k = F(v_0) \subsetneqq F(v_1) \subsetneqq \dots,$$

and $K = \bigcup_{j=0}^\infty F(v_j)$. Also, by [10, Theorem 9.1, Ch. VI], since all roots of unity in K belongs to F, we have $[F(v_{j+1}) : F(v_j)] = p$ for all $0 \le j < \infty$.

By induction on j we obtain that $\mathcal{R}_j(k, K) \supseteq F(v_j)$ for all $0 \le j < \infty$.

We now show that if $j \ge 0$, $u \in F(v_{j+1}) \setminus F(v_j)$ and $u^m \in F(v_j)$ for some $m \ge 1$, then $u = \alpha v_{j+1}^i$ for some $\alpha \in F(v_j)$ and $i \ge 0$; so, $u^p \in F(v_j)$. Assume that m is minimal. If q is a prime factor of m, we obtain by [10, Theorem 9.1, Ch. VI] that $[F(v_j, u^{\frac{m}{q}}) : F(v_j)] = q$. Thus $q = p$, and m is a power of p. Set $z = v_{j+1}$. Write $u = \dfrac{g_1(z)}{g_2(v_j)}$, where $g_1, g_2 \ne 0$ are polynomials over F. Replacing u by $g_1(z)$ we may assume that $u \in F[z]$: $u = \sum_{i=0}^n c_{r_i} z^{r_i}$, where all the coefficients c_{r_i} are nonzero. We also may assume that $r_0 = 0$. Assuming that $u \notin F[z^p] = F[v_j]$, let i be the least integer such that r_i is not divisible by p, thus $i > 0$. The coefficient of z^{r_i} in u^m is $m c_0^{m-1} c_{r_i} \ne 0$. But $u^m \in F[v_j] = F[z^p]$, and so r_i must be divisible by p, a contradiction. This proves our assertion.

Next, we show that if for some $j \ge 0$, $u \in F(v_{j+2}) \setminus F(v_{j+1})$, then u has no power in $F(v_j)$. Otherwise, as shown before, we have for some $\alpha \in F(v_{j+1})$ and $1 \le i < p$ that $u = \alpha v_{j+2}^i$, and $u^{p^2} \in F(v_j)$. Let $z = v_{j+1}$. As before, we may assume that $\alpha \in F[z]$. Also we may assume that z does not divide α in $F[z]$. We have $u^p = \alpha^p (1 + z)^i \notin F[z^p] = F[v_j]$, since the degree in z of $\alpha^p (1 + z)^i$ is not divisible by p. Since $u^p \in F[v_{j+1}] = F[z]$ and some power of u^p is in $F(v_j) = F(z^p)$, we see that some power of u^p is in $F[z^p]$. We infer that $u^p = \beta z^r$ for some $\beta \in F[z^p]$ and $r \ge 1$. Thus $\alpha^p (1 + z)^i = \beta z^r$, which implies that α is divisible by z in $F[z]$, a contradiction. This finishes the proof of our second assertion.

We now easily conclude that $\mathcal{R}_j(k, K) = F(v_j)$ for all finite j. □

THEOREM 2.11. *For any positive integer d there exists a d-dimensional noetherian domain A with the following properties:*

(1) *The root closure of A is not obtainable in finitely many steps.*

(2) *For any monoid S generated by primes the domain $\mathcal{R}^S(A)$ is noetherian, and $C(\mathcal{R}^S(A)) = S$.*

Proof. Let F be a field of zero characteristic. Let $\{t_p \mid p \text{ prime}\}$ be a set of independent indeterminates over F. For any prime p let v_p be defined as follows: set $u_{p,0} = t_p$, and by induction set $u_{p,i} = \sqrt[p]{1 + u_{p,i-1}}$ for $0 < i \le p$; set $v_p = u_{p,p}$, and $u_{p,i} = v_p$ for $i > p$. Let $k = F(t_p \mid p \text{ prime})$, and $K = F(v_p \mid p \text{ prime})$.

Let S be a monoid generated by primes.

For a prime $p \in S$ let M_p be the field

$$k(\{v_q \mid q \ne p \text{ a prime in } S\}).$$

By Lemma 2.10, we obtain for a prime $p \in S$

$$\mathcal{R}_i^S(k, K) \subseteq \mathcal{R}_i(M_p, K) = M_p(u_{p,i}).$$

Thus $\mathcal{R}_i^S(k, K) \subseteq \bigcap\limits_{p \text{ a prime in } S} M_p(u_{p,i}) = k(\{u_{p,i} \mid p \in S\})$.

The last equality is obtained from degree considerations replacing the set $k(\{u_{p,i} \mid p \in S\})$ by any of its finite subsets.

We have $\mathcal{R}_i^S(k, K) = k(\{u_{p,i} \mid p \in S\})$, and $\mathcal{R}^S(k, K) = k(\{v_p \mid p \in S\})$. Since $u_{q,1} = \sqrt[q]{1 + t_p} \notin R^S(k, K)$ for any prime $q \notin S$, we obtain that $C(\mathcal{R}^S(k, K), K) = S$.

We have $u_{p,i} \notin \mathcal{R}_{i-1}(M_p, K)$ for $0 < i \le p$. Hence

$$u_{p,i} \in \mathcal{R}_i(k, K) \setminus \mathcal{R}_{i-1}(k, K).$$

Thus

$$\mathcal{R}_0(k, K) \subsetneqq \mathcal{R}_1(k, K) \subsetneqq \cdots \subsetneqq \mathcal{R}_p(k, K)$$

for any prime p. It follows that the root closure of k in K is not obtainable in finitely many steps.

Let $\mathcal{L}$ be the family of subfields of K of the form $k(v_p)$ for p prime. By Theorem 2.8 the domain $A = \mathcal{A}(k, K, \mathcal{L})[Z_1, \ldots, Z_{d-1}]$ satisfies our requirements. $\qquad\square$

REFERENCES

[1] D.D. Anderson and D.F. Anderson, Divisibility properties in graded domains, *Canad. J. Math.* 34 (1982), 196-215.

[2] D.D. Anderson and D.F. Anderson, Divisorial ideals and invertible ideals in a graded integral domain, *J. Algebra* 76 (1982), 549-569.

[3] D.F. Anderson, Root closure in integral domains, *J. Algebra* 79 (1982), 51-59.

[4] D.F. Anderson, Root closure in integral domains, II, *Glasgow Math. J.* 31 (1989), 127-130.

[5] D.F. Anderson, Seminormal graded rings, *J. Pure Appl. Algebra* 21 (1981), 1-7.

[6] D.F. Anderson, D.E. Dobbs and M. Roitman, Root closure in commutative rings, *Sci. Univ. Clermont II, Sér. Math.* 26 (1990), 1-11.

[7] J.W. Brewer and D.L. Costa, Seminormality and projective modules over polynomial rings, *J. Algebra* 58 (1979), 208-216.

[8] J.W. Brewer, D.L. Costa and K. McCrimmon, Seminormality and root closure in polynomial rings and algebraic curves, *J. Algebra* 58 (1979), 217–226.

[9] R. Gilmer, *Multiplicative Ideal Theory*, Marcel Dekker, New York, 1972.

[10] S. Lang, *Algebra, Third Edition*, Reading Massachusetts, 1994.

[11] M. Nagata, *Local Rings*, Wiley, Interscience, New York, 1962.

Index